# Conservation Planning

# Conservation Planning

## Informed Decisions for a Healthier Planet

Craig R. Groves

Edward T. Game

ROBERTS AND COMPANY PUBLISHERS

Greenwood Village, Colorado

Library of Congress Cataloging-in-Publication Data

Names: Groves, Craig. | Game, Edward T.
Title: Conservation planning : informed decisions for a healthier planet / Craig R. Groves, Edward T. Game.
Description: Greenwood Village, Colorado : Roberts and Company Publishers, [2016]
Identifiers: LCCN 2015038114 | ISBN 9781936221516
Subjects: LCSH: Ecosystem management. | Nature conservation.
Classification: LCC QH75 .G728 2016 | DDC 333.72--dc23
LC record available at http://lccn.loc.gov/2015038114

Manufactured in Canada

10 9 8 7 6 5 4 3 2 1

Roberts and Company Publishers, Inc.
4950 South Yosemite Street, F2 #197
Greenwood Village, CO 80111 USA
Tel: (303) 221-3325
Fax: (303) 221-3326
Email: info@roberts-publishers.com

**Internet: www.roberts-publishers.com**

Publisher, Ben Roberts; production editor, Julianna Scott Fein; manuscript editor, Christianne Thillen; creative director, Emiko-Rose Paul; text designer, Jeanne Schreiber; cover designer, Emiko-Rose Paul; illustrator, Robin Mouat; proofreader, Jennifer McClain. The text was set in 10.5/13 New Century Schoolbook by Dovetail Publishing Services and printed on #45 Somerset Matte by Transcontinental Printing.

The views outlined in this book are the authors' own and not necessarily those of The Nature Conservancy.

Front cover: Pronghorn antelope (*Antilocapra americana*) graze in front of a natural gas drilling derrick in Jonah Field, Wyoming, USA. The Jonah Field area of sage and grass provides critical winter range for antelope, but it is also an economically important natural gas field whose exploitation has been made possible by new fracturing (fracking) techniques. *Conservation Planning* is about effectively navigating the increasing number of challenges like this so that we can live not only in a prosperous world, but in one where nature still thrives.

To Tasman and Atticus, who I hope, get as much pleasure from exploring this planet as I have. —E. G.

For my Mother and Father whose unflagging support provided me the opportunity to play a small role in conserving Nature. —C. G.

# Brief Contents

# Contents

## CHAPTER 10 Planning for Ecosystem Services: Building a Bridge to Human Well-Being 427

# Foreword

Conservation planning has had a checkered history. In some cases it is done poorly, while in many cases plans and strategies are never used and sit to rot on agency shelves. It is remarkable how many governments and environmental nongovernmental organizations continue to reinvent the field of conservation planning while ignoring fifty years of literature. Remarkably, some groups still present species richness maps as conservation plans. Worse still, plans are sometimes not made at all, and conservation opportunities are lost through ad hoc and impulsive decision making. *Conservation Planning: Informed Decisions for a Healthier Planet* is long overdue.

This is not just another dry academic tome describing tools and ideas that could work in theory. Groves and Game are practitioners, and consequently this is a very practical book. It is particularly strong in explaining all the socioeconomic contexts and forces essential to good planning. The content is broad and covers ideas not generally discussed in the dryer technical literature on conservation planning—such as the people and politics of conservation decisions. People in academia, NGOs, and government will learn a lot from the synthesis of the literature and also the wealth of practical experience the authors can use to explain real conservation planning situations.

While the famous saying "plans are useless but planning is essential" is a little trite, it makes a good point. Much of conservation planning is about process not the end product. This book covers the *process* of conservation planning much better than any other in its field. Groves and Game outline the steps in conservation planning—but not as a dull shopping list. They explain the crucial concepts in a palatable form with appropriate real-world examples and links to a diverse literature.

The book covers some important concepts that are not yet well understood or discussed in the traditional literature. For example:

- the idea of a counterfactual assumption (what would have happened if you did nothing) as a method of estimating the benefits of actions,
- the importance of good project management,
- adaptive management and double-loop learning,
- being clear about who a plan is for,
- implementation and the art of doing,
- the thorny issue of dealing with multiple objectives,

- the fact that robust decision-making tools mean that uncertainty is not an impediment to rational decision making, and
- an emphasis on calculating the costs of conservation actions.

This really is a book at the cutting edge of a discipline, determined to redirect, not merely summarize, the state of the art. And let there be no doubt, this is more art than science.

The surface of the planet is finite—we cannot make more of it. Deciding what actions to take on that finite surface, whether it be land or sea, swamp or stream, is both a privilege and a daunting responsibility. Arguably it is our greatest responsibility, because it affects not just present generations but future generations. Deciding that a valley will be a dam or a national park is a decision that will impose an environmental, social, and economic footprint for tens of generations.

Those who read this book will be more informed planners and decision makers. It is a timely contribution to an essential and expanding literature, and I commend it to anyone who cares about life.

Hugh Possingham
*The University of Queensland*
*Brisbane, Australia*

April 2015

# Preface

Over 45 years ago, Ian McHarg wrote the first edition of *Design with Nature*. And nearly 40 years ago, *A Nature Conservation Review* was first published in the United Kingdom as a means of evaluating sites for their ability to serve as representative natural areas. In many respects, these were two of the earliest efforts to outline approaches to conservation planning—the former from the perspective of an urban planner who placed importance on ecological processes and services, and the latter from that of conserving natural areas through a system of nature reserves. Fast forward to 2015, and an enormous amount of progress has been made both in the professional natural resource planning community and in the planning arena now associated with the discipline of conservation biology. And, of course, the natural world has changed dramatically during this period: accelerated losses and degradation of biological diversity and ecosystem processes, large-scale changes in land use practices that include dramatic reductions in natural land cover globally, and the worldwide impacts from global warming.

Beginning with a special issue of *Biological Conservation* in the late 1980s, in which a group of Australian biologists outlined a systematic approach to planning for nature reserves, hundreds of scientific papers have focused on where to establish nature reserves or conservation areas to conserve the greatest amount of biological diversity in the most efficient manner. We refer broadly to the approach of these papers as *spatial planning*, because the major emphasis of the papers was answering the question of where to locate nature reserves. But as Kent Redford and his colleagues aptly pointed out in a seminal paper in *Conservation Biology* entitled "Mapping the Conservation Landscape" (2003), conservation planning also seeks to resolve an equally important question: how will conservation get accomplished on the ground—or, in other words, what set of interventions will be necessary to achieve conservation in a particular landscape or seascape? Answering such a question is the realm of strategic planning, and separate planning methods have been promulgated that focus on *strategic planning* for conservation. The best known of these is the *Open Standards for the Practice of Conservation*, written and published by the Conservation Measures Partnership, a consortium of conservation organizations and conservation-oriented foundations.

One of us wrote a conservation planning book focused primarily on spatial planning aimed at conserving biodiversity, as the book title suggests—*Drafting a Conservation Blueprint: A Practitioner's Guide to Planning for Biodiversity* (2003). In this book, however, we take another tack—combining the methodologies of spatial and strategic planning into one integrated approach. We do so largely because we have learned how difficult it is to separate the decisions that conservation practitioners and managers need to make about where to take conservation action from the decisions about what strategies and actions need to be taken. It is all well and good to have a spatially explicit list of places that are important for conserving a particular set of species or ecosystems, but then what? At any one of these places, a variety of conservation actions might need to take place that collectively add up to a considerable amount of staffing and financial resources, all with certain risks and uncertainties as well as opportunities. Undertaking conservation at one particular place could leverage conservation actions at other similar sites, while in other situations the interventions taken at one site might be a one-off opportunity with limited chances for scaling up. In either case, it should be clear that setting conservation priorities involves making decisions about what actions to take where, and that is why we have integrated spatial and strategic planning.

Throughout the book, we have endeavored to highlight two other important themes. The first is that most conservation problems and the conservation plans that are designed to address them involve working to achieve multiple objectives. Whether it is deploying the tools of marine spatial planning to design a system of marine protected areas while maintaining local fishing communities, creating new protected or conservation areas in response to mitigation offsets from the energy sector, or implementing a new REDD mechanism (Reduced Emissions from Deforestation and Forest Degradation) to meet timber outputs, emission reductions, and biodiversity goals, there is a substantial need for, and much to be gained by, the conservation community investing in better understanding, planning, and balancing of the trade-offs between biodiversity and other objectives.

The third and final underlying theme of this book relates to making decisions. Our premise is simple. Many important decisions get made every day in natural resource agencies and conservation organizations about where and how to take various types of conservation and resource management actions. The principal aim of conservation plans should be to inform those actions. Thus, much of the guidance we provide in this book is steeped in decision science and structured decision making—all with the aim of helping planners and practitioners make more informed decisions (hence the book's subtitle) that we hope will lead to better outcomes for conservation.

We owe a considerable debt to our present employer—The Nature Conservancy—for providing us years of experience around the world in developing conservation planning methods and working with a variety of planning teams in both developed and developing nations. At the same time, this is decidedly not a Nature Conservancy book. The methods and experiences that we relied on to write this book span a range of disciplines, institutions, scientific literature, experiences, and expertise. Although we are both conservation biologists and planners by training, in writing this book we have drawn heavily from economics, decision science, sociology, and political science as well as conservation biology and ecology. Many of conservation's biggest challenges are indeed "wicked problems." At a minimum, they will take a multidisciplinary or transdisciplinary approach to solve—and that applies to planning as well as implementation.

We have written this book first and foremost for conservation practitioners—including planners, scientists, and natural resource managers who work for governmental agencies or nongovernmental organizations around the world—as well as for students who are interested in conservation practice. To make the concepts presented in this book accessible to a wide audience, we attempt to illustrate most ideas, methods, and tools with real-world examples, and often with graphical material. In addition, the text is written somewhat less formally than that in a scientific journal. Although we have cited the important literature, we are deliberately selective rather than exhaustive in use of references. This book is designed as a relatively complete and coherent guide to a process of conservation planning. Nevertheless, we have tried to write it so that each chapter can be read on its own. Someone interested, for example, in the process of establishing objectives for a planning exercise could just read Chapter 3.

Like many individuals who get involved in conservation or natural resources, we've spent significant times in our lives both working and recreating in some of the grandest natural areas the world has to offer. We are deeply committed to planning for and conserving these remaining natural areas. Yet, we appreciate that nature conservation and the management of natural resources comes in many forms. The conservation and natural resource community of planners, scientists, practitioners, and managers is and should be a "big tent," open to diverse approaches in a variety of places, from the urban environment to the wildest places left on Earth. We sincerely hope that this book will resonate with and be useful to this broader community.

## Acknowledgments

We are deeply grateful to many people for their role in making this book happen. Our thanks to The Nature Conservancy for the opportunity to both gain and share the knowledge and experience contained in this book. Peter Kareiva steadfastly believes in the value of sharing knowledge through books and encouraged us to write this one. Ben Roberts at Roberts and Company shares our vision of an accessible, well-designed, but affordable book written for practitioners. Thanks also to Julianna Scott Fein for her patience, attention to detail, and deft editorial skill in talking us out of our bad ideas and into her good ones. Christianne Thillen's copy editing made our writing so much better. Robin Mouat and Emiko Paul took our figures, maps, and other art and substantially upgraded their quality and appearance. Leonie Seabrook prepared the index for us; Joan Keyes typeset the book.

We are greatly indebted to the knowledgeable colleagues who kindly reviewed chapters of this book, provided importance guidance, or helped with the text in specific places: Tom Barrett, Leandro Baumgarten, Michael Bode, Kathy Boomer, Mark Burgman, Dick Cameron, Rob Campellone, Jamie Carr, Robin Cox, Molly Cross, Frank Davis, Amielle DeWan, Karl Didier, Nigel Dudley, Jon Fisher, Toby Gardner, Josh Goldstein, Doria Gordon, Sara Gottlieb, Hedley Grantham, David Hulse, Lise Hanners, Kim Hall, Jodi Hilty, Mark Humpert, Malcolm Hunter, Carter Ingram, Arlyne Johnson, Josh Lawler, Heather Leslie, Peter Kareiva, Katie Kennedy, Carissa Klein, Andrew Knight, Cristian Lasch, Craig Leisher, Richard Margoluis, Rob Marshall, Michael Mascia, Matt Miller, Jensen Montambault, John Morrison, Reed Noss, Sally Palmer, Bruce Stein, Tim Tear, and Sara Vickerman.

In addition to these reviewers, CG conducted research interviews for the book, and several individuals contributed valuable information to these interviews: Ross Alliston, Sandy Boyce, Rob Campellone, Bill Connelly, Daniel Juhn, John Morrison, Rachel Neugarten, and Mary Pfaffko. We are especially grateful to Kent Redford, who reviewed the entire book, challenged us on many points, and helped ensure that we did a better job of reaching target audiences. We have been inspired by the work of colleagues around the world. A great many of them are mentioned in this book and were kindly willing to share the examples, photos, figures, and tables that we use to illustrate the process and tools of conservation planning.

EG is grateful to Craig Groves for his friendship, generosity, and the opportunity to do and share the work contained in this book; and to Eve McDonald-Madden for her unwavering support and partnership on this and many other journeys. I am indebted to Eddie Game, whose family generously opened their home in Brisbane to me on numerous occasions, steeped me in decision science and multi-criteria analyses, and had the patience and calm to see this project through its various roadblocks and challenges.

Craig Groves
*Bozeman, Montana, USA*

Edward Game
*Brisbane, Queensland, Australia*

October 2015

# About the Authors

**Craig Groves** is the executive director of the Science for Nature and People Partnership (SNAPP)—a collaboration of The Nature Conservancy (TNC), the Wildlife Conservation Society (WCS), and the National Center for Ecological Analysis and Synthesis (NCEAS) (www.snap.is). SNAPP delivers evidence-based, scalable solutions through multidisciplinary working groups to global challenges at the interface of nature conservation, economic development, and human well-being. He also serves as the series editor for IUCN's World Commission on Protected Areas (WCPA) *Best Practice Guidelines.* In his 30-year career in nature conservation, Craig has worked as a conservation scientist and planner for TNC, WCS, and the Idaho Department of Fish and Game. Natural resource agencies and nature conservation organizations worldwide have used his first book on conservation planning, *Drafting a Conservation Blueprint* (Island Press, 2003). He has published numerous peer review articles on conservation planning, climate adaptation, monitoring, and evaluation, and on the ecology of at-risk species in the Rocky Mountains, USA. He resides in Bozeman, Montana, USA.

**Edward Game** is a lead scientist with The Nature Conservancy and editor in chief of the journal *Conservation Letters.* He has worked on conservation plans in over 15 countries and published more than 30 papers on aspects of conservation decision making. He authored the manual for the world's most widely used conservation planning software, Marxan, and received the Great Barrier Reef Foundation's inaugural prize for innovative concepts to conserve the reef in the face of climate change. He holds an adjunct faculty position at the University of Queensland. Eddie has explored some of the world's least visited destinations and has written for magazines including *Outdoor* and *Australian Geographic* on adventures such as mountain biking in Kyrgyzstan and kayaking in Greenland. He lives in Brisbane, Australia.

# 1

# The Why, Where, How, and What of Conservation Planning

## Overview

Conservation planning is the process of making informed conservation decisions. In this introductory chapter, we explore the rationale behind conservation planning and the many different forms that such planning can take. Scientific information and analyses can make major contributions to planning, but the values held by organizations and institutions are important and underappreciated drivers of planning processes. Spatial priority setting (addressing where conservation actions should take place) and strategic planning (what actions are needed) have traditionally been conducted separately, but we integrate the two forms in this book. Conservation planning is evolving and growing to meet increasingly complex conservation challenges. At the same time, there are areas within the current practice of planning that need to improve. We outline six key areas for improving the practice of conservation planning.

## Topics

- *Rationale for planning*
- *Values and science in planning*
- *History of conservation planning*
- *Organization of this book*
- *Areas for improvement*
- *Different types of conservation planning*

## The Importance of Conservation Planning

Imagine a watershed where the mountainous forests in the headwaters are rich in endemic species, and are also the primary source of drinking water for a major city. Native shrublands and grasslands in the foothills are slowly being fragmented by agricultural expansion that endangers several wildlife species. Poor logging practices are degrading water quality and the ecological integrity of the forested landscape. Downstream in the watershed, new dams are being proposed as climate change–induced periods of drought pose long-term threats to water supplies. A regional conservation plan has highlighted how critical this landscape is to conserving biodiversity and some ecosystem services. Scientists and planners are just beginning to wrestle with the myriad problems they need to confront to conserve this landscape and watershed. Although they have given considerable attention to the species and ecosystems of interest, they recognize that more of the planning effort will need to be directed at understanding the social, political, and economic context that exists in the watershed.

The watershed we have just described could be almost anywhere in the world, and although fictitious, the challenges presented by this hypothetical project probably sound all too familiar. At the same time, these are exactly the types of challenges that the methods and tools of conservation planning were designed to address. Twenty years ago, systematic planning efforts to conserve nature were rare in most conservation organizations and government agencies. Nature conservation was too often an ad hoc and inefficient enterprise that was largely focused on opportunities. Now that has all changed. Private donors, foundations, multilateral organizations, and governments have provided millions of dollars in funding to systematically develop conservation plans. These plans contain visions, goals, priority areas, conservation outcomes, and strategies and actions to achieve those outcomes. And those same institutions have provided even more funding to implement these plans because of the compelling and transparent visions and approaches the planners present (e.g., Conservation International's Biodiversity Hotspots and the World Wildlife Fund's Global 200 Ecoregions). Conservation planning, in all its different forms and fashions, is now both an expected and a valued practice in nongovernmental organizations (NGOs) and government agencies alike.

We plan because we always have limited resources, and planning helps us make decisions about where and how to deploy those resources. Planning can be a means of settling conflicts among multiple uses and users, as anyone who has participated in a public planning process that involves lands and waters managed by government can appreciate. National forest planning processes in Australia and the United States are prime examples of management planning that involves multiple and often conflicting management objectives. Sometimes planning is undertaken to better understand a problem and to begin to develop solutions, while at other times planning is a strategy unto itself to engage stakeholders and raise the profile of conservation. Finally, planning provides for a level of transparency to all involved that not only helps build trust among stakeholders and project team members but also allows for critiquing and improving of project strategies as they learn about the effectiveness of their efforts over time.

Like any industry or discipline, conservation planning needs to continuously improve to be effective. This is especially true in our era of global climate change and global economic change, both of which are having and will continue to have serious consequences for nature conservation efforts and contribute to the growing complexity of nature conservation work. In fact, many of today's conservation challenges are what one of us has referred to as "wicked problems": problems that lack clear solutions because each solution is linked to other problems.[1] The take-home message is that conservation planning needs to evolve and improve in critical ways to tackle today's conservation problems. At the same time, current practices in conservation planning have well-known shortcomings—for example, implementation often falls short, alternative strategies are too often left unexplored, and there is a strong reliance on expert opinion that is not always elicited in the most effective manner.

We will explore the evolution and growth of conservation planning, as well as its shortcomings, in more detail in the chapters ahead. We will also discuss and demonstrate the steps, methods, tools, and processes for developing and implementing conservation plans that work. But what do we mean by the expression "plans that work"? We mean that if the steps and methods outlined in this book are adopted as part of a planning process, the conservation project will have a greater chance of achieving its desired outcomes.

Planning that incorporates some aspect of nature conservation can arise in many different institutional contexts and may have various motivations. The planning may be mandated by law or policy, in the case of many government natural resource agencies. In other situations, planning may be related to achieving broader sustainability goals for society at scales from local to national.[2] Planning may be motivated by international conventions or treaties, such as the Convention on Biological Diversity, or by a broader natural resource policy framework, such as ecosystem-based management.[3] Whether it is planning to achieve the twin goals of conservation and development,[4] planning that is largely focused on ecological restoration,[5] nature conservation within a land use planning context,[6] or planning that is primarily focused on biodiversity conservation, we want this book to apply to a broad range of institutional settings and motivations for nature conservation. It is our intention that conservation practitioners, project managers, planners, and conservation scientists from around the globe, whether in nongovernmental organizations, government agencies, academic institutions, or the private sector, will find the methods, tools, and examples that we outline in the chapters ahead to be useful.

We hope that as you use this book, you will come away convinced that the principles of good planning and implementation are largely the same across a variety of spatial and temporal scales, institutional settings, and motivations for nature conservation, and that the information we have detailed in this book will help improve the practice of conservation planning across the board.

## Why Plan?

This is a book about planning for nature conservation, so a reasonable question to ask at the outset is, why plan? For starters, nature conservation, as we have already suggested, is an increasingly complicated business. Many, if not most, conservation projects and natural resource management programs involve solving complex problems that have ecological, social, political, and economic dimensions.[7] In part, these problems are complex because humans are increasingly the dominant force affecting the world's species and ecosystems—presenting a challenge that has been referred to as "conservation in the Anthropocene."[8] As Thomas Friedman has pointed out repeatedly,[9] even small changes in economic conditions in one part of the world can have large consequences for biodiversity, land use, and other natural resources in distant parts of the globe.

Whereas biodiversity conservation was once heavily focused on conserving species across landscapes, the Millennium Ecosystem Assessment, conducted under the auspices of the United Nations Environment Programme, awakened the conservation community to the importance of conserving ecosystem processes and services, such as water provision and nutrient cycling, and the benefits these processes can provide to people. In short, conservation has become such a challenging enterprise that developing adequate strategic responses that have a good chance of succeeding often requires a considerable amount of forethought and planning. Although there is no such thing as a *typical* conservation project, REDD (Reducing Emissions from Deforestation and Forest Degradation) projects demonstrate these challenges, as they may involve payments from a carbon market to a government in exchange for limiting timber harvest, an emissions reduction component, a forest restoration component, some aspects of

FIGURE 1.1 Tropical forests of Berau, Indonesia, on the island of Borneo. These forests are the location of a REDD+ (Reducing Emissions from Deforestation and Forest Degradation) project. REDD projects are only one example of the increasing complexity of nature conservation. In this case, carbon markets and financial incentives are used to reduce greenhouse gas emissions. In practice, REDD projects usually include multiple objectives such as biodiversity conservation, improved forest and protected area management, timber production, job opportunities, and improved governance, all while relying on relatively technical measurements of forest carbon stocks.

biodiversity conservation, and possibly some social development and employment objectives—all of which will involve trade-offs (FIGURE 1.1).

There are, of course, many reasons to plan. Conservation biologist Kent Redford and colleagues[10] have suggested that conservation plans are intended to answer two major questions: *where* on the ground (or in the water) are the most important places to undertake conservation activity to achieve the stated goals of a project, and *what* are the strategies, interventions, and actions that are best to implement? Conservation biologists and planners often refer to planning processes that answer the former question as *spatial* and those that answer the latter question as *strategic*. The "where" question has been the subject of hundreds of articles in scientific journals that broadly address the topic of conservation planning or spatial planning, but strategic planning for conservation has received far less attention. In the end, all plans, spatial or strategic, are about allocating resources to some sort of conservation action, so we see spatial planning simply as part of strategic planning. We will return to this theme later in the book.

Although there are many good reasons to engage in conservation planning, we believe that the most important reason is to develop effective strategies and actions that will lead us to better conservation outcomes than intuition alone would lead us to. This is the fundamental assumption of any type of planning—that a thoughtful, deliberate planning process will lead to better decisions and better conservation outcomes than those conducted with a less thoughtful process—and it is the premise of this book as well (as its subtitle makes clear). Having said that, we recognize that there is considerable latitude in what constitutes a thoughtful planning process, including how many resources should be invested in any particular planning effort (we discuss resources in more detail in Chapter 2). Not all conservation projects, initiatives, or management efforts are created equal, and it follows that not all planning efforts merit equivalent investments in staff time and financial resources.

Throughout this book, we make references to scientific information, data, analyses, and a range of planning tools and methodologies. It would be easy to give our readers the impression that conservation planning is a science-driven process. Certainly there is an element of truth in such a statement. Yet it is an underappreciated aspect of conservation planning that individual human and institutional values play a significant role in shaping conservation projects. Sometimes it is difficult to distinguish between science-driven approaches and tools and those that are more value laden. Understanding the distinction between the values underlying a conservation plan and the scientific methods, data, and analyses used to develop that plan will help ensure that conservation scientists and practitioners alike make the most of the advice we provide in the chapters ahead on planning approaches, methods, data, and tools (**Box 1.1**).

In the remainder of this chapter, we briefly review the history of conservation planning to provide the context for how and why it has changed over time, summarize the different types of conservation planning, articulate why and how conservation planning needs to improve and evolve, and describe how this book is organized to assist practitioners in developing, improving, and implementing conservation plans.

## A Brief History of Planning for Nature Conservation

Systematic efforts to plan for nature conservation got their start in the 1970s. In the United States, Natural Heritage programs—cooperative ventures between The Nature Conservancy (TNC) and state natural resource agencies designed to inventory and manage information on rare species and ecological communities—were initiated in the

## BOX 1.1 Values, Science, and Conservation Planning

The authors of this book have spent significant periods of our professional lives encouraging the use of scientific data and analyses in the design, implementation, and management of conservation projects and initiatives. At the same time, we need to recognize the important role that human values play in planning and managing conservation projects. More importantly, we need to understand that many conservation projects fail to acknowledge the differences between values, value judgments, and more data-driven or scientifically objective components of plans and projects.

The conservation biologist George Wilhere[a] has provided an excellent example of the differences between human values and scientific data in a discussion of a question that is critical to conservation planners: How much is enough? In other words, how many populations of a species, or how much of an ecosystem, do we need to conserve to avoid the extinction of that species or the loss of that ecosystem? Although there are data, analyses, and models that can inform an answer to this question for many features of biodiversity, considerable human judgments and values enter into it as well, in the form of what level of extinction risk is acceptable to one person versus another.

What are human values in the conservation context? The Western Australian government scientist Ken Wallace[b] has defined them as "enduring beliefs concerning ultimate, preferred end-states of existence," and he has outlined a planning framework that specifically includes human values and their relationship to goals, objectives, biodiversity assets, ecosystem services, and strategic actions that could be taken to conserve those assets and services. Within a conservation project, these "end-states of existence" might refer to the physical environment, the native species that occur within the project area, or the human community and what its members might prefer socially, economically, or politically. The fundamental objectives of a conservation project (see Chapter 3) are really statements that articulate the values relevant to the plan—they are the reason for caring about the outcome of the plan and its subsequent implementation. Chapter 3 also emphasizes the important role that values play in focusing our planning efforts on achieving what really matters.

On the conservation side of planning, values are usually tied to the mission of the agencies, institutions, or organizations involved in the project. For example, many projects around the world are focused on conserving coral reefs and the biodiversity and ecosystem functions associated with them. In some cases, local human communities depend on reef fish as either a source of protein (nutrition) or a source of income.

Marine protected areas are one commonly applied strategy for conserving these reef systems. The human values associated with a coral reef conservation project may be the long-term preservation of fish species for recreational, spiritual, or aesthetic reasons as well as for purposes of food security or income. Scientific data and analyses can inform what sorts of actions conservationists might take to conserve a reef, how large an area needs to be considered, what sorts of factors may be influencing the survival of reef fish and of the reefs themselves, the trade-offs involved in different human uses of the reef and their effects on the reef community, and whether various conservation actions are having the intended effect of conserving the reef system. Marine protected areas are a means for achieving a goal that was established on the basis of human and organizational values.

The point to remember here is that a consideration of human values has helped clarify what the conservation initiative cares about and wants to achieve, whereas scientific data, models, and analyses can inform the actions that will best help realize those values. Being clear and transparent about what human

*(continued)*

**BOX 1.1 Values, Science, and Conservation Planning** *(continued)*

values are at play in a conservation initiative will ultimately benefit any project by avoiding confusion between means and ends, by clarifying trade-offs between different values and their related goals and objectives, and by avoiding the common misunderstanding that objective, evidence-based science is the principal driver of most aspects of conservation projects.

---

[a] Wilhere, G. F., *The how-much-is-enough myth.* Conservation Biology, 2008. 22(3): 514–517.

[b] Wallace, K. J., *Values: Drivers for planning biodiversity management.* Environmental Science and Policy, 2012. 17: 1–11.

early 1970s. Information and data from these programs, which now cover most of the 50 U.S. states and Canadian provinces as well as a few locations outside North America, have many uses, but chief among them is the identification of important sites for conservation. In the early years of the Natural Heritage Network, the programs produced Natural Diversity Scorecards, which ranked potential conservation sites based on the elements of biodiversity contained within the sites and their conservation status. In turn, these scorecards were used by TNC and government agencies to help prioritize the places most in need of conservation attention. In 2001, NatureServe[11] was established to serve as an umbrella organization for the network of Natural Heritage programs and to help market and disseminate regional and national data products on the status of biodiversity to a variety of government and nongovernmental clients.

At about the same time that systematic efforts began in the United States, *A Nature Conservation Review* was published in the United Kingdom.[12] This report represented one of the first systematic efforts to use a set of criteria to evaluate sites for their potential as nature reserves. These criteria, some of the most important of which were rarity, species diversity, and degree of naturalness, were used to nominate National Nature Reserves. Today, a quasi-governmental body—Natural England—manages a network of National Nature Reserves in collaboration with many partners.

In the 1980s, Australian scientists initiated efforts to establish a comprehensive and ecologically representative set of nature reserves. They too applied criteria such as rarity, richness, and naturalness to help identify reserves. A special volume of the journal *Biological Conservation*, published in 1989, featured numerous papers on conservation planning in Australia and introduced the concept that would later be known as systematic conservation planning (described in the next section of this chapter). Systematic conservation planning continues to strongly influence the Australian government's efforts to establish an ecologically representative set of nature reserves throughout the country as part of the National Reserves Program. And as will be evident throughout this book, Australian expertise in conservation planning has had a profound global influence on the planning, design, and establishment of protected and conservation areas in all their various forms, as well as on the process of conservation planning more generally.

Since the late 1980s, the disciplines of conservation biology and landscape ecology have considerably influenced the development of conservation planning methods. The importance of population viability analyses, the concept that larger conservation areas allow for greater long-term persistence of populations and ecosystems, the underlying importance of ecological processes to species and ecosystems, and advances in thinking about connectivity and corridors are but a few of the most important contributions these fields have made to improvements in conservation planning. Arguably one of the greatest advances for conservation planning was the advent of **gap analysis**: the process of identifying which species, ecosystems, and other features of in-terest are already adequately conserved in an existing network of conservation areas, then identifying the remaining biodiversity features that have not been adequately conserved and determining where the best places are for filling in those "gaps" in protection.[13] Gap analyses (FIGURE 1.2) in various forms have now been incorporated into most of the regional conservation planning efforts globally.

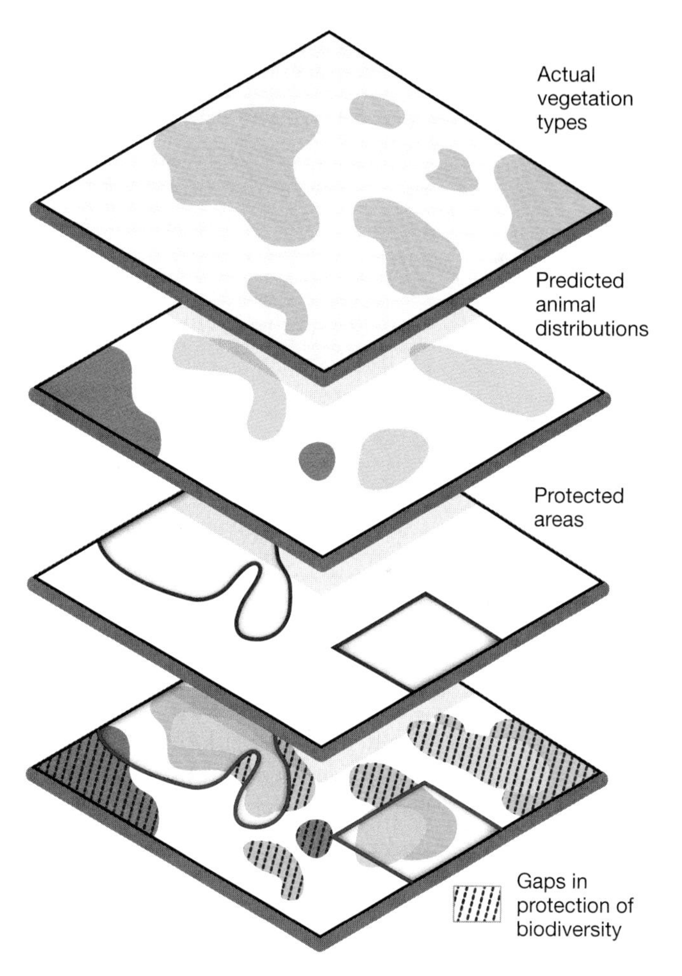


FIGURE 1.2 Gap analysis. Conservation organizations and natural resource agencies began using geographic information systems in the 1980s to identify conservation priorities. One of the first major efforts involved a process of overlaying several data layers for a planning region, such as existing protected areas, vegetation communities or types, and distributions of focal animal species, to identify gaps in protection. These gaps represented species or vegetation communities that were not adequately represented in protected areas. (Adapted from Scott et al., 1993.)

In 2002, scientists and practitioners from several major conservation organizations who were concerned with measuring progress in conservation projects convened a meeting. Out of that meeting the Conservation Measures Partnership was born, and soon after came one of its first products—the *Open Standards for the Practice of Conservation*.[14] Today, these standards are in their third generation, and the underlying adaptive management cycle on which they are based (described later in this chapter) has become standard practice in a number of conservation organizations and increasingly in natural resource agencies and foundations with an environmental or conservation portfolio. In essence, these standards represent the basis for strategic conservation planning approaches. The best known of these approaches is TNC's Conservation Action Planning (CAP), which in itself strongly influenced the *Open Standards* and helped contribute to the development of the desktop planning software—Miradi—that supports their application. CAP and its users also made significant contributions to the launching of a global network of practitioners and coaches who facilitate *Open Standards*–related planning processes: the Conservation Coaches Network, or CCNet.[15]

The United Nations Convention on Biological Diversity (CBD)—a multilateral treaty aimed at conserving the world's biodiversity and first signed at the Rio Earth Summit in 1992—has also contributed substantially to the development of conservation plans. All parties (nations) that are signatories to the CBD (nearly all nations in the world) are required to develop National Biodiversity Strategy and Action Plans (NBSAP). Most nations have developed such plans (184), and many are revising their plans during 2014–2015 in response to the strategic plan of the CBD (2011–2020) and its biodiversity targets.

While conservation biologists and planners have generally focused their planning efforts on relatively ecologically intact regions, landscapes, and watersheds, a parallel effort with many similarities has been undertaken by planners in more urban and developed areas. Their efforts are often referred to by terms such as local planning, land use planning, landscape planning, ecological planning, or urban planning. These planning disciplines were strongly influenced in their early years by a seminal book, *Design with Nature*,[16] which, according to one of today's leading landscape architects, also influenced the fields of environmental impact assessment, coastal zone management, zoo design, river corridor planning, and (ecological) sustainability.[17] Yet, the greatest contribution of *Design with Nature* and of its author, Ian McHarg, was probably the insertion of the principles of ecological planning into local government planning, whereby inventories and studies of the "biophysical and socio-cultural systems reveal where specific land uses may be practiced."[18] McHarg and other early landscape architects paved the way for the expansion of ecological thinking to urban and developed environments, including concepts such as greenways,[19] green infrastructure,[20] nature-friendly communities,[21] and landscape sustainability.[22]

## The Many Forms of Planning for Nature Conservation

Planning for nature conservation comes in many shapes and forms. We have adopted an approach for this book that includes a wide range of efforts to conserve, manage, or restore species, communities, and ecosystems in the terrestrial, freshwater, and marine realms as well as the underlying ecological processes that support these systems and the ecosystem services they provide. In some projects, conservation planning will be primarily focused on lands and waters that are in a natural or seminatural state, while in others, it will be focused on a wider range of lands and waters, including those that have been heavily modified by humans. We recognize that it is increasingly difficult to define what is "natural,"[23] a subject we take up in more detail in Chapter 4.

Because of the many situations in which it can be applied, conservation planning is difficult to define. Lance Craighead and Charles Convis[24] defined conservation planning as "the use of the best scientific information to ensure that natural systems are conserved as human-induced change takes place." Our view of conservation planning is slightly broader:

> *Conservation planning is a systematic process that is primarily focused on identifying, developing, and implementing strategies to conserve specific features of biological diversity, the ecological processes that sustain this diversity, and the ecological (ecosystem) services that are provided by it.*

Identifying the places on the ground or in the water where conservation actions can be most appropriately implemented is in itself a strategy, and we treat it as such. For many forms of planning, our definition of conservation planning will seem appropriate, and the methods and approach that are outlined in this book can be readily adopted. For others, however, such as some types of natural resource management or land use planning, only some elements of our definition may apply. In any case, our approach to conservation planning can be informed by, and can help inform, these well-established professional planning processes.

Each of these planning situations has developed its own flavor of planning, often drawing on different disciplines for inspiration. Nonetheless, there is a great deal of overlap in these different forms of planning, and they have all informed the guidance in this book to some degree. In the following section, we briefly describe several different types of planning and major planning efforts that have both contributed to and benefit from advances in conservation planning.

### Systematic Conservation Planning

**Systematic conservation planning** (**SCP**) can trace its roots to the 1970s in Australia, when biologists there made their first efforts to identify and establish a comprehensive and biologically representative set of nature reserves. In 2000, two Australian biologists and planners, Chris Margules and Bob Pressey, formally outlined a structured approach to conservation planning[25] that would achieve two broad objectives: to represent the full spectrum of biodiversity and the long-term persistence of that biodiversity in a set of reserves. The basic steps involved in their systematic approach range from compiling data on the biodiversity of the planning region to managing conservation areas (FIGURE 1.3). Many agencies and organizations, along with scores of scientific articles, have applied this approach to setting priorities for biodiversity conservation. A number of improvements in both methods and tools for conducting systematic conservation planning have been advanced.[26]

### Strategic Conservation Planning

Systematic conservation planning has been primarily focused on identifying places to conserve through the creation or expansion of conservation areas. Less emphasis has been placed on developing and implementing the range of management strategies needed to adequately conserve biodiversity in these areas. That is the realm of strategic planning. Whether in the nonprofit, government, or for-profit world, an enormous amount of strategic planning takes place. Much of that planning involves an analysis to identify an organization's or project's strengths, weaknesses, opportunities, and "threats" (known as a **SWOT analysis**), as well as activities such as developing a vision, an organizational goal, a broad set of strategies, and more specific objectives and actions to implement those strategies.

In the 1990s, several conservation organizations started developing their own brand of strategic planning, usually aimed at conserving some features of biodiversity in a particular region (known as **site conservation planning**). As we mentioned earlier in this chapter,

FIGURE 1.3 Major steps in systematic conservation planning. These steps, and this planning process, have significantly influenced conservation priority-setting exercises worldwide and have been the basis for scores of publications in the scientific literature. The dashed rectangle represents planning steps originally identified in Margules and Pressey (2000). (Data from Pressey and Bottrill, 2008.)

**1 Scoping and costing the planning process**
Deciding on the boundaries of the planning region, planning team, budget, required funds, and approach to each step in the process

**2 Identifying and involving stakeholders**
Involving, communicating with, and building capacity for stakeholders who will influence or be affected by conservation decisions and implementation of conservation action

**3 Identifying the context for conservation areas**
Assessing the social, economic, and political context for the planning process, including constraints on and opportunities for establishing conservation areas

**4 Identifying conservation goals**
Progressively refining the values of stakeholders from a broad vision statement to specific qualitative goals that shape the rest of the process

**5 Collecting socioeconomic data**
Collecting and evaluating spatially explicit data on tenure, extractive uses, costs, threats, and existing management as a basis for planning decisions

**6 Collecting data on biodiversity and other natural features**
Collecting and evaluating spatially explicit data on biodiversity pattern and process, ecosystem services, and previous disturbance to potential conservation areas

**7 Setting conservation targets**
Translating goals into quantitative targets that reflect the conservation requirements of biodiversity and other natural features

**8 Reviewing target achievement in existing conservation areas**
Assessing, by remote data or field survey, the achievement of targets in different types of conservation areas

**9 Selecting additional conservation areas**
With stakeholders, designing an expanded system of conservation areas that achieves targets while integrating commitments, exclusions, and preferences

**10 Applying conservation actions to selected areas**
Working through the technical and institutional tasks involved in applying effective conservation actions to areas identified in the conservation plan

**11 Maintaining and monitoring established conservation areas**
Applying and monitoring long-term management in established conservation areas to promote the persistence of the values for which they were identified

*Stages described by Margules and Pressey*

**Steps in stage 3**

3.1 Preparing a situation analysis
3.2 Assessing threats in the context of conservation areas
3.3 Identifying actions and mechanisms for identifying threats
3.4 Identifying urgent conservation needs on the advice of stakeholders
3.5 Reviewing the effectiveness of existing conservation areas
3.6 Assessing perceptions and attitudes to planning
3.7 Assessing the strength of governance systems
3.8 Identifying constraints on establishing conservation areas
3.9 Identifying opportunities for establishing conservation areas
3.10 Identifying actions necessary to complement conservation areas

**Steps in stage 10**

10.1 Developing a strategy for the day-to-day mechanics of implementation
10.2 Allocating conservation actions to specific areas
10.3 Deciding how to deal with areas outside the conservation plan
10.4 Estimating the cost of applying conservation actions
10.5 Developing a strategy for scheduling conservation actions
10.6 Mainstreaming the conservation plan for stakeholders
10.7 Applying conservation actions to specific areas
10.8 Identifying lessons for locating and designing conservation areas
10.9 Reviewing progress in applying conservation actions

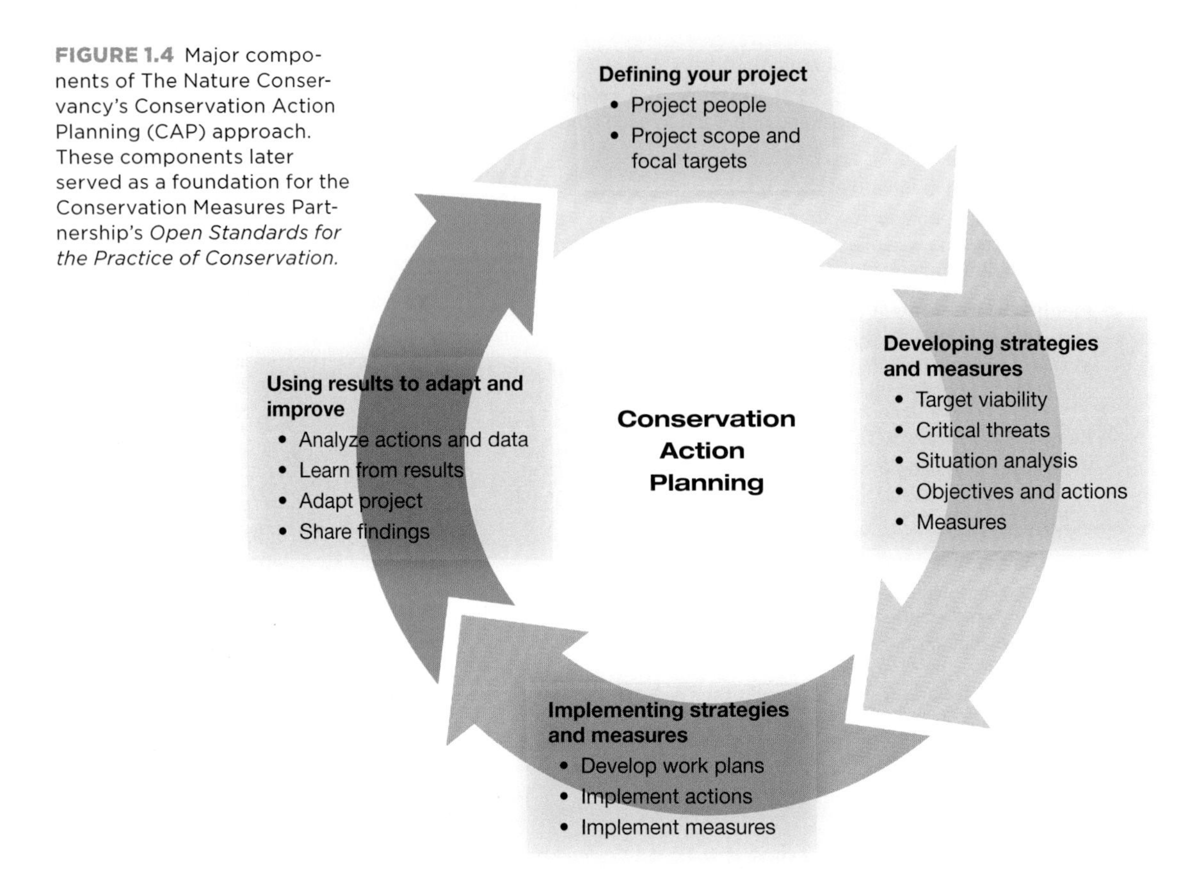


**FIGURE 1.4** Major components of The Nature Conservancy's Conservation Action Planning (CAP) approach. These components later served as a foundation for the Conservation Measures Partnership's *Open Standards for the Practice of Conservation.*

the best known of these strategic planning approaches is TNC's Conservation Action Planning, or CAP (**FIGURE 1.4**). Its predecessor 5-S system for site conservation planning included systems (the elements or features of biodiversity), stresses (threats to those systems), sources (of the stresses), strategies (for abating the threats), and success measures (for determining the effectiveness of the strategies). The Wildlife Conservation Society (WCS) promoted a similar form of site conservation planning, referred to as the Landscape Species Approach[27] (**FIGURE 1.5**). From these early efforts emerged the Conservation Measures Partnership's *Open Standards for the Practice of Conservation* (**FIGURE 1.6**), which includes steps for conceptualizing the project, planning actions and monitoring, implementing actions and monitoring, analyzing results and adapting strategies, and sharing knowledge.

## Natural Resource Management Planning

Most government natural resource agencies develop management plans in response to a set of laws and policies by which they are required to manage their lands and waters. The U.S. Forest Service, for example, is required to develop land management plans that cover multiple uses of forest land, as detailed in the National Forest Management

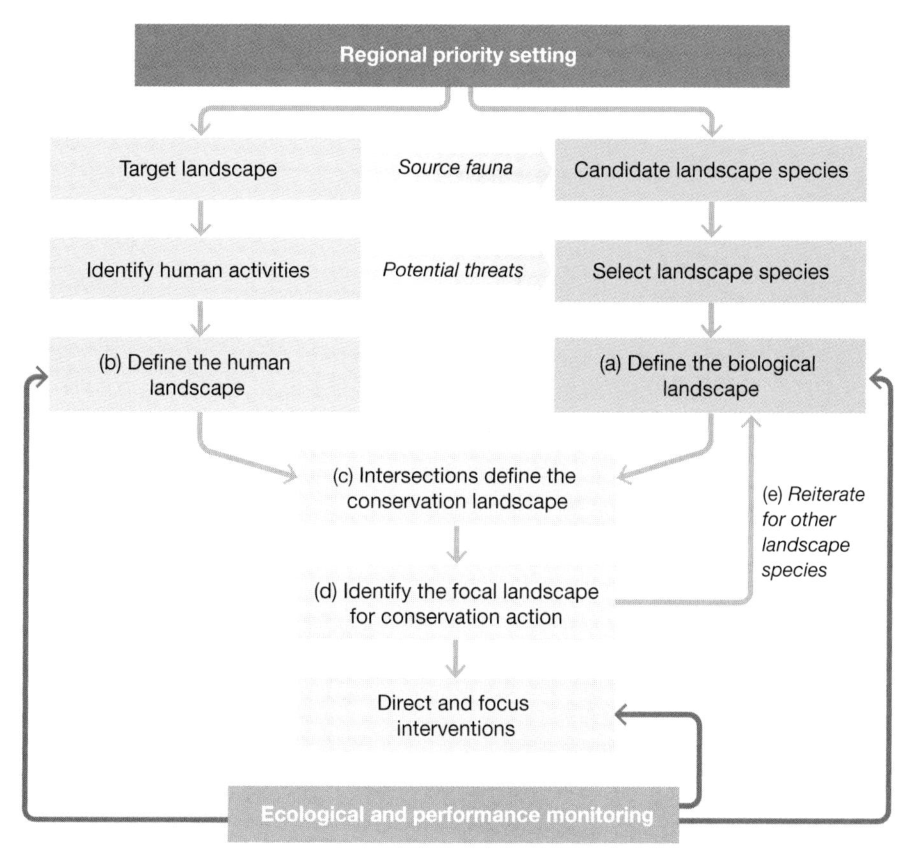


**FIGURE 1.5** The Landscape Species Approach of the Wildlife Conservation Society. Like Conservation Action Planning, the Landscape Species Approach assumes first that a regional planning exercise has established a place and species as priorities. The steps outlined here can be thought of as a form of site conservation planning because they are focused on a place or a site where conservation action should take place (known as the conservation landscape). (Adapted from Sanderson et al., 2002.)

Act. A 2012 planning rule governs the "development, amendment, and revision of land management plans designed to maintain and restore National Forest System land and water ecosystems while providing for ecosystem services and multiple uses."[28] The rule is intended to meet multiple needs: watershed protection, forest restoration, species conservation, and a "sustainable flow of benefits, services, and uses of NFS lands that provide jobs and contribute to the economic and social sustainability of communities."

Laws requiring what essentially amounts to a conservation plan are, of course, not restricted to developed nations. In Indonesia, spatial planning for both terrestrial and marine areas at district, provincial, and national levels is required by national laws (Laws 26 and 27/2007). Under these laws, districts and provinces are required to allocate space for the development and protection of natural resources and features. Guidelines accompanying the legislation outline a system of zones within which different activities are permitted, and they also emphasize the need for public participation in the planning process. Like the National Forest Management Act, Indonesian law explicitly prescribes a multiple-objective planning process to balance ecological considerations with development and resource exploitation

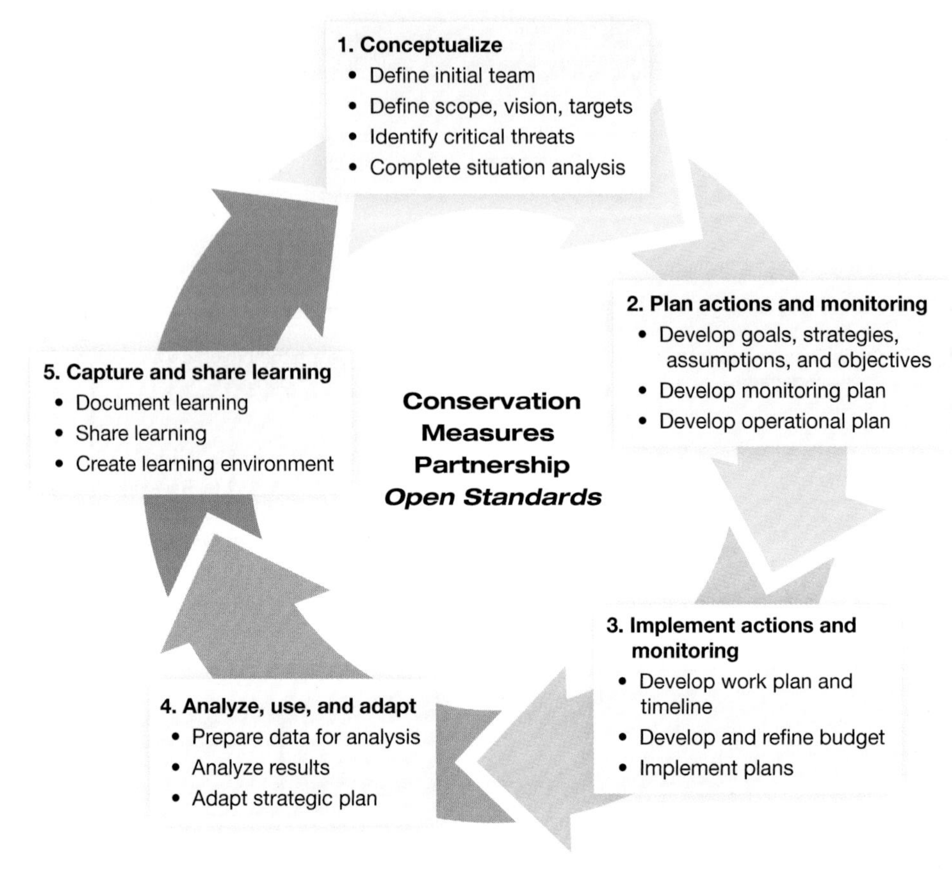


FIGURE 1.6 Major components or steps of the Conservation Measures Partnership *Open Standards*. The steps in this adaptive management cycle are intended to guide staff who are planning and implementing a conservation project. The CMP *Open Standards* are increasingly being adopted worldwide by a variety of government and nongovernmental organizations as well as foundations that support nature conservation efforts. (Adapted from Conservation Measures Partnership, 2013.)

objectives. This law has provided an avenue for conservation NGOs to assist district and provincial governments with planning, leading to valuable collaborations for both partners and important opportunities to influence natural resource outcomes at a large scale.

## Multiple-Use and Multiple-Objective Planning

Although some conservation organizations and government agencies have traditionally had a narrow preservation mandate, many agencies (such as the U.S. Forest Service, as described earlier) and conservation organizations are also engaged in collaborative planning efforts with other sectors of society that encompass multiple objectives, including those related to human use.

Some of the best-known examples of these efforts are marine spatial planning, mitigation planning,[29] and planning for an array of other types of projects that promote ecosystem services (e.g., water funds[30]). For example, the Federation of St. Kitts and Nevis, an island nation in the Caribbean, has developed a marine spatial plan with multiple objectives related to fisheries, ecotourism, ocean transportation, and biodiversity conservation (FIGURE 1.7). The methods and tools used for this sort of planning range from relatively simple spa-

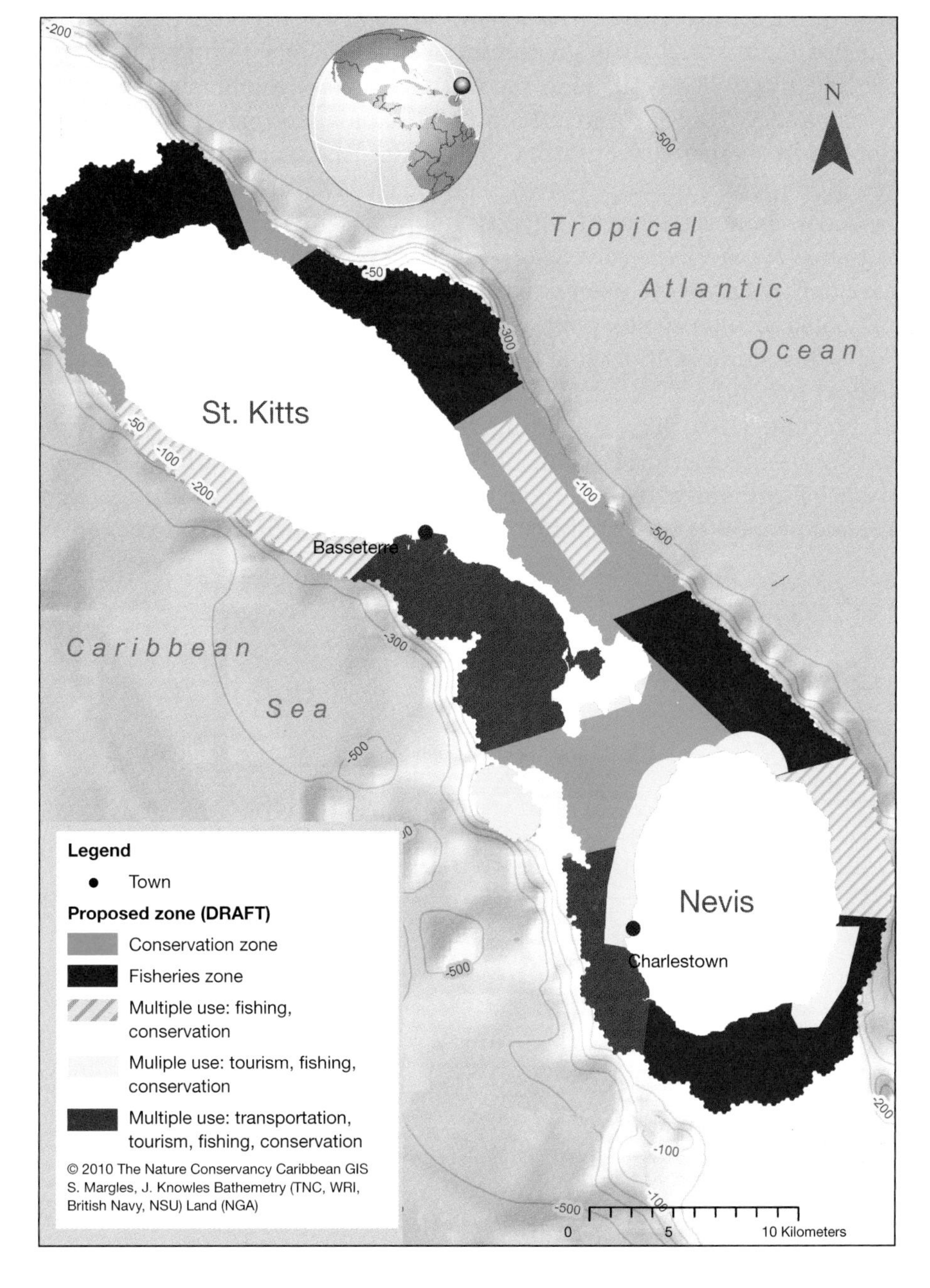


FIGURE 1.7 Results of a marine spatial planning exercise for St. Kitts and Nevis. Conservation planners are able to delineate zones for different uses such as fisheries, conservation, tourism, and transportation. (Adapted from V. N. Agostini et al., *Marine Zoning in Saint Kitts and Nevis: A Path towards Sustainable Management of Marine Resources.* 2010, Arlington, VA: The Nature Conservancy.)

tial mapping or zoning of multiple uses to programs that assist planners in optimizing and analyzing trade-offs among multiple objectives. Planning for mitigation purposes, often through mechanisms known as **biodiversity offsets**, would also fall under the umbrella of multi-objective planning.[31] More detailed information and examples of applying these methods are provided in Chapter 7.

## Ecological Network Planning

Beginning in the 1980s, the idea of ecological networks emerged in Europe (FIGURE 1.8) through the landscape ecology community[32] and in the United States through the work of conservation biologists Reed Noss, Larry Harris, and others.[33] Although the concept has been applied in a variety of situations and scales, the term **ecological network** loosely refers to a network of conservation areas that are ecologically connected for the purposes of promoting biodiversity conservation.[34] It usually involves some sort of core conservation areas and corridors designed to meet the landscape-level needs of species, but it may also include social, political, and cultural components.

Considerable effort has gone into the design and planning of ecological networks,[35] which now number over 150 globally.[36] The state of Florida, USA, has a shining example of a successful statewide network—the Florida Ecological Greenways Network—in which hundreds of millions of dollars have been spent since 2000 to acquire and protect over 900,000 ha of important habitats, which were identified as

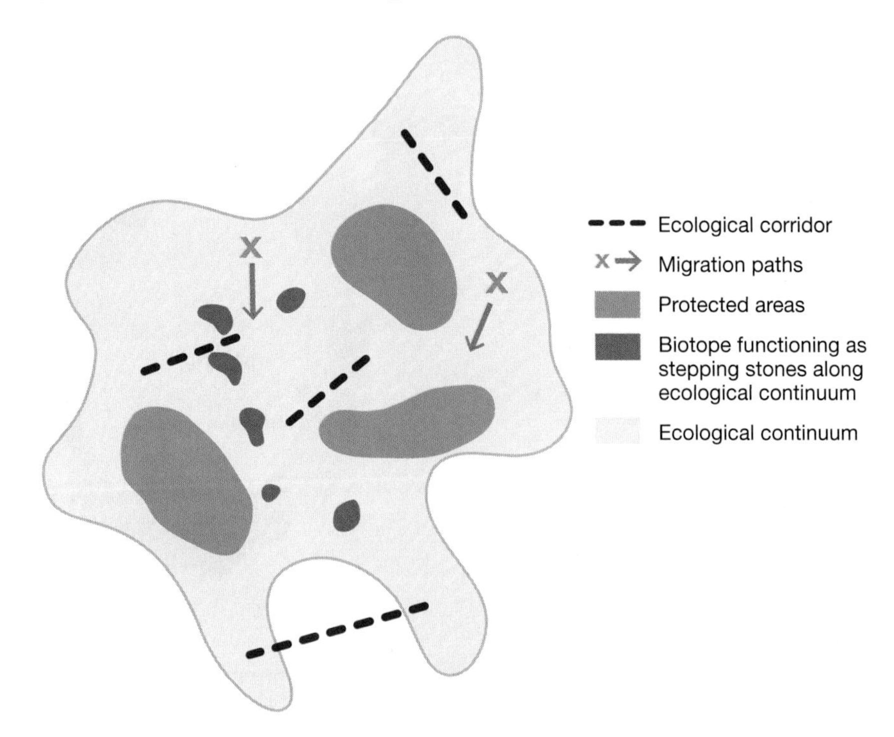


FIGURE 1.8 The major elements of an ecological network. These types of networks have been designed for many European countries as part of a green infrastructure strategy adopted by the European Commission. (Adapted from the Alpine Ecological Network [accessed May 6, 2014]. http://www.alpine-ecological-network.org/the-alpine-ecological-network/ecological-network.)

priorities through a scientifically based plan that included greenways and other critical data layers used to establish acquisition priorities.[37]

### Protected Area Planning

At the beginning of this chapter, we suggested that much of conservation planning has addressed two questions: where conservation areas should be located (spatial planning), and what actions should be taken to manage them (strategic planning). These questions apply quite well to conservation planning that has taken place in relation to protected areas, especially under the auspices of the IUCN World Commission on Protected Areas (WCPA).

The commission's early guidance focused on how to develop a national system of protected areas,[38] but its focus later shifted to management planning for individual protected areas.[39] Several new best practice guidelines and WCPA reports now cover a range of topics, including urban protected area management, the management of large-scale marine protected areas, transboundary conservation, and the identification of **Key Biodiversity Areas**.[40]

### Climate Adaptation Planning

Adaptation strategies (FIGURE 1.9) are now seen as paramount for conserving the world's biodiversity and ecosystem processes and services over the long term in the face of climate change. A number of recent papers and frameworks have outlined approaches for either

FIGURE 1.9 Sign in a U.S. National Forest. A recent U.S. Forest Service guidance document on adaptation to climate change suggests four general types of adaptation strategies: resistance, resilience, response, and realignment. Planning for these adaptation strategies and others will be discussed in detail in Chapter 9. (Adapted from D. L. Peterson et al., *Responding to Climate Change in National Forests: A Guidebook for Developing Adaptation Options.* General Technical Report PNW-GTR-855. 2011, Olympia, WA: U.S. Forest Service, Pacific Northwest Research

developing new conservation plans that incorporate adaptation strategies or modifying existing ones.[41] Most of these approaches rely in part on establishing corridors for plants and animals to move through in response to a changing climate, although a number of other strategies have been outlined as well.[42]

Practitioners who are well grounded in conservation planning recognize that establishing corridors between conservation areas has long been regarded as an important action in fragmented habitats. We will explore these adaptation strategies in more detail in Chapter 9.

### Species Conservation Planning

There are many different forms of species conservation plans, along with a wide variety of guidance to support them. The spectrum is broad—from IUCN Species Action Plans to range-wide species conservation plans[43] to habitat conservation and recovery plans under the U.S. Endangered Species Act. Most of these planning efforts include some sort of review of a species' status and threats to the species, a vision and goal, strategies for conservation, a specific objective, costs and budget, and a monitoring and evaluation component.

In a provocative 2011 paper, Kent Redford and colleagues from the Wildlife Conservation Society outlined the key attributes of successful species conservation efforts, which go well beyond avoiding extinction.[44]

### State Wildlife Action Planning

Each state and territory of the United States was required by the U.S. Congress to prepare a Comprehensive Wildlife Conservation Strategy[45] by 2005 to "assess the health of each state's wildlife and habitats, identify the problems they face, and outline the actions that are needed to conserve them." Although there is significant variation among the 56 state wildlife action plans (SWAPs) that were produced (all of which must be reviewed and updated by 2015), each plan contained eight basic elements, the most important of which were (1) the distribution and abundance of wildlife species, (2) habitat descriptions, (3) assessment of problems facing wildlife and habitats, and (4) proposed conservation actions.

Most of these SWAPs produced a map of important places in the state that need actions if the state's wildlife species and their habitats are to be conserved (FIGURE 1.10). The results of these plans collectively represent a national network for wildlife conservation, although considerable efforts are needed to improve consistency across plans, establish more regional conservation programs, integrate more effectively with federal wildlife programs, and identify gaps in conservation by looking across groups of regionalized state plans.[46]

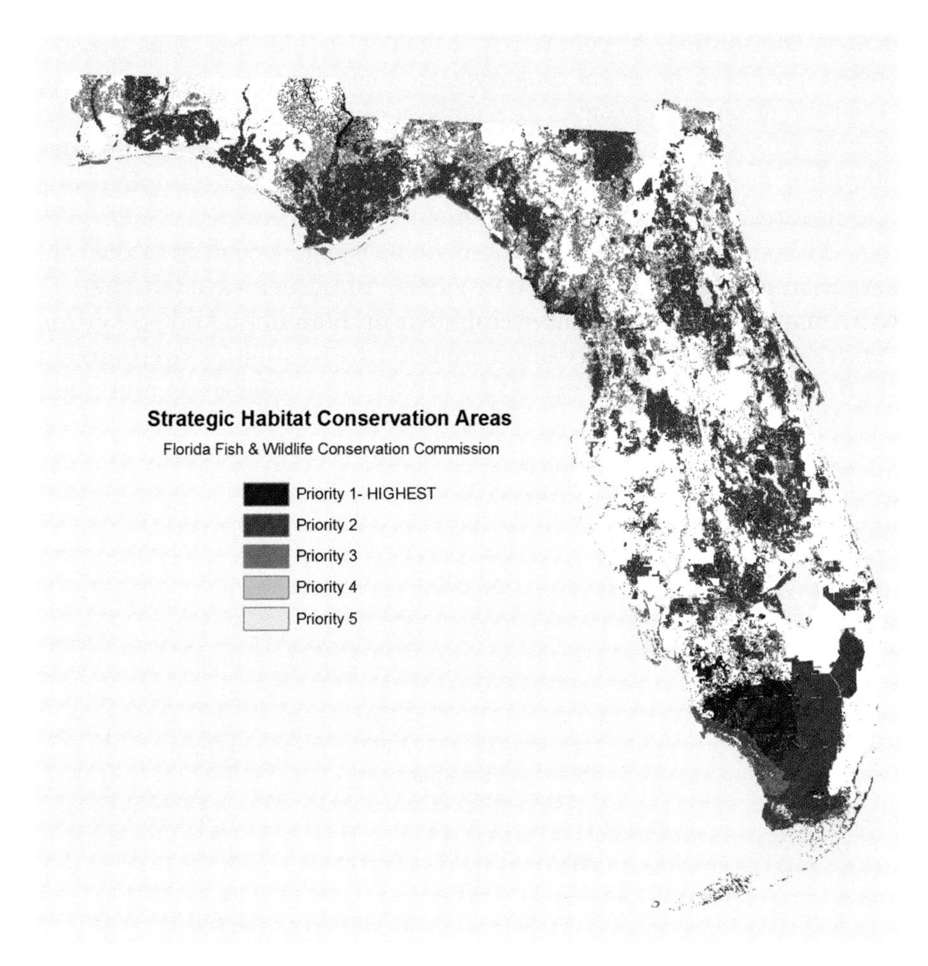


FIGURE 1.10 Strategic Habitat Conservation Areas identified by the Florida Fish and Wildlife Commission. Identifying these areas has been an important part of the creation of the Florida State Wildlife Action Plan, as it will help to identify gaps in the existing statewide system of wildlife conservation areas and will inform ongoing land acquisition and conservation efforts. (Adapted from Florida Fish and Wildlife Commission; map from the CLIP 3.0 database: J. Oetting, T. S. Hoctor, and M. Volk, *Critical Lands and Waters Identification Project (CLIP):Version 3.0.* Technical Report 2014, Tallahassee, FL: Florida Natural Areas Inventory.)

## Land Use and Related Forms of Local Planning

In many parts of the world, but especially in developed nations, a great deal of planning occurs at the level of local or regional governments, such as municipalities, townships, and counties. Nature conservation is sometimes a consideration in local planning, but it is rarely a major emphasis. Planning at the local government level is strongly influenced by graduates of professional planning and landscape architecture schools, and by textbooks and scientific articles published by academics in those schools;[47] this community of professional planners dwarfs the number of people who consider themselves to be conservation planners or even conservation biologists. As noted earlier, this type of planning comes in a variety of forms and falls under the diverse rubrics of green infrastructure, landscape planning, land use planning, ecological planning, greenway development, local planning, and nature-friendly communities, just to name a few.

Although professional planners do pay some attention to the conservation of natural areas, urban parks, and greenways (**FIGURE 1.11**), their plans often place greater emphasis on recreation and ecosystem services, and they are more often created in a regulatory environment than are those plans generated by conservation biologists and plan-

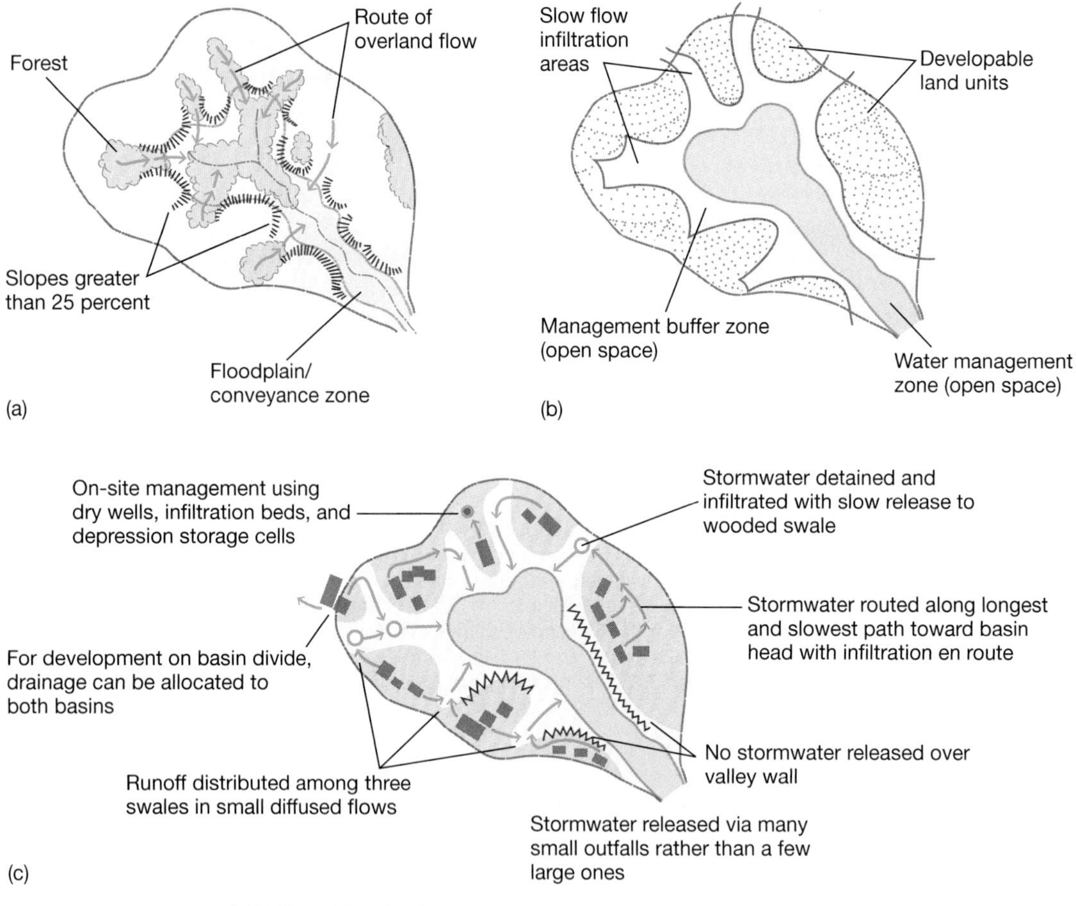


**FIGURE 1.11** A land use plan for a small drainage basin. The maps show (a) hydrologic zones, (b) land use units, and (c) guidance for site planning. There is considerable overlap between the fields of land use planning and conservation planning, although the former is most concerned about places that can be developed (such as the drainage basin shown in this figure) while the latter is often focused on places or strategies for conservation. Although land use planners have traditionally been most interested in ecological processes and services and conservation planners have been most focused on the patterns of biodiversity, these boundaries are now breaking down. (Adapted from Marsh, 2010.)

ners. One of the most widely used textbooks in environmental landscape planning,[48] for example, places considerable emphasis on geophysical features, soil types, wastewater and groundwater flow issues, stormwater management, water quality, and soil erosion, in addition to more biodiversity-oriented topics such as landscape ecology, vegetation, and habitat management.

### National Biodiversity Strategy and Action Planning

As mentioned briefly in the previous section, the Convention on Biological Diversity requires that all signatory nations prepare and maintain National Biodiversity Strategy and Action Plans (NBSAPs). As of May 2015, 184 countries had completed NBSAPs (92% of CBD signatories). Detailed information on each country's NBSAP can be found on the NBSAP portion of the Convention on Biological Diversity website.

In 2010, the United Nations University Institute of Advanced Studies published the only comprehensive analysis of NBSAPs.[49] This assessment found that NBSAPs are highly varied in their form and content. Most of these plans are viewed as comprehensive and were developed in a participatory manner. Although their original intent was to help countries in reducing the rate of biodiversity loss, this assessment concludes that, generally, the implementation of NBSAPs has been unable to achieve this goal. Newer or revised NBSAPs are viewed as being more effective than first-generation efforts, many of which are more than a decade old. Too few plans have been able to "mainstream"[50] the conservation of biodiversity into development sectors, and only a small portion consider climate change. All countries are being asked to revise and update their NBSAPs to reflect the CBD's revised strategic plan for 2011–2020 and the accompanying **Aichi Biodiversity Targets**. As of August 2015, 43 of the 186 nations had revised their NBSAPs.

From the global to the national, regional, and local scales, there are many different types of planning processes that to varying degrees incorporate some elements of what we refer to as conservation planning. In most cases, there is a body of peer-reviewed literature that reports on the strengths, weaknesses, methods, and tools of these different forms of planning. Although we have fallen short of an exhaustive review of all the planning literature, the approaches, methods, tools, and recommendations that we will make throughout this book have drawn on lessons learned from these diverse forms of planning and planning efforts. In turn, we are confident that the guidance we provide in the chapters ahead will be useful to planners and practitioners across this diverse field of planning. In the next section of this chapter, we draw on some of these lessons to focus on how to strengthen the conservation planning process to better position planning efforts to be successfully implemented.

## Improving Conservation Planning for Successful Implementation

Our brief overview of the different types of planning for nature conservation shows that approaches to planning will vary by place, scale, and institutional context. Our review of the lessons learned from this range of planning types, as well as from our own experiences, highlights the need for the methods and tools of conservation planning to grow and improve in the years ahead to meet the needs of conservation in the Anthropocene.

In this section, we outline six key areas for improvement that should be applicable to almost any planning situation and circumstance. Although we will cover these topics in more detail in later chapters, it will be useful for planners, scientists, and practitioners alike to have some understanding from the outset of a planning process of those components of conservation planning that are likely to be of disproportionate importance in crafting a successful plan.

1. *Multiple-objective planning:* Conservation organizations and government agencies have long worked with other sectors, such as the livestock, hydropower, and forestry industries, to achieve conservation objectives. In many cases, these collaborations have occurred on lands or waters that might be termed natural or seminatural. These types of collaborations are evolving in two significant ways. First, the conservation community is finding an increasing number of opportunities to plan for and achieve conservation with more development-oriented partners by planning and working in what some have termed "production landscapes,"[51] lands and waters that are primarily, as the term suggests, oriented toward producing an economic good or service (e.g., agricultural landscapes, forestry plantations, or even urban settings). Second, there are now more sophisticated tools for multiple-objective planning that allow teams to evaluate how to most effectively and efficiently achieve a variety of objectives and understand the trade-offs between what can be competing ends. We will explore some of these tools in more detail in Chapter 7. Critical to the success of these efforts is a willingness of all organizations involved to recognize the legitimacy of conservation and nonconservation objectives alike and to work collaboratively toward them. Ultimately, most if not all conservation planning projects can be treated as multi-objective problems to be solved.

2. *Greater emphasis on social, economic, and political science:* Economist Elinor Ostrom's work[52] on governance processes that lead to improvements in natural resource management

has contributed to the increasing recognition of the important role of social, economic, and political science in conservation planning.[53] So, too, has the renewed interest in the trade-offs between conservation and development,[54] the recent attention given to the limits of sustainability and the needs of sustainability science,[55] the debate over the linkages between the conservation of biological diversity and efforts to alleviate poverty,[56] and the overall importance of considering human well-being in conservation projects and initiatives.[57] The greater attention that we have focused on what many researchers and conservation practitioners have referred to as the **social-ecological system** is, in part, an acknowledgment that any conservation project must operate within social and economic systems that are embedded within ecological systems and that there are important linkages and feedback loops between these systems.[58] Although this book is focused on nature conservation, it is clear that ecological, social, economic, and political science all make important contributions to successful conservation projects and that they all need to be thoughtfully considered in the planning process (FIGURE 1.12).

FIGURE 1.12 Gathering and analyzing social data is an important component of the conservation planning process. Researchers are increasingly using automated technologies to gather such information. Here, participants from local communities in central Kenya are being interviewed in relation to a proposed Payment for Ecosystem Services project in the Tana River watershed. The information from the interviews is being recorded on an iPad for efficient downloading and analysis of interview results.

3. *Integration of spatial and strategic planning:* It may seem obvious that identification of the places that are important for conservation must be closely tied to the strategies that will be used to conserve those places. But in many conservation organizations and agencies, these two aspects of planning have been separate exercises that have been poorly linked and in many cases remain poorly integrated throughout the planning process.[59] Today, we appreciate that most conservation planning occurs at relatively large spatial scales at which it is necessary to identify the places within seascapes, landscapes, and watersheds that are most important for achieving conservation. As we identify these places, we are either implicitly or explicitly beginning to identify strategies and actions that will conserve them. Conversely, when we entertain a strategy (such as readjusting dam operations to provide water-flow patterns that more closely simulate flows that would have occurred in the absence of a dam), we nearly always have in mind places where that strategy will be undertaken. Thinking about both together will result in a more cost-effective and efficient conservation plan. We will review the critical steps of both strategic and spatial planning in Chapter 6, and then discuss how they can best be integrated, along with methods, tools, and examples for doing so, in Chapter 7. Importantly, we argue that all conservation projects should focus on setting priorities for what actions need to take place and where to undertake those actions.
4. *Evaluating alternative actions and strategies:* It is human nature to default to what we do best, and that is true in conservation organizations and natural resource agencies. For example, a conservation organization such as a land trust may have a lot of expertise in acquiring land or purchasing conservation easements. These land protection strategies are excellent solutions to many conservation problems. However, in a situation that requires, for example, improving watershed management, buying or otherwise protecting land may not be the most effective strategy, even if it is the primary expertise of a local conservation organization. To overcome this natural tendency, it is important for project teams to consider a range of possible actions that could help achieve the desired outcomes of a project and to document with some evidence[60] why one course of action may be preferred over another. Such an evaluation needs to consider the consequences of alternative courses of action as well as the risks and assumptions behind those alternatives. We will explore this evaluation of alternative actions and the impor-

tance of taking an evidence-based approach to conservation in more detail in Chapters 6 and 7.

5. *Risk analysis:* All conservation actions or projects have a chance of turning out worse than we want or expect. We know from bitter experience, for example, that conservation actions that involve eradicating an invasive pest often fail. Conservation actions may also turn out poorly because the people implementing the action fail (in extreme cases, the money is spent on other things), because sociopolitical forces (such as loss of community support or a change of government interest) thwart the action, or simply because the design of the action was based on incorrect assumptions about the way social-ecological systems function. We refer to these uncertain events that might have a negative effect on the outcome of conservation projects as **risks**. From a planning and decision-making viewpoint, risks should influence how we perceive the expected benefits and costs of a conservation action, yet frequently conservation plans contain no mention of risks at all. Indeed, in our experience, many practitioners regard a risk as something more akin to an obstacle that they need to work around, such as an action that could damage the credibility of an agency or organization. Ecologist and risk assessment expert Mark Burgman[61] and others have argued that risk is an area of judgment and intuition that is particularly prone to bias in conservation work. It has also been repeatedly demonstrated that people typically have a poor understanding of probabilities, which further detracts from our ability to intuitively assess risk.[62] Chapter 8 explores the role of risk in conservation planning in more detail.
6. *Planning-implementation gap:* Much has been written about the planning-implementation gap in conservation[63] (also sometimes referred to as the knowing-doing gap and the research-implementation gap). The peer-reviewed literature that addresses this gap tends to focus on regional priority-setting exercises and how few of them result in conservation action. Although there are certainly elements of truth to this view, it is also a reflection of the bias in the scientific literature that we have previously noted: only a few articles on conservation planning actually address strategic planning and implementation of conservation actions. In addition, many of the authors of conservation priority-setting papers never intended them to be avenues for implementation; they were primarily academic exercises. Still, even for organizations that invest heavily in strategic conservation planning, there can be serious shortcomings in

plan implementation. In the final chapter of this book, we dive into the details of why implementation—the process of making decisions and undertaking actions—often falls short, and we explore how it can be improved. At the same time, we appreciate that most successful conservation planning efforts consider implementation from their inception and throughout the planning effort. Our aim is to do the same in this book by providing small doses of advice about implementation in most chapters.

## How This Book Is Organized

The purpose of this book is to provide conservation practitioners, scientists, natural resource managers, and planners with the best possible guidance for developing and implementing an effective conservation plan anywhere in the world. To achieve this purpose, we have blended knowledge, practice, and methods from a wide array of planning disciplines and settings. Our approaches and tools have their origins in a variety of disciplines, including ecology, conservation biology, economics, sociology, and political science. We also draw heavily on **decision theory** and **decision science**. Along the way, we have endeavored to demonstrate methods and tools with examples of their application in real conservation projects while taking care to highlight some of the most important scientific literature. We make no attempt to be exhaustive in either case.

Although we must describe the different steps of conservation planning in a somewhat linear fashion here, the planning process itself is anything but linear (FIGURE 1.13). There are many points of feedback among what we see as the three major components of any planning process:

1. Framing conservation planning problems
2. Solving conservation planning problems
3. Planning implementation

We will remind readers of the many linkages between the different steps of the planning process, as described in Figure 1.13, throughout this book.

To assist the reader in better comprehending the development and implementation of a conservation plan, we have divided the book into three main parts:

Part One: Developing a Conservation Plan

Part Two: Special Topics in Conservation Planning

Part Three: Putting the Plan in Practice: Implementation

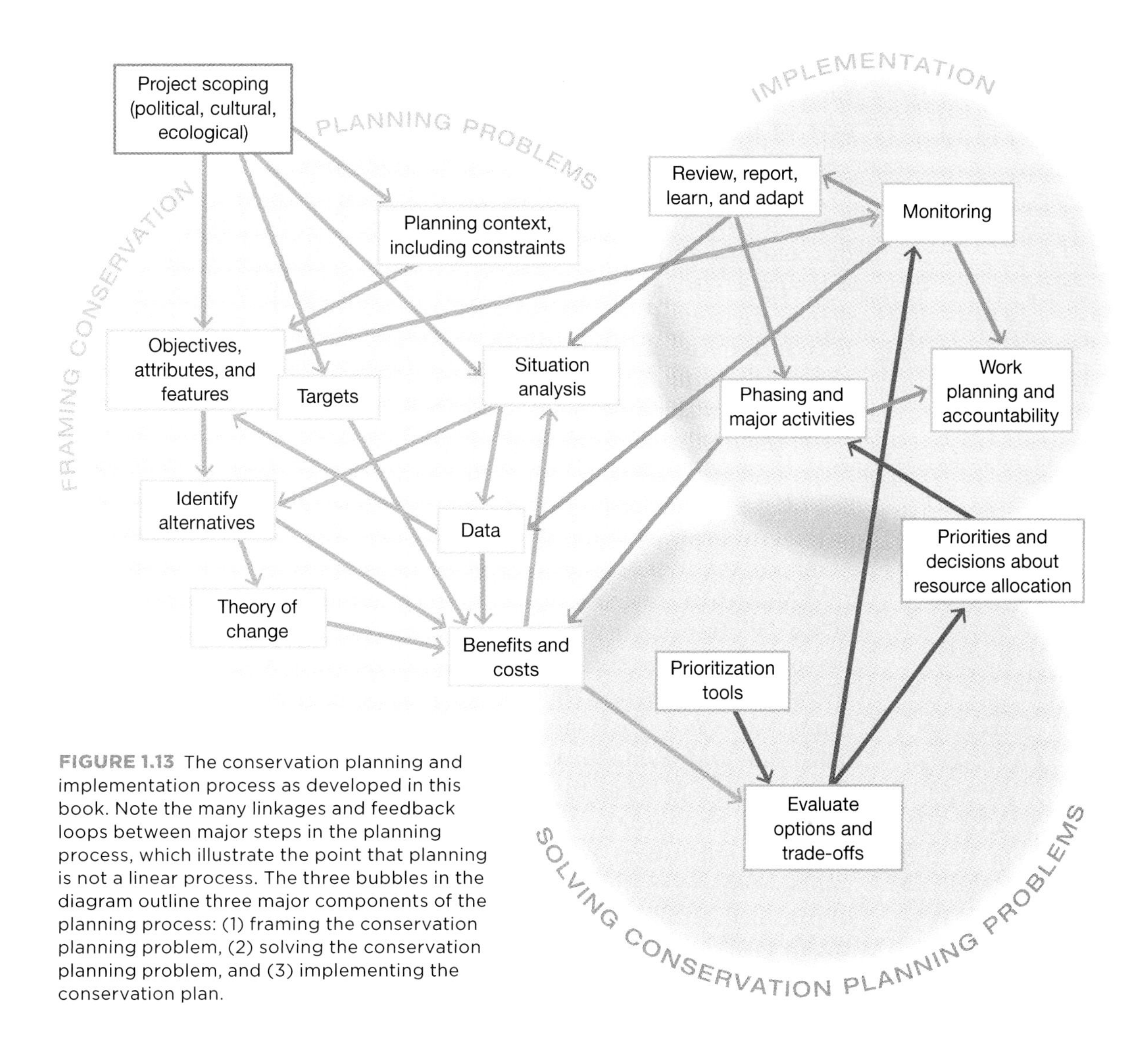


**FIGURE 1.13** The conservation planning and implementation process as developed in this book. Note the many linkages and feedback loops between major steps in the planning process, which illustrate the point that planning is not a linear process. The three bubbles in the diagram outline three major components of the planning process: (1) framing the conservation planning problem, (2) solving the conservation planning problem, and (3) implementing the conservation plan.

Part One is intended to provide scientists, planners, and other conservation professionals with the guidance needed to develop a conservation plan. Chapter 2 is focused on helping a planning team get started on the right foot by considering what questions a plan should answer, what decisions it will help make, and who the decision makers to actually put the plan into practice will be. It contains the nitty-gritty on such topics as assembling a team and managing a planning process to fruition, and it introduces the basic steps of any conservation planning process. Many plans are handicapped from the outset by a failure to develop clear objectives, so Part One devotes an entire chapter (Chapter 3) to identifying and structuring the objectives that will guide subsequent planning, including environmental, social, and economic objectives. Objectives are distinguished from targets (Chapter 4), which are specific quantitative statements about the desired level of achievement of those objectives. We think of targets as objectives viewed through the lens of available data, so in Chapter 5 we review the major types of information and data that can be used in plans and provide guidance on assessing threats. Chapter 6 describes how we can frame a conservation plan as a problem to solve by considering the range of strategic options available and developing a theory of change describing how strategies and actions will help deliver the conservation outcomes we desire. The intent of Chapter 7 is to introduce some of the most important tools for solving conservation planning problems, such as scenario analyses, return-on-investment analyses, and multi-criteria decision analyses. In Chapter 8, the final chapter of Part One, we devote attention to one of the most underappreciated topics in conservation planning: the risks related to undertaking certain actions, the uncertainty associated with many aspects of conservation projects, and approaches for understanding and confronting risk and uncertainty.

Many topics related to developing a conservation plan are worthy of special attention or emphasis. In Part Two of this book, we address two of these topics that are receiving considerable attention by conservation planners and that are significantly influencing many of the component steps of conservation planning that are addressed in this book. The first is adaptation to climate change. Although some of the first articles and books to address this topic were published in the late 1990s, the first decade of the twenty-first century has ushered in several academic papers, reports, and books[64] with broad guidance about adaptation, especially as hopes of curbing worldwide emissions have diminished. In Chapter 9 we'll review this guidance and suggest how it may best be incorporated into conservation planning.

In Chapter 10, we'll explore the developing field of ecosystem services and how it is being incorporated into conservation planning. Although conservation biologists have long been focused on biodiversity, landscape architects and urban planners have traditionally been more concerned with ecosystem services and their provisioning for human well-being. This chapter brings those concerns together, discusses some of the philosophical issues related to conserving ecosystem services, considers some of the challenges of planning for them, and highlights tools and examples that are moving the field of conservation planning forward to more effectively address both biodiversity and ecosystem services.

Even the best available scientific information and all the technical bells and whistles of conservation planning will not necessarily lead to implementation, good decisions, and conservation results. Part Three of this book focuses on implementation and monitoring. In

Chapter 11, we highlight some of the more effective practices for, tools for, and examples of plan implementation. Communication about the plan, engagement of constituents, sound project management and work planning, budgeting and fund-raising, scaling up results from individual projects for broader impacts, and our own "top 10" list for effective implementation will round out this chapter. Chapter 12 focuses on a question that has plagued conservation projects for decades: How do we know that we are being effective and making progress toward the desired outcomes of a conservation project or initiative? We discuss what some of the most important challenges to effective project monitoring and evaluation have been and why the conservation community has a notoriously poor track record in this arena. At the same time, good progress has been made in recent years, and we offer practical advice on how much to invest in monitoring and evaluation and under what circumstances, how to overcome some of the most important pitfalls, why adaptive management isn't the panacea many had hoped it would be, and how we can scale up learning across conservation projects to improve project performance.Finally, we close with a brief Chapter 13 in which we posit that the future of conservation planning will be framed around planning for nature and people, identify some of the challenges in doing this, and summarize the most important approaches and tools that will help meet these challenges.

## KEY MESSAGES

- Conservation planning has brought a systematic, strategic, and transparent approach to the enterprise of conservation over the last twenty years that has inspired donors and constituents for nature conservation and helped to inculcate planning as a routine practice in conservation.
- There are many good reasons to plan: to efficiently allocate limited resources, to resolve conflicts among stakeholders, to better understand problems, and to build support for certain strategies, to name only a few. But fundamentally, we invest in thoughtful planning processes because we believe they will help us reach better outcomes and make better decisions for conservation than we might reach in the absence of such efforts.
- Scientific information and analyses have an important role to play in conservation planning, but an institution's values are also an important driver of planning processes and are ultimately more important in shaping the fundamental goals and objectives of conservation projects and initiatives.

- Planning for nature conservation takes many different forms, occurs in different institutional contexts, and is driven by different motivating factors. The guidance in this book draws on a wide range of planning experiences and disciplines and should, in turn, be useful to practitioners, planners, managers, and scientists in government natural resource agencies and conservation organizations as well as to land use planners and industries engaged in natural resource extraction and production.
- Conservation planning is evolving and improving to meet the needs of conservation practitioners in the Anthropocene. Greater consideration of multiple-objective planning, integrating spatial and strategic planning, evaluating alternative strategies and actions, infusing more socioeconomic science into the planning process, and more attention to the risks and assumptions behind strategies are some of the most important areas where planning can be strengthened to improve the chances for successful implementation.

## References

1. Game, E. T. et al., *Conservation in a wicked complex world; challenges and solutions.* Conservation Letters, 2014. **7**(3): 271–277.
2. Rockström, J. et al., *Planetary boundaries: Exploring the safe operating space for humanity.* Ecology and Society, 2009. **14**(2): 1–33.
3. Levin, P. S. et al., *Integrated ecosystem assessments: Developing the scientific basis for ecosystem-based management of the ocean.* PLoS Biology, 2009. **7**(1): e1000014.
4. McShane, T. O. et al., *Hard choices: Making trade-offs between biodiversity conservation and human well-being.* Biological Conservation, 2011. **144**(3): 966–972.
5. Zedler, J. B., J. M. Doherty, and N. A. Miller, *Shifting restoration policy to address landscape change, novel ecosystems, and monitoring.* Ecology and Society, 2012. **17**(4): 36.
6. Steiner, F. et al., *The ecological imperative for environmental design and planning.* Frontiers in Ecology and the Environment, 2013. **11**(7): 355–361.
7. Game et al., *Conservation in a wicked complex world.* (See reference 1.)
8. Harden, C. et al., *Understanding human–landscape interactions in the "Anthropocene."* Environmental Management, 2014. **53**(1): 4–13.
9. Friedman, T., *The World Is Flat: A Brief History of the Twenty-First Century.* 2005, New York: Farrar, Strauss, and Giroux. 660.
10. Redford, K. H. et al., *Mapping the conservation landscape.* Conservation Biology, 2003. **17**(1): 116–131.
11. NatureServe [accessed September 3, 2013]. http://www.natureserve.org.

12. Margules, C., and M. B. Usher, *Criteria used in assessing wildlife conservation potential: A review.* Biological Conservation, 1981. **21**(2): 79–109.
13. Scott, J. M. et al., *Gap analysis: A geographic approach to protection of biological diversity.* Wildlife Monographs, 1993. **1993**(123): 3–41.
14. Conservation Measures Partnership. *Open Standards for the Practice of Conservation.* [accessed May 6, 2015]. http://www.conservationmeasures.org.
15. Conservation Coaches Network. *Conservation Gateway* [accessed April 23, 2014]. http://www.ccnetglobal.com.
16. McHarg, I. L., *Design with Nature.* 1969, Garden City, NY: Doubleday. 208.
17. Steiner, F., *Healing the earth: The relevance of Ian McHarg's work for the future.* Philosophy and Geography, February 2004. **7**: 141–149.
18. Steiner, F., *The Living Landscape: An Ecological Approach to Landscape Planning.* 2000, New York: McGraw-Hill. 477.
19. Hellmund, P. C., and D. S. Smith, *Designing Greenways: Sustainable Landscapes for Nature and People.* 2006, Washington, DC: Island Press. 270.
20. Benedict, M. A., and E. T. McMahon, *Green Infrastructure: Linking Landscapes and Communities.* 2006, Washington, DC: Island Press. 299.
21. Duerksen, C., and C. Snyder, *Nature-Friendly Communities: Habitat Protection and Land Use Planning.* 2005, Washington, DC: Island Press. 420.
22. Musacchio, L., *The scientific basis for the design of landscape sustainability: A conceptual framework for translational landscape research and practice of designed landscapes and the six Es of landscape sustainability.* Landscape Ecology, 2009. **24**(8): 993–1013.
23. Willis, K. J., and H. J. B. Birks, *What is natural? The need for a long-term perspective in biodiversity conservation.* Science, 2006. **314**(5803): 1261–1265.
24. Craighead, F. L., and C. L. Convis Jr., editors, *Conservation Planning: Shaping the Future.* 2013, Redlands, CA: Esri.
25. Margules, C. R., and R. L. Pressey, *Systematic conservation planning.* Nature, 2000. **405**(6783): 243–253.
26. Pressey, R. L., and M. C. Bottrill, *Approaches to landscape- and seascape-scale conservation planning: Convergence, contrasts and challenges.* Oryx, 2009. **43**(4): 464–475.
27. Sanderson, E. W. et al., *A conceptual model for conservation planning based on landscape species requirements.* Landscape and Urban Planning, 2002. **58**(1): 41–56.
28. Department of Agriculture, 36 Code of Federal Regulations Part 219. National Forest System Land Management Planning, Final Rule and Record of Decision.
29. Kiesecker, J. M. et al., *Development by design: Blending landscape-level planning with the mitigation hierarchy.* Frontiers in Ecology and the Environment, 2009. **8**(5): 261–266.

30. Goldman-Benner, R. L. et al., *Water funds and payments for ecosystem services: Practice learns from theory and theory can learn from practice.* Oryx, 2012. **46**(1): 55–63.

31. Kiesecker, J. M., et al., *Development by design: Blending landscape-level planning with the mitigation hierarchy.* Frontiers in Ecology and the Environment, 2009. **8**(5): 261–266[3]; and Gardner, T. A., et al., *Biodiversity offsets and the challenge of achieving no net loss.* Conservation Biology, 2013. **27**(6): 1254–1264.

32. Jongman, R. H. G., and I. Kristiansen, *National and Regional Approaches for Ecological Networks in Europe. Nature and Environment,* Vol. 110. 2001, Strausborg, France: Council of Europe Publishing.

33. Noss, R. F., and L. D. Harris, *Nodes, networks, and MUMs: Preserving diversity at all scales.* Environmental Management, 1986. **10**(3): 299–309.

34. Boitani, L. et al., *Ecological networks as conceptual frameworks or operational tools in conservation.* Conservation Biology, 2007. **21**(6): 1414–1422.

35. Jongman, R. H. G., and G. Pungetti, *Ecological Networks and Greenways: Concept, Design, Implementation.* 2004, Cambridge: Cambridge University Press. 368.

36. Bennett, G., *Linkages in Practice: A Review of Their Conservation Practice.* 2004, Gland, Switzerland, and Cambridge: IUCN. 28.

37. Hoctor, T. S., M. H. Carr, and P. D. Zwick, *Identifying a linked reserve system using a regional landscape approach: The Florida Ecological Network.* Conservation Biology, 2000. **14**(4): 984–1000; and Oetting, J., T. S. Hoctor, and M. Volk, *Critical Lands and Waters Identification Project (CLIP):Version 3.0.* Technical Report. 2014, Florida Natural Areas Inventory: Tallahassee, FL [accessed March 14, 2015]. http://fnai.org/pdf/CLIP_3_technical_report.pdf.

38. Davey, A. G., *National System Planning for Protected Areas.* 1998, Gland, Switzerland, and Cambridge: IUCN. x + 71 pp. [accessed March 14, 2015]. https://portals.iucn.org/library/efiles/edocs/PAG-001.pdf.

39. Thomas, L., and J. Middleton, *Guidelines for Management Planning of Protected Areas.* Best Practices Protected Areas Guidelines Series No. 10. 2003, Gland, Switzerland: IUCN. 79 [accessed March 14, 2015]. https://portals.iucn.org/library/efiles/documents/PAG-010.pdf.

40. IUCN, *Consultation Document on an IUCN Standard for the Identification of Key Biodiversity Areas.* World Commission on Protected Areas and Species Survival Commission. 2014, Gland, Switzerland [accessed March 4, 2015]. https://portals.iucn.org/union/sites/union/files/doc/consultation_document_iucn_kba_standard_01oct2014.pdf.

41. Groves, C. et al., *Incorporating climate change into systematic conservation planning.* Biodiversity and Conservation, 2012. **21**: 1651–1671.

42. Heller, N. E., and E. S. Zavaleta, *Biodiversity management in the face of climate change: A review of 22 years of recommendations.* Biological Conservation, 2009. **142**(1): 14–32.

43. Sanderson, E. W. et al., *The human footprint and the last of the wild.* BioScience, 2002. **52**(10): 891–904.
44. Redford, K. H. et al., *What does it mean to successfully conserve a (vertebrate) species?* BioScience, 2011. **16**(1): 39–48.
45. Association of Fish and Wildlife Agencies, *State Wildlife Action Plans (SWAPs)* [accessed September 3, 2013]. http://teaming.com/state-wildlife-action-plans-swaps.
46. Meretsky, V., and R. L. Fischman, *Learning from conservation planning for the U.S. National Wildlife Refuges.* Conservation Biology, 2014. **28**(5): 15–27.
47. Steiner, *The Living Landscape.* (See reference 20.)
48. Marsh, W. M., *Landscape Planning: Environmental Applications.* 5th ed. 2010, Hoboken, NJ: John Wiley and Sons. 528.
49. Prip, C. et al., *Biodiversity Planning: An Assessment of National Biodiversity Strategies and Actions Plans.* 2010, Yokohama, Japan: United Nations University Institute of Advanced Studies.
50. Huntley, B., and K. H. Redford, *Mainstreaming Biodiversity in Practice: A STAP Advisory Document. Scientific and Technical Advisory Panel.* 2014, Washington, DC: Global Environmental Facility [accessed March 14, 2015]. http://www.thegef.org/gef/sites/thegef.org/files/documents/GEF.STAP_.C.46.Inf_.04%20Mainstreaming%20Biodiversity%20in%20Practice_April%2025%202014%20rev.pdf.
51. Wilson, K. A. et al., *Conserving biodiversity in production landscapes.* Ecological Applications, 2010. **20**(6): 1721–1732.
52. Ostrom, E., *A general framework for analyzing sustainability of social-ecological systems.* Science, 2009. **325**(5939): 419–422.
53. Ban, N. C. et al., *A social–ecological approach to conservation planning: Embedding social considerations.* Frontiers in Ecology and the Environment, 2013. **11**(4): 194–202.
54. McShane et al., *Hard choices.* (See reference 4.)
55. Rockström et al., *Planetary boundaries.* (See reference 2.)
56. Roe, D. et al., *Linking biodiversity conservation and poverty reduction: De-polarizing the conservation-poverty debate.* Conservation Letters, 2013. **6**(3): 162–171.
57. Milner-Gulland, E. J. et al., *Accounting for the impact of conservation on human well-being.* Conservation Biology, 2014. **28**(5): 1160–1166.
58. Miller, B. W., S. C. Caplow, and P. W. Leslie, *Feedbacks between conservation and social-ecological systems.* Conservation Biology, 2012. **26**(2): 218–227; and Walker, B., and D. Salt, *Resilience Thinking: Sustaining Ecosystems and People in a Changing World.* 2006, Washington, DC: Island Press. 174.
59. Game, E. T., P. Kareiva, and H. P. Possingham, *Six common mistakes in conservation priority setting.* Conservation Biology, 2013. **27**(3): 480–485.

60. Cook, C. N., H. P. Possingham, and R. A. Fuller, *Contribution of systematic reviews to management decisions.* Conservation Biology, 2013. **27**(5): 902–915.
61. Burgman, M. A., *Risks and Decisions for Conservation and Environmental Management.* 2005, Cambridge: Cambridge University Press. 488.
62. Plous, S., *The Psychology of Judgment and Decision Making.* 1993, New York: McGraw-Hill. 352.
63. Knight, A. T. et al., *Knowing but not doing: Selecting priority conservation areas and the research–implementation gap.* Conservation Biology, 2008. **22**(3): 610–617.
64. Glick, P., B. A. Stein, and N. A. Edelson, *Scanning the Conservation Horizon: A Guide to Climate Change Vulnerability Assessment.* 2011, Washington, DC: National Wildlife Federation [accessed March 14, 2015]. http://www.nwf.org/~/media/pdfs/global-warming/climate-smart-conservation/nwfscanningtheconservationhorizonfinal92311.ashx; Hilty, J. A., C. C. Chester, and M. S. Cross, *Climate and Conservation: Landscape and Seascape Science, Planning, and Actions.* 2012, Washington, DC: Island Press. 416.

# Part One

## Developing a Conservation Plan

# 2

# Getting Started: Foundations, Planning Principles and Standards, and a Road Map to Conservation Planning

## Overview

Understanding the context for planning, including the decisions that will be made as a result of it, the constraints on those decisions, and the audience for the plan, is a critical foundational element of good planning. Equally important is the composition of the planning team and its leadership. Most conservation planning processes also use fundamental principles and standards that can trace their roots to the disciplines of conservation biology, program evaluation, and decision science. Although conservation planning methods have been strongly influenced by plans developed in the terrestrial realm, considerable progress has been made in marine and freshwater planning and in integration of plans across these realms. In this chapter, we incorporate all of these considerations into a question-based road map for conservation planning that can be used to guide scientists, planners, and practitioners through the conservation planning process.

## Topics

- *Planning context (purpose, scope, decisions, constraints, audience, investment)*
- *Planning team and team leadership*
- *Planning principles and standards*
- *Integration of freshwater-marine-terrestrial planning*
- *Road map to conservation planning*

A few years ago, on the way home from a conference, one of us was chatting with a colleague about a large conservation planning effort that she had been engaged in. The plan was prepared largely by scientists and planners, and project directors and managers had limited involvement. Although this situation was not necessarily unusual, our colleague was disappointed because not much was happening in terms of implementing the plan.

In hindsight, it was no surprise that this plan lacked the traction needed for implementation. Whether in an agency or conservation organization, experience has taught us that if program managers, directors, or other important decision makers are not engaged in plan development, exactly what decisions the plan is intended to support are often unclear, and decision makers are less likely to "own" the results and take the lead in implementing them. Having a clear understanding of the decisions that a plan is intended to support and engaging the decision makers in the planning process are essential to comprehending the planning context, and they help to get a plan started in the right direction. Doing so with a diverse planning team of planners, scientists, practitioners, and implementers as well as solid leadership is another foundational element of launching a planning process that is well positioned for success. We will explore both of these elements in more detail in the pages ahead.

Most agencies and organizations undertake new conservation initiatives, or new approaches to conservation or natural resource management, that require some form of planning (FIGURE 2.1). The Biodiversity Hotspots approach of Conservation International, the ecoregional visioning of the World Wildlife Fund (WWF), and the development

FIGURE 2.1 Planning workshop in the Northern Rangelands Trust project, Kenya. Workshops that involve the planning team and some stakeholders are critical to launching a conservation planning process. Early workshops need to help build an effective team and consider the context for the planning effort, including the overall purpose and scope of the plan and project, the decisions that the plan will help to make, constraints on those decisions and on the planning effort, and the audience for the plan.

of National Biodiversity Strategy and Action Plans (under the Convention on Biological Diversity) are three such examples. In these situations and others like them, some programs and staff will enthusiastically adopt a new approach, while others remain less seriously engaged, often treating the new approach as a mandate to "go through the motions" of meeting its intent. Some may lack enthusiasm because the purpose of the new approach was unclear, leaders failed to explain their expectations adequately, or the internal or external audience for the plan was never clarified. These situations can often result in initial planning efforts being disorganized and of insufficient quality to have much success or impact. We refer to these as situations in which the context for planning was unclear—a common occurrence in the initial stages of new conservation initiatives.

After a discussion of planning context and the importance of establishing a solid foundation for planning, this chapter also introduces a set of planning principles and standards that we will elaborate on in the chapters ahead. Finally, we conclude the chapter with a road map for conducting a conservation planning process from beginning to end, and an overview of how the chapters ahead help navigate that map. That road map has some nuances that depend on whether we are traveling through terrestrial, freshwater, or marine ecosystems, or sometimes all three in combination. Although most of the advice, tools, and methods in this book should apply to all three realms, we note a few of the unique circumstances posed by planning in aquatic realms.

## Planning Context

Understanding the **planning context** is one of the most important and, in our experience, underappreciated steps in planning. The context for planning has five major components:

1. Purpose and scope of the plan
2. Decisions to be made by the plan
3. Constraints on the planning process
4. Audience for the plan
5. Level of investment in the planning process

### Purpose and Scope

There are many reasons for developing a conservation plan. It might be, for instance, to identify important areas within a region where action can be taken to achieve a particular goal (for example, creating open space and trails within a land use plan for a local government). Alternatively, a plan might be focused solely on identifying strategies and actions to conserve or manage a particular place or to implement new or revised policies that a natural resource agency is promulgating. The plan may need to meet several, possibly competing, objectives (such as a national park established to conserve mountain gorillas and generate tourism revenues and jobs for local people), or it may be more singularly focused (a nature preserve managed for an endangered species). Some plans may be policy oriented, whereas others may be focused on concrete actions on the ground. For some conservation organizations or agencies, the purpose of a conservation plan may be to address a major environmental problem or some part of that problem. In the Lake Chad region of Africa, for example, climate change and poor water management practices have resulted in a dramatic decrease in the size of the lake, which in turn is causing desertification of farmland that relied on irrigation from the lake. The point, of course, is that any conservation plan absolutely must have a clear rationale for being conducted, and all those involved must understand and agree on that rationale and be comfortable with the planning process.

Establishing a purpose or rationale for developing a conservation plan is closely related to defining a problem; after all, most conservation efforts involve addressing a set of problems. One trap that is easy to fall into is to concentrate more on the solution than the problem. For many project directors or natural resource managers, this idea may seem to fly in the face of conventional wisdom. Most program managers would no doubt be pleased if their staff were focused on solutions (positive) instead of on problems (negative). But what we are referring to here is a tendency to jump to a solution before carefully thinking through the problem. Protecting riparian habitat downstream of a dam from development may be one solution to a problem, but that action may do little to solve a more important problem of disrupted flow regimes that are precluding reproduction in the dominant riparian tree species.

Albert Einstein was quoted as saying that if he had one hour to save the world, he would spend 55 minutes defining the problem and 5 minutes on the solution.* A bit extremely stated, perhaps, but the point is clear: we need to take time to fully understand problems so that we can craft appropriate solutions. These solutions will be influenced by the missions and mandates of the organizations and agencies involved, the strategies and tools at their disposal, and the differing views of stakeholders.

---

*Although this quote is widely attributed to Einstein, its exact source has not been determined.

Fortunately, there is plenty of good advice on how to define problems. In *Smart Choices: A Practical Guide to Making Better Life Decisions,*[1] John Hammond and colleagues devote an entire chapter to problem definition. Some of their best advice, sprinkled with a bit of our own wisdom, includes asking yourself or your team the following questions:

- What is the problem, and why are we concerned about it?
- What triggered the problem?
- Whose problem is it?
- What are the constraints around the problem? (For example, a set of dams is going to be constructed in a watershed by a certain date, but can we influence the locations and operating conditions of the dams?)
- What are the components of the problem? (Break the problem into smaller pieces.)
- What other problems or decisions are related to the current problem? (For example, two adjacent property owners have agreed to take out easements on their land if we secure an easement on our property *x*.)
- How big is the problem we trying to solve? (For example, desertification in Africa, or a portion of that problem.)
- Who else can help us solve this problem with fresh eyes and fresh thinking?

In reality, a strong approach to conservation planning involves thinking about these questions throughout the planning process, but they are most important at the outset to help the team clarify the purpose of the plan and what problems they hope to address through the planning process. Not all members of the planning team need to be engaged in addressing these problem-defining questions, but efforts should be made to reach consensus on the answers so that the whole team is focused on the same overall purpose and problems.

In addition to giving initial thought to the overall purpose of a plan, another useful early step in planning is to outline the scope of a potential project. **Scope** can refer to strategic scope (for example, who the major actors are; what sectors, such as water or energy, might be involved), geographic scope[2] (generally, where the project will be undertaken), and temporal scope (the potential duration of the project).[3] Time invested in this initial, albeit broad, scoping of a project can pay dividends later in helping a team decide how much to invest in planning as well as in preventing “scope creep,” in which the interests of a project go well beyond its original intent, often in unproductive ways.

## Decisions

Closely related to clearly understanding the purpose of the plan is knowing what decisions will be made as a result of it and, importantly, who will make those decisions. The decision makers, often a project director or a superintendent or other senior manager in a natural resource agency, are one of the key audiences for the plan (a topic we will discuss in more detail later in this chapter). After an initial discussion about purpose, we have found it effective to begin conservation planning exercises by drawing up a list of decisions and putting the decision maker's name next to each one. This process in itself often reveals important things about the purpose of the plan and how it needs to be structured if it is to support these decisions adequately. This list of decisions is extremely useful throughout the planning process for checking whether the process is on track.

From the inception of a conservation project, it is essential for those involved in the plan to agree on what decisions it will influence and what decisions have already been made that will influence the planning process before its start (we refer to these as **constraints**). A lack of such agreement will eventually plague the implementation of the plan, a topic we will take up in greater detail in Chapter 11. For example, an agency or nongovernmental organization (NGO) may go through some sort of conservation priority-setting exercise, but the senior leaders of that agency or NGO may not have agreed to act on the priorities that emerge from such an exercise. In the case of some conservation plans that are focused on land management actions and are underpinned by a regulatory framework, the decisions that such a plan will influence are of heightened importance because they may be challenged through appeals or litigation. In other situations, a planning team may have the responsibility to plan activities for a wide range of staff and even partners in different organizations, but limited authority for implementation. No matter what the situation, any conservation plan should strive to make clear from the beginning what decisions it intends to influence.

In government natural resource agencies, it is usually clear who has responsibility for implementing a conservation plan (for example, superintendents or directors of national parks), but in nongovernmental conservation organizations, decision-making authority is often much murkier. In a recent review of conservation planning that we conducted, we heard repeatedly that project managers were not involved or sufficiently engaged in developing a strategic plan and therefore not necessarily in agreement with the overall direction of the plan.

Another way to think about this situation is to ensure that conservation plans are addressing questions that project directors or key audiences want answers to. Having a group of scientists or planners develop a plan that either answers questions that project directors or key audiences are not interested in or doesn't answer questions they do want answers to is one sure path to wasted planning effort. For example, the managers of national parks and national forests are often perceived as important audiences of priority-setting plans that nongovernmental conservation organizations prepare and advance. Yet, these plans often address questions that don't immediately seem relevant to a natural resource manager or speak a different language than what natural resource managers are accustomed to. In addition, these plans may answer questions in a manner (identifying priority conservation areas) that is inconsistent with how most of these managers view their jobs (to manage for multiple uses) or think about problems. In some cases, project directors or program managers may not know what questions or issues a plan should address. In these cases, it is often time well invested for some members of the planning team to engage the directors or managers in a discussion of the most important issues, and then elicit their opinions on which of those issues are most important to them and identify the issues for which they will ultimately need to make a management-related decision.

In an ideal planning world, there is a clean pathway from (1) identifying the decisions to be made to (2) doing the planning that establishes outcomes and develops strategies and actions and ultimately to (3) making the decisions on which actions to advance. Almost inevitably, however, conservation planners will at some stage find themselves involved in planning exercises in which decision makers clearly have already made up their minds about a course of action, or in which the actions seem so obvious that a plan seems redundant. There might be no discernible "decisions" actually on the table. Yet often these same decision makers remain supportive of conducting a conservation planning exercise. Is this just expensive window dressing?

Not necessarily. We believe that some of the components of the planning process also play a critical role in legitimation and sense-making, boosting confidence in the choices made through presentation of the logic and evidence supporting them and ensuring that those choices have been rigorously thought about, even if somewhat retrospectively. We know this idea will be borderline offensive to many planners, but being flexible about the planning process and methods in these situations may bear useful fruit. For example, these sorts of situations may present opportunities to articulate and check the basic assumptions underpinning the decisions (a process that commonly highlights major weaknesses that at the very least make senior staff more cautious in their advocacy), and there is the distinct possibility of uncovering surprising results that might change people's minds about the best course of action.

## Constraints

Most conservation projects and planning efforts for them are affected to some degree by decisions that have been made at a higher organizational level. For example, a project in the Bozeman (Montana, USA) ranger district where one of us lives will be guided in part by the larger Gallatin National Forest Plan (of which the district is part). Similarly, what the Wildlife Conservation Society's country program in Gabon does will be influenced by its Africa strategic direction and by the strategic plan of the Global Conservation Program of WCS. Plans for conserving a new Natura 2000 site[4]—such as Dogger Bank in the United Kingdom, which provides critical spawning and nursery grounds for North Sea commercial fishes—are governed to some extent by the data standards established by the European Commission for nominating Natura 2000 sites. The take-home message is that planning for most conservation projects is influenced by policies, laws, or strategic directions that are determined at a larger spatial scale and need to be considered as part of the planning context at a project's inception.

The outcomes of a planning process are also likely to be constrained by the legacy of existing conservation actions. Plans are rarely developed on a blank slate; instead, they take shape against a background of ongoing work by the organization and its employees, either in a particular geography or in line with a particular strategy. Failure to consider the legacy of existing commitments, whether formal (such as deliverables on a grant) or informal (such as staff expertise) is a defeating weakness in many otherwise rational and defensible planning processes. The good news is that these commitments and the higher-level decisions mentioned earlier serve to bound conservation planning processes, which makes them far more tractable. In other words, as well as being vital pieces of context, these "constraints" are useful boundaries. The term *constraints* should not be read in a pejorative sense here. As one example of a constraint, conservation projects funded by the **Global Environment Facility** (**GEF**) are intended to represent "additional" efforts and are required to build on existing "baseline" projects that have their own co-financing.

The easiest way to get at what the constraints on a plan might be is to ask: What decisions have already been made that would affect this plan? It can take some prompting to get the list of decisions flowing. Typical responses might be items that have been included in funding proposals, geographies where investments need to be focused, commitments that have been reached through higher-level planning processes, or, for a conservation organization, agreements that may have been reached a priori with a donor.

An understanding of the constraints placed on decisions is useful, but it is equally important to ensure that these are indeed real constraints and not just perceived ones. For example, is the project really restricted to working on revegetation, or could that be changed if a compelling enough case were made? There is a fine line to navigate between a plan that comes up with solutions that will never be implemented and one that remains open to finding the best solution. In general, our advice is to be honest about constraints, but if there is any uncertainty, err on the side of not including them as constraints. A situation analysis or a conceptual model of a conservation project (described in Chapter 5) is an appropriate place to document constraints.

### Audience for the Plan

Few tasks are more important at the outset of planning than thinking through the audience for a conservation plan. Many factors will influence this audience, including the size of the organization, the scope and scale of the project or program for which the plan is being developed, whether there are legal requirements for public involvement, the degree to which the exercise is internal or will involve partner organizations, and the organizations and individuals that are funding the project.

For example, a small land trust or conservation organization that has an operating unit the size of a local government unit and an annual operating budget of perhaps a few hundred thousand dollars (US$) will probably have a small target audience for any planning effort, and that audience might include staff, a governing board, and a couple of key funders. At the other end of the spectrum could be a public agency such as a state or provincial wildlife agency that is responsible for developing a conservation plan for all wildlife species and habitats. Such a planning effort not only may have statutory public engagement requirements, but will need to engage a wide variety of stakeholders in order to be successfully implemented.

Multi-country conservation initiatives, such as Natura 2000 or the **Micronesia Challenge**, will also have potentially large and diverse audiences. From a different perspective, a plan focused internally on the staff of an organization might contain very frank detail, even on sensitive issues such as internal management risks and the chance of failure, while a plan with a significant external audience would naturally be vaguer on these topics. Whether small or large, internal or external, the audience for a plan should to some degree dictate who is involved in planning, how the plan is developed and written, and how it is communicated.

Fortunately, a growing number of tools and organizations (for example, the International Association for Public Participation) are available to assist planning teams in determining the audience(s) for a plan and those who should be engaged in the plan. Such tools generally go by the term **stakeholder analyses**, although both *stakeholders* and *stakeholder analysis* mean different things to different people. Mark Reed and colleagues[5] have provided an excellent review of stakeholder analysis in natural resource management, including definitions of stakeholders, approaches to stakeholder analysis, and a series of case studies that illustrate different approaches. For conservation planning purposes, **stakeholders** can be defined as "any individual, group, or institution that has a vested interest in the natural resources of the project area and/or who potentially will be affected by project activities and have something to gain or lose if conditions change or stay the same."

As the WWF Programme Standards[6] for planning suggest, stakeholder analyses have three important steps:

1. Identifying the key stakeholders
2. Evaluating the interest, potential role, and importance of each stakeholder
3. Determining how to best engage stakeholders

TABLE 2.1 summarizes different methods for identifying and evaluating stakeholders in natural resource management, along with the resources required and the strengths and weaknesses of each approach. The Nature Conservancy recently conducted a stakeholder analysis for investing in Water Fund projects in Latin America.[7] FIGURE 2.2 on p. 50 shows the results of their analysis.

For those who are new to conservation planning, a key piece of advice is warranted: do not underestimate the time and resources required to conduct a stakeholder analysis and to engage those that the analysis suggests are important. Depending on the scope of the plan and the project, it may or may not be necessary to conduct a formal stakeholder analysis. For some conservation projects, the scope may be limited, the audience may be limited, and such an analysis may be unnecessary. At the same time, for many projects the results of a stakeholder analysis will ultimately have considerable bearing on success or failure (FIGURE 2.3 on p. 51).

In a highly successful public planning process for marine protected areas (MPAs) along the northern coast of California, USA, a transparent and professional stakeholder engagement process was highlighted as one of the critical components that led to the successful designation of a new network of marine protected areas.[8] No small

**TABLE 2.1 Methods for analyzing stakeholder participation in a conservation project**

| Method | Description | Resources | Strengths | Weaknesses |
|---|---|---|---|---|
| Focus groups | A small group brainstorms stakeholders, their interests, influence, and other attributes and categorizes them | High-quality facilitation; room hire; food and drink; facilitation materials (e.g., flip-chart paper and sticky notes) | Rapid and hence cost-effective; adaptable; possible to reach group consensus over stakeholder categories; useful for generating data on complex issues that require discussion to develop understanding | Less structured than some alternatives; requires effective facilitation for good results |
| Semi-structured interviews | Interviews with a cross section of stakeholders to check/supplement focus group data | Interview time; transport between interviews; voice recorder | Useful for in-depth insights into stakeholder relationships and to triangulate data collected in focus groups | Time-consuming and hence costly; difficult to reach consensus over stakeholder categories |
| Snowball sampling | Individuals from initial stakeholder categories are interviewed to identify new stakeholder categories and contacts | As above: successive respondents in each stakeholder category are identified during interviews | Easy to secure interviews without data protection issues; fewer interviews declined | Sample may be biased by the social networks of the first individual in the snowball sample |
| Stakeholder-led stakeholder categorization | Stakeholders themselves categorize stakeholders into categories that they have created | Same as semi-structured interviews | Stakeholder categories are based on perceptions of stakeholders | Different stakeholders may be placed in the same categories by different respondents, making categories meaningless |
| Q methodology | Stakeholders sort statements drawn from a concourse according to how much they agree with them; analysis allows social discourses to be identified | Materials for statement sorting, interview time, transport between interviews | Different social discourses surrounding an issue can be identified and individuals can be categorized according to their "fit" within these discourses | Does not identify all possible discourses, only the ones exhibited by the interviewed stakeholders |

*(continued)*

part of this success was due to a well-thought-out stakeholder engagement strategy. Conservation projects and their plans can sometimes be contentious affairs, and anyone involved in conducting or facilitating a stakeholder meeting needs to be a trained facilitator who can

| Method | Description | Resources | Strengths | Weaknesses |
|---|---|---|---|---|
| Actor-linkage matrices | Stakeholders are tabulated in a two-dimensional matrix and their relationships described using codes | Can be done within focus group setting (see opposite), or individually by stakeholders during interviews (see opposite), or by researcher/ practitioner | Relatively easy, requiring few resources | Can become confusing and difficult to use if many linkages are described |
| Social network analysis | Used to identify the network of stakeholders and measure relational ties between stakeholders through use of structured interview/ questionnaire | Interviewer, questionnaire, training in the approach and analyses, time, software | Insight into the boundary of the stakeholder network and the structure of the network; identifies influential and peripheral stakeholders | Time-consuming; questionnaire is a bit tedious for respondents; need specialist in the method |
| Knowledge mapping | Used in conjunction with social network analysis; involves semi-structured interviews to identify interactions and types of knowledge | Same as semi-structured interviews | Identifies stakeholders that would work well together as well as those with power balances | Knowledge needs may still not be met due to differences in the types of knowledge held and needed by different stakeholders |
| Radical transactiveness | Snowball sampling to identify fringe stakeholders; development of strategies to address their concerns | Training in the approach, time | Identifies stakeholders and issues that might otherwise be missed and minimizes risks to future of project | Time-consuming and hence costly |

*Source:* Adapted from M. S. Reed et al., *Who's in and why? A typology of stakeholder analysis methods for natural resource management.* Journal of Environmental Management, 2009. **90**(5): 1933–1949.

manage meetings and stakeholder groups and the individuals who sometimes attempt to dominate such meetings.

Closely related to stakeholder analysis, and also deserving of attention, is a consideration of which organizations you may want to engage with directly as partners in both the planning process and the project implementation. Whether or not a potential partner has the resources to commit to a project, whether joint planning is a priority, and whether your organization will be relying on a partner to deliver the outcomes of a conservation project are among the most important considerations.[9] The Conservation Partnership Center[10] is a useful resource for thinking through all aspects of investing in a partnership related to the planning and implementation of a conservation project.

| Level of interest in the water fund area | Less influence | More influence |
|---|---|---|
| More | Private sector:<br>Soft-drink company<br>Livestock<br>Multilateral cooperation agencies<br>International NGOs | Water company<br>Hydropower plant<br>Agricultural sector with intensive water use<br>Local NGOs<br>Protected areas agency<br>Community-based organizations |
| Less | Private sector not related to water (chemicals, food)<br>Financial sector<br>Others | Local government<br>Regional government<br>Landowners<br>Environmental authorities<br>Indigenous communities and communities of African descent |

Level of influence in the water fund area
Less → More

FIGURE 2.2 An application of the actor-linkage method for stakeholder analysis as articulated in Table 2.1. In this case, the actors are potential stakeholders in TNC water fund projects in Latin America, a form of payment for ecosystem service projects (see Chapter 10). Data from Calvache, A., S. Benítez y A. Ramos. 2012. Fondos de Agua: Conservando la Infraestructura Verde. Guía de Diseño, Creación y Operación. Alianza Latinoamericana de Fondos de Agua. The Nature Conservancy, Fundación FEMSA y Banco Interamericano de Desarrollo. Bogotá, Colombia. 144p.

## Level of Investment in Planning

Not all conservation plans are, or should be, created equal. Most of us intuitively appreciate that a conservation plan for a noncontroversial land acquisition and development of a set of hiking trails by a local government "open space" agency should not require the same level of investment in planning as a large-scale proof-of-concept project related to conserving a large expanse of tropical forest with proposed timber concessions through a forest carbon mitigation scheme. This point of investing in different levels of planning depending on a project's scope has too often fallen on deaf ears. Perhaps it is the existence of planning standards (or appeals and lawsuits in some cases) that drives organizations and agencies to make similar investments in planning regardless of the scale or complexity of a project. Or maybe it's analogous to people's tendency to fill up our luggage regardless of whether we're packing for a one-night or a seven-day trip. In any event, it makes little sense to invest in conservation plans at a similar level for all projects. In fact, investing more than is warranted not only wastes valuable resources but also tarnishes the perceived value of planning for future initiatives.

FIGURE 2.3 WWF India stakeholder meeting. Stakeholder meetings are critical opportunities to gather opinions, data, and other information from key audiences for a conservation project as well as a chance to build community support for the project.

We have developed a few rules of thumb that may help in making decisions on investments in the planning process:

- *Risk:* Greater levels of financial, legal, ecological (for example, high extinction risk), or reputational risk in a project suggest the need for a greater investment in planning.
- *Replication:* For a *proof-of-concept* project intended to demonstrate a conservation strategy that can be replicated across other projects, a correspondingly greater investment in planning might be warranted. On the flip side, replicating a strategy that is known to be successful under a variety of circumstances (such as establishing marine protected areas) suggests the need for investing less time in planning.
- *Complexity:* Projects can be complex for many reasons. For example, they may involve multiple partners, large numbers of stakeholders, or complicated strategies, or they may depend on the success of multiple linked strategies at different organizational levels to achieve the overall conservation outcomes. As a general rule, greater complexity suggests the need for a greater investment in planning.
- *Planning as a strategy:* For some types of conservation projects in which stakeholder involvement is integral (for example, marine spatial planning or establishment of MPAs), planning

itself is a strategy that through the demarcation of conservation areas likely will lead directly to conservation outcomes. In other circumstances, planning is a mechanism for building buy-in to strategies by conducting the planning effort in collaboration with partner organizations. Both of these situations should necessitate a relatively greater investment in planning.

Certainly other factors will influence investment in planning. FIGURE 2.4 is a guide to some of the factors that should influence greater or lesser investments in staff time and resources directed to planning. Such guidance is all relative, of course.

You may be wondering how much a conservation planning effort really costs. Unfortunately, few organizations or agencies track such costs. Groves and colleagues estimated the median cost of 24 ecoregional plans developed by TNC to be US$234,000 in 2001 (including staff time and operating costs).[11] More recently, Madeleine Bottrill and Bob Pressey[12] provided cost estimates for conservation planning efforts that were of different durations, covered vastly different-sized planning regions, and were conducted by different types of organiza-

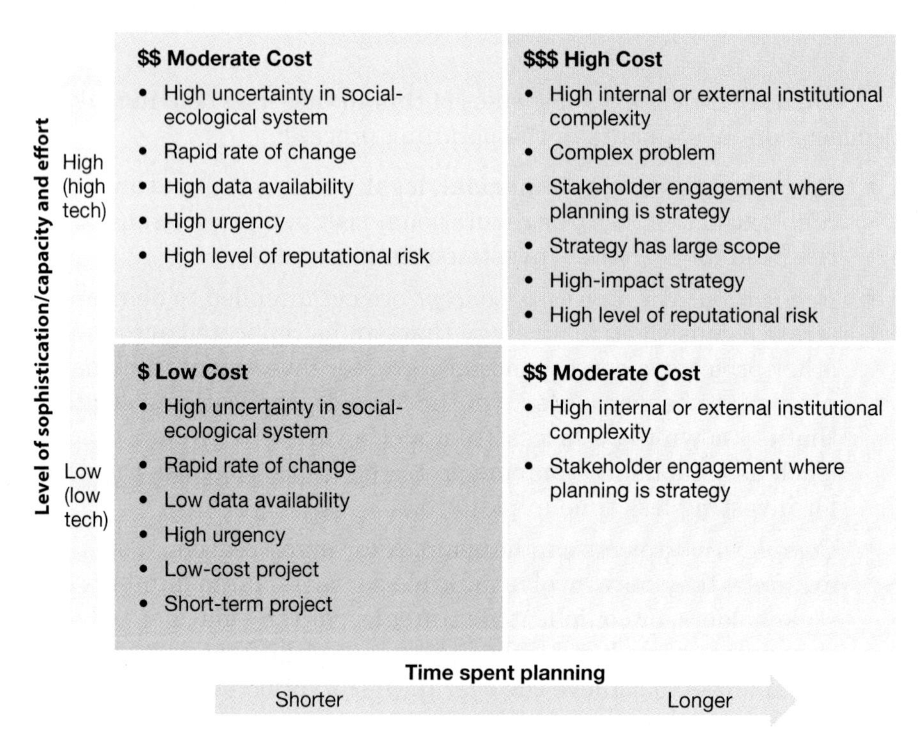


FIGURE 2.4 Recommended levels of staff and financial investment in conservation planning are influenced by various factors. These factors include but are not limited to urgency, project duration, complexity, data availability, and the degree to which planning itself is a strategy.

tions and agencies, which ranged from US$400,000 (a Conservation Action Plan by TNC in Papua New Guinea) to nearly US$8 million (a Great Barrier Reef Marine Park rezoning plan). These few data points suggest that it is not unusual for significant resources to be spent on conservation planning, which is all the more reason to give careful thought to that level of investment and to follow the guidance in this book to get the best return on that investment!

We've now considered several factors that collectively constitute what we refer to as the planning context. In the next section, we discuss the second major foundational element of conservation planning—putting together an effective planning team.

## Building an Effective Planning Team: The Power of a Multidisciplinary Approach

Over a decade ago, Dee Boersma of the University of Washington led a team of conservation biologists and graduate students in an evaluation of recovery plans under the U.S. Endangered Species Act (ESA).[13] One of their key findings was that having a diversity of planning team members, including government and nongovernmental participants, improved the likelihood that the species population would be trending upward as a result of plan implementation. According to Boersma and colleagues, there were at least two plausible explanations for this result. First, including planning team members from outside the government helped ensure that the most recent scientific data were incorporated into the effort. Second, the involvement of diverse participants may have encouraged greater investment in both the planning process and its implementation. In many cases, it makes sense to include key partners and stakeholders on the planning team for the same reasons suggested by the ESA study: they can bring added expertise, and their inclusion can bring additional support and credibility to the plan and project, which will make its successful implementation more likely.

A more recent synthesis of the conservation science literature on multidisciplinary approaches reveals that although these approaches pose consistent challenges, there is sage advice, based on years of experience, on how to best go about establishing a multidisciplinary team and project, especially one that contains both natural and social scientists. According to Simon Pooley and colleagues,[14] among the most significant challenges to these teams are (1) methodological challenges, including differing views on data quality, cultural differences between the social and natural sciences, and the need to invest team time in building mutual understanding and trust; (2) differing views of value judgments; (3) different theories of how knowledge is obtained;

(4) disciplinary prejudices (for example, some natural scientists think that the primary role of social scientists is to help change the behavior of target audiences); and (5) the different languages in which social and natural scientists communicate.

Fortunately, Pooley and colleagues have also suggested a set of steps, based on the lessons learned from 50 years of multidisciplinary efforts in conservation, for establishing an effective multidisciplinary team:

1. Recruit leaders and collaborative members from disciplines appropriate to the project at hand.
2. Refine the project's questions with everyone's input.
3. Develop mutual respect and trust, and clarify leadership.
4. Identify and determine how to best accommodate hidden values and conceptual differences.
5. Educate the team on theories, concepts, methods, and data from the range of disciplines.
6. Develop a shared language and good internal team communication.

Their paper provides detailed references and background discussion on each of these steps that many planning teams will find useful. Even though their synthesis was developed primarily for a research team, the findings of Pooley and colleagues apply equally well to planning teams.

Whether or not a team is multidisciplinary, it is critical to give considerable thought to team composition, management, and leadership. Anyone who has served on a poorly managed planning or project team that lacks solid leadership knows how unpleasant the experience can be. Much has been written about team management, many cartoons have made fun of the process (FIGURE 2.5), and there are many resources to draw on, although it is beyond the scope of this book to cover this topic extensively. Twenty years after it was first published, Jon Katzenbach and Douglas Smith's *The Wisdom of Teams*[15] still has much to offer, including the following sage advice:

- Keep your team small. Effective teams tend to be no larger than 10.
- Establish a separate advisory or steering committee as a helpful means of bringing additional support and expertise to the planning process.
- Carefully consider your team's composition in terms of its mix of technical, disciplinary, strategic, interpersonal, problem-solving, and decision-making skills.

"Before we get going, let's start with a quick pack-building exercise."

**FIGURE 2.5** Team building is a key component of most successful teams. Research from both business management and conservation concludes that when teams take the time to build trust and respect, they are more likely to succeed.

- Establish key roles and responsibilities for team members, and use a team charter if necessary to formalize them.
- Build a sense of trust and mutual accountability among team members.
- Leadership, of course, really matters. Teams can have more than one type of leader (for example, process leaders and intellectual leaders). A team leader needs to have strong facilitation abilities, interpersonal communication skills, project management experience, the ability to delegate, and the confidence to build trust and bring out the best in the talents of team members. It's a tall order, but many teams fail or succeed on the basis of leadership.

In our experience, we cannot overstate the importance of good leadership. Whether it is a planning effort or a conservation project or initiative, good leaders can make all the difference in the success of the effort. Simon Black and colleagues[16] examined the leadership of successful and unsuccessful species conservation programs. They found that successful conservation leaders could establish shared, long-term visions and work toward them with short-term goals; they were able to provide hands-on management skills, such as understanding the strengths and weaknesses of individual team members and how to manage those effectively; they could manage at both the detailed and big-picture levels; and, perhaps most importantly, they were skilled at encouraging learning and improvement by individual members while remaining open to alternative solutions to problems. Black and colleagues' detailed findings on 10 common problems in conservation programs and the 10 positive traits of strong leaders are worthwhile reading for any conservation project or planning team.

In a best-selling management book, Patrick Lencioni[17] outlines five dysfunctions of teams that he witnessed regularly in Fortune 500 companies. The issues he describes appear to apply equally well to teams in government agencies and NGOs: absence of trust, fear of conflict, lack of commitment, avoidance of accountability, and inattention to results. Fortunately, Lencioni also provides a considerable dose of advice for overcoming these dysfunctions, especially for the team leadership.

For example, for teams that lack trust, it is critical for the leader to create an environment in which team members can openly admit to weaknesses or failures. For teams that fear conflict, it can be helpful to identify a key team member as a "miner of conflict"—a person who has the courage to point out sensitive issues and force team members to work through them. Committed teams make clear decisions and have buy-in to those decisions from all team members, but it often falls to the team leader to ensure that such decisions get made (even without all the ideal data) and to work with individual team members to develop the support for those decisions. Perhaps no dysfunction is more difficult to confront than a team that lacks accountability to the tasks it set out to undertake, but sometimes there is a simple remedy: announce publicly and clearly what the team needs to achieve, what each person's responsibilities and expectations are, and how everyone has to perform to get the job done. Finally, team members may care about other issues more than they do about the team's goals. Lencioni calls this inattention to results the ultimate dysfunction and suggests that the cure is a selfless and objective leader who rewards team members who contribute to team goals. Most readers who have served on planning or conservation teams will recognize these dysfunctions. The good news is that Lencioni provides both team members and leaders with advice, tools, and exercises for helping to overcome these hurdles.

For those interested in more detailed guidance on team composition, operations, and leadership, the WWF Programme Standards are an excellent reference source on this topic.[18] Their guidance covers the different types of knowledge and skill sets needed by most conservation teams; the importance of defining roles, responsibilities, and a general team process at the outset; how a team or project charter can help communication and accountability; and some of the key attributes of effective and ineffective teams. As we noted, there is no shortage of guidance on establishing and operating an effective team. Understanding the basics from *The Wisdom of Teams*; heeding Lencioni's advice on dysfunction; paying attention to some of the conservation-specific recommendations by Pooley, Black, and their colleagues; and using the WWF standards for team operations will have any conservation planning team fully moving in the right direction.

The observation that conservation problems are increasingly complex and require multiple disciplines to effectively confront them is not new. Conservation projects that successfully integrate multiple objectives, such as those focused on marine spatial planning, those focused on mitigation associated with energy development, or any of a host of projects related to ecosystem services are increasingly "testing the waters" of how to best develop multi-objective conservation plans and improving the multidisciplinary process. One of the earliest conservation successes in marine spatial planning was the establishment of representative no-take zones in the Great Barrier Reef, Australia. Key to this success was the wide engagement of teams with a diversity of areas of expertise, including heritage values, tourism and recreation, fisheries, indigenous values, and social impact assessment in addition to a variety of marine resources.[19]

Going beyond the use of multidisciplinary planning and project teams, South African conservation planner Belinda Reyers and her colleagues[20] have put forth the provocative notion that for conservation to be more effective, it needs to be more of a transdisciplinary process. They propose that because nature conservation is a messy business, conservation practitioners, scientists, and planners have to get more involved in the social processes of engaging a variety of stakeholders, understanding different value systems, and changing behaviors; and until this happens, the conservation projects are not likely to be as relevant and influential as they could be.

## Principles and Standards for Conservation Planning

We have just considered some important foundational elements for launching a successful planning effort. Equally important to this solid foundation is grounding in a set of principles and standards that will guide the planning effort. In *Drafting a Conservation Blueprint*,[21] one of us reviewed the different principles for conservation planning that had been advanced up to that point. Most of those principles applied primarily to selecting and designing networks of conservation or protected areas (in other words, spatial planning). They can be synthesized into three overarching principles:

1. *Representation:* The need to represent different ecosystem types and the environmental gradients in which they occur across a system of conservation or protected areas

2. *Persistence or adequacy:* The need to pay attention to the viability and integrity of biodiversity features that occur in conservation areas to ensure their long-term persistence
3. *Efficiency:* The need to achieve conservation goals in conservation areas of the smallest number or size

These principles remain relevant to conservation planning today. For example, representation underlies our discussion of features in Chapter 3 and is specifically referenced in a section on spatial prioritization in Chapter 7. The principles of persistence and efficiency are, in part, the foundation for setting targets in Chapter 4.

At the same time, relatively newer concepts and methods are now contributing to what will likely turn out to be new planning principles, especially in regard to our expanded view of what constitutes conservation planning and the diversity of conservation strategies in use. We suggest that at least three additional planning principles may be emerging today: (1) a focus on the *socio-ecological system,* (2) using a *theory of change* to develop strategies and actions, and (3) incorporating consideration of *risk and uncertainty* in the planning process.

As we suggested in Chapter 1, a greater focus on the linkages of conservation planning with social-ecological systems (SES) and **human well-being** (**HWB**) has emerged as a new principle for planning. The notion of social-ecological systems (also called socio-ecological systems) has been around for quite some time and has been variously defined. Fikret Berkes from the University of Manitoba and Swedish scientist Carl Folke have published numerous papers on the topic. They define SES as nested multilevel systems that provide essential services to society, such as supplying food, fiber, energy, and drinking water. These nested multilevel systems are essentially sets of social and ecological factors that are interacting to produce a set of goods and services for society.[22] The concept of SES has also been linked to the concepts of ecosystem services (see Chapter 10), resilience (see Chapter 9), and sustainability.[23]

There are, of course, both positive and negative feedbacks between social systems and ecological systems that conservation planners need to be aware of. Although the interactions between conservation and development have received the most attention as an example of these feedback loops, Brian Miller and colleagues from the University of North Carolina[24] have provided examples of several different types of these feedback loops, such as conflicts between wild and domesticated animals, human displacement, and comanagement of natural resources (**TABLE 2.2**).

Likewise, human (or social) well-being has been interpreted broadly. Social scientists Sheri Stephanson and Michael Mascia[25] have

**TABLE 2.2** Feedback loops between social systems and ecological systems

| Feedback | Example narrative |
|---|---|
| Conflict between wild animals and humans | Conservation initiatives increase probabilities of persistence of wild animals, which increases interactions and conflicts with humans, who develop negative attitudes toward conservation and may respond with actions that reduce probabilities of persistence of wild animals. |
| Social movements | Conservation initiatives marginalize local resource users and suppress human rights, which engenders discontent and leads to collective protest, subversion, or resistance. |
| Adaptive comanagement | Local people are integrated into conservation planning and implementation, so effects on humans and the environment are mediated by iterative decision-making processes. This fosters positive attitudes and cooperation. |
| Loss of traditional management | Indigenous resource management institutions achieve conservation objectives; however, top-down conservation initiatives can undermine local traditional management structures and decrease management capacity. |
| Traditional ecological knowledge | Conservation practices are altered to include traditional practices and knowledge, and the probability of achieving conservation objectives increases. |
| Human displacement and risks to livelihoods | Human communities are relocated or human access to resources is restricted or threatened, which results in a variety of undesirable effects on humans. People develop alternative livelihood strategies that may reduce the probability of achieving conservation objectives. |
| Conservation and development | Human communities benefit economically (e.g., jobs, income, resource protection) via mechanisms such as ecotourism, integrated conservation and development, and community-based management. These incentives increase the probability of achieving conservation objectives. |

*Source:* Adapted from B. W. Miller, S. C. Caplow, and P. W. Leslie, *Feedbacks between conservation and social-ecological systems.* Conservation Biology, 2012. **26**(2): 218–227.

defined social well-being as "the human capabilities and conditions associated with a productive fulfilling life," a definition they felt was consistent with that of the Millennium Ecosystem Assessment and various human development policies and goals (for example, the United Nations Millennium Development Goals). EJ Milner-Gulland and colleagues[26] have provided an excellent overview of the use of the human well-being concept in conservation and have suggested a framework (well-being in developing countries, or WED) for implementing it. This framework lists three conditions necessary for individual human well-being: (1) human needs are met, (2) humans can act meaningfully to pursue goals, and (3) humans are able to experience a satisfactory quality of life. The authors outline some of the challenges involved in applying such frameworks, including the need to develop case studies under a variety of circumstances. The most recent version of the Conservation Measures Partnership's *Open Standards*, which we introduced in Chapter 1 and discuss further in this chapter, also provides a treatment of human well-being in conservation planning (see Chapter 10 for a more thorough discussion of this topic).

Although the work of Elinor Ostrom and colleagues has garnered the most attention as a framework for SES, conservationists could use many different frameworks to analyze SES.[27] Claudia Binder and colleagues[28] have reviewed these frameworks in detail (**TABLE 2.3**) and have suggested several prompting questions that conservation planners and practitioners can use when selecting a framework. Not included in their review is the recent work by Stephanson and Mascia,[29] who have developed a hierarchical framework for incorporating and mapping data on social well-being across five domains: economic well-being, health, education, political empowerment, and culture. We address these social domains in more detail as part of the situation analysis in Chapter 5.

Using the concept and methods of **theory of change** (**TOC**) constitutes what we believe to be a second emerging principle for conservation planning. Originally conceived in the field of program evaluation, TOC involves mapping a set of causally linked steps in a pathway believed to be necessary to achieve a long-term conservation outcome through the deployment of one or more strategies or interventions. Using TOC methods has become widespread in many major conservation organizations and natural resource agencies as they develop strategies to achieve certain outcomes and consider the short-term and intermediate steps needed to achieve those outcomes as well as the evidence and assumptions underlying the practicality and feasibility of those steps. Indeed, TOC methods are central to the *Open Standards for the Practice of Conservation*, which make frequent use of results chains, a tool for visualizing a theory of change. In Chapter 6, we dive into the details of using TOC methods.

Understanding climate change impacts and adapting strategies to address these impacts have made most conservation scientists and planners much more aware of the role of uncertainty in conservation planning. At the same time, understanding and managing risks in conservation planning and implementation has also become more commonplace. Together, incorporating a consideration of risk and uncertainty represents our third addition to important planning principles. We delve into the details of how to most appropriately consider risk and uncertainty in the planning process in Chapters 8 and 9.

While representation, persistence, and efficiency can be viewed as planning principles that emerged from the spatial planning focus of con-

**TABLE 2.3** Analyzing social-ecological systems: Ten proposed frameworks and their major purposes

| Framework | Purpose |
|---|---|
| Driver, Pressure, State, Impact, Response (DPSIR) | Develop an improved understanding of, indicators for, and appropriate responses to impacts of human activities on the environment along the causal chain-drivers-pressure-state-impact-responses. |
| Earth Systems Analysis (ESA) | Understand the global interactions in and dynamics of the earth system as well as its sustainable evolutions. |
| Ecosystem Services (ES) | Analyze the integral, dynamic, and complex interactions of biotic and abiotic components of an ecosystem in relation to the supply of services this system provides to support life on Earth. |
| Human Environment Systems Framework (HES) | Provide a methodological guide or template for analyzing the structure of social-ecological systems and understanding the processes and dynamics between the social and ecological systems as well as within different scales of the social system. |
| Material and Energy Flow Analysis (MEFA) | Analyze the metabolic profiles of societies. Analyze the material and energy flows as representing the metabolism of a society, region, or nation. |
| Management and Transition Framework (MTF) | Support the understanding of water systems, management regimes, and transition processes toward more adaptive management; enable comparative analyses of a wide range of diverse case studies; and facilitate the development of simulation models based on empirical evidence. |
| Social-Ecological Systems Framework (SESF) | Provide a common language for case comparison for organizing the many variables relevant in the analysis of SES into a multitiered hierarchy that can be unfolded when needed, and for facilitating the selection of variables in a case study. |
| Sustainable Livelihood Approach (SLA) | Analyze which combination of livelihood assets enables the following of what combination of livelihood strategies with sustainable outcomes. |
| The Natural Step (TNS) | Provide a framework for planning toward sustainability based on constitutional principles (how the system is constituted); outcome (principles for sustainability); and process to reach this outcome (principles for sustainable development). |
| Vulnerability Framework (TVUL) | Analyze who and what are vulnerable to multiple environmental and human changes, and what can be done to reduce these vulnerabilities. |

*Source:* Adapted from C. R. Binder et al., *Comparison of frameworks for analyzing social-ecological systems.* Ecology and Society, 2013. **18**(4).

servation planners in the 1990s and early 2000s, the social-ecological system, theory of change, and risk and uncertainty have arguably gained traction in conservation planning largely as a result of strategic considerations. We can think of these principles as rules, beliefs, or simply good ideas that should help guide and underpin the conservation planning process. In contrast, standards—although closely related to principles—are entities that represent an approved way of doing something and can be used in comparative evaluations.

In nature conservation work, the CMP *Open Standards* is a widely recognized set of standards for conservation planning. It heavily emphasizes the strategic aspects of conservation planning and the use of visual, conceptual models and theories of change. Indeed, the five basic steps (or stages) of the *Open Standards* (see Figure 1.6)—conceptualizing the project, planning actions and monitoring, implementing actions and monitoring, analyzing results and adapting strategies, and sharing knowledge—as well as the detailed outputs underlying these steps and the desktop planning software Miradi (**Box 2.1**), which helps support the standards, have significantly affected planning in conservation organizations and government agencies worldwide. Although several organizations have developed their own versions of the *Open Standards*, even among those different versions there is consistency in terminology that allows practitioners from different organizations and different parts of the world to speak the same language. The conservation logic of the standards, the common language, the availability of training and software for applying the standards, and their widespread use speak to the strengths of this approach. One of its important limitations is the absence of a clear link to setting spatial priorities: the standards are largely focused on the question of how conservation should be accomplished.

These principles and standards of conservation planning should prove useful in a variety of planning settings, from protected areas to "working landscapes and seascapes" and to what landscape architect and ecologist Laura Musacchio has referred to as **designed landscapes** or "places where people have reshaped the spatial and functional heterogeneity of ecosystems for the benefit of themselves and sometimes nature."[30] As conservation biologists, planners, and practitioners, we increasingly appreciate that humanity is having a growing influence over much of the Earth's surface, and nature conservation needs to span the gamut from urban areas to protected areas. Our argument remains that good conservation planning is critical to conservation successes anywhere along this continuum and that the planning process we have outlined in this book will help lead plan-

### BOX 2.1 Miradi

Miradi is the name for a popular desktop program that helps conservation scientists and practitioners plan and manage their projects. The Conservation Measures Partnership and a private company, Benetech, collaborate to develop, maintain, and regularly update and improve the software. Its step-by-step approach is meant to assist users in implementing the *Open Standards for the Practice of Conservation*. The familiar conceptual models and results chains are but two of the many capabilities of Miradi. Besides guiding practitioners through the familiar steps of visioning, setting objectives, selecting targets, and developing strategies, Miradi also helps with the management, budgeting, and work planning involved in actually implementing projects. See https://www.miradi.org for details.

ners, scientists, and practitioners to successful results. In that spirit, as we delve into the details of conservation planning in the chapters ahead, the principles and standards outlined earlier, whether from the leaders of conservation biology or landscape architecture and sustainability science, will be foremost on our minds.

## It's Not Just about Land Conservation: Special Considerations for Planning in Freshwater and Marine Realms

No foundational chapter on conservation planning would be complete without at least an introduction to some of the nuances involved in planning for freshwater and marine realms[31] and the importance of integrating these efforts with land-based planning. Planning for freshwater and marine ecosystems necessitates that we consider how these realms are different from terrestrial ecosystems and how those differences can be accounted for through particular planning approaches while at the same time capitalizing on advances in terrestrial planning approaches that can be applied to their more watery counterparts.

For planning in freshwater ecosystems, four characteristics related to the information about them, ecology of them, pressures on them, and actions to conserve them stand out:[32]

1. Compared with terrestrial biodiversity, there is a real dearth of information on the status and distribution of aquatic biodiversity.
2. Connectivity—longitudinally (downstream and upstream), laterally (into terrestrial systems), and vertically (surface to groundwater)—has profound effects on the life histories of biota in freshwater systems and on the impacts of threats to these systems.
3. There are enormous social and economic pressures on water resources for human uses, and in contrast to at least some terrestrial ecosystems, concerns over the needs of people for freshwater resources cannot be divorced from conservation efforts directed toward the biodiversity of these resources. At the same time, streams are used in fewer total ways than lands, which may, in turn, make it easier to define the conservation problems that we must address for freshwater ecosystems.
4. Implementation of conservation actions will involve large numbers of stakeholders who may not yet appreciate that conserving aquatic ecosystems often requires actions that must

take place over large watersheds. Furthermore, there are only a few good examples of effective implementation of such actions in freshwater ecosystems.

A little over a decade ago, WWF aquatic biologist Robin Abell wrote a compelling essay on freshwater biodiversity conservation.[33] She convincingly outlined the case that freshwater ecosystems and species are among the most imperiled in the world, generally far more so than their terrestrial counterparts. She correctly observed that most conservation planning efforts tended to overlook freshwater or balkanize it into a separate effort. Fortunately, there has been steady progress over the last decade in freshwater conservation planning and in the attention being paid to the plight of freshwater ecosystems.

In 2011, the journal *Freshwater Biology* devoted a special issue to conservation planning for freshwater ecosystems that summarized many of the key issues and articulated many of the advances as well. This journal issue and other related papers demonstrate that systematic conservation planning can be and is being applied to freshwater ecosystems; that standing waters are receiving some attention, but far less than stream systems; that it is possible to do planning in data-poor regions (of which there are many for freshwater ecosystems); and that incorporation of ecosystem functions into planning in freshwater systems is critically important and is beginning to happen.[34]

There are three primary threats to freshwater ecosystems: land use change at the watershed level, alterations of hydrological regimes and wholesale modifications of physical riverine systems (for example, via channelization), and introductions of non-native species[35] (**FIGURE 2.6**). In turn, there are three major strategies to confront these threats: managing whole watersheds; restoring natural flow regimes; and exclusion, eradication, and control of non-native species. Although these strategies have been associated with aquatic protected or conservation areas (for example, beaver [*Castor canadensis*] removal in Tierra del Fuego, South America), they are by no means exclusive to a protected area approach, nor are they intended to be an exhaustive list of freshwater-associated strategies.

Freshwater planning will continue to advance. Some of the leading freshwater planners and applied conservation biologists have recommended key areas for improvement.[36] Among these, we would highlight the need to strengthen planning in non-riverine systems, improve the effectiveness of freshwater biodiversity surrogates (see Chapter 3 on surrogates), better incorporate freshwater connectivity into software planning tools, and develop good examples of conservation plans that span and integrate freshwater, marine, and terrestrial realms while also considering the trade-offs and benefits of this integration.[37]

(a)

(b)

(c)

FIGURE 2.6 The three most significant threats to freshwater ecosystems. Alteration of hydrological regimes by (a) dams and diversions, (b) invasive species (such as zebra mussel, *Dreissena polymorpha*), and (c) poor land management practices over entire watersheds (e.g., poor timber harvest practices and sedimentation from roads) represent the greatest threats to these ecosystems. As a result, strategies such as reestablishing ecological flow regimes below dams, controlling or removing invasive species, and improved land use planning and management over entire watersheds are needed to restore many aquatic ecosystems.

Conservation planning in marine ecosystems must take into account two important characteristics of marine life. First, many marine species have multiple life history stages, including planktonic phases early in life; the movements, population dynamics, and genetics of these planktonic forms are all important considerations for conservation. Second, like many terrestrial species, many marine fishes both migrate and aggregate as adults, and as a result, their migration routes and aggregation areas are critical components of marine conservation planning.

Attention to marine conservation and advances in marine conservation planning have grown nearly exponentially over the last decade. In 2004, for example, the Great Barrier Reef Marine Park in Australia implemented a new zoning plan, based on systematic conservation planning processes, that resulted in it being recognized as the largest marine protected system in the world. One-third of the reef is under strict protection, and other areas are tightly zoned and managed for other uses. In the same year, the state of California, USA, implemented the Marine Life Protection Act Initiative, which put in place a task force, a science advisory group, and a wide range of stakeholders to develop the first marine protected area system for California's Central Coast. Other results of this legislation and subsequent marine conservation planning were the establishment of 22 new marine protected areas along the north-central California coastal region and documentation of how and why an inclusive and transparent public planning process worked in achieving these conservation outcomes.[38]

In 2007, the first global classification of marine ecoregions was published, and it has subsequently seen considerable use in various conservation planning efforts.[39] In 2012, a group of scientists, policy makers, and conservationists developed and published the first Ocean Health Index,[40] which takes into account the status of regional marine ecosystems and the major factors influencing their quality. In the same year, the World Risk Report for 2012[41] focused on the importance of natural habitats along the coast in reducing risks and hazards for human communities. These are just a few highlights of the recent progress in marine conservation and planning efforts.

Over the last decade, marine conservation planning has advanced first from a focus on individual protected areas to systems of marine protected areas, and subsequently from marine ecosystem-based management (EBM) to marine spatial planning (MSP). In the first instance, the focus was primarily on conservation objectives, whereas both EBM and MSP involve planning for multiple objectives. Although most of this planning has occurred within coastal marine ecosystems and reefs, one of us has introduced these approaches to open-ocean conservation in the pelagic zone,[42] while others are bringing more consideration to marine ecosystem services in their planning efforts. Precise definitions of marine ecosystem-based management have proven elusive,[43] and differences between scientists and managers as to what it is and what it includes abound and have sometimes been a barrier to advancing conservation. To help overcome this barrier, a group of over 220 marine scientists and policy makers produced a consensus statement on ecosystem-based management:

> *Ecosystem-based management is an integrated approach to management that considers the entire ecosystem, including humans. The goal of ecosystem-based management is to maintain an ecosystem in a healthy, productive and resilient condition so that it can provide the services humans want and need. Ecosystem-based management differs from current approaches that usually focus on a single species, sector, activity or concern; it considers the cumulative impacts of different sectors.*[44]

**Marine spatial planning** (**MSP**) brings a spatial or zoning component to the principles of marine EBM (FIGURE 2.7). One definition of marine spatial planning is "a public process of analyzing and allocating the spatial and temporal distribution of the human activities in marine areas to achieve economic, ecological, and social objectives that are usually specified through a political process."[45] It is essentially the marine counterpart of land use planning and zoning.

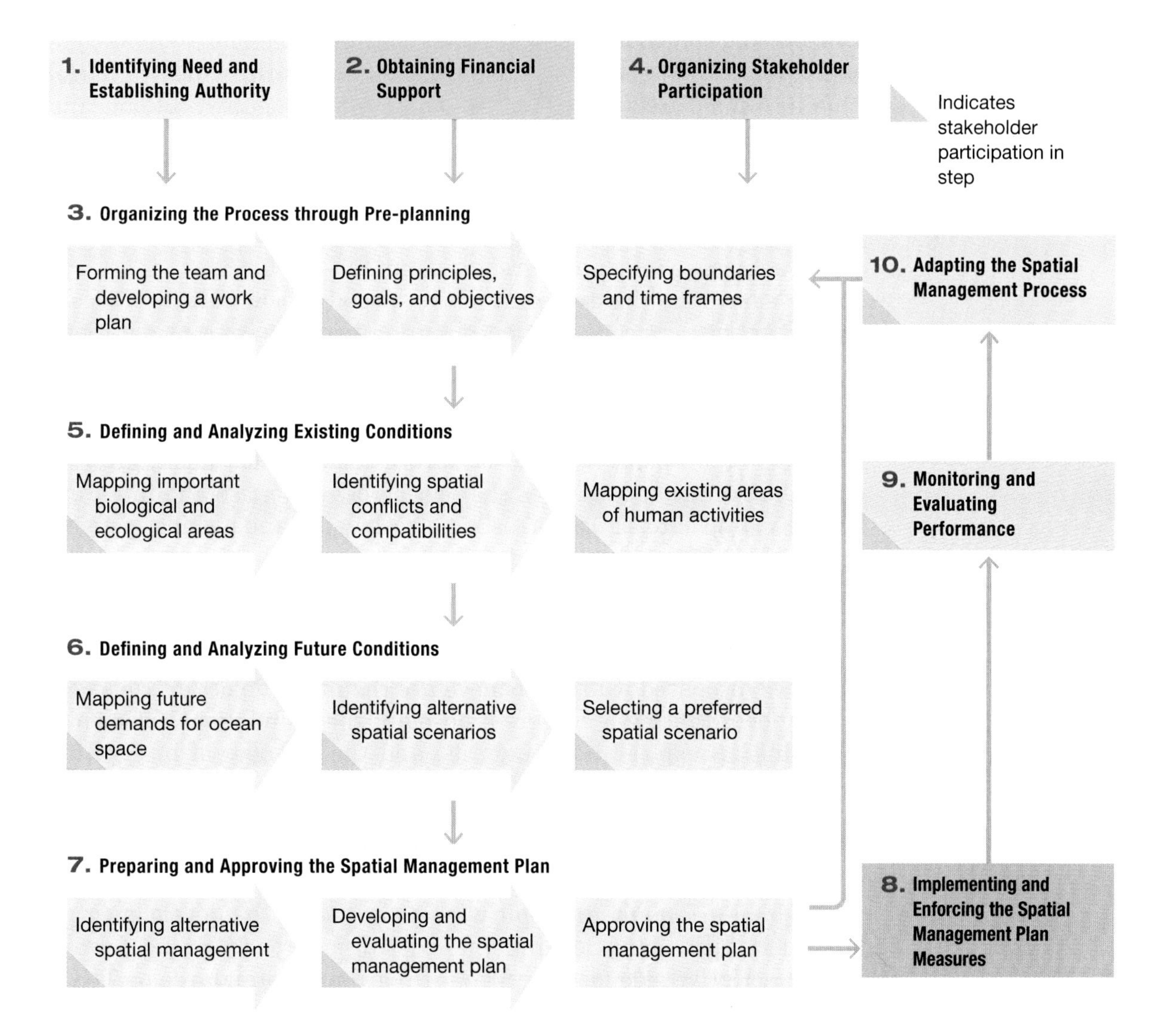


FIGURE 2.7 The major steps of marine spatial planning. These steps extend from defining the purpose of the planning process to designating spatial management zones such as marine protected areas. Adapted from Intergovernmental Oceanographic Commission, *Marine Spatial Planning: A Step-by-Step Approach toward Ecosystem-Based Management*, R. Dahl, editor, 2009. Paris: UNESCO.

As we have already mentioned, conservation planning in terrestrial, freshwater, and marine ecosystems has until recently advanced largely on separate tracks. The lack of integration of planning across these boundaries, and the fact that many threats to biodiversity in one realm originate in another, creates real problems due to the interconnectedness of ecosystems (for example, soil sedimentation on land negatively affects stream ecosystems, and pollution from freshwater and terrestrial ecosystems degrades marine systems). Although many projects have begun to integrate terrestrial and marine conservation planning,[46] there remain challenges to doing this well. Jorge Alvarez-Romero and colleagues[47] suggest that in some cases there will be competing objectives across realms, a lack of data with which to model threats, and fundamental social barriers among institutions and organizations that manage resources in these three realms, but rarely across them. The interdisciplinary approach that we discussed earlier in this chapter will almost certainly be required, but so too will be planning for multiple objectives across multiple natural resource sectors—a topic we take up in greater detail in Chapter 6.

We have highlighted some of the special considerations of freshwater and marine conservation planning here. These considerations notwithstanding, the basic steps and methods we outline for conservation planning throughout this book should apply equally well in these contexts. This approach is consistent with our view that the future of conservation planning, and of nature conservation itself, involves multidisciplinary approaches that cut across expertise and integrate where appropriate across terrestrial, freshwater, and marine systems.

## A Road Map for Conservation Planning

We have discussed what we believe are the most important considerations for initiating a conservation plan: the foundational elements, planning principles and standards, and the nuances of integrating across the terrestrial, freshwater, and marine realms. In this final section of the chapter, we outline a road map (FIGURE 2.8) for conducting a conservation planning process from beginning to end, including its implementation and monitoring and evaluation of progress.

There are many different ways to approach the development of a conservation plan. Some of the existing approaches tend toward the prescriptive side (see the WWF Programme Standards or TNC's Conservation Action Planning methods), while others provide more general guidelines (for example, National Biodiversity Strategy and Action Plans). Some approaches emphasize strategic planning (*Open Standards*), others emphasize spatial planning (systematic conservation planning), and some make an effort to integrate both.

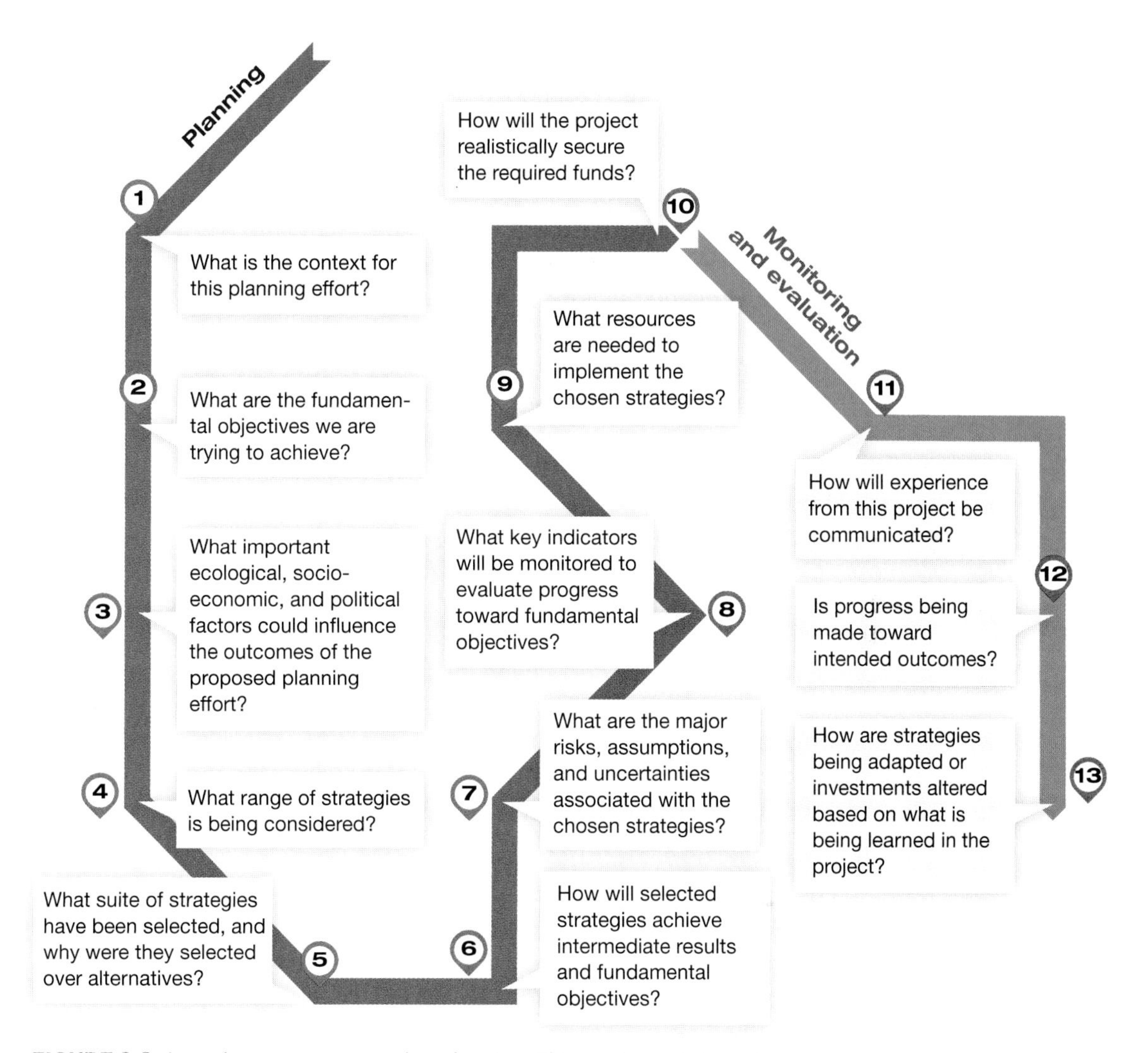


**FIGURE 2.8** A road map to conservation planning. This road map is based on a series of questions that project managers, planners, and scientists should address as part of any conservation project. Leaders and managers may be more receptive to planning efforts that they perceive to be guided by questions that need to be answered than to a planning process described by a series of steps that may be steeped in jargon. Adapted from The Nature Conservancy, *Conservation Business Planning Guidance.* 2013, Arlington, VA: The Nature Conservancy.

Our road map* to conservation planning, which integrates strategic and spatial planning, consists of a series of questions. How, in what sequence, and to what depth these questions are answered can be tailored to the specific project or program at hand. We have taken this approach for two reasons. First, it is our experience that many

* Much of the thinking behind this road map evolved from an internal TNC effort to evaluate and improve conservation planning as well as from existing TNC guidance on Conservation Business Planning.[48]

conservation project leaders, who may not have either a planning or a science background, respond more effectively to a planning effort when it addresses questions that they can relate to and generally agree need answering than to a set of planning steps that may be steeped in jargon. Second, the challenges that conservation projects face are highly variable around the globe in terms of the problems, the available information, and the skills and capacity of the project team. What's needed is a flexible planning approach that can draw on a wide variety of planning tools and scientific information and data. The question-based road map to planning provides the needed flexibility and should influence how project teams think about conservation project design without being overly prescriptive. It also helps emphasize the well-known adage that in planning, the process of answering questions is at least as important as the answers or the plan itself.

Although, for simplicity and display purposes, we have organized our road map in a linear fashion, we fully appreciate that many of these questions will be addressed in parallel, and not necessarily in the order in which we have placed them. In reality, this road map should have many smaller "roads" connecting the various questions in any number of feedback loops. As we observed in Chapter 1, conservation planning and adaptive management of a conservation project are closely linked. Here, we outline this road map in more detail and provide a rationale for answering each question.

1. What is the context for this planning effort?

   *Why answer this question?* To ensure that plans are focused on providing the answers actually needed to make decisions relevant to natural resource managers and conservation program directors, that the purpose of the planning exercise is clear, that the decisions made as a result of the plan are clear and transparent, and that the project team is in agreement as to the audience for the plan. Earlier in this chapter, we addressed this question as one of the foundational elements of good planning.

2. What are the fundamental objectives we are trying to achieve?

   *Why answer this question?* Being clear about the fundamental objectives or the ultimate outcomes of a conservation project will focus strategies on what actually matters—ends, not means. Ends are where we want to go. They are what we care about and value. Means are ways to get there. Clearly stated fundamental objectives provide a basis for consistently and transparently comparing and prioritizing alternative strategies and exposing trade-offs between them. Chapter 3 explores the development of fundamental objectives in more detail as well as their relationship to specific biodiversity or social features and the goals or quantitative targets that may be established for them.

3. What are the important ecological, socioeconomic, and political factors that could influence the outcomes of the proposed planning effort?

   *Why answer this question?* Setting the stage for identifying good strategies and potential points of intervention requires looking at the full social-ecological system in which conservation work takes place. This step, which is generally known as the situation analysis, will be outlined in more detail in Chapter 5.

4. What range of strategies is being considered?

   *Why answer this question?* It is easy for any agency or conservation organization to fall back on what it knows and does best, and too often that leads to a failure to explore a variety of possible strategies. Addressing this question transparently is more likely to lead to innovative strategies. It incorporates the selection of spatial priorities (the areas on the ground or in the water in which conservation actions take place), which we consider to be a subset of the overall strategies of any conservation project or initiative. Chapter 6 explores this question in depth.

5. What suite of strategies have been selected, and why were they selected over alternatives?

   *Why answer this question?* For many conservation problems, there are probably a variety of solutions. For practical reasons, most project teams need to focus on a small number of these solutions and implement them well. Selecting these strategies and actions is arguably the most important step in planning, yet many conservation projects rarely explore alternatives in any meaningful way, often falling back on what an organization has always done or is good at. In addition, it's important for planners and practitioners to document the evidence base for why a particular set of strategies is likely to succeed. Doing so builds confidence in the actions of the project, makes it easier to make adjustments, and allows other projects to learn from successes and failures. As we outlined in Chapter 1, exploring different avenues to address conservation problems and challenges is one of the most important ways in which conservation planning can improve. Chapter 6 delves further into this question.

6. How will the selected strategies achieve intermediate results and fundamental objectives?

   *Why answer this question?* At its core, this question forces planners and practitioners to link the strategies and actions they are proposing to the project's fundamental objectives. This is often accomplished through a theory of change that will help validate the realities of any specific strategy by carefully thinking through the steps, and the results of those steps, by which the strategy will help achieve project outcomes. Chapters 6 and 7 explain how planners may best answer this question.

7. What are the major risks, assumptions, and uncertainties associated with the chosen strategies?

   *Why answer this question?* Too few conservation projects seriously consider the risks they take and the assumptions they make in implementing various strategies and actions. That is unfortunate, because a thoughtful consideration of both can often substantially improve a project's chances of success and expose strategies that are unnecessarily risky. A transparent and explicit statement of risks allows us to mitigate them, makes us more risk tolerant in designing strategies, and establishes a clear connection between risks and strategy selection. We discuss tools for evaluating project risks and uncertainty in detail in Chapter 8.

8. What key indicators will be monitored to evaluate progress toward the fundamental objectives?

   *Why answer this question?* As we mentioned at the start of this road map, monitoring and evaluation are inextricably linked to planning. It is part of good planning practice to think clearly about how we will measure the progress of various strategies and actions and make adjustments to strategies as appropriate (see Chapter 12). Conservation practitioners have been notoriously poor at evaluating the effectiveness of strategies and actions, and part of getting better at this involves thinking about measuring effectiveness from the outset of a project.

9. What resources are needed to implement the chosen strategies? (See answer under question 10.)

10. How will the project realistically secure the required funds?

    *Why answer these questions?* In most conservation organizations and institutions, there is a tendency to underestimate the real costs of conducting a conservation project. And the bigger and more complicated a project is, the more difficult this task becomes. The costs are usually dominated by staff salaries, and opportunity costs are frequently involved in deploying staff on particular projects. An even greater challenge is being realistic about fund-raising, especially in challenging financial and economic times. Making an investment in answering these questions will increase the likelihood that a conservation project will have the necessary resources to advance successful strategies. We will probe these questions in more detail in Chapter 11 on implementation.

11. How will experience from this project be communicated?

    *Why answer this question?* Whether in a government agency or an NGO, an enormous amount of reinventing the wheel takes place in terms of applying good planning practices, implementing various conservation strategies, and evaluating the effectiveness of strategies through monitoring. In an era when it is both easy and increasingly an expected practice to share data, information, and lessons learned in conservation, it is critical that all conservation projects improve at communicating and sharing their methods and results. We'll explore this communication issue in Chapter 11.

12. Is progress being made toward intended outcomes? (See answer under question 13.)
13. How are strategies being adapted or investments altered based on what is being learned in the project?

    *Why answer these questions?* As we noted earlier, conservation projects have a long history of poor performance related to evaluation of effectiveness. For many practical reasons, not every project can or should have a highly inferential monitoring program to evaluate its effectiveness. Although this topic of monitoring and evaluation could itself be the subject of a book, we will address its most important aspects in Chapter 12 as well as touch on the subject throughout the book. Even when there is information to suggest a needed change in approach or strategy, it can be difficult to make that happen. Too few conservation projects take the time, or make it part of their day-to-day management of the project, to routinely consider changing their approach even when the current set of activities isn't working to the degree that the team had hoped.

## KEY MESSAGES

- Understanding the context for planning a conservation project (purpose, decisions, constraints, audience, and investment) and putting together an effective multidisciplinary team are two critical components that will serve as a strong foundation to any conservation plan.
- Different principles and standards have been advanced to assist practitioners in developing a conservation plan. We have summarized the most important of these principles (representation, persistence, efficiency), highlighted three emerging ones (social-ecological systems; theory of change; risk and uncertainty)

and shown how they have been incorporated into the road map for planning on which this book is based.

- Conservation planning has long emphasized terrestrial systems, yet substantial advances in conservation planning methods have been made in the marine and freshwater realms over the last decade. The best planning efforts will be those that successfully integrate themselves across the realms.
- The diversity of planning situations around the world suggests that a flexible approach is needed for developing conservation plans. We present a road map for planning that underpins the overall structure of this book. It provides the needed flexibility by asking questions that managers, leaders, and project directors are likely to find useful to answer and that may help engender broader support for any planning effort.

## References

1. Hammond, J. S., R. L. Keeney, and H. Raiffa, *Smart Choices: A Practical Guide to Making Better Life Decisions.* 1999, New York: Broadway Books. 242.
2. Dallimer, M., and N. Strange, *Why socio-political borders and boundaries matter in conservation.* Trends in Ecology and Evolution, 2015. **30**(3): 132–139.
3. The Nature Conservancy, *Conservation Business Planning Guidance.* 2013, Arlington, VA: The Nature Conservancy.
4. European Commission, *Natura 2000 Network* [accessed July 10, 2014]. http://ec.europa.eu/environment/nature/natura2000.
5. Reed, M. S. et al., *Who's in and why? A typology of stakeholder analysis methods for natural resource management.* Journal of Environmental Management, 2009. **90**(5): 1933–1949.
6. Golder, B., and M. Gawler, *Cross-cutting tool: Stakeholder analysis*, in *WWF Standards of Conservation Project and Programme Management Standards.* 2005, Gland, Switzerland: WWF International.
7. Calvache, A., S. Benítez y A. Ramos. 2012. Fondos de Agua: Conservando la Infraestructura Verde. Guía de Diseño, Creación y Operación. Alianza Latinoamericana de Fondos de Agua. The Nature Conservancy, Fundación FEMSA y Banco Interamericano de Desarrollo. Bogotá, Colombia. 144p.
8. Gleason, M. et al., *Science-based and stakeholder-driven marine protected area network planning: A successful case study from north central California.* Ocean & Coastal Management, 2010. **53**(2): 52–68.
9. The Nature Conservancy, *Conservation Business Planning Guidance.* 2013, Arlington, VA: The Nature Conservancy.
10. The Nature Conservancy, *Conservation Partnership Center* [accessed September 6, 2013]. http://www.conservationgateway.org/Conservation Planning/partnering/cpc/Pages/default.aspx.

11. Groves, C. R. et al., *Planning for biodiversity conservation: Putting conservation science into practice.* BioScience, 2002. **52**(6): 499–512.
12. Bottrill, M. C., and R. L. Pressey, *The effectiveness and evaluation of conservation planning.* Conservation Letters, 2012. **5**: 407–420.
13. Boersma, D. et al., *How good are endangered species recovery plans?* BioScience, 2001. **51**(8): 643–649.
14. Pooley, S. P., J. A. Mendelsohn, and E. J. Milner-Gulland, *Hunting down the chimera of multiple disciplinarity in conservation science.* Conservation Biology, 2014. **28**(1): 22–32.
15. Katzenbach, J. R., and D. K. Smith, *The Wisdom of Teams.* 1993, Boston: Harper Business. 291.
16. Black, S. A., J. J. Groombridge, and C. G. Jones, *Leadership and conservation effectiveness: Finding a better way to lead.* Conservation Letters, 2011. **4**(5): 329–339.
17. Lencioni, P., *The Five Dysfunctions of a Team: A Leadership Fable.* 2002, San Francisco: Jossey-Bass, Wiley. 228.
18. Beale, W., *Step 1.1: Define team composition and operations*, in *Resources for Implementing the WWF Project and Programme Standards.* 2006, Gland, Switzerland: WWF International.
19. Fernandes, L. et al., *Establishing representative no-take areas in the Great Barrier Reef: Large-scale implementation of theory on marine protected areas.* Conservation Biology, 2005. **19**(6): 1733–1744.
20. Reyers, B. et al., *Conservation planning as a transdisciplinary process.* Conservation Biology, 2010. **24**(4): 957–965.
21. Groves, C., *Drafting a Conservation Blueprint: A Practitioner's Guide to Planning for Biodiversity.* 2003, Washington, DC: Island Press. 457.
22. Berkes, F., and C. Folke, editors, *Linking Social and Ecological Systems: Management Practices and Social Mechanisms for Building Resilience.* 1998, Cambridge: Cambridge University Press. 476.
23. Walker, B., and D. Salt, *Resilience Thinking: Sustaining Ecosystems and People in a Changing World.* 2006, Washington, DC: Island Press. 174.
24. Miller, B. W., S. C. Caplow, and P. W. Leslie, *Feedbacks between conservation and social-ecological systems.* Conservation Biology, 2012. **26**(2): 218–227.
25. Stephanson, S., and M. B. Mascia, *Putting people on the map through an approach to integrating social data in conservation planning.* Conservation Biology, 2014. **28**(5): 1236–1248.
26. Milner-Gulland, E. J. et al., *Accounting for the impact of conservation on human well-being.* Conservation Biology, 2014. **28**(5): 1160–1166.
27. Ostrom, E., *A general framework for analyzing sustainability of social-ecological systems.* Science, 2009. **325**(5939): 419–422.
28. Binder, C. R. et al., *Comparison of frameworks for analyzing social-ecological systems.* Ecology and Society, 2013. **18**(4): 26.

29. Stephanson and Mascia, *Putting people on the map.* (See reference 25.)
30. Musacchio, L., *The scientific basis for the design of landscape sustainability: A conceptual framework for translational landscape research and practice of designed landscapes and the six Es of landscape sustainability.* Landscape Ecology, 2009. **24**(8): 993–1013.
31. Ban, N. C. et al., *Marine and freshwater conservation planning: From representation to persistence,* in *Conservation Planning: Shaping the Future,* F. L. Craighead and C. L. Convis Jr., editors. 2013, Redlands, CA: Esri. 175–217.
32. Nel, J. L. et al., *Progress and challenges in freshwater conservation planning.* Aquatic Conservation: Marine and Freshwater Ecosystems, 2009. **19**(4): 474–485; Linke, S., E. Turak, and J. Nel, *Freshwater conservation planning: The case for systematic approaches.* Freshwater Biology, 2011. **56**(1): 6–20.
33. Abell, R., *Conservation biology for the biodiversity crisis: A freshwater follow-up.* Conservation Biology, 2002. **16**: 1435–1437.
34. Turak, E., and S. Linke, *Freshwater conservation planning: An introduction.* Freshwater Biology, 2011. **56**(1): 1–5.
35. Saunders, D. L., J. J. Meeuwig, and A. C. J. Vincent, *Freshwater protected areas: Strategies for conservation.* Conservation Biology, 2002. **16**(1): 30–41.
36. Nel et al., *Progress and challenges in freshwater conservation planning.* (See reference 32.)
37. Adams, V. M. et al., *Planning across freshwater and terrestrial realms: Cobenefits and tradeoffs between conservation actions.* Conservation Letters, 2013. **7**(5): 425–440.
38. Gleason et al., *Science-based and stakeholder-driven marine protected area network planning.* (See reference 8.)
39. Spalding, M. D. et al., *Marine ecoregions of the world: A bioregionalization of coastal and shelf areas.* BioScience, 2007. **57**(7): 573–583.
40. Halpern, B. S. et al., *An index to assess the health and benefits of the global ocean.* Nature, 2012. **488**(7413): 615–620.
41. Beck, M. W. et al., *World Risk Report 2012.* 2012, Bonn, Germany: United Nations University.
42. Game, E. T. et al., *Pelagic protected areas: The missing dimension in ocean conservation.* Trends in Ecology & Evolution, 2009. **24**(7): 360–369.
43. Arkema, K. K., S. C. Abramson, and B. M. Dewsbury, *Marine ecosystem-based management: From characterization to implementation.* Frontiers in Ecology and the Environment, 2006. **4**(10): 525–532.
44. McCleod, K. L. et al., *Scientific consensus statement on marine ecosystem-based management.* 2005, Communication Partnership for Science and the Sea. Corvallis: Oregon State University.
45. Intergovernmental Oceanographic Commission, *Marine Spatial Planning: A Step-by-Step Approach toward Ecosystem-Based Management,* R. Dahl, editor. 2009, Paris: UNESCO.

46. Makino, A. et al., *Integrated planning for land–sea ecosystem connectivity to protect coral reefs.* Biological Conservation, 2013. **165**: 35–42.
47. Alvarez-Romero, J. G. et al., *Integrated land-sea conservation planning: The missing links.* Annual Review of Ecology, Evolution, and Systematics, 2011. **42**(1): 381–409.
48. The Nature Conservancy Planning Evolution Team, *Planning for Tomorrow's Challenges: Recommendations of the Planning Evolution Team.* 2011, Arlington, VA [accessed 4 March 2015]. https://www.conservationgateway.org/Files/Pages/report-planning-tomorrow%E2%80%99.aspx; and The Nature Conservancy, *Conservation Business Planning Guidance, Version 1.3.* July 10, 2013, Arlington, VA. [accessed March 4, 2015]. http://www.conservationgateway.org/ConservationPlanning/BusinessPlanning/Documents/CBP_Guidance.pdf.

# 3

# Which Way Paradise? Establishing Objectives

## Overview

Conservation planning is a process intended to achieve desirable outcomes for nature and people. The objectives of a plan are versions of these desirable outcomes. Defining objectives represents a crucial foundation for good planning. In this chapter we introduce the notion of a fundamental objective based on values, and we highlight the distinction between ends (where we want to go) and means (how we want to get there) that must be made when establishing and structuring objectives. We distinguish environmental, social, and economic objectives, and we offer guidance on understanding and articulating objectives and on deciding which of these are relevant to the plan.

## Topics

- *Terminology*
- *Importance of fundamental objectives*
- *Means versus ends*
- *Types of objectives*
- *Eliciting objectives from stakeholders*
- *Conservation features*
- *Species*
- *Community and abiotic features*
- *Surrogacy*
- *Ecosystem services*
- *Ecological processes*
- *Human well-being and other non-biodiversity objectives*

## Terminology

For cultural, historical, and institutional reasons, conservation planners have adopted a diversity of terms to describe the same concepts. The Conservation Measures Partnership (CMP) first described this diversity of terms in an initiative in 2004, known as the Rosetta Stone. It was a comparison of the project management systems and terminology used by various conservation organizations along with some proposed standard terminology. Since then the CMP, and more recently the Conservation Coaches Network, have facilitated a movement to use a standard terminology for management of conservation projects.

The standard terminology of CMP as articulated in the most recent version of the *Open Standards* (Version 3.0) follows the familiar strategic planning hierarchy of terms: *vision*, *goals*, *objectives*, and *actions*. We depart somewhat from this path, both in our use of the terms and the sequence in which we approach them, most noticeably with regard to goals and objectives (TABLE 3.1). Most significantly, we emphasize *fundamental objectives* as those things that we ultimately want our conservation actions to achieve, and we distinguish these from *intermediate objectives*, which relate to specific activities undertaken toward these fundamental outcomes. This departure is firmly intentional and draws on the use of these terms in decision science (**Box 3.1**). We believe that the terminology we have adopted helps us to clearly recognize, and distinguish, the roles of science and values in the planning process, and that this distinction ultimately makes our approach more powerful.

We will use our chosen terms throughout this book, but we are not puritanical about this terminology. If, as a reader, you prefer a

**TABLE 3.1** Key planning terminology used in this book

| Term | Definition |
|---|---|
| Vision statement | A brief and inspirational statement about what we hope the future will look like. |
| Objectives | The desirable outcomes of a planning process, stated as something stakeholders care about and the direction in which they want it to go. Objectives can be fundamental objectives, intermediate objectives, process objectives, or strategic objectives. |
| Features | Specific attributes of broad concepts such as *biodiversity* for inclusion in a conservation plan. |
| Indicator | A measurable property of an objective or feature. |
| Target | A quantitative statement of the desired level of achievement for an indicator. Targets can be achieved or not. |
| Actions | The set of intervention or strategy options available to choose from. |

## BOX 3.1 A Rosetta Stone for Fundamental Objectives

As this introductory section of Chapter 3 suggests, we believe it is important to start any conservation planning process with a discussion of what we ultimately want to achieve; these are sometimes referred to as outcomes or ultimate outcomes, but we refer to them by the decision science term—*fundamental objectives.* There is no clear synonym for this term in the *Open Standards (OS)* terminology. Fundamental objectives are related to both goals and objectives in the OS classification.

In OS terminology, goals are "formal statements of a desired impact of a project such as the desired future status of a target." As such they are intended to be measurable, time-bound, specific, and linked to a conservation target or biodiversity feature. Objectives, in OS terminology, are also formal statements; but they are focused on outcomes or intermediate results necessary to achieve goals. They, too, are intended to be measurable, time-bound, and specific but also practical within the social and political context of a conservation project. In contrast, fundamental objectives are not intended to be specific, time-bound statements about what we want to achieve that are measurable in their own right. Rather, fundamental objectives represent statements about the things we value, such as fish, or jobs, or safety, and what we hope will happen to these things, such as increase.

In Chapter 4, we introduce the terminology of targets that are intended to be measurable, time-bound, and linked to specific features of the plan. Our use of *targets* appears to be nearly synonymous with the OS use of *goals.*

different term for one of the concepts we describe, feel free to substitute terms as you see appropriate. In all cases, the terminology used is not as important as understanding the concepts and their purpose.

It is also worth noting that, throughout the book, we use the term *landscape* for simplicity; but unless specified otherwise, it should be interpreted broadly to include *landscapes*, *seascapes*, or *freshwater systems*.

## Vision Statement

Conservation planning conventionally starts with a **vision statement**: a brief and inspirational statement about what we hope the future will look like—for example, "To manage land use in a way that maintains resilient local communities and healthy natural ecosystems." Rather than offering guidance on constructing a compelling vision statement, we urge you to put aside conventionality and get started on objectives. To be clear, we think it is perfectly reasonable to construct a vision statement—but do it later in the planning process.

We do not see vision statements as an effective way to elicit useful input or buy-in from any group before planning. Understanding objectives, as described in this chapter, is a far more effective starting point for a conservation plan. We often see the construction of a vision statement become a time-consuming exercise in wordsmithing for no discernible gain. Do not spend valuable time, or burn valuable credit with partners and stakeholders, for the sake of a vision statement. Vision statements can make great communication pieces that can also be helpful in fund-raising efforts if they are constructed with that purpose clearly in mind—but separate this role from strategic planning.

## Fundamental Objectives

### Ends versus Means

> *"Would you tell me, please, which way I ought to go from here?"*
>
> *"That depends a good deal on where you want to get to," said the Cat.*
>
> *"I don't much care where—" said Alice.*
>
> *"Then it doesn't matter which way you go," said the Cat.*
>
> *"—so long as I get SOMEWHERE," Alice added as an explanation.*
>
> *"Oh, you're sure to do that," said the Cat, "if you only walk long enough."*
>
> — Lewis Carroll, *Alice in Wonderland*

Conservation planning is an exercise in figuring out the best route to desirable outcomes for people and nature. As Alice's conversation suggests, planning a route requires a destination: Where are we headed? What do we want to achieve?

In conservation, our destination is those things that we ultimately want our conservation actions to achieve. Which outcomes are most important to those who will be influenced by the planning process? Think of these as the reasons for caring. More abundant fish, fewer species lost, culture sustained, well-being improved. What is our view of paradise? We refer to these outcomes as a plan's **fundamental objectives**. This term is commonly used in the decision science literature.

As obvious and logical as it might seem to start a planning process with a clear understanding of the outcomes we ultimately desire, doing so frequently proves more challenging than expected (although insight from behavioral psychology suggests that this challenge should be the expectation rather than the exception). Even those who have been involved in conservation planning for many years are often surprised at having the question of fundamental objectives posed. One reason this question is challenging is that people find it difficult to separate where they want to go from how they want to get there—in other words, to separate *ends* from *means*. A second reason is that a focus on value-based fundamental objectives is likely to include outcomes that seemingly do not have a conservation focus (such as the number of jobs in a region) but that may be affected by, or affect, conservation outcomes.

Conservation plans are filled with examples of substituting means for ends when defining what we want to achieve. One of the most common examples we see is a stated objective to establish a new protected area or park when doing so is clearly a means to a more fundamental end: the conservation of a particular species, ecosystem, or ecological process. Even when explicitly asked about fundamental objectives, planning teams often come up with objectives along the lines of "Improve the connectivity of forests" (or substitute any other habitat type). Did you notice that a *means* (improved connectivity) has been substituted for the fundamental *end* (probably something to do with a healthy and extensive forest ecosystem)? Would we ultimately be satisfied with well-connected forestlands that were overrun with invasive species and had suffered many local extirpations? Probably not. It is likely that connectivity was listed as a desired outcome because those involved in the plan believed habitat fragmentation to be a major barrier to a healthy, functioning forest ecosystem, or perhaps because connectivity was an easy outcome to communicate and one that captured many of the activities the program was likely to undertake.

Research in behavioral psychology suggests that substitution of means for ends is the default mode for most of us. Means (in other words, the actions and strategies we are going to use to get somewhere) are generally easier to think of because they are what we spend our days doing—they are more available to us.[1] Similarly, human minds are hungry for richness in descriptions and are likely to have trouble thinking of an outcome without simultaneously thinking of mechanisms to get there.[2] Identifying fundamental objectives requires deliberate, effortful thinking. There are good reasons to make this effort at the start of a conservation planning process.

## Why Focus on Fundamental Objectives?

Fundamental objectives are legitimately matters of opinion, not science. They reflect *values*. As we mentioned in Chapter 1, values are what we care about. They are the basis by which we evaluate the desirability of different actions and outcomes. For conservation planners and ecologists accustomed to emphasizing the scientific basis of our work, this focus on values has often been a hard pill to swallow.[3] However, we see acknowledging the value-driven nature of conservation objectives as a strength for good planning. Whereas the science behind conservation decisions can be legitimately questioned (and often is), there is some sanctity around the notion that we are free to develop and seek support for our own values.

The focus on value-driven objectives in decision support has been heavily influenced by the work of decision scientist Ralph Keeney. Keeney and colleagues have written extensively on *value-focused thinking*[4] (**Box 3.2**) and have been instrumental in developing the planning approach known as **structured decision making**.[5] Although their work deals with decision making more generally, it provides reference material that is comfortably applicable to conservation planning. They argue that values can be identified through careful and deliberate thinking. Fundamental objectives translate those values into clear statements by which to judge alternative courses of action. In reflecting the values of stakeholders, fundamental objectives can also play a strong role in justifying the decisions and actions that result from a conservation plan—ideally, we wish to be able to say that these

### BOX 3.2 Value-Focused Thinking

Decision scientist Ralph Keeney has proposed that values should be the driving force in any decision-making process. Keeney's logic is that your interest in any decision is the hope of getting a desirable outcome. What makes an outcome desirable? The relative desirability of an outcome is based on what you value.

A colleague of ours, Mike Runge, illustrates this idea neatly with an example: If a friend asks you where the two of you should go out to dinner tonight, what is your typical response? For most people, it is "Well, what are the alternatives?" Your friend then lists a few restaurants, and you choose one. But why did you choose it? It might be that you want to eat something different from what you cook at home, or that the restaurant has a good atmosphere and you can talk easily, or is a place you don't have to travel far to, or a place with reliably good service. These are all things that you value and that, perhaps unconsciously, have influenced your choice. Explicitly articulating these values for a casual dinner out with friends is probably too much planning, especially given your ability to do this almost unconsciously.

For more important decisions, however, with substantial resource, social, or environmental consequences, taking time and using a structured process to understand and articulate these values is time well spent. As Keeney himself says, "Focusing early and deeply on values when facing difficult problems will lead to more desirable consequences." Planning and decision support is then an exercise in evaluating the consequences of alternative courses of action against values. For an authoritative source on value-focused thinking and its role in planning and decision making, readers should look directly to Keeney's own work, but we hope you will see the threads of this way of thinking throughout this book.

## BOX 3.3 Support for Fundamental Objectives

A focus on fundamental objectives is likely to engender broader support for conservation planning and actions than would be possible if planners focused on particular strategies or actions. Consider an objective that is commonly stated in conservation plans: establishing a certain number or extent of marine protected areas (MPAs).

If those involved are questioned about what the plan is ultimately trying to achieve, most likely it will not be to establish protected areas per se, but rather to adequately protect marine habitat and preserve important marine stocks or species. MPAs are simply a favored and proven means of accomplishing this. However, a sector of the stakeholder community almost certainly is uncomfortable with the notion of protected areas. Starting with the means as the stated objective immediately places the planning process in opposition to these stakeholders. There is likely to be far greater agreement and broader support for a planning process with the stated objective of sustaining fish populations.

As a more general example, it can be reasonably speculated that resistance to climate change action in the United States is at least partly due to fundamental opposition to the proposed solutions (generally taxes). If the proposed solutions had come secondarily to discussion about the fundamental objectives—namely, protecting American values and lifestyles in a changing climate—broader support for action might well have been galvanized.

actions represent the best way to balance and achieve the set of values held by all affected parties (**Box 3.3**).

Being clear about fundamental objectives also helps keep the planning process open to new ideas. For example, if planning is focused too early on specific actions, such as creating protected areas (and the myriad approaches to this task rather than fundamental objectives), there is a risk that a better solution for achieving what we actually care about might not be noticed. It might well be that some sort of market-based incentive would be equally effective and less costly than MPAs in conserving marine stocks and habitats. The benefit is twofold here. We want a planning process that not only keeps our minds open to alternative solutions but also keeps us focused on what we care about.

Fundamental objectives form the basis by which we should judge the desirability of possible conservation actions or solutions. They should also influence the data and information we use in subsequent parts of the planning process: we want data that will help predict the influence of actions on our objectives, and data that will help us track the performance of our chosen actions in achieving our objectives.

### Getting to Fundamental Objectives

Determining the set of fundamental objectives that will guide the subsequent planning process is rarely as simple as asking a group to identify them in a workshop and then writing them down. The influence of the psychological biases mentioned earlier makes it challenging to think of outcomes without actions. The psychologist Daniel Kahneman refers to our need for complete stories.[6] We like rich details; an end stripped of its means often appears less compelling and therefore does not spring to mind as easily.

The good news is that with a deliberate effort, we can quite easily get to an understanding of values and fundamental objectives, and we can even use our preference for stories to help us get there. We have found the following procedure useful in eliciting fundamental objectives in group workshop settings. The points can be used in any order.

1. Clearly define what fundamental objectives are. Describing the distinction between ends and means usually helps.
2. Ask participants what they care most about in the landscape in question. What do they really want or need out of that landscape? **Box 3.4** gives an example of how participants in a planning workshop for the Pacific coasts of Washington and Alaska, USA, answered this question.
3. Use prompting questions. The following are all useful:

   What, in your view, is wrong with the current situation?

   What, in your view, would be the best possible outcome? What makes this outcome great?

   What, in your view, would be the worst possible outcome? What makes this outcome terrible?

   If no action is taken, what do you think will happen? What is good or bad about this outcome?

**BOX 3.4** Value Statements

These answers were given by participants in a planning workshop for the Pacific coasts of Washington and Alaska, USA, to the question of what they cared most about in the landscape and seascape of the region. Such answers are great starting points for determining the fundamental objectives of a planning process.

- Access for recreation
- Preservation of cultural values
- Increased fish abundance for sport fishing
- Wildness
- Aesthetic value of natural landscape
- Identity of indigenous and nonindigenous people from the region
- Economic opportunities associated with resource extraction
- Self-determination of indigenous people
- Lands that allow for quiet solitude
- Sustainability of revenue
- Abundant wildlife
- Jobs
- Government revenue
- Old-growth forests

4. Play the *Why* game. One way to get at fundamental objectives is to work from participants' initial statements of what they think should be achieved by asking, "Why is that important?" Consider these statements and questions:

   An offset program is successfully implemented in all new energy developments.

   *Why is that important?*
   Because energy developments are likely to damage habitats they occur in.

   *Why is that important?*
   Because often they occur in rare or sensitive habitats.

   *Why is that important?*
   Because we believe that these habitats constitute an important part of saving biodiversity.

   *Why is that important?*
   Well, because it just is!

   At this point, we could say that a fundamental objective is actually to *minimize the loss of biodiversity.*

   Depending on where you sit, an answer to the last "Why is that important?" question might equally be, "Because the legislation says we need to protect this habitat," in which case the fundamental objective might be to *maximize the extent of the habitat identified in the legislation.*

5. As the list of objectives builds up, think about grouping them. Often there will be many closely related objectives, and grouping them can provide a good indication of what is ultimately important. At the highest level, objectives might be grouped into themes representing environmental, social, and economic objectives. Beneath these there might be multiple objectives relating to populations of species or the extent of habitat, which could also be grouped.

6. Consider turning the groupings from point 5 into what is generally referred to as an **objective hierarchy**.[7] Inevitably, some objectives will be more specific than others, and some will largely be clarifications of more general objectives that might also be listed (we'll call them "sub-objectives"). For example, the broad objective "minimizing biodiversity loss" might have the sub-objective of "minimizing loss of threatened species." Higher-level objectives can have many possible sub-objectives that help clarify them. An objective hierarchy simply involves

arranging objectives on a board or computer so that they are grouped appropriately to minimize the number of high-level objectives, and so that objectives that appear alongside one another are at a similar level of explicitness. Hierarchies can rapidly help us make sense of objectives that vary widely in scope and explicitness. FIGURE 3.1 illustrates how the values identified in Box 3.4 were subsequently turned into an objective hierarchy.

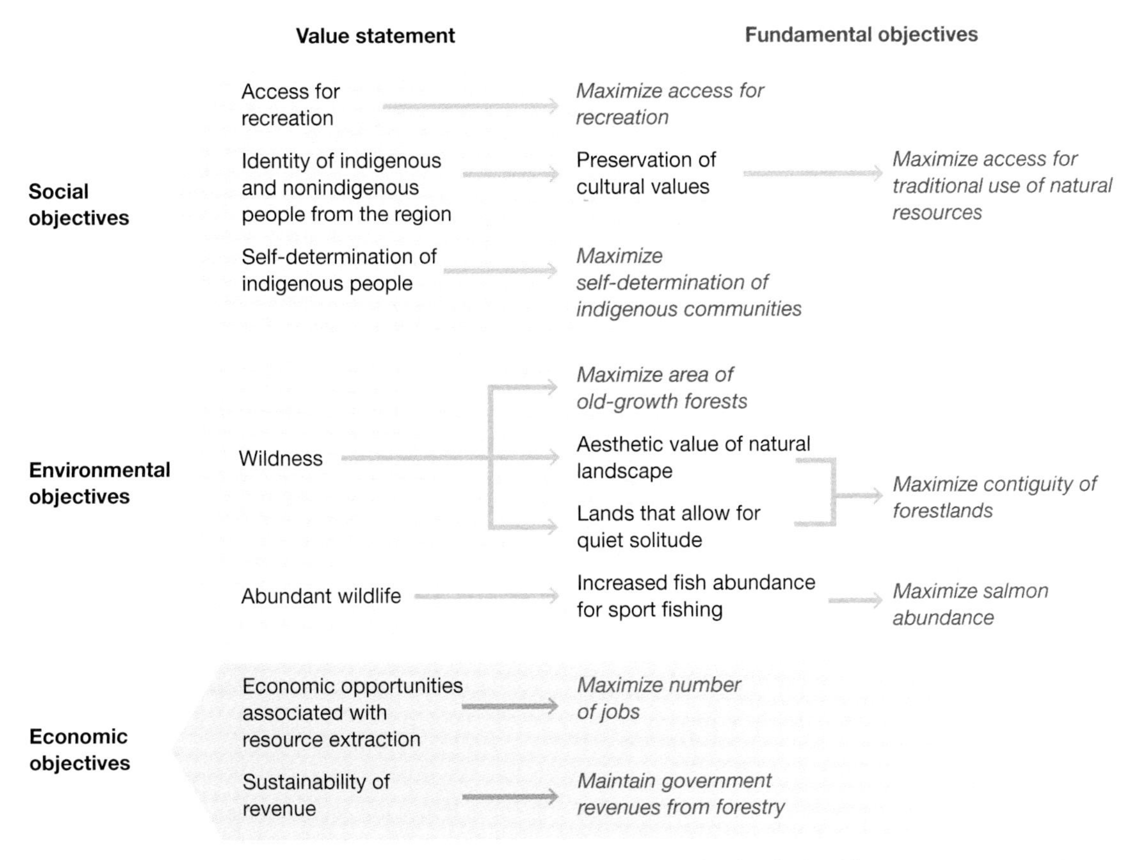


FIGURE 3.1 Example of an objective hierarchy for a conservation plan for the Pacific coasts of Washington and Alaska, USA. At the highest level (left side), the objectives are grouped into three classes: social, environmental, and economic. Such high-level groupings are not a necessary part of an objective hierarchy, but are used here to show the diversity of objectives found in many modern conservation plans. To the right of these general groupings are the value statements taken from Box 3.4. Some of these statements have sub-objectives that clarify the important elements of the main objective. For example, the general value of "wildness" has sub-objectives for the aesthetic value of natural landscapes and the potential for quiet solitude. The value statements are then expressed as fundamental objectives (italic type)—wildness has the dual objectives of maximizing old-growth forest and maximizing the contiguity of forestland. Each fundamental objective takes a value and expresses it as the thing that is important (e.g., old-growth forest, salmon, jobs) and the direction we want it to go (e.g., maximize, maintain, minimize).

Procedures like this one work best when politics are kept out of the discussion. To the extent possible, elicitation of objectives requires a situation in which people can be honest. It is perfectly fine for a range of objectives to sit alongside one another, even if they appear to be in opposition to one another (as economic and conservation objectives often do). There will be plenty of time later to look for mutually beneficial solutions and navigate the necessary trade-offs. Similarly, we think it is important to start by being inclusive: when someone suggests something, write it down. Group or remove objectives later. This procedure is both good for morale and good for keeping our minds and options open.

## What Makes a Good Fundamental Objective?

According to Ralph Keeney's classic text, *Value-Focused Thinking*,[8] good fundamental objectives should contain just two components: the thing that matters and the direction it should move in. Direction is generally specified as some form of *increase* or *decrease*, such as increased revenue or decreased loss of biodiversity. Anything else is superfluous. **TABLE 3.2** contains a set of fundamental objectives that demonstrate these two components.

Simple objective statements like these are almost certain to contain terms that could be considered vague (such as *biodiversity* or *revenue*). Vagueness is acceptable at this stage because it is the role of features (discussed later in this chapter), together with indicators and targets (discussed in Chapter 4), to more carefully define how these objectives should be interpreted.

Shouldn't objectives be SMART (specific, measurable, attainable, realistic, time-bound)? Many conservation planners are familiar with the notion of SMART objectives[9] that are quantitative and time-bound,

**TABLE 3.2** A set of fundamental objectives from a conservation plan for the mid-Atlantic seascape in the United States

1. Maximize populations of threatened and endangered marine mammals.
2. Maximize biomass of apex predators (large pelagics such as sharks, tuna, and marlin as well as seabirds such as northern gannets and loons).
3. Maximize the extent of cold-water coral and other biogenic habitats (for example, live-bottom patch habitats) from near shore (20 m) to the foot of the continental shelf–continental slope break (2500 m).
4. Increase total biomass of diadromous fishes (shad, river herring, eels, and sturgeons) that use mid-Atlantic habitats and resources.
5. Maintain extent of soft-bottom benthic habitats.

including statements of how much and by when. We agree that these are important parts of a plan, and in our road map they appear as the indicators and targets that help define each objective (see Chapter 4). It is simply our emphasis on value-focused objectives and consistency with the use of the term *objective* in the decision sciences that makes the SMART acronym inappropriate for this first step in the planning process.

Fundamental objectives simply say what and in which direction. It is safe to assume that, in general, the faster and more substantially an objective is achieved, the better. Situations where that is not the case are probably easily navigated. It might be important to reassure those tasked with implementing a conservation plan that fundamental objectives worded as we describe here will provide useful starting points for defining objectives that are more operational.

## Challenges in Identifying Fundamental Objectives

### We Care about It All

Conservation planners, and many of those involved in conservation plans, care about nature. They also know a lot about nature and clearly recognize the many species, habitats, and communities that exist in a planning region. This wealth of knowledge can sometimes seem problematic when identifying objectives, which can quickly turn into long lists of species and habitats.

How these different features of biodiversity are captured and represented in the planning process is discussed in detail below, but for the purpose of defining fundamental objectives, it is not necessary to list every element of nature we care about; sometimes phrases such as "native habitats" or "important species" will suffice. Some decision scientists use a simple test of importance to help them decide whether an objective belongs in the hierarchy: they ask participants to describe how excluding that objective would substantially alter a decision about a solution. If the objective is unlikely to change decisions in a substantive way, it is not included.

### How Many Fundamental Objectives Should There Be?

There is no "right" number of fundamental objectives for a plan. Of course, the more objectives, the more difficult it can be to explore the effects of potential actions on them. While each listed objective does not need to be of equal importance (and they rarely will be), they should at least be roughly similar; if an objective seems insignificant relative to others, then consider whether it is truly fundamental. If something seems vastly less important, it might still be a consideration, but not necessarily a fundamental objective.

## Distinguishing Types of Objectives

When surveying conservation groups about why they undertook particular planning exercises, we were surprised at how frequently we came across instances in which planning itself was a conservation strategy—a way to engage partners and stakeholders, build trust, and establish relationships. While we firmly believe that planning must be directed toward actual decisions and implementation (otherwise, there is a high likelihood of wasting resources and undermining future planning efforts), we also acknowledge its important role as an avenue for collaboration. Outcomes such as building collaborative relationships can be important enough in planning exercises that we feel there is good reason to understand the different types of objectives that influence conservation plans.

### Process Objectives

**Process objectives** are guides to how planning and decision making should take place, rather than objectives related to the desired outcomes of conservation action—think of them as something like planning ground rules. Establishing relationships and building trust are not fundamental conservation objectives, but they might be important objectives for a conservation planning process, and if so, they should be identified as such so that the planning process can adequately deliver the desired outcomes. For example, a common process objective we've seen in our work is to acknowledge and reinforce self-determination in indigenous communities. Process objectives might also include conservation planning skill development and transfer.

### Intermediate Objectives

We have emphasized fundamental objectives for the reasons articulated earlier in this chapter: we believe it is critical that planning teams be aware of and clear about the ultimate ends their plans are working toward. However, the context of a plan (see Chapter 2) might dictate that something other than these end points be the principal focus. For instance, a statewide or region-wide planning effort might already have identified improving habitat connectivity as a critical action to minimize the loss of native species. If you are part of a team that is charged with planning for greater habitat connectivity, then despite this clearly being a *means* to achieve something larger, it is an essential objective for the purpose of your plan. Many conservation planners know these sorts of objectives as **intermediate objectives**, *means objectives*, or perhaps *project objectives*.

Any planning process aimed at implementing a particular strategy (perhaps identified at an organizational level) will involve intermediate objectives. For a given project, intermediate objectives can fall anywhere on a sliding scale of end points that all contribute to a larger, more fundamental objective. For example, **TABLE 3.3** contains a set of objectives for a project led by the U.S. Fish and Wildlife Service to combat the spread of white-nose syndrome (a fungal disease) in bats. These objectives are articulated in the same way as the fundamental objectives in Table 3.2, but they are focused on the management of a particular threat—white-nose syndrome—rather than the more fundamental objective of maximizing the persistence of bat populations in the United States. The topics of intermediate objectives and their role in conservation planning appear again in Chapters 6 and 11.

### Strategic Objectives

Conservation planning is about supporting decisions, and politics is an unavoidable part of decision making. Public and private organizations involved in conservation and natural resource management generally develop statements about their strategic role that speak to both the mission of the organization and how they intend to go about achieving it. For example, in pursuing its mission to "conserve the lands and waters on which all life depends," The Nature Conservancy emphasizes its adherence to a nonconfrontational, collaborative, and science-based approach to conservation. Certain conservation strategies, such as public name-and-shame advocacy campaigns, might be considered very effective for particular issues but are obviously inconsistent with the stated values and approach of The Nature Conservancy. Mounting such a campaign would therefore represent a risk to public and partner perceptions of The Nature Conservancy. Concerns about the effects of a decision on the organizations involved in a plan

**TABLE 3.3** A set of intermediate objectives identified in a plan to develop a strategic response to white-nose syndrome (WNS) affecting bats in the United States

1. Prevent the spread of WNS within Area 3.
2. Avoid unacceptable risks to endemic cave-obligate biota resulting from WNS or management strategies directed at WNS.
3. Minimize public health risks from management strategies related to WNS and from moribund bats.
4. Minimize restrictions on the affected public and land managers.
5. Minimize the total number of bats directly killed as a result of management strategies.
6. Provide research opportunities to accelerate the search for solutions to manage this disease.

*Source:* Data from J. A. Szymanski et al., *White-nose syndrome management: Report on structured decision making initiative.* 2009, Fort Snelling, MN: Department of Interior, U.S. Fish and Wildlife Service.

can be captured as **strategic objectives**. Common strategic objectives might be to avoid negative effects on the public perception of an organization, or simply to ensure consistency with an organization's mission.

### Three Classes of Fundamental Objectives

When discussing the types of objectives that appear in conservation plans, we've found it helpful to talk about three general classes of fundamental objectives: environmental, social, and economic. Environmental objectives are those classically associated with conservation plans, including statements about biodiversity, ecosystem processes, and ecosystem services. Economic objectives speak directly to the profitability of key economic activities in the planning region and their drivers. Social objectives reflect the many nonfinancial things that contribute to human well-being (for example, health, education, cultural heritage).

There are many points of overlap between the three classes, and many objectives that could fit in more than one class: for example, many socio-cultural values (such as natural areas for recreation) could constitute ecosystem services. We also recognize that we are using a very narrow definition of the term *economic*, and that it could reasonably be argued that all three classes of objectives we have listed are related to economics. The three classes of objectives described here are not formal divisions but, rather, helpful categories to consider when establishing the objectives for a conservation plan.

Objectives of all three classes are likely to be present in most conservation plans, so if they are not in the initial list, it is worth asking whether they should be. In the remainder of this chapter, we have structured our discussion of conservation planning objectives around these three classes.

## Environmental Objectives

Conservation planners and biologists have thought and written a lot about environmental objectives and how they should be represented in conservation plans. Conservation planning has traditionally been motivated by a desire to conserve the natural environment. It makes sense, then, that at the center of our planning are objectives relating to nature. Nature might not take primacy in planning exercises involving other sectors, or even in the minds of many stakeholders in conservation plans, but this simply makes it even more imperative that conservation planners think carefully about how and what aspects of nature are represented in a planning process. For the purposes of this discussion, we will divide these aspects into *biodiversity*, *ecological processes*, and *ecosystem services*.

## Biodiversity

No matter which way you look at it, nature is vast. In any planning region there is likely to be a staggering diversity of life—what we call **biodiversity**. The conservation community began widely using the term *biodiversity* in the late 1980s, and it has been a signature term of conservation planners ever since, perhaps partly because it gave a "sciency" feel to the vast concept of nature. In any case, it has been a useful concept in conservation. It is a bit like the Swiss Army knife of the conservation toolkit: it can be nearly anything it needs to be. It can be interpreted either as strictly as the diversity of living forms (taxonomic, phenotypic, behavioral, etc.) or as loosely as required to encompass all ecological systems and their processes. The success of biodiversity as a concept is so widespread that an international convention is named after it (the Convention on Biological Diversity), and it has been influential in policy at all levels of government.

We use *biodiversity* here in its broadest sense—the diversity of ecological systems together with their components and processes—because we find it useful. However, the term has also been criticized as overly technical and failing to resonate with the public. In a widely reported survey (although we have never been able to find the original source for it) about the meaning of the word *biodiversity*, a common answer was apparently "a brand of washing powder." Due partly to communication challenges like this, many conservation groups and agencies have moved away from emphasizing biodiversity. Our usage of *biodiversity* is equally consistent with the term *natural heritage*.

Conserving biodiversity can be considered a fundamental objective of most conservation plans. However, because of the complexity, ambiguity, and differences in interpretation of the term, biodiversity as an objective should be broken down into its component parts in the planning process so that it can be functionally represented in a plan.

### Conservation Features

In this book, we refer to the representations of biodiversity in a conservation plan as **conservation features**. This terminology owes its genesis to the systematic conservation planning field pioneered by Bob Pressey and colleagues in Australia.[10] Although widely used, *conservation features* is not a universal term. One of the most popular alternative terms for the pieces of biodiversity that constitute the core of a conservation plan is *targets*. For example, targets formed the basis of the Ecoregional Assessments and Conservation Action Plans conducted by The Nature Conservancy. The use of different terms has contributed to some confusion, as systematic conservation planning typically uses the term *target*, as we do in this book, to refer to the quantitative amount of each conservation feature that is desired within a solution.

Regardless of terminology, biodiversity in its entirety is simply impossible to fully represent—it is either too vast or too unknown. At some point, a successful conservation plan depends on breaking down biodiversity into more explicit components that reflect the aspects of nature we particularly care about and that enable us to realistically assemble data.

### Categories of Biodiversity Features

Three categories of features are used to represent biodiversity in conservation plans: (1) species, (2) communities or ecosystems, and (3) abiotic units. This section introduces these three types of features and sets the stage for the next two chapters on targets and data. One of us has written an earlier book[11] that provides a more thorough introduction to biodiversity features and their genesis, identification, selection, and role in conservation planning. Here we cover each of them in only enough detail so that planners will know what to look for and decision makers will know what they are looking at.

In providing information on the sorts of features in each of these three categories that are commonly included in conservation plans, we are not suggesting that these are necessary components of a plan. Which features are appropriate depends entirely on the context of a plan. A conservation plan focused on Asian wild cattle species has a narrow and clearly identified biodiversity focus, so it is unlikely to need landscape-scale abiotic conservation features to represent biodiversity.

The three categories of features have acquired many names over the development of conservation planning, and they have been lumped and divided in various ways. One of the most widely embraced categorizations is the concept of conservation features as either coarse filters or fine filters.[12] **Coarse filters** are intended to compensate for our incomplete knowledge of all biodiversity by reflecting diversity at a higher level of ecological organization that we can more readily observe. Communities, ecosystems, and abiotic units can all be considered coarse-filter features because conserving representative examples of these features would also conserve most of the species in a region. In other words, these features serve as **surrogates** for biodiversity generally (we consider the concept of surrogacy more thoroughly in **Box 3.5**).

The "filters" metaphor is appropriate because some important aspects of biodiversity will inevitably slip through these coarse filters and need to be picked up by other features—the **fine filters**. The

## BOX 3.5 The Concept of Surrogacy

Surrogate features are one of the most useful tools in the conservation planner's toolkit. In the context of biodiversity as a whole, we have spatial distribution data for almost comically few species. Changing this is not easy: reliably understanding the distribution of even a single species is a significant research undertaking. And yet it is hard to be rigorous in our attempts to conserve biodiversity if we cannot measure it. To overcome this limitation, conservation planners commonly make use of *surrogates*: to represent biodiversity (or parts of it), they use something else that is easier to observe, map, or measure.

The conservation planner Hedley Grantham and colleagues describe surrogates as either taxonomic or environmental.[a] *Taxonomic surrogates* are typically well-known taxa used as surrogates for other, less well-documented taxa. Birds and mammals are probably the two most widely used taxonomic surrogates for the simple reason that they are relatively easy to observe. A typical application might involve assuming that the pattern of species richness in birds across a region is representative of species richness patterns for other faunal taxa more generally, or that the response of bird communities to activities in the landscape is indicative of the responses of other taxa as well. *Environmental surrogates* are either ecological communities, such as vegetation types, or abiotic classifications of the landscape. Both are commonly used as features in conservation plans, as described in this chapter. The use of environmental surrogates is predicated on the notion that environmental variation drives species variation, and that conserving the diversity of environments in a region will therefore conserve its underlying biodiversity.

[a]Grantham, H. S. et al., *Effectiveness of biodiversity surrogates for conservation planning: Different measures of effectiveness generate a kaleidoscope of variation.* PLoS One, 2010. **5**(7): e11430.

Related to the notion of surrogacy just described are individual species used as features or indicators in conservation plans because of their perceived ability to act as surrogates for the conservation needs of other species or for broader landscape conservation objectives. These species are variously termed *umbrella species*, *landscape* or *focal* species,[b] *keystone species*, *indicator species*, or simply *surrogate species*. Ecologist Tim Caro[c] has compiled an excellent summary of types of surrogate species and their uses. Typically, these sorts of surrogates are large-bodied animals with large area requirements (see Figure 3.2), and there is a good chance that stakeholders will care about many of these species in their own right.

The validity of using surrogates in a conservation plan has been the source of significant academic investigation and debate. To quote Caro, "There are very few arenas in which we can say that studies have unequivocally demonstrated that certain surrogate species or species groups represent the distribution of other taxa or the responses of other taxa to environmental change." Numerous methods have been proposed to test the effectiveness of surrogates at doing what they are intended to: represent biodiversity, either in its entirety or specific elements of biodiversity. Unfortunately, the many studies that have investigated the performance of surrogates have revealed few reliable generalities. Still, the following come close:

- Surrogates tend to perform better in a *complementarity* sense than a *hotspot* sense (FIGURE 3.5.1). In other words, a set of sites that captured all the species of a surrogate

[b]Caro, T. M., and G. O'Doherty, *On the use of surrogate species in conservation biology.* Conservation Biology, 1999. **13**(4): 805–814.

[c]Caro, T., *Conservation by Proxy: Indicator, Umbrella, Keystone, Flagship, and Other Surrogate Species.* 2010, Washington, DC: Island Press.

*(continued)*

## BOX 3.5 The Concept of Surrogacy *(continued)*

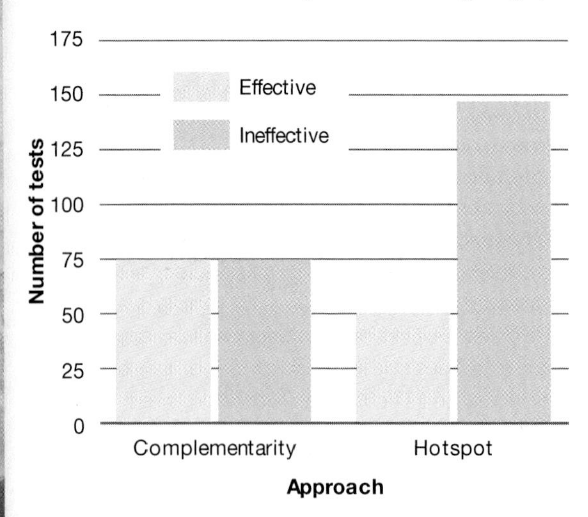


FIGURE 3.5.1 Number of effective and ineffective surrogates in tests of surrogates reported in the conservation literature identified with complementarity or with hotspot approaches. Effectiveness indicates significant prediction of target taxon richness from surrogate richness ($\times 2$, $p < 0.001$). (Data from Lewandowski et al., 2010.)

taxon, such as birds, would be likely to capture a good deal of the species diversity in another taxon, such as mammals. But patterns in the richness of bird species would be unlikely to reflect patterns in the species richness of mammals or of any other taxa.[d]

- The larger the region being considered, the better surrogates tend to perform[e]—which is convenient, because the larger the region, the lower the likelihood of our having extensive species-level data. As you might expect, taxa that have many species and are distributed throughout the entire landscape, such as birds, are generally the best-performing surrogates.[f] Probably many invertebrate groups also fit this bill, but they are not as easy to observe or as well known.
- Surrogates tend to be less effective at capturing rare or threatened species. For example, using short-grass prairie as a surrogate would probably not be effective at representing a rare grassland sparrow, which might be found in only a small number of these grasslands and under very specific habitat conditions.
- Adding more surrogate taxa improves performance in representing a region's biodiversity. Basically, the more decent data sets you have, the better!
- There is no consistently superior taxonomic surrogate, nor are environmental or taxonomic surrogates consistently better or worse than each other.

From a pragmatic viewpoint, these last two points are actually quite useful. They suggest that (1) it pays to combine taxonomic and environmental surrogates in planning, and (2) if representing biodiversity is your aim, you should use all available data. You will rarely be swamped with choices of surrogates, and this effectively means one less decision that planners need to make—just go with what is available.

The creative use of surrogates is a useful tool that can help in moving a plan forward in a relatively objective fashion. We have discussed surrogacy here in the context of biodiversity features, but the principle of surrogacy can apply equally well to threat (for example, proximity to roads), cost (productivity of the land), and even the probability of project success (see Chapters 5 and 6). In a plan to help The Nature Conservancy prioritize expansion of its work in Africa, we used the Ibrahim Index of African Governance as a surrogate for the probability that a project in a particular country could be successfully implemented in the medium term (**FIGURE 3.5.2**).

*(continued)*

---

[d]Lewandowski, A. S., R. F. Noss, and D. R. Parsons, *The effectiveness of surrogate taxa for the representation of biodiversity.* Conservation Biology, 2010. **24**(5): 1367–1377.

[e]Lewandowski et al., *The effectiveness of surrogate taxa.*

[f]Larsen, F. W. et al., *Birds as biodiversity surrogates: Will supplementing birds with other taxa improve effectiveness?* Journal of Applied Ecology, 2012. **49**(2): 349–356.

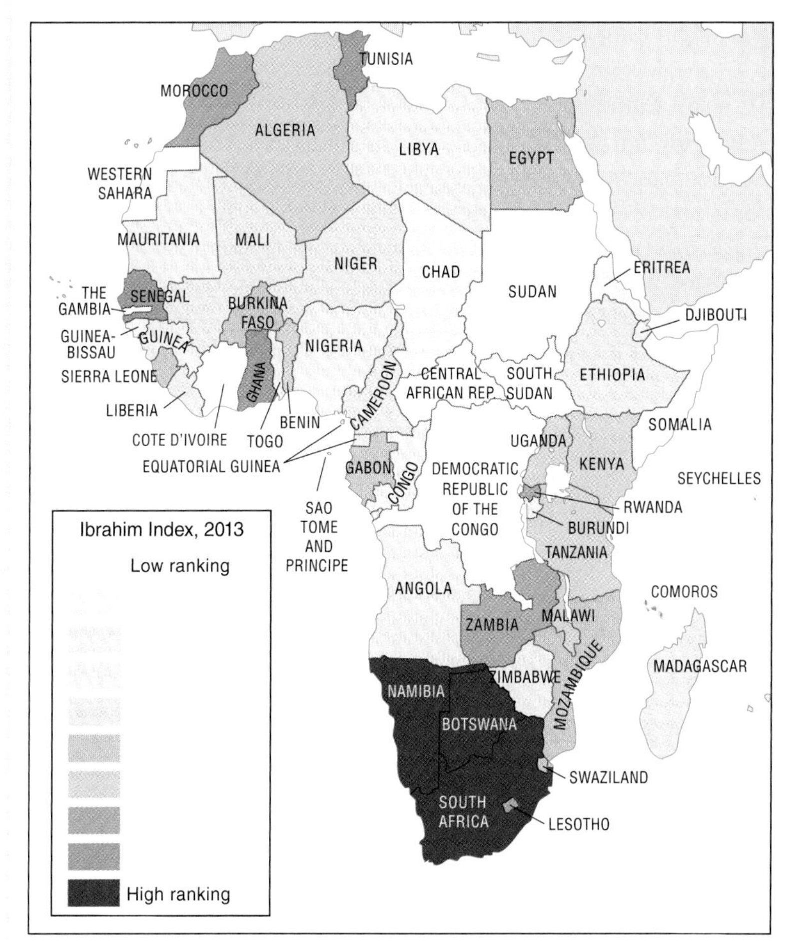


**FIGURE 3.5.2** The Ibrahim Index of African Governance, used as a surrogate in a continental conservation plan to reflect the likelihood of successfully implementing a project in a given country in the medium term. The index scores countries based on a range of criteria that include rule of law, human rights, sustainable economic opportunity, and human development. (Adapted from Mo Ibrahim Foundation http://www.moibrahimfoundation.org/.)

types of biodiversity that are particularly likely to slip through the coarse filters are threatened or endangered species, endemic species with a restricted range, or small ecological features like vernal pools.

All three categories of biodiversity features exist on a spectrum and are not mutually exclusive; sometimes particular species, communities, or abiotic units serve as surrogates for broader biodiversity and as direct features of importance at the same time. Furthermore, the choice of features should be matched to the scale of the planning exercise. A plan for a small geographic area requires a more highly resolved set of features than would make sense for a large area. For example, a marine protected area design process for the coast of KwaZulu-Natal Province in South Africa identified 121 habitat features (ecological communities or abiotic features) along the province's 640-km coastline. Compare that with a national coastal assessment that classified South Africa's 3100-km coastline into 37 habitat features. Both worked well in the relevant context.

### Influence of Data Availability on Choice of Conservation Features

In an ideal world, we would think of the features that best represented our conception of biodiversity and the natural world and then use those features in our planning. In the world we live in, however, it would be optimistic to the point of foolishness to suggest that data availability is not a factor that influences our choice of conservation features. Hence, some of the discussion that follows about selecting the most appropriate features will make more sense after the discussion of data in Chapter 5.

### How Many Features Can You Have?

How long is a piece of string? There is no wrong answer to this question, nor to the question of how many features should be used in conservation planning. There is, however, an obvious trade-off between the number of conservation features in a plan and the richness of information it is possible to have about them. The context of the plan, particularly its scale, will also dictate how many features it makes sense to include.

If the plan is aimed at establishing spatial priorities for a country or province based largely on the distribution of each feature, there is almost no limit to the number of features that could reasonably be considered. Powerful geographic information systems (GISs) combined with highly efficient conservation planning software means that even thousands of features are not difficult to handle. If, on the other hand, the plan requires establishing the locations and extent of different types of land use or management across a district, some knowledge about how those different uses will affect different conservation features is needed, but the trade-off in data analyses will mean that fewer features will be considered.

Plans principally focused on developing and prioritizing strategies at a particular site require even more data about each feature (about the consequences of actions for the features of interest and their effects on other objectives), thus further reducing the number of features that it is realistic to consider. For example, the CMP's *Open Standards* strongly encourage a maximum of eight features, simply out of pragmatism, given the number of judgments that must be made for each of those features. Features can sometimes be kept to a practical number by lumping similar features under broad categories such as *alpine communities*. Restricting the number of features that represent biodiversity will undoubtedly require hard choices.

### Species

Species are the most easily recognized unit of biodiversity. When people think of biodiversity, they think of species, and when people think about loss of biodiversity, they think about the loss of particular species. People have studied, recorded, and collected species in every region of the planet; there is probably no region where something is not known about some of the species that live there. As a natural unit of biodiversity, species are an obvious choice as conservation features for plans focused on the conservation of biodiversity. However, not all species are equally important or useful as conservation features. Much effort has been expended in debating what sorts of species can best represent biodiversity. This section introduces some of that debate and our own opinions about the types of species that might best be included as features in a conservation plan.

***Threatened and Endangered Species*** People often care a great deal about threatened and endangered (T & E) species. Many members of the public, and some scientists, perceive conservation and conservation planning to be primarily about T & E species. Whether or not this perception is correct, it has created an expectation: threatened species are the tangible end of biodiversity loss and commonly need to be acknowledged as features in a conservation plan.

In addition to upholding public expectations of the function of conservation, protection of T & E species is often mandated by endangered species legislation, such that one or more of the parties or stakeholders involved in a conservation plan may have some legal responsibility for the conservation of T & E species in the region. Regardless of legal obligation, T & E species can be important conservation features for at least three additional reasons. First, they often have the potential to leverage resources and actions unavailable for biodiversity conservation more broadly. Second, T & E species are less likely than most other species to be adequately captured through the use of surrogates or coarse-filter features. Third, nearly by definition, some distributional and population data are available for T & E species, as such data are necessary to establish a species' status as threatened in the first place (FIGURE 3.2).

Many countries, and even states and provinces, have established criteria for identifying threatened and endangered species, but by far the most widely used and influential classification of threatened species is the IUCN Red List (TABLE 3.4). The details of this classification system and the criteria for it are well documented,[13] but in general, a species is considered threatened if it has undergone a severe contraction of distribution or population size, it exists only in a very small or

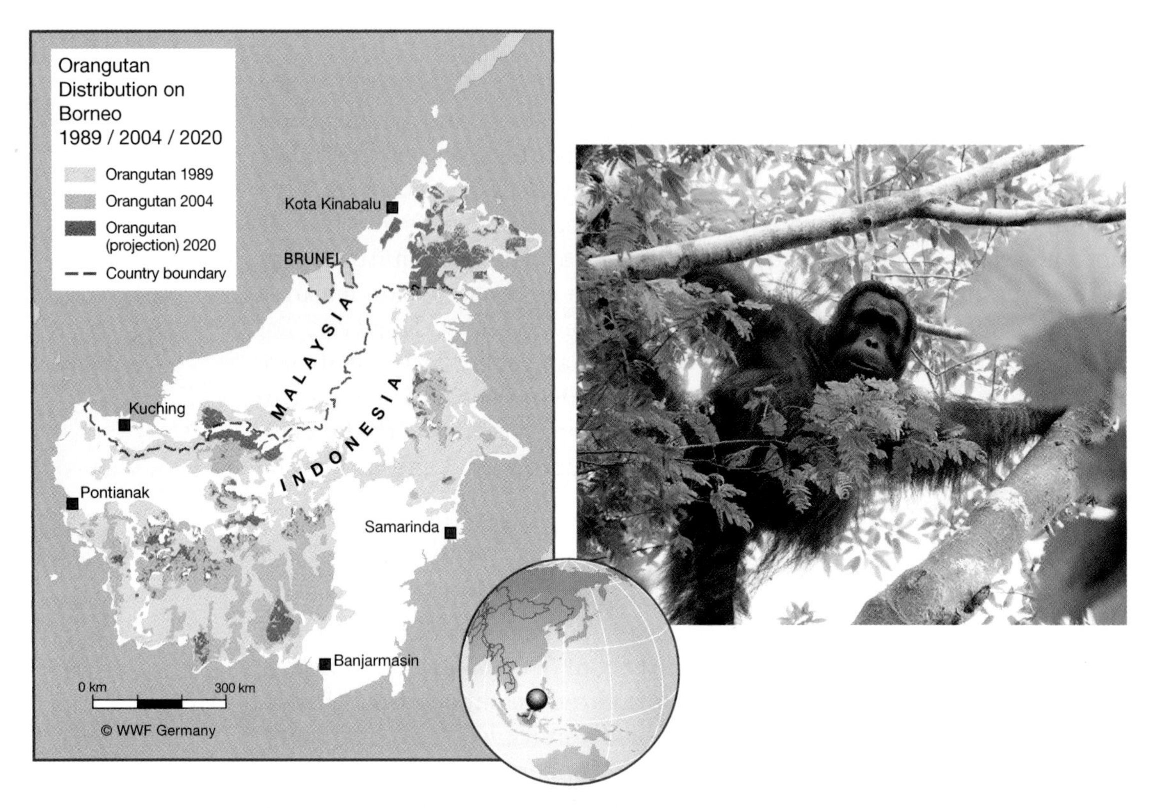


**FIGURE 3.2** An endangered species. The Bornean orangutan (*Pongo pygmaeus*), which has undergone enormous population reduction and range contraction, is a good example of an endangered species that has been a prominent feature in conservation plans.

highly restricted population, or a quantitative analysis has indicated that the species is at high risk of extinction.

***Endemic Species*** In all but the most localized of planning regions, there are likely to be species existing in that region and nowhere else—species that are **endemic** to that region. Endemic species may also be T & E species. The presence of endemic species is significant from a planning point of view because if these species are not conserved adequately in the region, global biodiversity will be irretrievably diminished. In many cases, endemic species are also distinctive species and sources of regional pride and identification with nature—obviously valuable qualities for a conservation feature.

The number of endemics in a region has long been interpreted as an indication of the global importance of the region for biodiversity conservation—hence the notion of biodiversity **hotspots**.[14] If the planning area contains many endemic species, as might happen in plans at a national scale or larger, it might be reasonable to include

**TABLE 3.4** Criteria for the IUCN Red List Endangered category

A taxon is Endangered when the best available evidence indicates that it meets any of the following criteria (A to E), and it is therefore considered to be facing a very high risk of extinction in the wild:

A. Reduction in population size based on any of the following:

1. An observed, estimated, inferred, or suspected population size reduction of ≥70% over the last 10 years or three generations, whichever is longer, where the causes of the reduction are clearly reversible *and* understood *and* ceased, based on (and specifying) any of the following:
   - (a) Direct observation
   - (b) An index of abundance appropriate to the taxon
   - (c) A decline in area of occupancy, extent of occurrence, and/or quality of habitat
   - (d) Actual or potential levels of exploitation
   - (e) The effects of introduced taxa, hybridization, pathogens, pollutants, competitors, or parasites
2. An observed, estimated, inferred, or suspected population size reduction of ≥50% over the last 10 years or three generations, whichever is longer, where the reduction or its causes may not have ceased *or* may not be understood *or* may not be reversible, based on (and specifying) any of (a) to (e) under A1
3. A population size reduction of ≥50%, projected or suspected to be met within the next 10 years or three generations, whichever is longer (up to a maximum of 100 years), based on (and specifying) any of (b) to (e) under A1
4. An observed, estimated, inferred, projected, or suspected population size reduction of ≥50% over any 10-year or three-generation period, whichever is longer (up to a maximum of 100 years in the future), where the time period must include both the past and the future, and where the reduction or its causes may not have ceased *or* may not be understood *or* may not be reversible, based on (and specifying) any of (a) to (e) under A1

B. Geographic range in the form of either B1 (extent of occurrence) *or* B2 (area of occupancy) *or* both:

1. Extent of occurrence estimated to be less than 5000 km and estimates indicating at least two of (a) to (c):
   - (a) Severely fragmented or known to exist at no more than five locations
   - (b) Continuing decline, observed, inferred, or projected, in any of the following:
     - i. Extent of occurrence
     - ii. Area of occupancy
     - iii. Area, extent, and/or quality of habitat
     - iv. Number of locations or subpopulations
     - v. Number of mature individuals
   - (c) Extreme fluctuations in any of the following:
     - i. Extent of occurrence
     - ii. Area of occupancy
     - iii. Number of locations or subpopulations
     - iv. Number of mature individuals
2. Area of occupancy estimated to be less than 500 km and estimates indicating at least two of (a) to (c):
   - (a) Severely fragmented or known to exist at no more than five locations
   - (b) Continuing decline, observed, inferred, or projected, in any of the following:
     - i. Extent of occurrence
     - ii. Area of occupancy
     - iii. Area, extent, and/or quality of habitat
     - iv. Number of locations or subpopulations
     - v. Number of mature individuals
   - (c) Extreme fluctuations in any of the following:
     - i. Extent of occurrence
     - ii. Area of occupancy
     - iii. Number of locations or subpopulations
     - iv. Number of mature individuals

*(continued)*

**TABLE 3.4 Criteria for the IUCN Red List Endangered category (*continued*)**

C. Population size estimated to number fewer than 2500 mature individuals and either:
   1. An estimated continuing decline of at least 20% within five years or two generations, whichever is longer (up to a maximum of 100 years in the future) *or*
   2. A continuing decline, observed, projected, or inferred, in numbers of mature individuals *and* at least one of the following (a–b):
      (a) Population structure in the form of one of the following:
         i. No subpopulation estimated to contain more than 250 mature individuals,
         *or*
         ii. At least 95% of mature individuals in one subpopulation
      (b) Extreme fluctuations in number of mature individuals

D. Population size estimated to number fewer than 250 mature individuals

E. Quantitative analysis showing the probability of extinction in the wild is at least 20% within 20 years or five generations, whichever is longer (up to a maximum of 100 years)

*Source:* Data from IUCN, *The IUCN Red List of Threatened Species. Categories & Criteria.* Version 3.1. 2014, Cambridge, UK.

only those endemic species with a highly restricted range—commonly referred to as **restricted-range endemics** (FIGURE 3.3).

***Apex Consumers*** Predators at the tops of food chains have declined or disappeared from ecosystems around the world.[15] Often the best-known examples of these **apex consumers** are large to medium-sized mammalian predators such as African lions (*Panthera leo*), gray

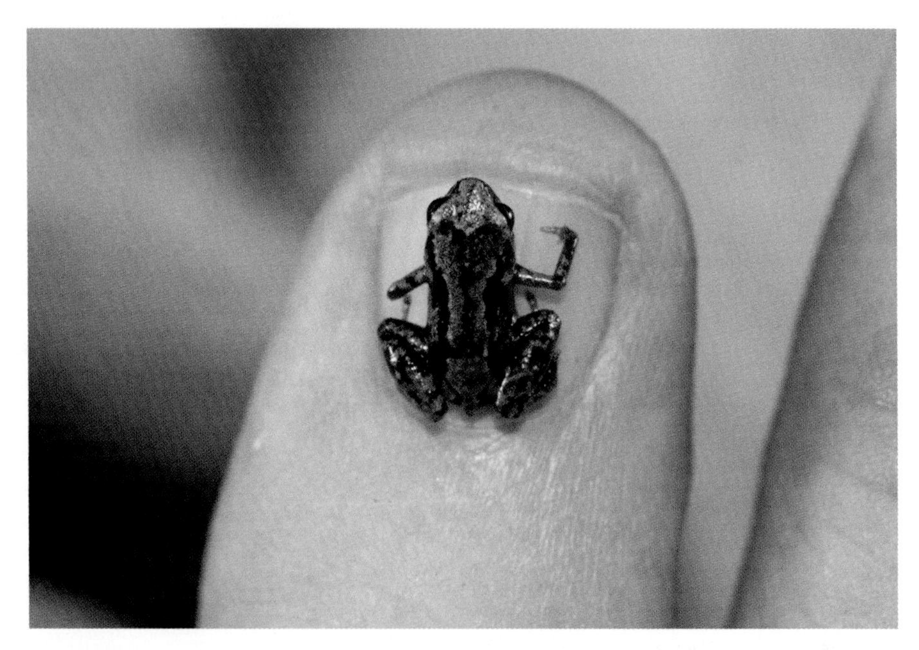

FIGURE 3.3 *Paedophryne swiftorum*, a restricted-range endemic from Papua New Guinea. One of the world's smallest frogs (adult males are ca. 8 mm SVL), the species is only known from an area that is around 3–4 km in diameter within the Kamiali Biological station on the north coast of Papua New Guinea.

wolves (*Canis lupus*), and Eurasian lynx (*Lynx lynx*). These types of species have often been included as features in conservation plans because of their threatened or endangered status or because they have large area requirements that require landscape-scale conservation actions.

More recently, ecologists have become aware that removal of apex consumers from nature—a process known as *trophic downgrading*[16]—can have serious and negative consequences for ecosystem structure, composition, and function. Top-down control of ecosystems by apex consumers occurs in terrestrial, freshwater, and marine systems, and trophic downgrading can be triggered by the loss of species from a variety of positions in food webs (herbivores, omnivores, carnivores).[17] There is now both a sufficient list of species known to function as apex consumers and a sufficient body of evidence of their disproportionate influence on ecosystem conservation to support their consideration as conservation features.

***Other Important Species*** Species do not need to be threatened or endemic to be considered as potential features in a conservation plan. Species typically falling in this category are particularly charismatic, are declining in a significant portion of their range but are not yet categorized as threatened or endangered (often referred to as "species at risk" or some similar nomenclature), or are economically or culturally important in the region.

In south east Queensland, Australia, where one of us lives, it is nearly inconceivable that a terrestrial conservation plan would be developed without including the koala (*Phascolarctos cinereus*) as a feature, even though it is neither threatened nor endemic to the region. Similarly, communities in the Solomon Islands often identify ironwood trees (generally Myrtaceae species) as specific conservation features because of their importance in making walking sticks, canoe paddles, and other decorative pieces. Stakeholders would not be satisfied with basing the conservation of such species on a surrogate or indirect measure.

Often these species represent important components of a region's cultural heritage. Ann Garibaldi and Nancy Turner have identified **cultural keystone species**,[18] which might be considered important inclusions in a conservation plan because they can act as a bridge between environmental objectives and social objectives (discussed in more detail later in this chapter). Any species for which it is desirable to be able to say explicitly how its conservation is being addressed should be included as a feature in the plan.

Some species can be economically important because they are charismatic and thereby help raise funds for conservation in the region. Such species are generally known as **flagship species** (and are often threatened or endangered species). Orangutans in Borneo (see Figure 3.2) are a good example, as they are relied on heavily to help raise money for forest conservation in the region and to raise local awareness of conservation plans and subsequent actions. Orangutans are also exemplary of the anthropomorphic characteristics (such as forward-facing eyes) that conservation planner Robert Smith and colleagues have shown make popular choices for flagship species.[19] Smith and colleagues have called for conservation planners to work closely with marketing professionals to better target the selection of flagship species.[20] Given that flagship species are often the public face of conservation, they clearly should be included as features in a plan.

***Rarity*** One of the most challenging issues in determining how to represent the species of a region in a conservation plan is rarity. Superficially, you might expect rare species to be important conservation features, but in practice the story is far more complicated. Some rare species will be included as features because they are either threatened or endemic, but other rare species might be relatively unimportant or even misleading to include because some species are naturally rare.

Although there is a substantial discourse on what constitutes rarity,[21] the most typically challenging situation is species that are rare in the region in question but common somewhere else. These "apparent" rarities might be close to the edge of their distribution, or they may be vagrant or possibly introduced species. For example, the burrowing owl (*Athene cunicularia*) is listed as a threatened species in Canada, but is a widespread species in the western United States (**FIGURE 3.4**).

### Communities and Ecosystems

Defining communities is an approach to classifying a landscape in a manner that represents the diversity of species that live there. Ecological **communities** can be considered relatively distinct assemblages of species that co-occur in space. In contrast, the terms **ecosystem** and **habitat** are generally taken to embody both species assemblages and their interactions with the geophysical environment; a desert ecosystem, for example, might include a distinct species assemblage, but it is also defined by the physical characteristics of the environment.

Although ecologists are usually clear about the differences between communities and ecosystems, conservation planners have tended to muddy those waters. The conservation planning literature often uses the terms *ecosystem*, *community*, *ecological system*, and *habitat* rather interchangeably. Ultimately, all of them can be useful and appropriate terms of classification when defining features for a conservation plan.

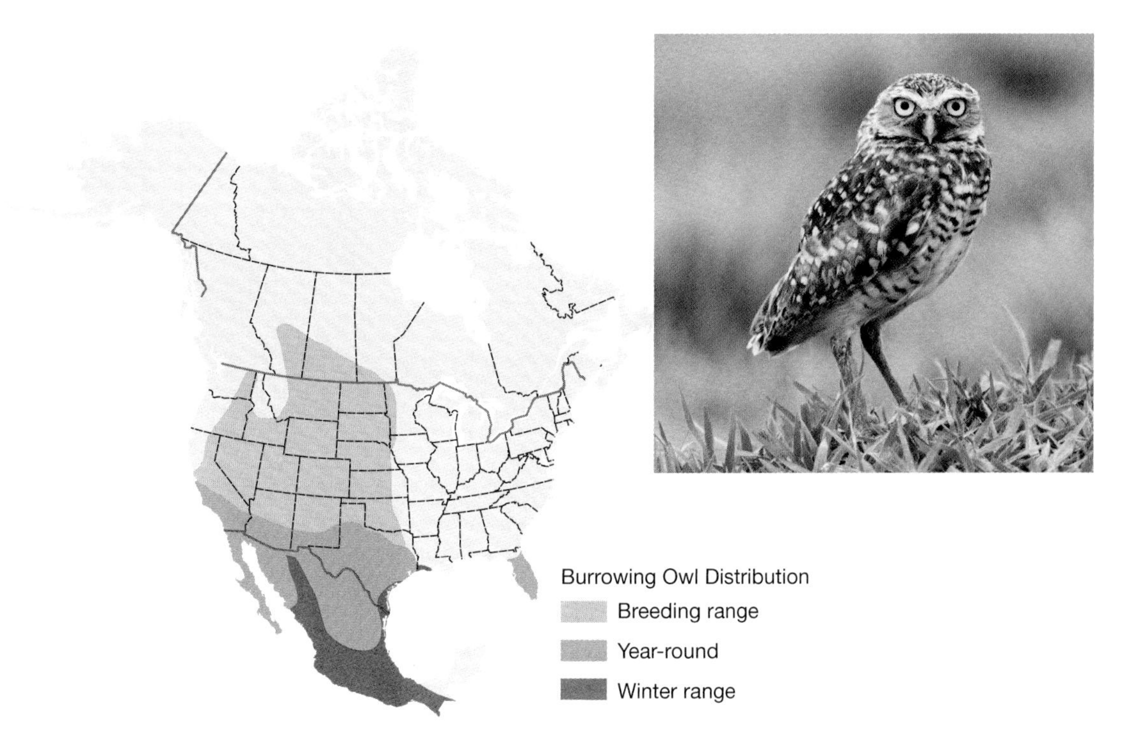


FIGURE 3.4 The tricky issue of local rarity. A small part of the distribution of the burrowing owl (*Athene cunicularia*) lies in Canada, where it is listed as a threatened species, but the species is widespread and reasonably common in the United States and Mexico.

***Approaches to Delineating Communities and Ecosystems*** There is no universal approach to delineating ecological communities or ecosystems. Hundreds of community and ecosystem classification schemes of various types exist depending on where a person works in the world. One of the most widely used classification schemes for terrestrial systems was developed by NatureServe and the U.S. Geological Survey for use in the National Gap Analysis Program. It is based on classification units known as *ecological systems*, which are defined as groups of vegetation communities that occur together within similar physical environments and are influenced by similar ecological processes, substrates, and environmental gradients. Ecological systems have been mapped for the entire United States and Latin America, and a national land cover database is now available online for use in conservation planning (FIGURE 3.5).

In marine and freshwater systems, communities are frequently defined based on a combination of geophysical characteristics and species observations. For example, the freshwater ecologist and conservation planner Jonathan Higgins and colleagues developed a system for defining freshwater communities based on both biological (for example,

FIGURE 3.5 Land cover data for the United States compiled by the USGS National Gap Analysis Program. Land cover categories are based on the NatureServe ecological systems classification.

fish distributions) and abiotic features.[22] Similarly, a National Marine Bioregionalization developed for Australia classifies marine ecosystems based on, among other things, fish assemblages and geomorphic features (FIGURE 3.6). This classification scheme has been extensively used in marine conservation planning exercises in Australia.

Which classification schemes may be most useful for defining features in a conservation plan depends largely on the plan's objectives and context. In 1999, ecologist and planner Mark Anderson and colleagues[23] suggested a number of points to consider that still make a great deal of practical sense today:

- Is the scale of the classification scheme well matched to the scale of the plan, or can the classification scheme be used at multiple scales? A national classification, for instance, might be too coarse to be useful for a local planning exercise because it will not adequately represent the variation in biodiversity across a local region.
- Can the units of the classification scheme be mapped, and to what extent is that mapping complete for the planning region? Classification schemes based largely on remotely sensed data

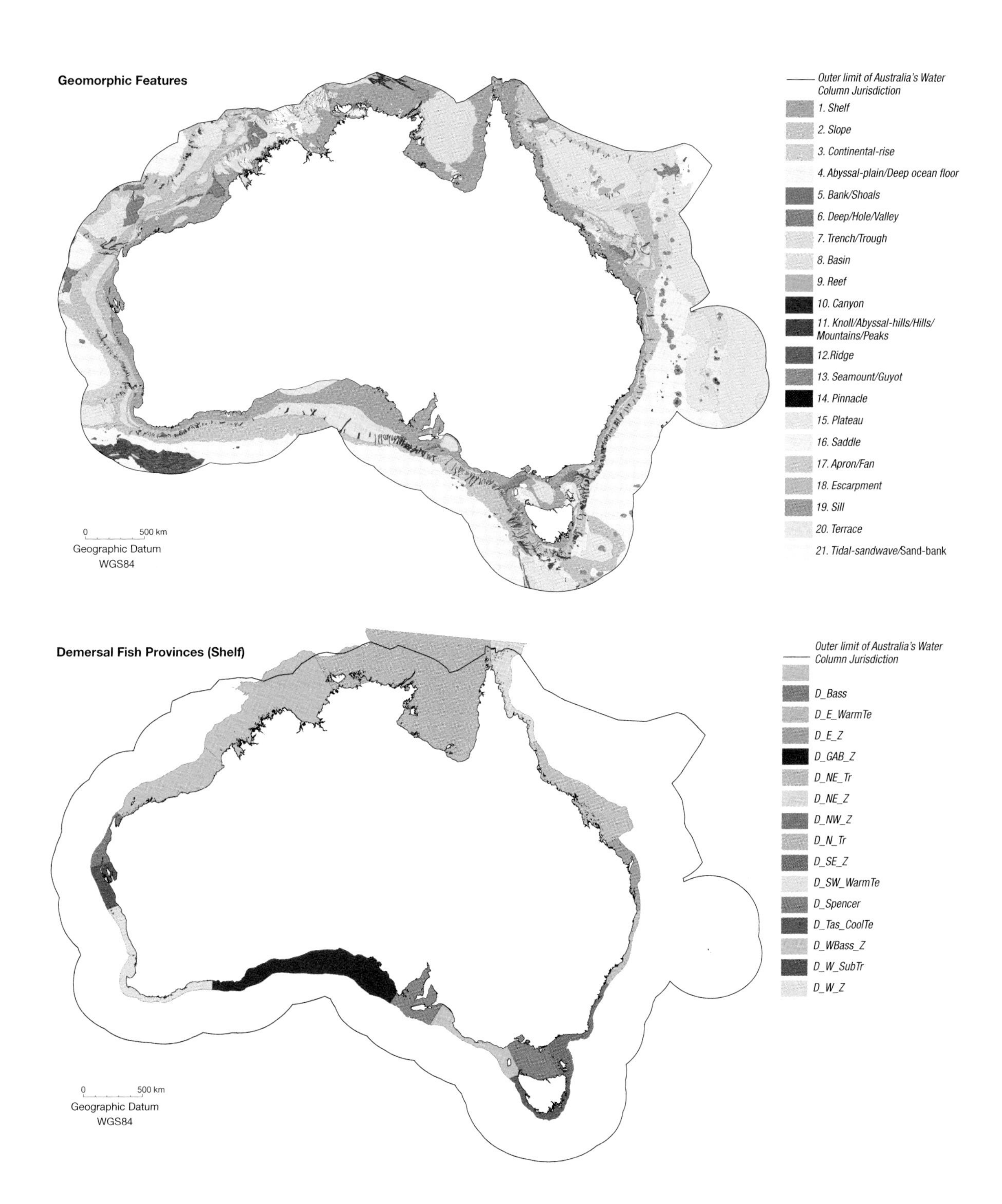


FIGURE 3.6 Australia's National Marine Bioregionalization. Two elements (demersal fish communities and geomorphic features) that contribute to a broad-scale classification of Australia's marine environments.

might be easily extended beyond their current coverage, but that is likely to be more challenging for those based on species assemblages.

- For what geography was the classification scheme developed? We will have the most confidence that a classification will capture the underlying variation in biodiversity in the geography where it was developed.
- To what degree has disturbance and its effect on the classification scheme been considered? A system that has experienced a high degree of anthropogenic disturbance will require that elements be added to the classification, at least, and care should be taken that a classification of communities or ecosystems actually reflects the biodiversity that still remains.

Whether their focus is on terrestrial, freshwater, or marine systems, many conservation planners will be working with community and ecosystem classification schemes that are already established. This does not mean that, where practical, conservation planners should not make sure that the classification scheme being used reflects a view of biodiversity consistent with the desired outcomes of the plan. A colleague of ours described the experience of driving across the fynbos vegetation of South Africa's Little Karoo region with his laptop and GIS open, calling out when the team was passing into a different "community" and seeing if they could detect a visual difference. Few better ways exist to check on whether your features are defensible.

As a way to represent biodiversity in a conservation plan, ecosystems and communities have a number of strong points. From an anthropocentric point of view, communities are a level of diversity that is reasonably well aligned with the scale at which humans view the natural world—even without ecological training, we can generally detect when we move from one community or ecosystem to another. Consequently, we tend to feel a certain level of comfort with the idea that conserving communities will conserve our conception of biodiversity. Communities and ecosystems are also closely related to people's sense of place, whether they live in that place or not (think of the moors of Scotland or the pampas of Argentina). Pragmatically, the delineation of communities and ecosystems can usually make use of remotely sensed data (see Chapter 5), which means that the classifications can be more comprehensive than is often possible with species. It also means that these classifications are more widely available than good species-level data. As mentioned previously, communities and ecosystems can serve as surrogates for all those elements of biodiversity we do not know.

***Threatened Communities and Ecosystems*** The previous discussion of species as features introduced the important role that the IUCN Red List has played in establishing features for conservation plans. A comparable effort, led by the Venezuelan conservation ecologist Jon Paul Rodriguez,[24] is under way to develop Red List criteria for ecosystems.[25] Just as communities as conservation features are considered better able to represent biodiversity than species alone, it is hoped that a Red List of threatened ecosystems will help provide a more accurate picture of threats to the diversity of life and compensate somewhat for the clear taxonomic bias in the Red List of species (proportionally, mammals are hugely overrepresented).

The strengths listed earlier—data availability, potential speed and extent of assessment, and resonance with the public—are all markers of value for a Red List focused on communities and ecosystems. In most cases, losses of communities or ecosystems reflect losses that are more immediately apparent to society than are losses of species: ecosystems are generally easier to interact with than threatened species.

Some governments have already established the concept of threatened communities under their relevant national legislation. Australia currently lists 55 threatened ecological communities under the federal Environmental Protection and Biodiversity Conservation Act. This threatened status reflects the risk of loss of these communities and establishes certain legal protections for them. Examples include Natural Temperate Grassland of the Southern Tablelands of New South Wales and the Australian Capital Territory (endangered) and Alpine Sphagnum Bogs (also endangered) (FIGURE 3.7). South Africa has over 200 nationally listed threatened ecosystems. Where they have been identified, threatened communities can be important conservation features for the same reasons that threatened species are.

## Abiotic Units

The distribution of flora and fauna is strongly influenced by the **abiotic** (nonliving) properties of the landscape. For example, even with no ecological training, it is easy to see that shaded spots have different plants than places with full sun. Insolation (the amount of solar energy a location receives) is just one of many abiotic factors that structure ecological communities. When we are interested in conserving the full range of biodiversity in a region, this range can be approximated by the range of abiotic units present in the region—in other words, by using abiotic features as **surrogates** for biodiversity. It is an intuitively appealing idea, especially as many important abiotic variables can be consistently mapped across a landscape using remotely sensed data (FIGURE 3.8).

(a)

(b)

**FIGURE 3.7** Some threatened ecological communities of Australia. (a) An alpine sphagnum bog in Tasmania. (b) A natural temperate grassland in the Southern Tablelands of New South Wales. Both of these community types are listed nationally as endangered.

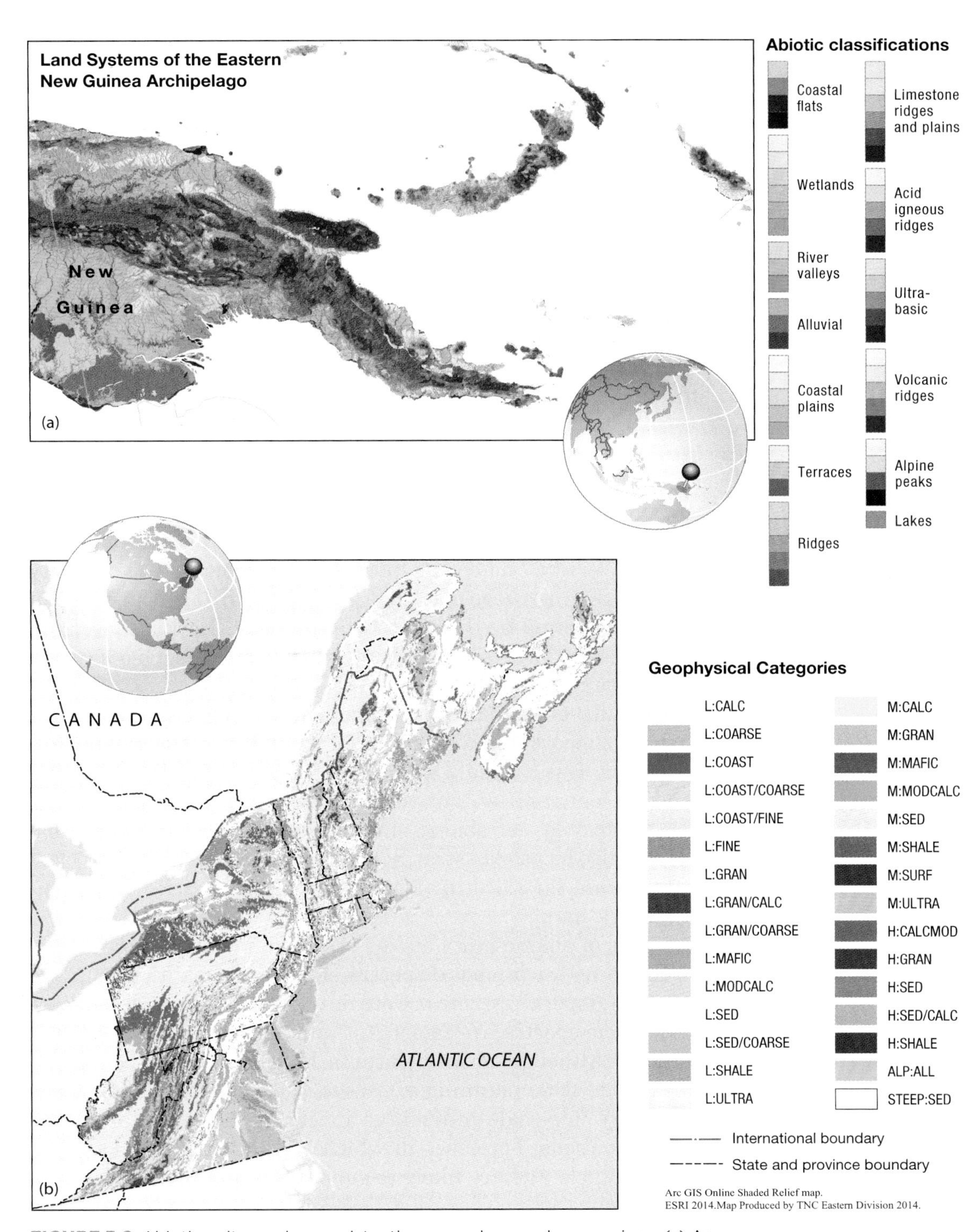


FIGURE 3.8 Abiotic units can be consistently mapped across large regions. (a) An abiotic classification scheme for Papua New Guinea, used as part of a national conservation assessment. (b) Geophysical settings used as features for conservation planning in the northeastern United States. The settings are classified by elevation zone and geology.

This advantage gives abiotic units a pragmatic strength of application, especially in regions with poor biological data. Abiotic units have been widely used in conservation planning, appearing under various names such as environmental features or surrogates, land facets, land systems, geophysical settings, or simply abiotic features. In marine and freshwater ecosystems, abiotic features are especially important in conservation planning due to the limitations of data on aquatic biota. The demand for climate change adaptation has renewed enthusiasm for the use of abiotic units in conservation planning—they are seen as enduring features of the landscape able to represent regional diversity regardless of distributional changes in floral and faunal communities. Their use in this context is covered in Chapter 9, on planning for climate change adaptation.

***Why Include Abiotic Features If You Already Have Biotic Features?*** If you already have data on ecological communities and some species that you care about, why consider using abiotic features? Their usefulness is predicated on the fact that environmental variation drives regional biodiversity, and on the likelihood that community diversity that we are either unaware of or that lies below our taxonomic resolution is also structured by the abiotic environment. In this sense, abiotic features can provide a more comprehensive picture of biodiversity than species or communities can.

The results of spatial conservation plans that use abiotic features tend to support this conclusion; for example, across Papua New Guinea, abiotic features do a great job of representing the diversity of vegetation communities, but vegetation communities alone do not capture the diversity of abiotic environments. Abiotic classification schemes can also be used to stratify coarser structural or floristic classifications so that the resulting features are a combination of community and abiotic data.

The logic of abiotic units makes the assumption that other species see the same fundamental spectra of physical conditions that we do (altitude, temperature, etc.). Certainly many studies do suggest that these same spectra are important in determining other species' distributions. Although abiotic features have been most commonly applied in large-scale planning exercises, there is plenty of evidence that even very fine-scale differences in abiotic conditions influence species distributions. For cases in which both biotic and abiotic features are available and are likely to support the plan's objectives, we recommend using both.

The use of abiotic units is a point of some contention among ecologists. They generally work better for vegetation than for fauna, largely because there are often many historical reasons (such as glacial or sea-level history) that fauna do not occupy habitats in which they might be predicted to survive. Similarly, even where abiotic units represent diversity well, they do not reveal anything about the condition of ecological communities. These observations suggest that they should be used in conjunction with more direct ecological data. Interestingly, opposition to the use of abiotic units is largely confined to the terrestrial realm.

For marine and freshwater conservation biologists, abiotic classification has been a standard and accepted practice for many years simply due to the paucity of comprehensive data on species distributions in marine and freshwater environments. For example, conservation plans for coral reef environments have routinely used conservation features based on an abiotic classification of reef types developed as part of the Millennium Coral Reef Mapping Project[26] (FIGURE 3.9).

***Generating Abiotic Features*** The global availability of remotely sensed data on environmental variables means that abiotic features can be generated for nearly all planning regions. Three decisions define the task of generating conservation features based on abiotic data: which variables are used, how they are classified, and how they are assigned to features. These three elements, along with important sources of data, are covered in detail in Chapter 5.

***How Do You Know You Have the Right Abiotic Features?*** Although the use of abiotic features as surrogates is supported by attractive logic and some evidence, planners will want confidence that the set of features they have chosen does indeed reflect biodiversity in a way they value. Any of the general approaches to testing surrogacy described in Box 3.5 are reasonable.

A simple test that we have used is to look at what percentage of other conservation features, communities, or species are captured in a set of areas that represent the diversity of environmental variables. FIGURE 3.10 illustrates just such a test applied to abiotic units, which in this case represented well over 90% of the vegetation features, but did a poor job of representing restricted-range endemic fauna.

## Ecological Processes

Ten years ago, one of us wrote in an earlier conservation planning book that ecological processes have "received short shrift from conservation planners." From the standpoint of planners who were primarily focused on biodiversity conservation, this was an accurate statement. From the broader view of conservation planning that we are advancing in this book, which includes land use planning, it was an incorrect statement. In fact, landscape planners and landscape architects who

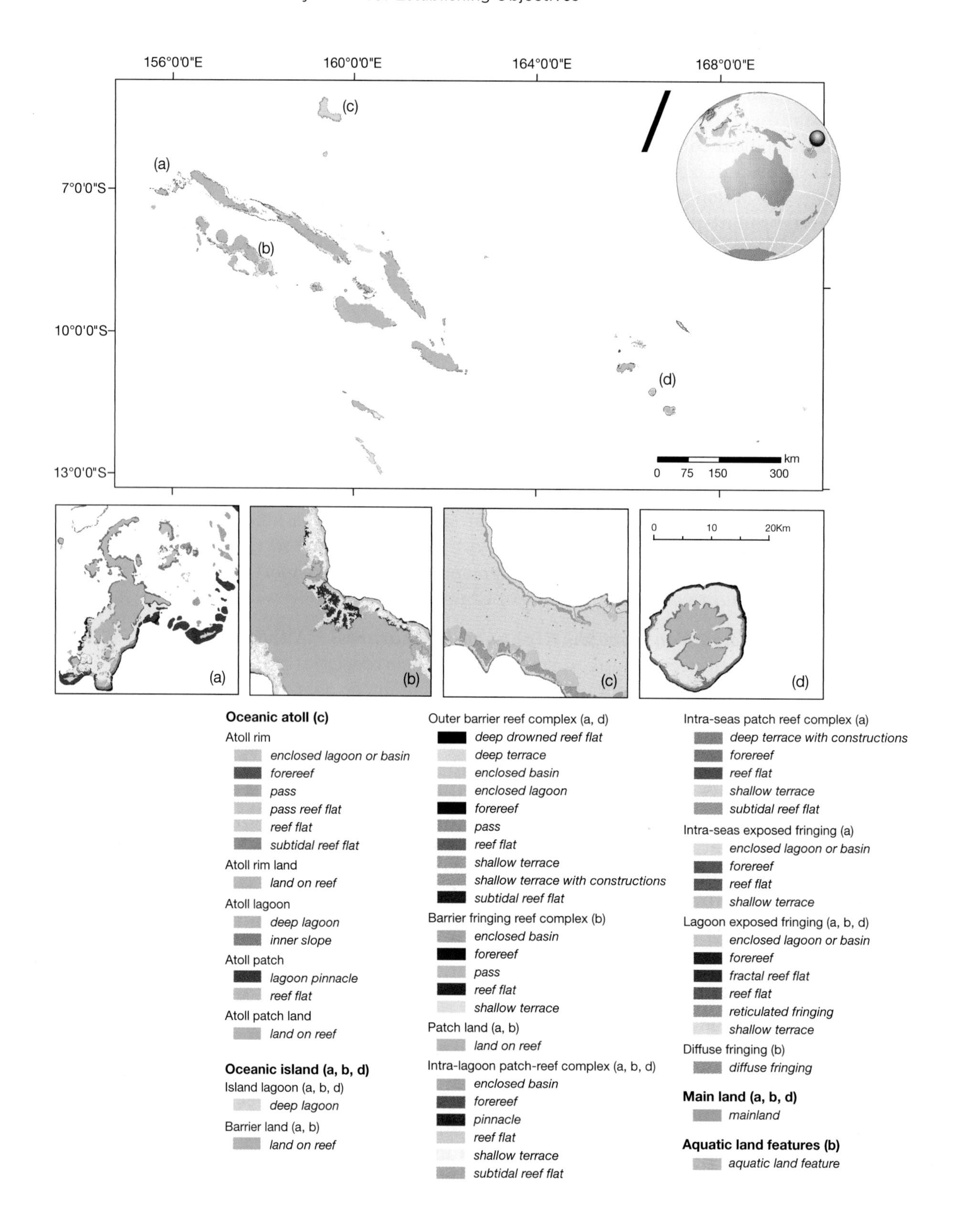
156°0'0"E
160°0'0"E
164°0'0"E
168°0'0"E
7°0'0"S
10°0'0"S
13°0'0"S
(a)
(b)
(c)
(d)
km
0 75 150 300
0 10 20Km
Oceanic atoll (c)
Atoll rim
enclosed lagoon or basin
forereef
pass
pass reef flat
reef flat
subtidal reef flat
Atoll rim land
land on reef
Atoll lagoon
deep lagoon
inner slope
Atoll patch
lagoon pinnacle
reef flat
Atoll patch land
land on reef
Oceanic island (a, b, d)
Island lagoon (a, b, d)
deep lagoon
Barrier land (a, b)
land on reef
Outer barrier reef complex (a, d)
deep drowned reef flat
deep terrace
enclosed basin
enclosed lagoon
forereef
pass
reef flat
shallow terrace
shallow terrace with constructions
subtidal reef flat
Barrier fringing reef complex (b)
enclosed basin
forereef
pass
reef flat
shallow terrace
Patch land (a, b)
land on reef
Intra-lagoon patch-reef complex (a, b, d)
enclosed basin
forereef
pinnacle
reef flat
shallow terrace
subtidal reef flat
Intra-seas patch reef complex (a)
deep terrace with constructions
forereef
reef flat
shallow terrace
subtidal reef flat
Intra-seas exposed fringing (a)
enclosed lagoon or basin
forereef
reef flat
shallow terrace
Lagoon exposed fringing (a, b, d)
enclosed lagoon or basin
forereef
fractal reef flat
reef flat
reticulated fringing
shallow terrace
Diffuse fringing (b)
diffuse fringing
Main land (a, b, d)
mainland
Aquatic land features (b)
aquatic land feature

FIGURE 3.9 Coral reef types in the Solomon Islands. These reef types were based on abiotic variables classified as part of the Millennium Coral Reef Mapping Project.

have taken an ecological approach to planning in more human-dominated landscapes have long been concerned about ecological processes and functions. Frederick Steiner's 1990 textbook, *The Living Landscape: An Ecological Approach to Landscape Planning*,[27] epitomizes this approach, which is focused on inventory and analysis of the biophysical environment and on suitability analyses (that is, of suitability for development) that emphasize hydrological and geomorphological processes, among other ecological processes and functions.

Over the last decade, planners primarily concerned with biodiversity conservation have gained a greater awareness of ecological processes and functions. The Millennium Ecosystem Assessment,[28] for example, conclusively demonstrated losses of ecosystem functions and services around the globe and their probable consequences for human well-being in the future. The popular media, too, have helped draw more attention to the importance of ecological processes, such as fire. Severe and large-scale fire events in Australia, Europe, and the United States over the last decade, exacerbated in part by climate change, have had real consequences for thousands of people who have lost their homes or been displaced. Other extreme weather events, such as flooding, heat waves, tornadoes, and hurricanes, and the disturbances that these events bring to natural and human communities and systems, have

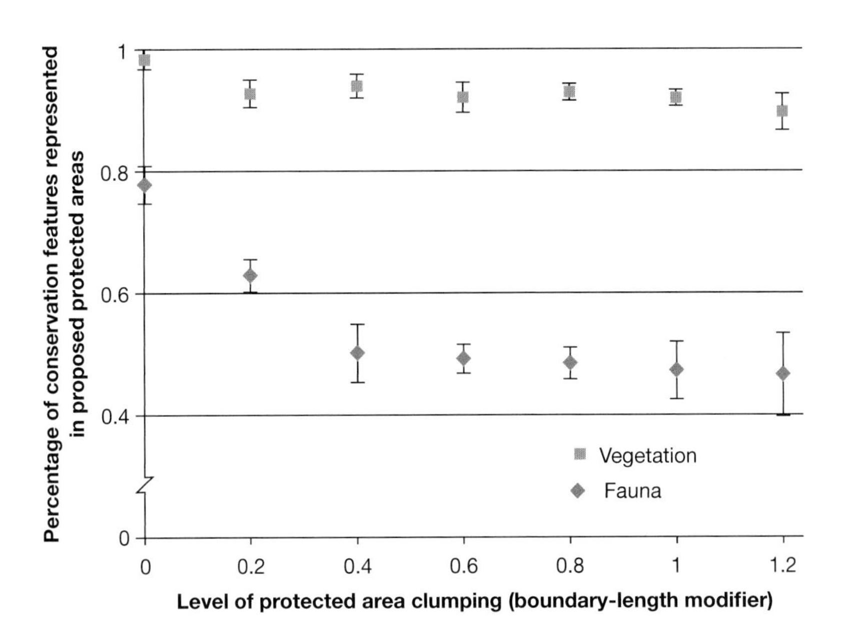


FIGURE 3.10 A test of the effectiveness of land systems as surrogates for biodiversity. The graph shows the effectiveness of proposed protected area networks, based on land systems, at representing diversity in vegetation communities and in restricted-range endemic fauna across Papua New Guinea. Representation is compared across different levels of protected area clumping (increased clumping results in fewer but larger protected areas). Error bars show standard deviation across networks.

also helped elevate the importance of ecological processes. As scientists and conservationists have turned their attention to developing and implementing climate adaptation strategies, conservation of ecological processes and functions is now being advocated as a major approach to climate adaptation[29] (see Chapter 9 for details).

The motivation to better include ecological processes in conservation planning is driven in part by individual species, ecosystems, and some human communities, which are sustained by ecological processes that have been substantially altered in many parts of the world and will continue to be altered by the effects of climate change. Yet there is another more subtle and less obvious reason for focusing on ecological processes. A recent visit to a small fen near Cambridge, England, helped illuminate this reason. Wicken Fen National Reserve is today a 170-ha remnant of what was once a vast peatland nearly 390,000 ha[30] in size (FIGURE 3.11).

Hydrologist Francine Hughes and colleagues are leading an innovative, if not controversial, restoration effort in Wicken Fen today that is focused on restoring dynamic ecological processes as its primary goal.[31] Like such efforts in many human-dominated landscapes around the globe, restoration efforts in Wicken Fen are challenged by the lack of evidence on this ecosystem's structure and function before its substantial alteration by human activities, by widespread non-native species that would be enormously costly to eradicate, and by climate change, which has influenced the underlying distribution of species. In the case of Wicken Fen, this means that community-supported efforts are under way to restore wetland habitats and functions that have no present-day analogs. Such cases are prompting restoration ecologists to call for an open-ended approach to restoring nature that focuses on the physical processes and structure of ecosystems and less on the species composition or patterns of biodiversity.[32] Broader-scale efforts to restore natural flow regimes and floodplain forests,[33] or other ecological processes such as fire, may similarly focus more on ecological processes than on patterns of biodiversity.

## Ecosystem Services

Society values natural systems both for biodiversity and for the services those systems provide. **Ecosystem services**, as they are most conventionally known, are distinguished from other ecological processes because they explicitly benefit human populations. An argument could be made that the maintenance of ecosystem services should be classified as a social or economic objective, but we have chosen to include it in environmental objectives because of the overlap and linkage between ecosystem services and ecological processes.

**FIGURE 3.11** Wicken Fen Preserve, United Kingdom. This preserve is being actively managed for ecosystem processes rather than for the species present.

Increasingly, the sustenance of ecosystem services is a desired outcome of conservation action. Explicitly linking the conservation of natural systems with human well-being, as we saw in Chapter 2, increases the immediate relevance of conservation and conservation planning to a broader swath of society. Ecosystem services now feature prominently in the mission statements of large conservation NGOs such as The Nature Conservancy and Conservation International. The argument that conserving natural systems is an effective way to ensure provision of services that would otherwise be costly, or sometimes impossible, to replace has also opened new sources of conservation funding and become a requirement for many existing ones, such as large philanthropic foundations and aid agencies. It seems likely that resources for ecosystem service–related conservation will continue to increase to the point of becoming the dominant source of funding for conservation action.

In this section, we briefly introduce the types of ecosystem services that might serve as features in a conservation plan. More information on measuring, mapping, and planning the conservation of ecosystem services is contained in the subsequent chapters on data and goals and in Chapter 10 of this book, which is devoted to planning for ecosystem services.

The influential Millennium Ecosystem Assessment distinguished four general types of ecosystem services: *provisioning services* (e.g., provision of seafood, fresh water, timber), *regulating services* (e.g., flood regulation, climate regulation), *supporting services* (e.g., pollination, soil formation), and *cultural, spiritual, and recreational services* (e.g., serenity, inspiration, mythology). All four types can be included as features in a conservation plan.

As with many other elements of conservation planning, South African planners have been instrumental in advancing the practice with regard to ecosystem services. In identifying priority sites for grassland conservation across South Africa, Belinda Reyers and colleagues[34] included the ecosystem services of water production, soil protection, carbon sequestration, and grazing provision in addition to conservation features based on vegetation classification and individual species. Ecosystem services are also the point of most substantial overlap between environmental objectives and the social and economic objectives dealt with in the next section. This overlap has allowed them to act as a sort of bridge for conservation practitioners who wish to consider improvement of human well-being as part of their work.

Conservation plans frequently include objectives related to cultural ecosystem services (for example, recreation, ecotourism, cultural heritage), often without realizing it. Terry Daniel and colleagues provide a good summary of cultural ecosystem services[35] and note that they have not been well incorporated into the ecosystem service framework.

While it is certainly true that valuing cultural services is extremely difficult, they have, nonetheless, appeared as features in conservation plans far more frequently than any other type of ecosystem service.

In developed nations, maintaining landscape aesthetics and providing recreational opportunities are important motivations, and therefore objectives, for much conservation action (as in the Pacific coast example noted earlier). And in all nations, but especially among indigenous communities, the protection of natural and cultural heritage is an important motivation for working with conservation groups.

In Melanesia, where one of us has been involved in numerous conservation planning exercises, natural features, such as particular tree species from which wood carvings are made, are frequently associated with the identity of communities, and maintaining this identity and the ability to share it across generations provides a strong incentive for conservation. Even features that are primarily social constructs, such as sacred or *tambu* (meaning taboo) areas, have been important elements of conservation plans in the region. Similarly, communities, indigenous or not, that have a close association with natural resources (for example, fishing communities) will often identify culturally significant species whose conservation would constitute an ecosystem service—a cultural heritage that is explicitly linked to a feature of the natural world (FIGURE 3.12). The cultural keystone species mentioned earlier would fall into this category.

The use of ecosystem services as objectives and features in conservation plans has been limited in part by difficulty in accounting for

FIGURE 3.12 Two members of Alaska's indigenous community carry freshly caught salmon ready for drying. Catching, drying, and eating salmon are considered essential parts of Alaskan Native culture. Communities, indigenous or not, that have a close association with natural resources (e.g., fishing communities) often identify culturally significant species whose conservation would constitute a cultural ecosystem service.

them. The relatively casual use of surrogates permissible in the case of biodiversity is seen as less acceptable when direct benefit to humans is concerned. The Natural Capital Project, a collaboration between WWF, TNC, Stanford University, and the University of Minnesota, is an attempt to address these limitations by providing tools to map and value ecosystem services (we will come back to these tools in subsequent chapters). This collaboration is significant for conservation planning because it has had a conservation focus from the outset and is designed to help account for ecosystem services in a manner compatible with conservation planning. For example, the Natural Capital Project has been working to support a number of conservation planning exercises that include such objectives as maximizing the ecosystem services of erosion control, flood mitigation, nutrient retention, and groundwater recharge. More information on the Natural Capital Project and the tools it has developed are provided in Chapter 10.

## Social and Economic Objectives

Conservation is a rapidly changing field, and conservation planning must shift with it. Helping conservation planning better reflect the realities of today's conservation challenges is a major motivator for writing this book. Many conservation organizations have shifted their strategic focus away from biodiversity and toward sustainable and equitable natural resource use and its importance for human well-being and economic prosperity. By legislative mandate and policy, many natural resource agencies already manage with this orientation. This shift partly reflects the changing interests of conservation funders, both public and private, but it also reflects an increasing recognition that the scale and wickedness of environmental problems mean that they will never be solved through biodiversity-focused strategies alone. Opportunities to leverage greater conservation impacts may be found through collaborative engagements with a wide range of partners outside the traditional conservation sector. Such engagements are not simply attempts at deepening the conservation resources pool, but attempts to do things better by finding solutions and actions on all sides that are as complementary and sustainable as possible.

What does this shift in conservation outlook mean for conservation planning in practice? The short answer is that conservation planning must be a multiple-objective process. Many institutions involved in conservation planning have used this approach for a long time. Agencies that manage national parks, national forests, or various types of wildlife management areas have long planned not only for the conservation of natural features or biodiversity but also for various human uses, including environmental education, hunting and fishing, grazing, watershed use, and extractive uses such as mining. Indeed, we find it hard to think of situations when conservation plans should not involve multiple objectives that recognize human activities and needs in the landscape and the effects conservation actions can have on them.

It would be trite and incorrect to suggest that conservation practitioners do not recognize the importance of human dimensions in their work; almost without fail, those we interact with are acutely aware of human interests and have long considered this aspect of conservation in their work (consider, for example, the rise of community-based conservation in the 1990s). However, with the exception of government agencies, which are often required to consider multiple objectives, this awareness does not necessarily translate into planning for conservation in a truly multiple-objective fashion. Effective and sustainable conservation outcomes require balancing multiple, often competing, objectives.

Conservation solutions will be durable only so long as people support the objectives that drive them. Plans and planning processes that acknowledge and clearly work toward achieving multiple objectives will almost certainly generate more interest as well as more realistic and well-supported solutions, and will therefore have an increased likelihood of implementation. We also see an emerging role for conservation planners as solution brokers for wider natural resource issues whereby they participate in, or even lead, planning efforts aimed primarily at strategies and decisions for other sectors such as mining, agriculture, forestry, or fishing. This role will require being deft at multiple-objective planning.

Many of the methods we describe in Chapter 7 are directly related to planning with multiple objectives and exploring and navigating the trade-offs between those objectives. But to do so efficiently, the relevant objectives need to be on the planning table from the outset. This means thinking about them and establishing them at the same time as we think about environmental objectives and the features that represent them. In this section, we attempt to encourage this level of integration and to consider how it can be approached.

### What Social and Economic Objectives Belong in a Conservation Plan?

The objectives that should be considered for inclusion in a conservation plan are any *relevant* objectives that are important to people who might be *impacted* by the outcome of the plan. This guideline obviously casts a wide net. Whether someone or some group will potentially be impacted by the outcome of a plan is a good benchmark by which to determine who should be considered a stakeholder in the planning process. Impact can be interpreted in this context as any change (negative or positive) in rights to use of or access to nature in any way—everything from simply where you can walk to what you can remove or exploit.

The question of what is a *relevant* objective is perhaps best approached by considering whether an objective is controllable in the context of the current plan. For example, improved health care is an obviously desirable objective for many communities (in both developing and developed nations) and yet is unlikely to be controllable given the set of decisions a conservation plan can generally inform. On the other hand, improved access to communities (which can be related to health care provision), be it by road, boat, air, or virtually, may indeed be an objective that a conservation plan and subsequent activities can influence.

### Equal or Not?

Planning teams starting to consider social and economic objectives as part of a conservation plan often spend time debating whether these objectives should be equal to conservation objectives or desirable but secondary concerns. This debate frequently raises the concern that conservation objectives will be marginalized or that the prominence of social and economic objectives will be alienating to some conservation supporters. Our advice: Don't worry about weighting or prioritizing objectives at this stage. If the freedom exists to weight objectives differently, it can be done when navigating trade-offs between solutions (see Chapter 7). At this stage of planning, it is best to focus on establishing the objectives, rather than on their relative importance.

### Eliciting Social and Economic Objectives

How do you find out what social and economic objectives are relevant to the plan? Just ask. There is no substitute for engaging stakeholders directly to find out what objectives belong in a plan. The hiring of social scientists by an organization or agency does not negate the need to engage directly with stakeholders to elicit the important objectives in each planning process, although it should make the organization or agency more effective at doing so. However, we know that "just asking" is more complicated than it sounds, and that getting the right objectives requires a careful and deliberate process. All the advice about eliciting objectives given earlier in this chapter applies to social and economic objectives as well.

### Communication

Perhaps the biggest challenge in establishing social and economic objectives in a conservation plan is convincing stakeholders that a conservation plan is a forum that might help them achieve their socioeconomic objectives. We expect that it will take some time to overcome a legacy of conservation planning that has considered many livelihoods as "threats" to be defended against and has viewed the planning process as something of a strategic skirmish against other interests and natural resource uses. Initial steps forward have involved conservation plans that engage stakeholders to ensure that conservation actions minimize inconvenience to them.

Ultimately, however, we believe that conservation planners need to communicate their approach not only as a way to reduce the impact of conservation actions on stakeholders' objectives, but as a tool that can actually help deliver those objectives—and deliver them in a potentially secure fashion. The pressure from groups and governments on individuals to conserve the environment is not going to disappear, so solutions that navigate satisfactory compromises now and find win-win outcomes where they exist are likely to be more secure for stakeholders in the long run. This is the direction of relevancy and sustainability for the field of conservation, but it cannot happen without good planning. In a grand vision, we see conservation planning not simply as a way to ensure that conservation resources are used to maximum effect, but as a method to guide equitable and sustainable natural resource management—and it starts here with objectives.

In this book we have tried to illustrate and guide a planning process that makes explicit space for projects with objectives other than biodiversity conservation to contribute in a meaningful way. As such, this book does not list specific economic and social objectives that might be part of a conservation plan, but rather focuses on the types of objectives to look for in a plan and on how to conduct a planning exercise that does these objectives justice.

## Types of Social and Economic Objectives

Conservation planners should be on the lookout for relevant economic objectives, social objectives, and process objectives. Economic objectives speak to the outcomes of key business activities in the planning region. Profitability is a core pursuit, the key consideration for most business sectors, and for public companies, a legally required objective. To the extent that the outcomes of a conservation plan may affect profitability (either negatively or positively), it is likely to be a relevant objective in the conservation plan. For example, a conservation plan for a small island state would need to consider the profitability of fisheries as an economic objective.

We recognize that including objectives relating to making money for industries will be anathema to many who have chosen a career in conservation. Including economic objectives in a plan does not mean that environmental objectives will be subjugated in pursuit of profitability; rather, they act as a flag to ensure that part of the planning process is to understand and consider the effects of conservation options on the economics of a region, and vice versa. Furthermore, conservation should have its own economic objective of improving its cost-effectiveness (or maximizing financial sustainability); we rarely have as much money as we would like, so we should rightly be concerned about achieving the most with the funding that we do have. We acknowledge that this is a narrow interpretation of the term *economic*.

Social objectives reflect the many elements other than money that contribute to quality of life: food availability, security, health, cultural customs, and so forth. The feasibility of conservation solutions is likely to be poorly reflected by economics alone because people tend to have strong preferences for how even their basic needs are met. Successful conservation relies on these "way of life" preferences, perhaps more than we often acknowledge.

In recent interviews with indigenous communities in Borneo related to a REDD demonstration project that one of us was involved in, some community members expressed strong preferences for resolving land tenure rights, for example, over livelihood or health issues. In another example, in Colombia, conservation groups working on silvo-pastoral systems with the ranchers of the llanos grasslands are counting on the sustenance of a ranching way of life being a strong motivator for local communities to participate in conservation; if financial return was all the ranchers cared about, they might well be expected to convert or sell their land for intensive cropping. In places in Melanesia where we've worked, social objectives commonly reflect the value placed on cultural heritage. Important objectives in plans we have worked on in Melanesia include protection of customs and spiritual areas as well as conservation of resources such as ironwood or megapode eggs that are more about culture than subsistence or economics.

In developed countries, social objectives are likely to be associated with things such as access to recreational opportunities (FIGURE 3.13) or the sustainability of local natural resource–based industries, such as fishing, that will not necessarily be reflected in purely economic objectives. Where one of us lives near Yellowstone National Park, USA, the debates between user groups on access to motorized versus non-motorized recreation are virtually endless and can reach uncomfortably confrontational levels.

Although the distinction between economic and social objectives is a blurry one, their division can be helpful in prompting consideration of the range of objectives that will probably exist. There is also overlap between economic and social objectives and the ecosystem service objectives already mentioned: both are focused on the needs

FIGURE 3.13 A hiker in the Beartooth Wilderness, Montana. Recreational use is an important cultural ecosystem service provided by many conservation areas.

of people, but the anthropogenic objectives described here are not necessarily dependent on maintaining ecosystem services (although that might still be the best approach).

Process objectives, introduced earlier in this chapter, focus on the way conservation planning exercises are executed and decisions made. Frequently, what appear to be social objectives might best be captured as process objectives, especially in the case of social values around rights and self-determination. These are objectives for the way natural resource decisions are made, rather than the reason they are made. For example, it might be important that final decisions about the siting of conservation projects be determined by the communities involved; therefore, the planning process must be designed to facilitate this (FIGURE 3.14).[36]

In Chapter 2, we introduced the concept of human well-being and noted its increasing prominence in conservation projects. We consider human well-being to be a broad term that encompasses elements related to social, economic, and environmental objectives. We, and many others in conservation, believe that everyone's life is improved by healthy natural systems. Improving human well-being might be thought of as the top of a fundamental objective hierarchy: something

FIGURE 3.14 Community-based planning. In the Solomon Islands, The Nature Conservancy explicitly designed a planning process that enabled local communities to have the final say in the location of conservation areas.

that needs more specific objectives beneath it in order to be useful in a planning process.

While there is no universal or consistent definition of what constitutes well-being, there are a number of established frameworks that identify constituent dimensions or focal areas of well-being. Among the most widely recognized of these are the United Nations Development Programme's Human Development Index[37] and the framework developed as part of the Millennium Ecosystem Assessment. The conservation social scientist Craig Leisher and his colleagues[38] have usefully summarized the most common dimensions of human well-being contained in a range of well-being indices currently in use around the world. This summary represents a good place to start when thinking about social objectives in a conservation plan.

## Building a Hierarchy of Social and Economic Objectives

Start by identifying the fundamental economic and social objectives and build a hierarchy as appropriate. What are the main economic or social activities happening in the planning region or landscape of interest? What is the main driver of satisfaction behind these activities? As we advised earlier in the chapter, start simple, with just the thing

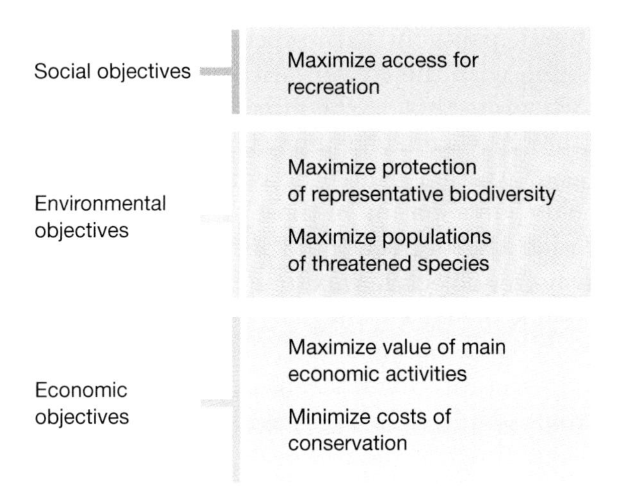


FIGURE 3.15 A simple and generic objective hierarchy for a conservation plan might contain a set of objectives that looks like this.

and the direction. Maximize crab catch. Maximize mineral resources available for extraction. Maximize the amount of power generated. More specific outcomes, such as the amount of wind energy or crab harvest required to ensure that industries remain profitable, will be taken care of in targets (see Chapter 4).

A simple and generic objective hierarchy for a conservation plan might contain a set of fundamental objectives, as in FIGURE 3.15. A set like this makes a great starting point for a planning process that aims to work for all stakeholders. Again, it is fine if these objectives appear to be in opposition to one another initially. There is plenty of time to explore possible trade-offs later.

Just as the broad objective of conserving biodiversity can be decomposed to more accurately reflect the outcomes we desire, so can the broad objective of economic prosperity. Doing so is critically important: how seriously the plan is received will depend on its objectives resonating closely with stakeholders. Starting with a hierarchy of objectives like those in Figure 3.15, it will be important to probe each objective to make sure that delivering on it would be the desired outcome.

We were involved in a conservation planning process for the rangelands of northern Kenya that identified the following set of fundamental objectives:

- Increase community income
- Improve security
- Improve rangeland condition
- Increase wildlife populations

Community participants in this process identified an increase in income associated with the conservation initiative as a key objective for the plan. We wrote this on the flip chart, but then asked the participants whether they would be happy if the total value of community income from conservation-related activities increased, but the number of households participating in those activities remained the same. Without exception, they agreed that this was not a desirable outcome, and that a far better objective was to increase the number of families receiving economic benefits from the program, thus increasing prosperity in a more equitable fashion. This seemingly small change in objective led to substantially different thinking about strategies—for example, expanding a livestock program previously focused on cattle to also include small stock such as goats and chickens so that households without cattle could also participate.

## KEY MESSAGES

- Conservation planning is an exercise in figuring out the best route to conservation success. What success looks like should be driven by our values—the things we care about. We refer to value-driven statements about desirable outcomes as *fundamental objectives* and urge that all plans begin with their definition.
- Fundamental objectives can be identified through deliberate, value-focused thinking with the assistance of some simple tools. Most critical is to separate ends (where we want to go) from means (how we want to get there).
- Conservation planners should be on the lookout for three general classes of fundamental objectives: environmental, social, and economic. Increasingly, conservation planning will involve navigating objectives in all three groups.
- Environmental objectives may include a focus on one or more of the following: biodiversity, ecological processes, and ecosystem services. Of these, the conservation community has gone to the greatest effort to develop approaches to represent biodiversity in conservation plans. These representations, referred to as conservation features, include species, classifications of communities or ecosystems, and abiotic units. Because there will never be data on all the elements of biodiversity, the use of surrogate measures is critical when defining conservation features.
- The social and economic objectives that belong in a conservation plan are any *relevant* objectives that belong to people who might be *impacted* by the outcome of the plan. *Impact* can be interpreted

in this context as any change (negative or positive) in rights to use or have access to nature in any way, while a *relevant* objective is an objective that is controllable in the context of the current plan.

## References


1. Kahneman, D., P. Slovic, and A. Tversky, *Judgement under Uncertainty: Heuristics and Biases.* 1982, Cambridge: Cambridge University Press. 544.
2. Kahneman, D., *Thinking Fast and Slow.* 2011, New York: Farrar, Straus and Giroux. 512.
3. Wilhere, G. F. et al., *Conflation of values and science: Response to Noss et al.* Conservation Biology, 2012. **26**(5): 943–944.
4. Keeney, R. L., *Value-Focused Thinking.* 1992, Cambridge, MA: Harvard University Press. 416.
5. Gregory, R. et al., *Structured Decision Making: A Practical Guide to Environmental Management Choices.* 2012, Oxford: Wiley-Blackwell. 299.
6. Kahneman, *Thinking Fast and Slow.* (See reference 2.)
7. Gregory et al., *Structured Decision Making.* (See reference 5.)
8. Keeney, *Value-Focused Thinking.* (See reference 4.)
9. Doran, G. T., *There's a S.M.A.R.T. way to write management's goals and objectives.* Management Review, 1981. **70**: 35–36.
10. Margules, C. R., and R. L. Pressey, *Systematic conservation planning.* Nature, 2000. **405**(6783): 243–253.
11. Groves, C. R., *Drafting a Conservation Blueprint: A Practitioner's Guide to Planning for Biodiversity.* 2003, Washington, DC: Island Press. 457.
12. Groves, *Drafting a Conservation Blueprint.* (See reference 11.)
13. IUCN, *IUCN Red List Categories and Criteria: Version 3.1.* 2001, Gland, Switzerland, and Cambridge: IUCN Species Survival Commission. 30.
14. Myers, N. et al., *Biodiversity hotspots for conservation priorities.* Nature, 2000. **403**: 853–858.
15. Ripple, W. J. et al., *Status and ecological effects of the world's largest carnivores.* Science, 2014. **343**(6167): 1241484.
16. Estes, J. A. et al., *Trophic downgrading of Planet Earth.* Science, 2011. **333**(6040): 301–306.
17. Estes et al., *Trophic downgrading of Planet Earth.* (See reference 16.)
18. Garibaldi, A., and N. Turner, *Cultural keystone species: Implications for ecological conservation and restoration.* Ecology and Society, 2004. **9**(3): 1.
19. Smith, R. J. et al., *Identifying Cinderella species: Uncovering mammals with conservation flagship appeal.* Conservation Letters, 2012. **5** (3): 205–212.
20. Verissimo, D., D. C. MacMillan, and R. J. Smith, *Toward a systematic approach for identifying conservation flagships.* Conservation Letters, 2011. **4**(1): 1–8.

21. Rabinowitz, D., *Seven forms of rarity*, in *The Biological Aspects of Rare Plant Conservation*, H. Synge, editor. 1981, New York: Wiley. 205–217.
22. Higgins, J. V. et al., *A freshwater classification approach for biodiversity conservation planning.* Conservation Biology, 2005. **19**: 432–445.
23. Anderson, M. et al., *Guidelines for Representing Ecological Communities in Ecoregional Conservation Plans.* 1999, Washington, DC: The Nature Conservancy.
24. Rodriguez, J. P. et al., *Establishing IUCN Red List Criteria for threatened ecosystems.* Conservation Biology, 2011. **25**(1): 21–29.
25. Keith, D. A. et al., *Scientific foundations for an IUCN Red List of ecosystems.* PLoS One, 2013. **8**(5): e62111.
26. Dalleau, M. et al., *Use of habitats as surrogates of biodiversity for efficient coral reef conservation planning in Pacific Ocean islands.* Conservation Biology, 2010. **24**(2): 541–552.
27. Steiner, F., *The Living Landscape: An Ecological Approach to Landscape Planning.* 1990, New York: McGraw-Hill. 338; and Groves, C., *The conservation biologist's toolbox for planners.* Landscape Journal, 2008. **27**(1): 81–96.
28. Millennium Ecosystem Assessment, *Ecosystems and Human Well-Being: A Framework for Assessment.* 2003, Washington, DC: Island Press. 245.
29. Groves, C. et al., *Incorporating climate change into systematic conservation planning.* Biodiversity and Conservation, 2012. **21**(7): 1651–1671.
30. Kahneman, *Thinking Fast and Slow.* (See reference 2.)
31. Hughes, F. M. R. et al., *Monitoring and evaluating large-scale, "open-ended" habitat creation projects: A journey rather than a destination.* Journal for Nature Conservation, 2011. **19**(4): 245–253.
32. Hughes, F. M. R., W. M. Adams, and P. A. Stroh, *When is open-endedness desirable in restoration projects?* Restoration Ecology, 2012. **20**(3): 291–295.
33. Rood, S. B. et al., *Managing river flows to restore floodplain forests.* Frontiers in Ecology and the Environment, 2005. **3**(4): 193–201.
34. Reyers, B. et al., *National Grasslands Biodiversity Program: Grassland Biodiversity Profile and Spatial Biodiversity Priority Assessment*, 2005, Council for Scientific and Industrial Research. Pretoria, South Africa.
35. Daniel, T. C. et al., *Contributions of cultural services to the ecosystem services agenda.* Proceedings of the National Academy of Sciences of the United States of America, 2012. **109**(23): 8812–8819.
36. Game, E. T. et al., *Informed opportunism for conservation planning in the Solomon Islands.* Conservation Letters, 2011. **4**: 38–46.
37. UNDP, *Human Development Report 2007/2008: Fighting Climate Change: Human Solidarity in a Divided World.* 2007, New York: United Nations Development Program. 399.
38. Leisher, C. et al., *Focal areas for measuring the human well-being impacts of a conservation initiative.* Sustainability, 2013. **5**(3): 997–1010.

# 4

# Making Objectives Measurable: Indicators and Targets

## Overview

Indicators are the things we measure to inform us of the extent to which we are achieving our objectives. Targets identify a particular level of achievement that we are striving for. Targets are useful, but not essential, elements of a conservation plan; indicators, on the other hand, are critical. In this chapter we provide guidance on how to identify good indicators: ones that faithfully reflect the intention of the objective and are practically measurable. We discuss the three different types of scales (natural, proxy, and constructed) that can be used to measure indicators. We acknowledge debate around the use of targets in conservation planning, and we identify the context in which they can be useful tools. Finally, we describe common approaches to establishing targets for ecological, economic, and social objectives.

## Topics

- *Identifying measurable indicators*
- *Natural scales*
- *Proxy scales*
- *Constructed scales*
- *Quantitative targets*
- *Trade-offs versus targets*
- *Setting targets*
- *Convention on Biological Diversity*
- *Historical baselines*
- *Naturalness*
- *Species-area relationship*
- *Population viability analysis*
- *Minimum viable population*
- *Precautionary principle*
- *Stakeholder-established targets*
- *Ecological thresholds*
- *Economic activity targets*
- *Social targets*

## Identifying Measurable Indicators

Objectives are what we care about. **Indicators** are what we plan to measure to tell us what is happening in regard to our objectives. Without measurable indicators, there is no way to know whether our objectives are being achieved, or to estimate the consequences of alternative courses of action. Identifying indicators is a critical aspect of conservation planning.

Consider the fundamental objective of maximizing salmon abundance in a region. While the intention of this objective is clear, how it should be measured is not. Is *salmon abundance* the number of breeding fish in all the rivers in the region? Would it matter if the same number of fish were all in one river or spread out across all the rivers? Can abundance be approximated by the commercial harvest, or does it require independent surveys? Bringing this sort of clarity to the objective is what it means to define indicators.

Indicators, as we use the term, are also sometimes referred to as **attributes** or criteria. Conventionally, the term *criteria* is most commonly associated with discriminating between alternative actions (in other words, decision criteria), while *indicators* is most commonly used in a results-monitoring context (as in "performance indicators"). Both of these contexts are covered in later chapters of the book; in Chapter 7 we look at the use of indicators in the context of deciding among alternative strategies, and in Chapter 12 we focus on evaluating performance.

### How Do Indicators Relate to Features?

In Chapter 3 we introduced *conservation features* as a way of representing biodiversity in a conservation plan, and we described three general types of features: species, ecological communities, and abiotic units. If features are to be actionable as part of a conservation plan, they require indicators and, in many cases, *targets*: **quantitative** (numerical) statements of the outcomes we want to achieve. In practice, our identification of conservation features often includes some implicit (or even explicit) association with indicators. For example, using abiotic units, such as geological types, as conservation features almost always involves the indicator "area protected" because few other aspects of these features can be measured—geological types don't have populations or condition. As we will see in the section on targets that follows, however, even a seemingly straightforward indicator such as area protected can be open to wide interpretation.

### What Makes a Good Indicator?

A good indicator has two equally important properties: it is *representative* and *measurable*. An indicator that is *representative* faithfully and fully reflects the intention of the objective. The best test of

FIGURE 4.1 A Family group of Yunnan golden monkeys (also known as Yunnan snub-nosed monkey) (*Rhinopithecus bieti*). If a conservation project had the objective of maximizing the population of Yunnan golden monkeys, the number of family groups in the area might be a good indicator.

representativeness is if, upon observing a positive change in the indicator, you would feel satisfied that things are going in the right direction. As a simplistic example, if our objective were to maximize the population of the Yunnan golden monkey (also known as the Yunnan snub-nosed monkey) in China (FIGURE 4.1), and our indicator was the number of family groups, we would most likely be happy talking about conservation success if we documented an increase in the number of family groups.

Straightforward cases like this one are probably rare. Consider another objective, that of maximizing the health of a grassland ecosystem. As an indicator, you could measure the total productivity of the grassland. However, even if you confidently documented an increase in productivity, you might feel that the indicator did not capture important parts of the objective's intention. For instance, what if the increase in productivity was due mainly to an increase in the abundance of an exotic grass species (FIGURE 4.2)?Representativeness might require considering an additional indicator related to the percentage of exotic grasses.

It is perfectly reasonable for objectives to be described by more than one indicator. However, it is important not to descend into simply listing large numbers of indicators for each objective. Doing so can be

(a)

(b)

FIGURE 4.2 Using productivity as an indicator of grassland health might mask undesirable properties like the presence of invasive species. (a) Grassland in Dinosaur National Monument, Utah, USA. (b) Grassland on Bluebell Ranch, South Dakota, USA. Both photos show what appear to be productive grasslands, but the grassland in (a) is dominated by cheat grass (Bromus tectorum), an invasive species that is considered a pest.

particularly tempting for objectives concerning biodiversity conservation, where, just as when we are identifying the objective hierarchy, we have a tendency to want to acknowledge the uniqueness of particular habitat types or populations or the many ecological processes that influence them. For example, when working with colleagues on conservation planning for coral reefs, one of us was presented with 61 indicators for an objective related to coral reef resilience. Our experience, as well as that of others in the monitoring field, is that trying to measure too many indicators is one of the reasons that monitoring efforts in conservation programs often fall short (we discuss this problem further in Chapter 12).

The decision scientists Lee Failing and Robin Gregory have pointed out that, unlike the fields of economics, development, or health and welfare, biodiversity conservation has largely resisted using **summary indicators**—essentially, a small basket of indicators that are combined into a single indicator.[1] Examples of summary indicators that will be familiar to most readers are the Dow Jones Industrial Average (an index of the New York Stock Exchange based on the performance of 30 large companies) and quality-adjusted life years, or QALYs (an index of human health based on five indicators: mobility, self-care, usual activities, pain/discomfort, and anxiety/depression). The objective of maximizing human health is nearly as broad as that of maximizing biodiversity, and yet there is reasonably broad agreement about useful summary indicators for quantifying the performance of health-related interventions. Summary indicators require a good dose of humility and altruism, and presumably ecologists and conservation biologists have not so far been convinced that their value outweighs the risk that they might not be representative of the particular biodiversity they are interested in or concerned about. One summary indicator that has been used in the United States for conservation planning that involves freshwater systems is James Karr's Index of Biotic Integrity.[2]

The second important property of any indicator is *measurability. Measurable* in this case does not mean simply that measurement is theoretically possible, but rather that measurement is possible without an inordinate amount of time, cost, or effort. Think practically. In an ideal world, data availability would not bias our selection of indicators. In the real world, it must. Indicators need not be complete prisoners to currently available data, but they should be accompanied by thoughtful consideration of what data are available or obtainable. We believe it is generally helpful to think of indicators as our objectives viewed through the lens of available data.

The issue of measurability becomes a little cloudy when we are dealing with indicators that are intended to be estimated through expert judgment rather than through more or less objective data. For example, let's say we chose an objective of maximizing biodiversity and an indicator of total species number. Yes, an expert could make a crude guess at the total number of species in a region at any time, but would you consider total species number reasonably measurable? Probably not. Any expert making such a guess would almost certainly be substituting an assessment based on a small number of taxa for an assessment of biodiversity as a whole (a phenomenon known as *indicator substitution*),[3] in which case it would be better to use a taxon that is more reliably measurable as the indicator.

The more clearly and specifically the objectives are formulated, the easier it is to identify good indicators. There is also important feedback between the two: frequently, the act of thinking about indicators can cause planners to recognize important properties that actually deserve to be objectives.

## Natural, Proxy, and Constructed Indicators and Scales

Because their primary role is to make objectives measurable, indicators must be associated with a scale that makes it possible to determine different levels of achievement. Another way to think of this sense of the term *scale* is as units of measurement. Indicators and their scales can be one of three kinds: natural, proxy, or constructed (**FIGURE 4.3**).

More objective → More subjective

| | Natural | Proxy | Constructed |
|---|---|---|---|
| Maximize wildlife population | Population size | Amount of suitable habitat | Index of threat severity |
| Increase community well-being | Household wealth | Number of local business ventures | Happiness index |
| Increase coral reef health | Percentage of live coral cover | Annual number of coral reef bleaching events | Index of reef resilience |

Indicator scale type

**FIGURE 4.3** Natural, proxy, and constructed indicators and scales span a spectrum from objective to subjective. Examples of possible scales of each type for three different fundamental objectives are given.

**Natural scales** are units of measurement that are obvious and preexisting ways of measuring an indicator. For example, if an objective was to increase stream flow, then an appropriate indicator might be base flow (the portion of stream flow that is not the result of runoff), and flow is conventionally measured in volume ($m^3$/second). Similarly, objectives relating to species conservation might have an indicator of population size, conventionally measured by the number of individuals or some variation thereof. Assuming that indicators are measurable, natural scales have the strength of the potential to measure them objectively—in other words, with empirical data rather than just good judgment. There are, however, likely to be situations in which natural scales are better, or necessarily, estimated by experts rather than measured from observational data.

Frequently, objectives will have indicators that have no obvious natural scale. For example, we might have an objective of maximizing recreational value in a landscape. There is no obvious, conventionally applied scale that we could use to measure recreational value. This leaves us two options: we can measure something else that has an obvious existing scale, but is likely to reflect changes in the objective we care about, and use that scale as a **proxy scale** for the indicator; or we can use a **constructed scale**. Because many ecological objectives, especially general properties such as ecosystem health, integrity, resilience, or even viability, do not have natural scales, they will require proxy scales, or even constructed scales.

In many cases, proxy indicators that can be measured objectively can be found with a little creativity. Among the objectives of a program we assisted in northern Australia was to maximize conservation management capacity among indigenous people, pastoralists, and private reserve holders. There is no natural scale for measuring management capacity, especially not across three very different groups of people; however, the program staff came up with the proxy indicator of the percentage of lands held by these three groups that were subject to early dry season burning at least once every three years. The logic was that burning in the early dry season reflected both the capacity and the intent to manage for conservation benefits and could be measured rapidly and objectively.

Where no satisfactory proxy scale can be found, a constructed scale must be used. Constructed scales are typically arbitrary scales that cover the spectrum from excellent to terrible. In their simplest form, constructed scales might be verbal interpretations, such as rating the level of threat to an area as high, medium, or low. More typically, however, constructed scales are subjective numerical representations, such as a score from 1 to 10. Constructed scales might also be summary or composite scales that combine a number of indicators. It is certainly possible that summary indicators might combine some indicators that can be measured on a natural scale and some that are measured on proxy or constructed scales (**Box 4.1**).

Constructed scales have the strength of being able to reflect any desired indicator without relying on what can be measured naturally

### BOX 4.1 Reporting on "True" Conservation Progress

The conservation decision scientist Eve McDonald-Madden and colleagues[a] developed a constructed indicator from two natural indicators for the fundamental objective of maximizing the protection of ecosystem types in a region. This objective has conventionally used the amount of each ecosystem type in conservation areas as an indicator, but McDonald-Madden and colleagues pointed out that our satisfaction with the objective of protection is also influenced by the amount of loss that is happening outside conservation areas. They proposed an indicator that is the net change in ecosystem protection, combining the area of the ecosystem that has been protected with the area of the unprotected ecosystem that has been cleared over the same period. Both elements of this indicator are measured on natural scales. Combining any number of variables into a constructed indicator requires great care, however, because it is easy to make mistakes that can compromise the validity of the scale. We return to this topic in detail in Chapter 7.

[a] McDonald-Madden, E. et al., *"True" conservation progress.* Science, 2009. **323**(5910): 43–44.

or through proxies. This means that the indicator can be perfectly tailored to be representative of the objective. However, constructed scales have the weakness of ambiguity around what the values in the scale actually mean and the need to either estimate or combine values using expert opinion, thus introducing additional subjectivity and bias. In Chapters 5 and 6, we cover issues of bias in expert judgment and how to minimize them.

As scientists, our preference is strongly for seeking scales that can be used relatively objectively—and to us that usually means natural or proxy scales. We find these scales more compelling and more reliable. If decision makers or those involved in the planning process do not find a particular indicator and scale convincing, it is often worth revisiting the objectives, because doing so might reveal additional objectives or help refine existing ones.

## Targets

> *It is easier to inspire people to climb a mountain when it has a summit than when there is none.*
>
> — Ralph Keeney and Howard Raiffa

We use the term **targets** to refer to quantitative statements of the outcomes we want to achieve for each objective: for example, the number of hectares to be protected, the number of populations to be conserved, the amount of carbon to be sequestered, and so forth. As noted in Chapter 3, our use of *targets* is different from some other conservation planning frameworks; but see our planning language "Rosetta Stone" for a crosswalk between the terms used in a range of planning frameworks. Unlike objectives (introduced in Chapter 3), which simply identify the things we care about and the direction in which we want them to go, targets identify a particular level of achievement that we are striving for—a target can be achieved or not.

The establishment and use of targets in conservation planning has been one of the most vigorously debated aspects of the entire systematic conservation planning field. The level of debate around targets occurs partly because they represent the pointed end of substantial scientific and philosophical debate about how much conservation is enough conservation. But also, we believe, the debate is partly a result of widespread misunderstanding about the role of targets. Targets are tools; nothing more, nothing less. They should not be imbued with responsibility for conservation salvation, but neither can they simply be dismissed. Pragmatically speaking, with regard to the outcomes we care about, more is generally better. In planning terms, however, more of one outcome nearly always comes at a cost to another outcome. Targets help conservation practitioners manage this balancing act.

Conservation is often a game of numbers. Targets, such as protecting 10% of all habitat types, have come to be the backbone of both international and national conservation policy.[4] Philosophically, these numerical targets are generally considered to represent the minimum amount of biodiversity that we want to effectively conserve. But despite their widespread use and endorsement in conservation plans and policies around the globe, targets are never far from criticism. Here are some common criticisms of targets:

- They are arbitrary and lack scientific credibility.
- They provide tacit acceptance of the loss of everything other than the target amount.
- They are too low and do not provide adequate protection for nature.
- They are too high to be achievable, and yet we are supposed to evaluate our progress against them.
- They are inflexible.
- They are too simplistic: conservation benefits are delivered in different ways, and targets cannot account for this complexity—for example, the amount of forest requiring protection to secure biodiversity is influenced by how forest is managed outside protected areas.

There are probably instances in which each of these criticisms has been true, at least to some extent. However, the conservation planner Josie Carwardine and colleagues[5] evaluated these criticisms carefully and concluded that most of them result from misunderstandings, misconceptions, and miscommunication. Their observations are worth summarizing:

- In many cases, targets should be viewed as sociopolitical rather than ecological. Rarely are targets such as 20% protection intended to reflect a deep ecological understanding. Rather, their principal aim is to ensure equity across biodiversity and help avoid mistakes of the past in which economically valuable habitats were grossly underrepresented in conservation areas (conservation largely got the pieces others didn't want!).
- Targets can be easy to convey. Politicians like, and can remember, numbers.
- Targets are intended to be aspirational and inspirational (as suggested by the quote at the beginning of this chapter). And while we should track progress against them, our conservation mission is not necessarily a failure if we do not reach them, nor necessarily complete if we do.

- Because our conservation work proceeds on many levels, areas above and beyond our targets are not simply sacrificed.
- Targets can be as simple or as complex as they need to be, and measuring achievement toward them can be deeply nuanced.

Too often, discussion around conservation targets is focused solely on the question of how much is enough. While this question is an important and challenging one, it has also distracted us from the practical roles that numerical targets play in conservation. Imagine instead that you were planning for public health outcomes in a developing nation—it would seem absurd in that context to have a discussion about how many children should have access to clean water. Of course, the more the better, but achievement of this outcome must be balanced with cost and other uses of resources. Thus, targets are used in health planning, for the same reasons outlined in the list above.

Targets should perhaps be seen as stars by which conservation plans can navigate, rather than the destinations themselves. After all, if we don't know where we are headed (in terms of what conservation we want to achieve for a feature), it is difficult to assess progress or know when we have arrived (achieving an objective). Targets do not deserve slavish adherence and should not be viewed as immutable. They can evolve. Sensible targets will always be a combination of values, science, and politics, none of which are static. Seen in this light, the need for conservation planners to have convincing ecological support for targets is diminished. This should not, however, detract from the need to understand how different levels of investment in conservation activities will translate into achievement of conservation objectives and the maintenance of those things we value.

Now that we have outlined the important role of targets in conservation planning, a significant caveat is required: conservation planning does not have to be target based. The alternative is for a conservation plan to look at levels of achievement of and trade-offs between outcomes across a range of scenarios[6] (**Box 4.2**). This approach applies equally to spatial planning (the Zonation conservation planning software[7] is a good example) and strategic planning. The role of science under this approach is to communicate clearly the consequences of different actions and to help stakeholders find the balance that works best. The question "How much is enough?" or more accurately, "What outcome are we willing to pay for?" is then determined by society rather than by scientists, which is probably how it should be.

In the remainder of this section, we introduce the origins of targets commonly used in conservation planning as well as some of the issues to be aware of when using different approaches to establishing targets.

## BOX 4.2 Trade-Offs or Targets?

Anna Roberts and colleagues have described the evaluation of trade-offs in managing the water quality of the Gippsland Lakes in Victoria, Australia.[a] The Gippsland Lakes are an important area for migratory birds, water-based recreation, and tourism, but they have been badly impacted by eutrophication due to phosphorus inputs from the surrounding farmland. A target of 40% reduction in phosphorus inputs had been agreed to by the government and stakeholders, but without a clear understanding of what the economic impact of this target would be on farming communities. When it became clear that meeting the 40% target would not be feasible with current environmental budgets, Roberts and colleagues constructed a range of scenarios that included phosphorus reductions between 10% and 40%. They then carefully evaluated both the environmental outcomes and economic costs of these scenarios and illustrated the trade-offs between those outcomes and costs (FIGURE 4.2.1). These scenarios and trade-offs then formed the basis of discussion with stakeholders about the best solution.

One of us conducted some research looking at the cost of reducing the risk of the catastrophic coral bleaching events that are wiping out reefs within protected areas on Australia's Great Barrier Reef.[b] The easiest way to reduce this risk is by protecting more reefs, but doing so costs money, both for protection itself and for compensating the fishing industry for lost revenue. Rather than setting a target for risk reduction, we looked at how the level of risk reduction was related to cost. We found that for only 2% more than the minimum amount required to meet conservation targets, risk could be substantially reduced (FIGURE 4.2.2). This is the sort of trade-off that the public might well be willing to accept, but that might be missed if conservation were driven entirely by targets.

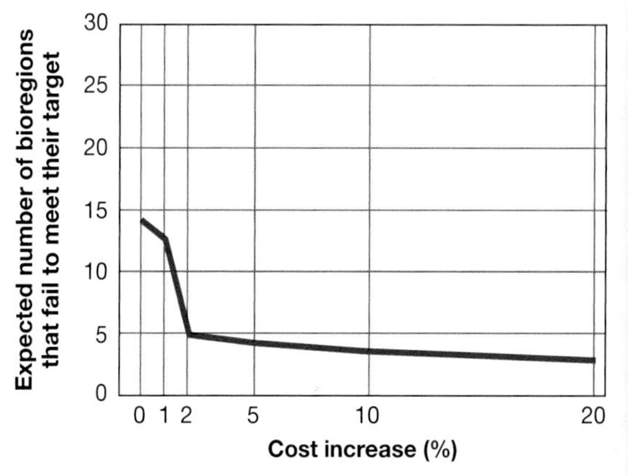


FIGURE 4.2.2 Effects of increasing the available budget on the risk that a reserve network on the Great Barrier Reef will fail to meet bioregional conservation targets as a result of catastrophic coral bleaching events. The 95% CI range is too small to be visible on the plot. (Data from Game et al., 2008.)

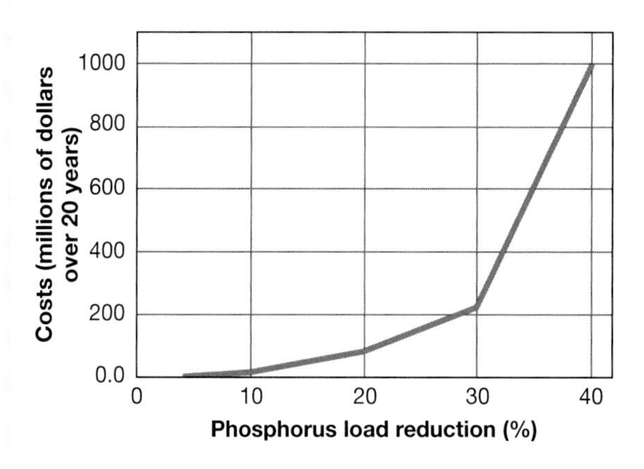


FIGURE 4.2.1 Trade-offs between phosphorus reduction and cost. (Data from Roberts et al., 2012.)

---

[a] Roberts, A. M. et al., *Agricultural land management strategies to reduce phosphorus loads in the Gippsland Lakes, Australia.* Agricultural Systems, 2012. **106**(1): 11–22.

[b] Game, E. T. et al., *Planning for persistence in marine reserves: A question of catastrophic importance.* Ecological Applications, 2008. **18**: 670–680.

## Where Do Targets Come From?

Targets specify the level of achievement we require or desire of our objectives. Targets should be expressed in terms of the indicators chosen for our objectives; otherwise, there is no way to account for expected outcomes or track progress toward our targets.

Where do targets come from? There is no single "right" way to establish conservation planning targets. In reality, most targets reflect a combination of both scientific and sociopolitical considerations,[8] a point well made by our colleague Tim Tear and others in a seminal article on the topic.[9] In fact, as the conservation biologist George Wilhere has pointed out eloquently,[10] it is erroneous and misleading to suggest that targets for conservation can be entirely based on objective scientific assessment. And yet such claims frequently grace the text of conservation plans and conservation planning studies.

Instead, targets are strongly, perhaps primarily, influenced by value judgments. To illustrate this, consider targets as reflecting the level of risk we are placing a species or habitat at: the higher the conservation target, the lower the risk. What is an acceptable level of risk to a species or habitat is a matter of values. Reliance on values to set targets is not unreasonable—in fact, it is inescapable—especially when targets are intended to speak to our objectives, which are themselves based on values (see the discussion of values in Chapter 1). It is critical, however, to acknowledge the influence of value judgments and distinguish them from scientific assessment. Unfortunately, conservation planning targets are often established in a way that precludes transparent interrogation of their appropriateness with respect to stakeholder values. In most nations' legal systems, value judgments do not need to be established for every individual assessment; some values can achieve a level of general support from society, and are translated into various forms of law. Conservation is no different.

### Convention on Biological Diversity

In Chapter 1 we introduced the Convention on Biological Diversity (CBD), which was established explicitly to promote the conservation and sustainable use of biological diversity and has influenced the discipline of conservation planning in a number of ways. It occupies the remarkable position of the most ratified international convention in history; as of early 2015, of the world's 198 recognized nations, only the United States and the Holy See are not parties to the convention.

Best known of the many targets set by the CBD are the 20 Aichi Biodiversity Targets (named after the city in Japan where they were agreed on) listed in the current CBD agreement. Readers of this book will recognize these "targets" as a combination of fundamental objectives, means, and targets as defined here—all prefaced with "By 2020" (or 2015 in some cases). The Aichi target that has been most useful to conservation planners in setting targets is Target 11, which reads:

> *By 2020, at least 17 per-cent of terrestrial and inland water areas, and 10 per-cent of coastal and marine areas, especially areas of particular importance for biodiversity and ecosystem services, are conserved through effectively and equitably managed, ecologically representative and well-connected systems of protected areas and other effective area-based conservation measures, and integrated into the wider landscapes and seascapes.*

Before 2010, the equivalent CBD targets were 10% protection of all terrestrial, inland, coastal, and marine areas. It is easy to see why these clear numerical targets have been enthusiastically used by conservation planners and in some cases, as in Australia and Costa Rica, have become a focus of national conservation policy (FIGURE 4.4).

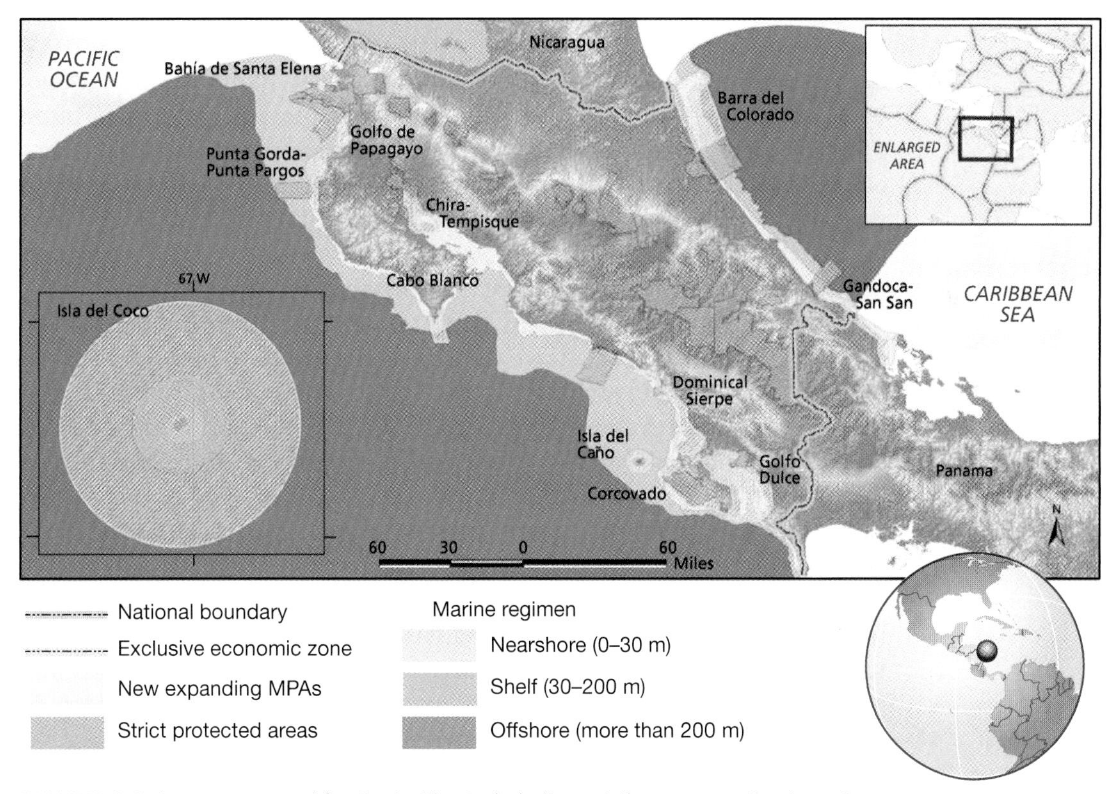


FIGURE 4.4 A map prepared for Costa Rica to help it meet the conservation targets established under the Convention on Biological Diversity. The combination of pink and yellow areas would result in a protected area network that would meet the CBD conservation targets.

From a pragmatic point of view, these CBD targets are a great starting point for conservation planning targets, especially for plans that involve establishing protected areas or other types of conservation areas; governments are nearly always receptive to CBD-based targets because they are in line with their obligations and because they leave plenty of room for interpretation. Often the CBD numbers are thought of as the default minimum targets for conservation planning exercises. The percentage targets are conventionally thought of as the percentage of each habitat type that should be protected, although "habitat type" can be defined as appropriate. We have used 10% of the area of all conservation features as a baseline in conservation plans for projects ranging from developing small, community-based conservation areas to designing protected area networks for entire countries.

### Historical Baselines and Naturalness

"One of the penalties of an ecological education is that one lives in a world of wounds." If this was true when Aldo Leopold wrote *A Sand County Almanac* in 1949,[11] it is even truer today. What does it mean to conserve 17% of a habitat's current distribution if this represents only a tiny fraction of what once existed? And how can that percentage be compared with a percentage of another habitat that has not been extensively cleared? To help level this playing field, it is often proposed that conservation planning targets should refer to baselines of historical distribution. As a simple example, this means that if we wanted to conserve 10% of a particular habitat, and we knew that 50% of that habitat had already been lost or cleared, the effective target would actually be 20% of the remaining habitat.

In many places, historical maps of habitat are available, often as a result of earlier aerial surveys or early attempts to map habitats or inventory public lands in order to determine future uses. In the United States, historical habitat data based on U.S. General Land Office original public land survey (PLS) records are widely available.[12] The historical distribution of a habitat is commonly referred to as the *pre-European* (though only outside of Europe) or *pre-clearing* habitat extent, reflecting a reference point before large-scale habitat conversion (FIGURE 4.5). However, the use of pre-European habitat extent as a baseline is being increasingly criticized as we learn more about the extent to which earlier human populations modified the landscapes around them. Even reference points that are more recent, but still historical, are useful baselines for conservation planning targets. Obviously, the extent to which historical baselines can be used in conservation planning is dictated by the available historical evidence. Establishing historical habitat distributions with the resolution required for conservation planning is rarely possible for periods earlier than 1930.

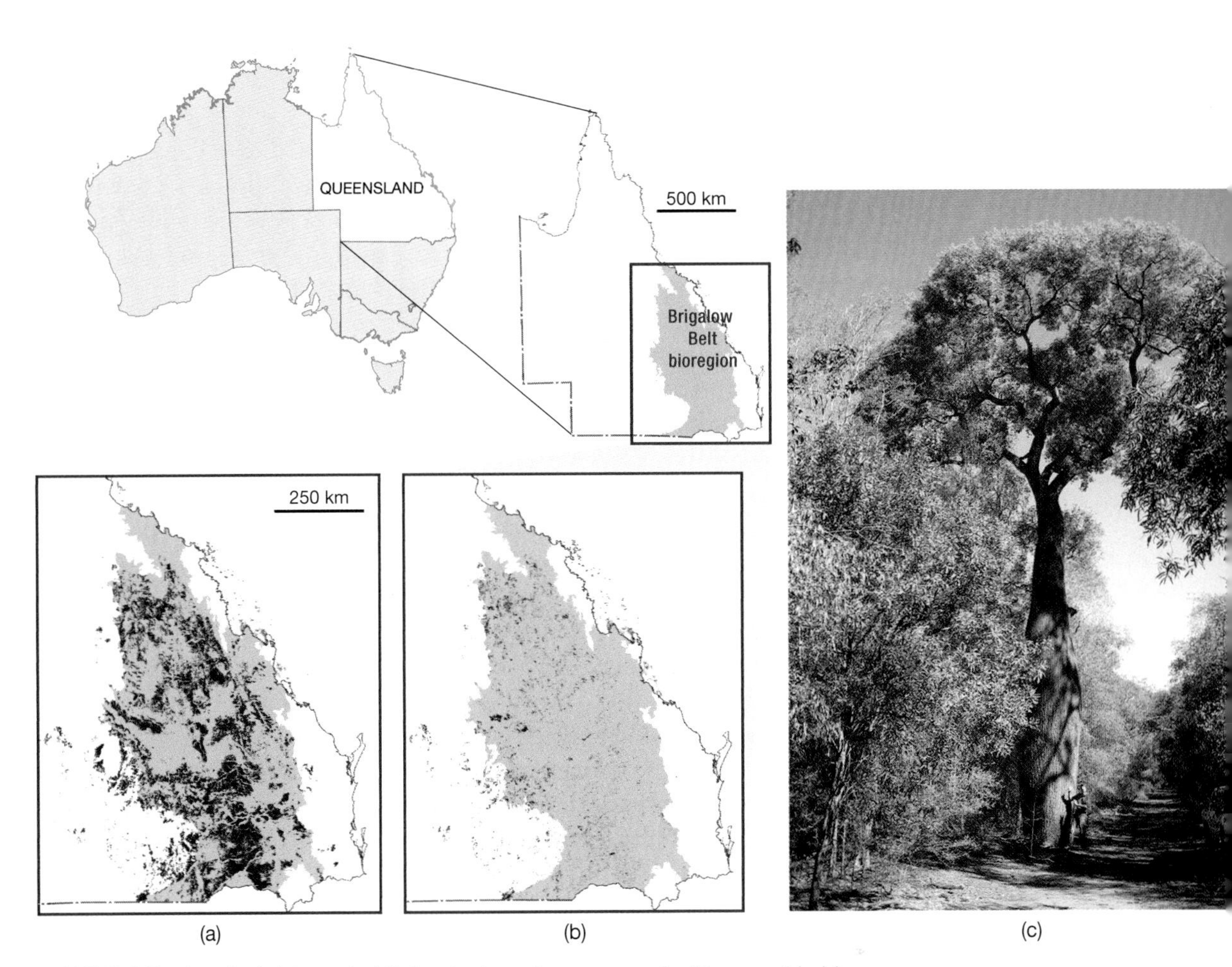


FIGURE 4.5 Historical data on habitat extent can be compared with current habitat distributions. (a) The pre-European extent of mature Brigalow ecosystems within Queensland, Australia. (b) The current extent of mature Brigalow ecosystems within Queensland. (Adapted from J. M. Dwyer et al., *Carbon for conservation: Assessing the potential for win-win investment in an extensive Australian regrowth ecosystem.* Agriculture, Ecosystems & Environment, 2009. **134**(1): 1–7.) (c) Brigalow ecosystem. Person is standing next to a Queensland bottle tree (Brachychiton rupestris), an iconic tree species found in Brigalow ecosystems.

Determining historical baselines for the abundance of species is an even more challenging task. Exception for some species whose harvest has been managed for a long time, especially commercial fisheries in Western nations, most estimates of historical abundance are purely anecdotal. This does not mean that they cannot be used to help establish targets, but it does make them especially prone to **sliding baselines**: the temptation to believe that memories of the near past reflect what natural abundance was, such that the perception of what is natural is progressively diminished.[13] For example, the Mexican conservation biologist Andrea Sáenz-Arroyo has clearly documented how sliding baselines influence current views of whether a population is "depleted."[14] In a study of three generations of fishermen in the Gulf of California, Sáenz-Arroyo and colleagues showed that, despite their catching 25 times fewer (as well as smaller) Gulf groupers (*Mycteroperca jordani*) than their grandfathers' generation, younger fishermen did not perceive the reefs to be as depleted as did older fishermen, who remember a far more productive system.

Historical baselines refer to more than the distribution of habitats or the abundance of species. They also refer to the structure, appearance, and functions of habitats and ecosystems. Conservation planning targets frequently make reference to maintaining or restoring an area or population to "natural" conditions. Scientists have made many efforts over the last century to define what *natural* means. For example, ecologist Jay Anderson, in a widely cited paper[15] developed while working in the Greater Yellowstone Ecosystem in the western United States, defined *naturalness* as having three components: the degree to which an ecosystem would change if humans were removed, the amount of cultural energy required to maintain an ecosystem in its current condition, and the suite of native species currently in the system compared with the species composition before human settlement.

However, as the conservation biologist Nigel Dudley has pointed out in a well-researched and well-reasoned book on the topic,[16] use of the terms *natural* or *naturalness* makes at least three assumptions that fail to hold up in many parts of the world today: (1) that there are places in the world without people, (2) that ecosystems will continue to function in the future as they have in the past, and (3) that it is possible to find baseline conditions by which we can measure changes in naturalness. Dudley advocates for a more contemporary view of naturalness based on today's rapidly changing world, in which most ecosystems have already been dramatically altered and finding baselines by which to measure the natural world away from human influence is increasingly difficult. He has used the term *authenticity* to denote this more contemporary view and has defined an authentic ecosystem as "a resilient ecosystem with the level of biodiversity and range of ecological interactions that can be predicted as a result of the combination of historic, geographic, and climatic conditions in a particular location."

Dudley suggests that the most important implication of focusing on authenticity, rather than naturalness, for conservation planning is the greater emphasis placed on managing for ecological process, a point that we have made earlier in this book. At face value, this means placing more importance on how an ecosystem is functioning and less importance on whether all the native components are in place. To be clear, Dudley does not dismiss the importance of native species, but takes a more relaxed view of judging an ecosystem's value for conservation based largely on some arbitrary measure of baseline species composition. He takes a similar view on invasive species, suggesting that we worry about them only when they significantly disrupt ecosystem function.

Leaving aside philosophy for the moment to take a pragmatic point of view, targets that reflect a desire to return ecosystems to their "natural" structure or function require some "natural" reference to measure contemporary ecosystems against. This need has led to the development of two related concepts: *historical range of variability* (HRV) and the notion of ecological departure. HRV has been proposed as a useful approach to establishing historical baselines for the purposes of conservation and management. The HRV concept has an associated methodology that has been well summarized by Robert Keane and colleagues at the U.S. Forest Service,[17] but its most important property is recognition that there is no single baseline condition for an ecosystem, but rather an often broad envelope.

Related to HRV is the idea of *ecological departure*, which is essentially a measure of how far the current ecosystem is from this notional natural baseline. As concepts, both HRV and ecological departure have been most prevalent in the conservation of terrestrial ecosystems, particularly for establishing targets related to the management of fire regimes. This prevalence might be partly explained by the finding that past fire regimes can often be established with reasonable confidence using signals such as charcoal deposits in the soil or tree ring counts. In many conservation plans, especially in developing nations for which data are less extensive than in the United States or Australia, we have adjusted target achievement by using simple proxies, such as logging history, as measures of ecological departure or degradation.

In a world where a changing climate is driving rapid ecosystem changes, however, and in which the vast majority of ecosystems are fragmented, the validity of HRV as a basis for conservation targets is increasingly questioned (**Box 4.3**). For example, where forests have been cleared and fragmented, a large fire might alter the entire ecosystem, whereas historically the forest may have been extensive enough to absorb a fire of the same magnitude. In such cases, there will probably be pressure to maintain prescribed burns, regardless of historical fidelity, to minimize the risk of dramatic fire impacts.

Ecological systems are naturally dynamic, and many of the properties that we might set targets for, such as the distribution of habitats or the abundance of species, change continually. However, because the pace of this dynamism has generally been slow relative to the temporal scales of conservation management, targets in conservation planning have typically been interpreted based on a static view of biodiversity. Climate change is accelerating the pace of change in many

## BOX 4.3 Moving Forward in an Anthropogenic World

There are significant philosophical and practical concerns about determining historical or natural baselines, given the pace of change that many ecosystems are experiencing in the Anthropocene. The environmental journalist and writer Emma Marris has written an account of conservation's obsession with historical and supposedly "natural" conditions.[a] Marris has suggested that conservation also needs to reflect current landscapes and psychology—perhaps today's baseline is as good as any. She asks whether we want zoos or functioning landscapes. Similar sentiments have been expressed by Hugh Safford and colleagues, who have suggested that we are likely "to have more management success focusing on function rather than form, process rather than pattern, and ecological resilience and integrity rather than historical fidelity."[b]

A rapidly changing climate is accelerating the pace of ecological changes in many ecosystems, further calling into question the utility of historical baseline conditions and the concept of HRV itself. In their book *Historical Environmental Variation* (2012),[c] the landscape ecologist John Wiens and colleagues suggest that there is now consensus that using past reference conditions as a blueprint for the future is inappropriate. Wiens and colleagues offer a set of recommendations to managers, restorationists, and conservation practitioners for the use of historical ecological information in a rapidly changing world. Among their most sage pieces of advice are the following: develop ideas of desired future conditions of ecosystems based on ranges of ecological structures, processes, and likely future trends; use information from both historical ecology and contemporary reference systems to understand the ecosystems of interest; appreciate that environmental conditions that are common on one part of a landscape shift geographically and topographically over time to other parts of a landscape and will continue to do so with climate change; and evaluate a range of possible future ecological conditions and different management practices that can be implemented in response to those varied conditions.

[a] Marris, E., *Rambunctious Garden: Saving Nature in a Post-Wild World.* 2011, New York: Bloomsbury.

[b] Safford, H. D. et al., *Historical ecology, climate change, and resource management: Can the past still inform the future?* in *Historical Environmental Variation in Conservation and Natural Resource Management,* J. A. Wiens et al. 2012, Hoboken, NJ: Wiley. Chapter 4.

[c] Wiens, J. A. et al., *Historical Environmental Variation in Conservation and Natural Resource Management.* 2012, Hoboken, NJ: Wiley.

ecological systems. For species and ecosystems that are being affected now, or in which models suggest there will be significant effects in the future, any effort to establish quantitative targets needs to carefully consider whether such targets will be realistic or possible under various climate change scenarios. We delve into this topic in more detail in Chapter 9.

Viewing ecological systems and species assemblages as dynamic is related to what we consider native versus non-native species and how these are treated in a conservation plan.[18] A currently topical example is the case of the dingo in Australia (FIGURE 4.6), which was

FIGURE 4.6 The dingo (*Canis lupus dingo*) is technically an introduced species in Australia but has been present there for at least 3500 years and is considered an integral part of Australian fauna, from both an ecological and cultural perspective.

introduced to the continent by humans. Dingoes are not native to Australia in the conventional sense, but having been present in Australia for at least 3500 years, they are now considered an integral part of Australian fauna, from both an ecological and cultural perspective.[19]

## Approaches to Setting Targets

In this section we cover some of the common approaches to establishing targets for use in conservation plans. Irrespective of the approach used to set targets, or whether the targets are for ecological or social objectives, three general tips are worth keeping in mind:

1. Avoid ambiguities—for example, the term *sustainable*. Make sure these terms are defined. This is the time for clarity.
2. State how the targets will be measured (using what indicators) and what the sources of these data will be.
3. Check to be sure whether achieving the targets would actually satisfy the objective they reflect.

### Species-Area Relationship

The **species-area relationship** (**SAR**) is about as close to a "law" as ecology gets. Put very simply, the SAR states that the larger the area, the more species will exist there. In a conservation context, this principle can be rephrased as follows: the more area you protect, the more species you protect. The relationship between species and area, however, is not linear (**FIGURE 4.7**). Initially, as you protect more area, you accumulate species faster than you accumulate area. As the amount of area protected grows, you accumulate new species at a progressively slower rate, until even large increases in the area protected add only a small number of new species.

Some of the early empirical work on the SAR suggested that as a general rule, it might be possible to "conserve" (read, "represent part of the distribution of") 50% of a region's species in as little as 10% of its area.[20] This observation almost certainly influenced the Convention on Biological Diversity's original 10% protection targets. Additional studies on the same topic suggest that this observation is a crude but generally accurate one, and that it applies to marine as well as terrestrial environments[21] (**TABLE 4.1**).

Conservation biologists like the SAR because it can be used to determine the proportion of a region's species likely to be retained in a given proportion of the region's habitat. This calculation most commonly takes the form

$$\log S' = z * \log A'$$

The value of $z$ reflects the rate at which species accumulate in a particular type of habitat. An average $z$-value of 0.3 is commonly cited; estimates are typically a little higher on islands, due to the number of local endemics, and a little lower on continents.

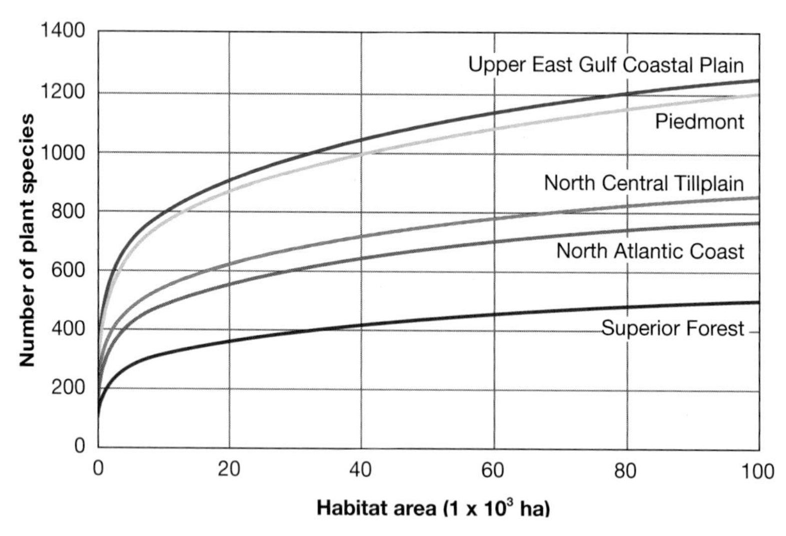


**FIGURE 4.7** Species-area curves for plants in five U.S. ecoregions. Note that the rate of accumulation of new species slows as the area grows larger. (Data from W. Murdoch et al., *Maximizing return on investment in conservation.* Biological Conservation, 2007. **139**(3-4): 375-388.)

**TABLE 4.1** Estimates of the percentage of species protected under a 10% conservation target in various marine habitats in two depth bands in the North Atlantic Ocean

| Depth band (m) | Substratum | *z*-value | 10% conservation target (% species protected) |
|---|---|---|---|
| 200–1100 | All data combined | 0.16 | 69 |
| | Bedrock | 0.39 | 41 |
| | Bedrock with carbonate veneer | 0.16 | 69 |
| | Gravel (biogenic; not coral) | 0.30 | 50 |
| | Gravel (boulders and cobbles) | 0.25 | 56 |
| | Gravel (coral rubble) | 0.22 | 60 |
| | Gravelly sand (pebbles) | 0.22 | 60 |
| | Mud | 0.39 | 41 |
| | Sand | 0.20 | 63 |
| | Sandy gravel (biogenic; not coral) | 0.40 | 40 |
| | Sandy gravel (pebbles and cobbles) | 0.25 | 56 |
| 1100–1800 | All data combined | 0.25 | 56 |
| | Bedrock | 0.27 | 54 |
| | Bedrock with carbonate veneer | 0.49 | 32 |
| | Gravel (biogenic; not coral) | 0.47 | 34 |
| | Gravel (boulders and cobbles) | 0.38 | 42 |
| | Gravel (coral rubble) | 0.32 | 48 |
| | Gravelly sand (pebbles) | 0.43 | 37 |
| | Mud | 0.93 | 12 |
| | Sand | 0.31 | 49 |
| | Sandy gravel (pebbles and cobbles) | 0.44 | 36 |

*Source:* Data from N. L. Foster, A. Foggo, and K. L. Howell, *Using species-area relationships to inform baseline conservation targets for the deep north east Atlantic.* PLoS One, 2013. **8**(3): e58941.

This same equation can easily be rearranged to work out the proportion of habitat needed to protect a desired proportion of species. This elegant property of the SAR suggests that it might sensibly be used to determine appropriate targets for the protection of different habitats. The South African ecologists Richard Cowling and Philip Desmet demonstrated how, with knowledge of plant species accumulation rates in different vegetation types, it was possible to establish habitat-specific conservation planning targets to represent 75% of the region's species[22] (**TABLE 4.2**). Of course, the 75% representation requirement is a social choice, but Cowling and Desmet were able to show exactly how much land would be required for different levels of representation, allowing the stakeholders to make an informed choice about representation with knowledge of the consequences of that choice.

Although calculating habitat-specific targets based on the SAR is attractive, it does require substantial data (extensive surveys of species diversity at many sites within each habitat) that are not commonly available, meaning that generalizations will inevitably need to be made. The level of sampling for species diversity has been shown to alter the resulting target calculations significantly.[23] It is also important to note that the SAR is influenced by habitat clearing and fragmentation (which essentially turns contiguous habitat into a series of islands) such that in highly transformed habitats, more area will be required to represent the same number of species than in untransformed habitats.

Another use of the SAR in conservation is predicting how many species are likely to be lost from a region as habitat is progressively cleared. If 10% of the area can represent 50% of the species, then it can be assumed that clearing 90% of a region's habitat would lead to loss of 50% of that region's species. The trouble is that this does not seem to be the case in practice; heavily cleared regions tend to retain most of their known species, even if they are restricted to tiny, isolated populations. The difference between the predicted species loss and the

**TABLE 4.2** Percentage of Succulent Karoo habitat type required to be protected in order to represent different proportions of plant species under the observed range of *z*-values

| | Target proportion of species | | | | |
|---|---|---|---|---|---|
| *z* | 0.5 | 0.6 | 0.7 | 0.8 | 0.9 |
| 0.1 | 1 | 1 | 3 | 11 | 35 |
| 0.125 | 1 | 2 | 6 | 17 | 43 |
| 0.15 | 1 | 3 | 9 | 23 | 50 |
| 0.175 | 2 | 5 | 13 | 28 | 55 |
| 0.2 | 3 | 8 | 17 | 33 | 59 |
| 0.225 | 5 | 10 | 20 | 37 | 63 |
| 0.25 | 6 | 13 | 24 | 41 | 66 |
| 0.3 | 10 | 18 | 30 | 48 | 70 |

*Source:* Data from P. Desmet and R. Cowling, *Using the species-area relationship to set baseline targets for conservation*. Ecology and Society, 2004. **9**(2).

observed loss has been accounted for by a concept known as **extinction debt**, which essentially assumes that small remaining populations of many species are just hanging on, but are not truly viable and will go extinct in the not too distant future. In 2011, the ecologists Fangliang He and Stephen Hubbell made a convincing argument that the SAR is not a good model for how extinctions happen, because we should really expect only species endemic to the lost areas to go extinct immediately.[24] They propose that when estimating expected species loss due to habitat clearing, an endemics-area relationship (EAR) should be used instead.

Despite its acknowledged weaknesses, the SAR equation is a useful one for conservation planners to be familiar with. It provides an important link between biodiversity and the targets for coarse-filter habitat-type conservation features, and it is also commonly used in return-on-investment prioritization methods (introduced in Chapter 7). Perhaps the most important lesson from the debate around the SAR is that caution is required when talking about what targets such as 10% of a habitat's area are actually achieving with regard to biodiversity conservation.

### Population Viability Analyses

**Population viability analyses** (**PVAs**) are models that project the fate of populations by considering changes in abundance as a result of demographic and environmental influences. They essentially provide a quantitative assessment of extinction risk, usually represented by a probability and a time frame. Green turtles (*Chelonia mydas*), for example, at the current levels of harvest, have a 15% chance of going extinct in the Indo-Pacific in the next 100 years[25] (FIGURE 4.8).

For many people, species extinction gets at the heart of conservation, and avoiding species extinctions is commonly an explicit or implicit objective of conservation initiatives. For this reason, PVA has long been a popular tool in the conservation toolkit. Because extinction risk is correlated with population size, it is intuitively attractive to use PVA to help set conservation targets by establishing a **minimum viable population** (**MVP**) for the species in question: the size at which survival of a population could be considered reasonably certain. Population viability is a subjective and rather arbitrary concept, but academic convention generally interprets viability as at least 90% certainty of persistence for at least 100 years.

The use of PVA to set targets for conservation planning has been the source of strong debate. Critics of its use point out the wide intervals and high uncertainty attached to predictions for most species, as well as to the fact that PVAs are often conducted with poor data and produce estimates that, however desirable, might be unrealistic

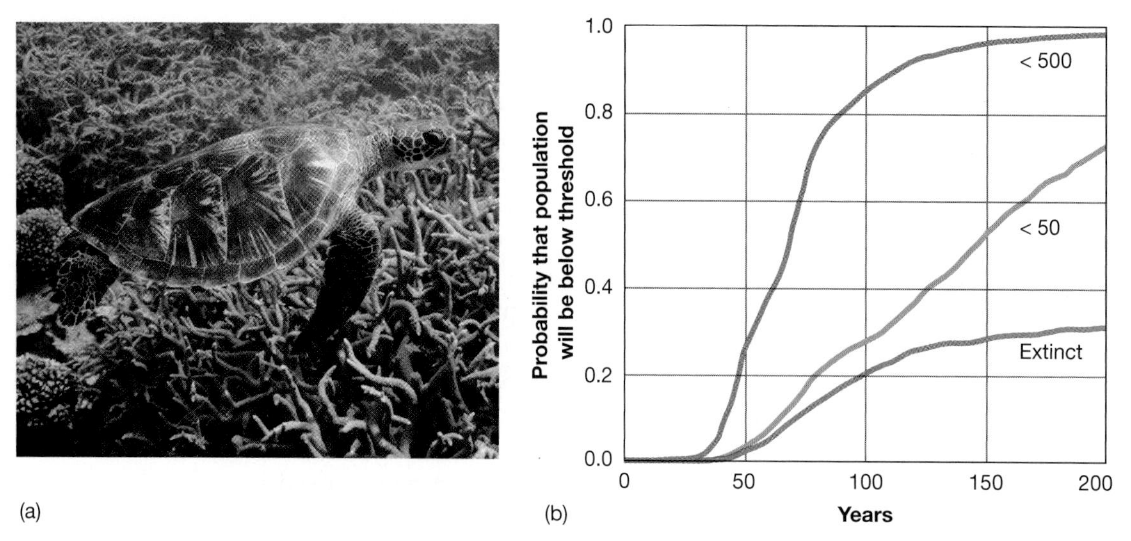


FIGURE 4.8 Population viability analysis. (a) Green turtle (*Chelonia mydas*). (b) Results of a population viability analysis of a genetically discrete green turtle population in the Indo-Pacific. The graph shows the probability of the population falling below thresholds of 500 individuals, 50 individuals, and zero individuals (extinct) over time under a certain set of management actions. (Data from K. E. M. Dethmers and P. W. J. Baxter, *Extinction risk analysis of exploited green turtle stocks in the Indo-Pacific.* Animal Conservation, 2011. **14**(2): 140–150.)

for management purposes. Certainly, assembling the data needed to conduct a PVA is either impossible or impractical for many species in many places. In response, there has been an active search for general rules that might be used instead, with different groups advocating for MVP sizes ranging from 500 to 5000 individuals.[26] While the scientific community does not agree on whether there is any validity to generalizations such as 5000 individuals[27] (small populations of island endemics are an obvious exception), it is clear that small populations have a high risk of extinction in the medium to long term. A great deal of caution is warranted in using the term *viability* in the context of objectives, targets, and outcomes of conservation planning efforts.

The use of PVA and MVP to set conservation targets can be problematic in other ways as well. Although 100 years is a short time in the history of a population, it is a long time for policy makers, who generally consider time frames of 5–25 years. Over these shorter time frames, extinction risks for most populations tend to be low enough as to not be compelling. If you presented the case that under current management a population had a 5% chance of going extinct in 50 years, most policy makers would be tempted to take this bet rather than increasing a conservation target, which would require more money. The use of MVP as targets has also been criticized for contributing to the creation of *conservation-reliant species*: species that require ongoing conservation investment in order to remain viable.[28]

In many situations, establishing sensible targets for conservation planning requires more than choosing a population size to aim for. The total area a species occupies, and how it is distributed within the region, are also important in determining its conservation requirements. These properties can be captured in a sophisticated, spatially explicit PVA.[29] Predictions of species persistence can be greatly improved by integrating spatial PVA with spatial conservation planning. This integration is eminently feasible from a technical planning point of view—see, for example, the work of Emily Nicholson and colleagues[30]—but it requires such a large amount of high-quality data that few planning exercises can seriously consider it. Closely related to the concept of spatially explicit PVA is that of *minimum area requirements* (MARs), or the amount of habitat that is required for long-term persistence of a population. MARs are often based on PVAs and are at times used in conservation planning to help set targets.[31]

Our somewhat negative view of the role of PVA in setting conservation planning targets should be tempered by its useful role in another area of conservation planning: choosing among strategies and actions. PVA has proved valuable for comparing the expected results of different potential strategies in a biologically relevant way during conservation planning. For example, Josie Carwardine and colleagues[32] have described the use of PVA to compare different strategies to conserve vertebrate biodiversity in the Kimberly region of Australia. When PVA is used in this way, some of the problems listed earlier can be ameliorated by testing the robustness of decisions to the uncertainty that is present in PVA estimates. One downside of this approach is that the uncertainty or noise present in PVA estimates can make them challenging to use as an indicator to evaluate the performance of strategies once they have been implemented. The process of developing a PVA can in itself lead to many useful insights about drivers of population change that can be helpful in developing conservation strategies and defining data collection and research priorities. Model building in general is also a great way to identify and reconcile the ambiguities and varying interpretations that commonly exist in conservation targets.

### Precautionary Principle

Despite a huge amount of study, there is an enormous amount we do not know about ecological systems and their conservation. Ecological systems are by any definition *complex systems*,[33] a term generally used to refer to systems with a vast number of interacting elements (think species, environment, people, geography, management, money, legislation, etc.) with no central control and many nonlinear relationships (which essentially means that seemingly minor alterations can lead to disproportionate effects). Conservation planning requires working within both complex ecological systems and complex social systems—or extremely complex social-ecological systems. The practical consequence is that the outcome of any intervention (whether conservation or threat) is, to a nontrivial extent, unpredictable. Clear scientific evidence is not always (in fact rarely) available.

The relationship between targets and the inherent uncertainty of social-ecological systems has long concerned conservation biologists; this concern stems from the awkward relationship between hypothesis testing (which defines the training of so many conservation biologists with an ecological background) and the need to make management decisions. For example, stating confidently (in a statistical sense) how much habitat is needed to sustain a community or population would require such an enormous amount of data as to frighten even the most optimistic graduate student.

Perhaps in response to the unavailability of certainty, conservationists often appeal to the **precautionary principle** or *precautionary approach*, especially in the domain of targets. In essence, the precautionary approach simply means not allowing scientific uncertainty to prevent conservation action. The widespread use of the precautionary principle was sanctioned by its appearance at the Rio Earth Summit in 1992. Since then it has made its way into the CBD, the United Nations Framework Convention on Climate Change (UNFCCC), and numerous national pieces of conservation legislation. Australia has relied on it very publicly to support conservation targets that require a significant increase in marine protected areas. The policy makers who engineered the expansion of the no-take areas within the Great Barrier Reef Marine Park, one of the seminal pieces of conservation planning action, unashamedly acknowledged that lack of scientific support for its conservation targets would not be used as justification to stall conservation action.[34] This approach is laudable, and the ability of this project to proceed without perfect knowledge serves to underscore clearly the role of targets.

Appealing to the precautionary principle comes with responsibility, however. It does not negate the necessity to assess strategies, actions, and the trade-offs attached to them. The original wording from the Rio Earth Summit that refers to the precautionary principle states that "lack of full scientific certainty shall not be used as a reason for postponing *cost-effective measures* to prevent environmental degradation" [italics ours]. *Cost-effectiveness* is a relative quality and thus presumes some assessment of alternatives, which might be associated with varying degrees of uncertainty. Rigidly applied, the precautionary principle also risks being inequitable and unrealistic; pragmatism is required. Conservation is not immune to the conservatism demanded by the precautionary principle; innovative conservation mechanisms risk being rejected in favor of conventional strategies, and we tend to overweight the status quo.

The discourse on appropriate use of the precautionary principle is substantial.[35] For the purposes of the conservation planner, however, the key points are these:

- Do not let scientific uncertainty around targets hinder a planning process.
- Do not be too rigid about targets; always assess trade-offs and compare options.
- Targets can evolve as we gather new information that informs them.

### Stakeholder-Established Targets

All "science"-based targets are, in theory, attempts to support objectives that are believed to belong to stakeholders in the broadest sense. A desire to sustain the diversity of habitats and species in a region is born out of a belief that (1) this diversity represents our collective natural heritage and therefore enriches our lives, (2) this diversity sustains a natural system that, when functioning in the way we believe it is intended to, improves our lives, or (3) we have a moral obligation to all species. These statements require interpreting *stakeholders* in the broadest sense possible—almost a global sense—to the point where it is hard to identify and point to one of *these* stakeholders, let alone involve them in a planning process. Conservation and natural resource professionals often act in what they believe to be the broader public's interest (although conservation has also been accused of being a "special interest" cause). Public conservation legislation and its accompanying regulations and policies, which often include targets, do the same.

In many conservation planning exercises, however, a relatively local contingent of stakeholders is likely to value some elements of nature, and some of the services it provides, more than others. Just think of how much conservation has been catalyzed by local concern about the imminent or potential loss of some local element of biodiversity that appears particularly important to local people. If you think back to the purpose of targets in a conservation plan, there are good reasons for local stakeholders to be heavily involved in setting these targets—none more so than the importance of community support, and of a plan that addresses the things communities care about, for the successful implementation of conservation actions.[36] Stakeholders with a strong interest in particular pieces of biodiversity or other landscape properties do not usually prevent conservation scientists and practitioners from using science to help inform targets. It is our experience that people are generally supportive of science informing them about what is required to achieve what they want.

The following approach to establishing targets with local stakeholders has worked well for us:

1. Develop an understanding of the features in the landscape or seascape that stakeholders particularly care about (refer to the sections on fundamental objectives and conservation features in Chapter 3).
2. Map these features using available data, then get local stakeholders to check these maps and supplement them as they see fit (update, improve accuracy, provide information about conditions, etc.).
3. Ask the stakeholders whether it would be acceptable to lose 50% of the feature they care about. If they say yes, adjust the percentage upward until they feel uncomfortable with the loss; if they say no, adjust percentage downward until they are comfortable.

As an example, one of us used this approach in a planning exercise with communities in the Solomon Islands, which resulted in establishing targets ranging from 50% to 95% representation for features considered particularly important for local communities: turtle nesting areas, fish spawning areas, habitat for megapodes whose eggs they collect (FIGURE 4.9), and ironwood trees used for making canoes and paddles. All other targets in the planning process were in line with the CBD targets,[37] so the final set of targets represented a

(a)

(b)

**FIGURE 4.9**
Some conservation features are particularly important to local communities. (a) A Melanesian megapode (*Megapodius eremita*). (b) This bird is an important conservation feature for local stakeholders in the Solomon Islands who collect its eggs, which are a highly prized food source.

combination of what was believed to be in the "broader" public interest and what was in local interests.

It will not always be the case that a conservation planning process can call on community stakeholders who are interested in establishing targets for conservation features. Indeed, in many cases local stakeholders will be actively opposed to the establishment of conservation targets. However, they might have an interest in targets that relate to ecosystem service objectives, particularly provisioning services such as the ability to harvest particular natural resources. These interests, which represent equally important and useful community-based targets, are discussed further in "Targets for Economic and Social Objectives."

### Threshold-Based Targets

Regulatory agencies and some conservation organizations (such as the Environmental Law Institute[38]) have long called for use of the **ecological thresholds** concept in conservation planning. One view of ecological thresholds defines them on the basis of anthropogenic effects on ecological features; for example, "the point at which a non-linear or substantive change in the dynamics or distribution of an individual organism, population, or community is observed relative to some level of (human) disturbance."[39] Another interpretation of ecological thresholds is the size and extent of habitat (for example, riparian buffers or habitat patches) needed to sustain a population of a species.[40]

Although such thresholds have found some use in agency land management plans, Canadian scientist Chris Johnson suggests that their use in conservation planning and actions has numerous limitations. Some of the most important of these are the lack of a clear and consistent definition of a threshold, the lack of consistent methods for identifying thresholds, the inability to generalize thresholds across species or ecosystems, and the paucity of examples of how thresholds have been effectively used in a planning or regulatory environment. Despite these rather serious limitations, at least some conservation scientists have suggested that identifying and ultimately employing thresholds in conservation decision making is one direction that future conservation efforts need to take.[41]

## Targets for Social and Economic Objectives

We do not expect that conservation planners will often be responsible for developing targets for social and economic objectives. We do, however, expect that most conservation planners will, in the course of their jobs, be involved in situations in which stakeholders or partners have important social or economic objectives that need to be considered when planning conservation actions. This section is a brief introduction to the sorts of social and economic targets that conservation planners are likely to encounter. Some partners or stakeholders may never even have thought about targets in the context of how conservation might affect their interests, in which case conservation planners might find themselves guiding on this issue.

### Targets for Economic Activities

Multiple-objective spatial planning frequently involves setting targets for economic activities potentially affected by changes in where different activities will be allowed to occur. The experience of marine spatial planning in European nations is a good example. In addition to targets for biodiversity conservation, European marine spatial plans have involved targets that reflect the spatial needs of activities including shipping, cables and pipelines, offshore wind farm installations, and fishing[42] (FIGURE 4.10). These targets reflected not only present uses and economic viability but also future projections of spatial needs under different scenarios of economic growth. In Belgium, for example, it was calculated that meeting the national target for renewable energy generation would require 100 km of offshore wind farms; this became a target in Belgium's marine spatial plan.

Establishing targets that allow for the economic viability of natural resource sectors has been commonplace for many natural resource management agencies, some of which are legally required to plan for multiple objectives. For example, the U.S. Forest Service's plan for the Tongass National Forest in Alaska included the objective of allowing for a sustained, economically efficient timber harvest, with a corresponding target to "provide an economic timber supply sufficient to meet the annual market demand for Tongass National Forest timber, and the market demand for the planning cycle, up to a ceiling of this Plan's allowable sale quantity, which is 2.67 billion board feet in the first decade."[43] *Economic timber supply* was clearly defined in the plan as the amount of timber at which "the average purchaser can meet all contractual obligations, harvest and transport the timber to the purchaser's site, and have a reasonable certainty of realizing a profit from the sale." Economic objectives such as this one were balanced with ecological objectives such as "maintain the eligibility of the total miles of river for the following recommended classifications: Wild 359.5 miles; Scenic 87.5 miles; Recreational 89.0 miles."

There is a belief that providing industries and stakeholders with the security of establishing their own "how much is enough" targets

FIGURE 4.10 Marine spatial plan for the German portion of the North Sea. As well as nature conservation, the plan explicitly allocates areas for shipping, pipelines and cables, wind energy development, and military use.

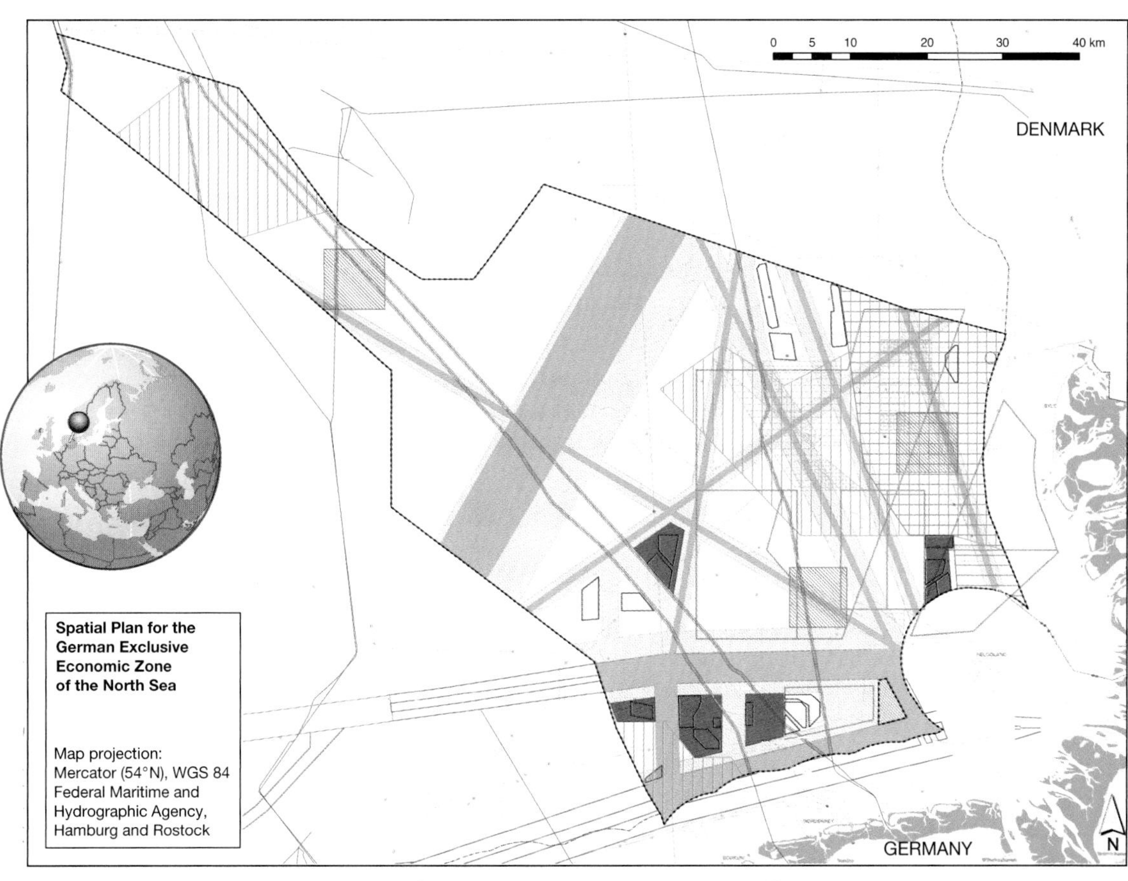


**Shipping**

Priority Area Shipping

Reservation Area Shipping

**Pipelines**

Priority Area Pipelines

Reservation Area Pipelines

**Submarine Cables**

Gate (Details A–C)

**Research**

Reservation Area Research

**Energy**

Priority Areas for Offshore Wind Energy

**Shipping**

Traffic Separation Scheme

Precautionary Area

Deep Water Road

Anchoring Area

**Exploitation of Natural Resources**

Sand / Gravel - Plan approved

Sand / Gravel - Project in Approval Procedure

Sand / Gravel - License

Natural Gas - License

**Pipelines**

Natural Gas (in use)

**Submarine Cables**

High Voltage Cable (in use)

High Voltage Cable (approved)

Data Cable (in use)

Data Cable (out of use)

**Energy**

Offshore Wind Farms appr. by 2009

Reference Area Offshore Wind Energy

**Nature Conservation**

Natura 2000 - SCI Habitat Directive

Natura 2000 - SPA Bird Directive

**Military Use**

Military Exercise Area

**Miscellaneous**

Previous Ammunition Area

Measurement / Converter Platform

**Boundaries**

Continental Shelf/EEZ

Territorial Sea/12 nm Zone

International Border

**Planning Area**

Boundaries of Planning Area

will increase their willingness to engage constructively in joint planning exercises. It allows planning exercises to look for solutions that genuinely meet economic needs while minimizing the impact on the environment. For example, some of our colleagues are working with an industry partner in Brazil that is looking to develop a sugarcane-to-ethanol plant. There is clear knowledge of how much sugarcane production is required for the plant to be profitable, and the conservation plan is being developed to see how this target could be met in a way that delivers the best possible outcome for natural ecosystems.

In practice, however, the stakeholder groups likely to be involved in these sorts of plans—with interests in fisheries, forestry, agriculture, tourism, mining, and so forth—can be reluctant to state some minimum economic goal, in case this is all they get. Their concern is understandable. Both industry and governments, which are often ultimately responsible for economic development and decisions about land and sea use, are likely to want to see the trade-offs between conservation gains and the impacts on the industry (for example, employment) for a range of scenarios. To use the European marine spatial planning example again, in the Belgian portion of the North Sea, planners developed a series of six alternative spatial designs that emphasized different values (**FIGURE 4.11**). None of these designs was to be selected in its entirety; instead, they offered a way to explore the trade-offs between different activities.

A consistent exception to this reluctance to set minimum targets might be industries not yet developed in a region, but projected to require a certain amount of resources to be viable—for example, new wind energy or even mining developments in a region. The targets of interest to these industries might look like minimum amounts of production that are required in an area. For example, one of us has been involved in a plan for a mining development in Isabel Province, Solomon Islands, that required a certain volume of ore to make its processing financially viable, which in turn limited the company's flexibility in negotiating with communities about conservation concerns. The extent of the profit margin involved in these calculations was not revealed, and the amount needed for financial viability could not easily be verified independently. These are real challenges that clearly illustrate the amount of good faith required for collaborative planning exercises.

Economic targets might also relate to the viability of activities undertaken as part of a conservation initiative. An example is the level of income required by a tourism concession for it to benefit the local community or be financially sustainable. Economic targets around the financial sustainability of conservation initiatives are increasingly common. *Sustainability* can be a fairly ambiguous term, but in an exercise we were involved in, it was clearly defined as the "operating budget guaranteed for at least the subsequent 10 years."

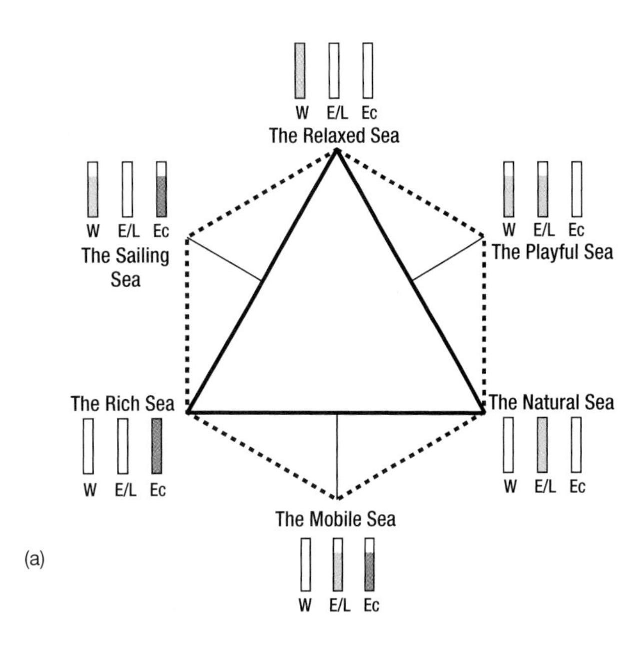


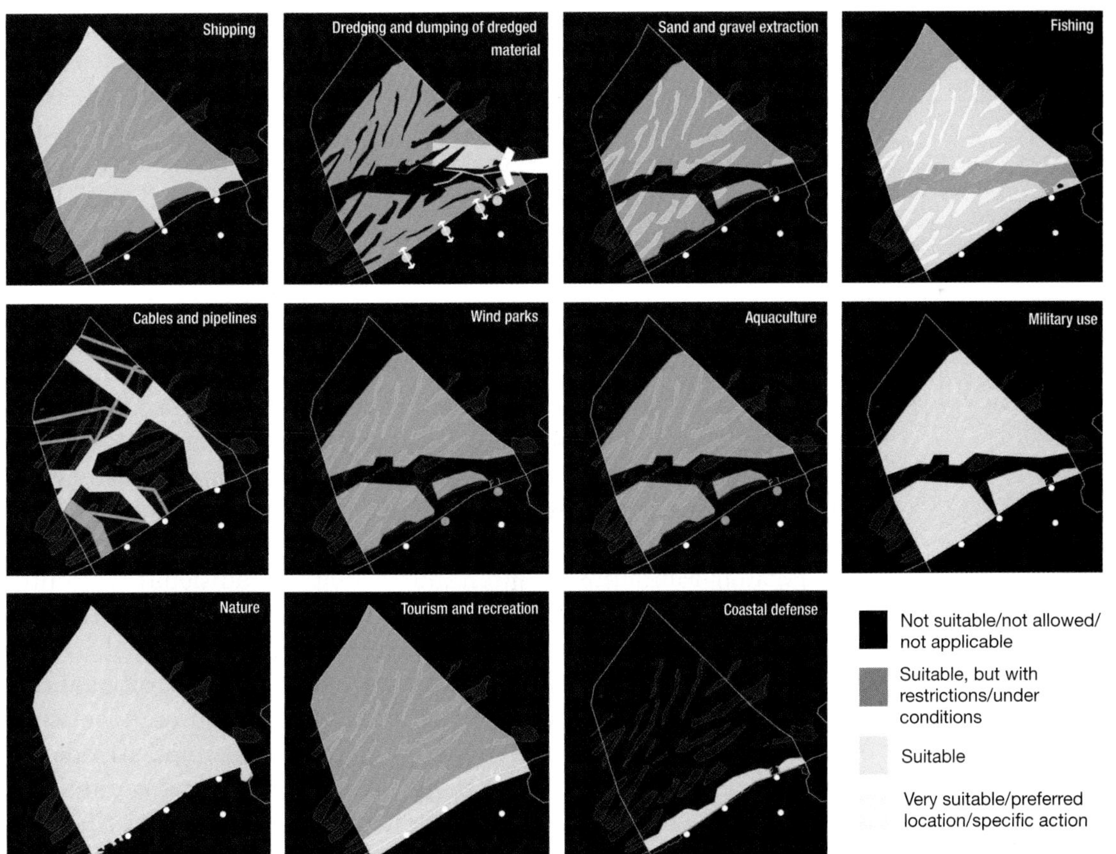


FIGURE 4.11 Six different visions for the Belgian portion of the North Sea, each emphasizing a different set of values. (a) Illustration of the extent to which each of the six scenarios deliver on three different fundamental objectives: well-being (W); ecology and landscape (E/L), and economic (Ec). (b) Maps showing what one of these six scenarios (the Mobile Sea) would look like for key marine activities. The blue shading indicates the level of suitability for the activity listed, and the light gray areas in the background represent underlying bathymetric features. These illustrations were used as a way to explore the trade-offs between different activities. (Adapted from F. Maes et al., *A flood of space: Towards a spatial structure plan for sustainable management of the North Sea*. Belgian Science Policy, Brussels, 2005.)

### Social Targets

For many conservation planners, working with and establishing targets for social objectives might seem like an unfamiliar task. Our experience, however, is that most conservation planners have encountered social targets in some form or another—after all, it could be argued that most environmental targets are social targets of a sort. Nevertheless, establishing appropriate targets for social objectives is likely to require involvement by someone with specialized expertise in this area.

It is important to emphasize again that social targets should be considered only for objectives that are relevant to the decisions at hand. As we saw in Chapter 3, relevant objectives are those that have the potential to be influenced by decisions made as a result of implementing the conservation plan. As you can easily imagine, we generally want the things that constitute social objectives, such as access to food, clean water, and safe homes, to get better continuously. This goal of continuous improvement is known as the principle of *progressive realization*.[44] Nevertheless, specific targets for these sorts of objectives can still be useful not just because of their role in planning but also politically, to highlight obligations to provide, at the least, minimum essential levels of human well-being or to make sure progress occurs at a reasonable rate.

In developing or disadvantaged communities, targets for social objectives might be based on bridging the gap between conditions experienced in those communities and the expectation in the wider national or even international communities. Frequently, governments will have established targets (sometimes by adopting international treaties) for major goals such as poverty reduction or gender equality. For example, many nations set explicit targets for different aspects of poverty reduction under an International Monetary Fund (IMF) and World Bank scheme known as Poverty Reduction Strategy Papers (PRSP). Countries as different as Benin and Nepal have established specific quantitative targets for things such as access to clean drinking water and primary school enrollment, which may be relevant to conservation plans, and if so, can serve as useful starting points for social targets.

Conservation projects with social objectives tend to be focused on particular communities, often indigenous or traditional communities. There is a risk that national-scale targets for human well-being could actually have negative consequences in the context of indigenous measures of well-being (for example, carrying on traditional cultural ways).[45] It has been suggested that focusing on gaps between indigenous and mainstream majority populations implicitly downplays the significance of unique indigenous priorities and worldviews. Regardless of where they come from, it is clearly important that targets reflect local conceptions of well-being.

Targets for social objectives related to the health of or access to natural environments can also rely heavily on comparisons. In contrast to some aspects of human well-being (such as security), there are generally few established "rights" in regard to the environment, so targets commonly draw on issues of equality and environmental justice. The obvious exception is where environmental issues overlap with human health and safety. Occasionally, there will be existing legally mandated targets or requirements that are relevant to conservation plans, such as limiting exposure to certain risks—either risks arising from the environment (such as reducing the risk of fire damage in Australia or in forest-urban interfaces in the United States by clearing vegetation near homes and creating firebreaks) or risks buffered by the environment (such as recognizing the role of vegetation in helping cities meet their targets for air quality). These sorts of targets can be particularly relevant for urban and landscape planning.

TABLE 4.3 illustrates some social indicators and associated targets that were established for a forest conservation project in Berau, Indonesia. These targets are aligned with some of the domains of human well-being we introduced in Chapter 3. It should be evident from this brief section that intuitive and useful targets can be established for social objectives in much the same way as they are for ecological objectives: by drawing on a combination of stakeholder input and relevant policy frameworks.

**TABLE 4.3** Example of social indicators and targets established for a forest conservation and carbon mitigation project in Indonesia

| Human well-being domain | Intermediate objective | Indicator and target |
|---|---|---|
| Security | Increase formal management rights for local communities | Increase of forest area (in ha) under formal management by communities (target: 5% increase or around 20,000 ha by 2015) |
| Opportunity | Increase income and livelihood options | Percent of households in participating villages that have increased income and livelihood options (target: at least 50% by 2015) |
| | Improve basic services | Percent of households in participating villages that have increased access to basic services attributable to improved forest management practices (target: at least 50% by 2015) |
| Empowerment | Increase the ability of community and village government to mobilize financial and human resources | Percent increase in financial resources mobilized by participating villages (target: 30% increase by 2015) |
| | Increase the capability of local institutions to manage financial resources | Local institutions in participating villages capable of financial and natural resource management (target: at least 50% of local institutions in participating villages have good financial management systems and natural resource management plans by 2015) |

## KEY MESSAGES

- To track outcomes and evaluate the consequences of alternative actions, it is necessary to establish measurable indicators. Good indicators are representative of an objective's intention and are, in a practical sense, measurable.
- Indicators can be measured in natural, proxy, or constructed scales. Constructed scales are useful for indicators that are hard to measure, but they also introduce more subjectivity than natural scales. Our preference is for using natural or proxy scales where possible.
- Targets are quantitative expressions of the level of achievement we are seeking for an objective. They can be achieved or not.
- Targets often form the backbone of national and international conservation policy, but they are also the subject of much debate and criticism. A lot of criticism of targets stems from a misunderstanding about the practical role targets can play in managing the balancing act between different objectives.
- Targets are never objective statements of science; they are always influenced by value judgments and risk tolerances. Targets typically reflect a combination of scientific and sociopolitical considerations. Numerous approaches are available to help support decisions about appropriate targets.
- In addition to ecological targets, a range of social and economic targets will often be relevant to a conservation plan. We encourage conservation planners to ensure that people with expertise in these areas are involved in the planning process.

## References

1. Failing, L., and R. Gregory, *Ten common mistakes in designing biodiversity indicators for forest policy.* Journal of Environmental Management, 2003. **68**: 121–132.
2. Karr, J. R., *Biological integrity: A long-neglected aspect of water resource management.* Ecological Applications, 1991. **1**(1): 66–84.
3. Kahneman, D., and S. Frederick, *Representativeness revisited: Attribute substitution in intuitive judgment*, in *Heuristics and Biases: The Psychology of Intuitive Judgment.* 2002, Cambridge: Cambridge University Press. 49–81.
4. Svancara, L. K. et al., *Policy-driven versus evidence-based conservation: A review of political targets and biological needs.* BioScience, 2005. **55**(11): 989–995.

5. Carwardine, J. et al., *Hitting the target and missing the point: Target-based conservation planning in context.* Conservation Letters, 2009. **2**(1): 4–11.
6. McShane, T. O. et al., *Hard choices: Making trade-offs between biodiversity conservation and human well-being.* Biological Conservation, 2011. **144**(3): 966–972.
7. Moilanen, A., H. Kujala, and J. Leathwick, *The Zonation framework and software for conservation prioritization*, in *Spatial Conservation Prioritization*, A. Moilanen, K. A. Wilson, and H. P. Possingham, editors. 2009, Oxford: Oxford University Press. 196–210.
8. Svancara et al., *Policy-driven versus evidence-based conservation.* (See reference 4.)
9. Tear, T. H. et al., *How much is enough? The recurrent problem of setting measurable objectives in conservation.* BioScience, 2005. **55**(10): 835–849.
10. Wilhere, G. F., *The how-much-is-enough myth.* Conservation Biology, 2008. **22**(3): 514–517.
11. Leopold, A., *A Sand County Almanac, and Sketches Here and There*, illus. by Charles W. Schwartz. 1950, New York: Oxford University Press.
12. Schulte, L. A., and D. J. Mladenoff, *The original US public land survey records: Their use and limitations in reconstructing presettlement vegetation.* Journal of Forestry, 2001. **99**(10): 5–10.
13. Pauly, D., *Anecdotes and the shifting baseline syndrome of fisheries.* Trends in Ecology & Evolution, 1995. **10**(10): 430.
14. Sáenz-Arroyo, A. et al., *Rapidly shifting environmental baselines among fishers of the Gulf of California.* Proceedings of the Royal Society B: Biological Sciences, 2005. **272**(1575): 1957–1962.
15. Anderson, J. E., *A conceptual framework for evaluating and quantifying naturalness.* Conservation Biology, 1991. **5**(3): 347–352.
16. Dudley, N., *Authenticity in Nature.* 2012, London: Earthscan. 244.
17. Keane, R. E. et al., *The use of historical range and variability (HRV) in landscape management.* Forest Ecology and Management, 2009. **258**(7): 1025–1037.
18. Schlaepfer, M. A., D. F. Sax, and J. D. Olden, *The Potential Conservation Value of Non-Native Species.* Conservation Biology, 2011. **25**(3): 428–437; and Vitule, J. R. S. et al., *Revisiting the potential conservation value of non-native species.* Conservation Biology, 2012. **26**(6): 1153–1155.
19. Letnic, M. et al., *Demonising the dingo: How much wild dogma is enough?* Current Zoology, 2011. **57**(5): 668–670.
20. Diamond, J. M., and R. May, *Island biogeography and the design of natural reserves*, in *Theoretical Ecology: Principles and Applications*, R. May, editor. 1976, Oxford: Blackwell Scientific Publications. 163–186.
21. Foster, N. L., A. Foggo, and K. L. Howell, *Using species-area relationships to inform baseline conservation targets for the deep north east Atlantic.* PLoS One, 2013. **8**(3) e58941.

22. Desmet, P., and R. Cowling, *Using the species-area relationship to set baseline targets for conservation.* Ecology and Society, 2004. **9**(2).
23. Metcalfe, K. et al., *Impacts of data quality on the setting of conservation planning targets using the species-area relationship.* Diversity and Distributions, 2013. **19**(1): 1–13.
24. He, F. L., and S. P. Hubbell, *Species-area relationships always overestimate extinction rates from habitat loss.* Nature, 2011. **473**(7347): 368–371.
25. Dethmers, K. E. M., and P. W. J. Baxter, *Extinction risk analysis of exploited green turtle stocks in the Indo-Pacific.* Animal Conservation, 2011. **14**(2): 140–150.
26. Traill, L. W. et al., *Pragmatic population viability targets in a rapidly changing world.* Biological Conservation, 2010. **143**(1): 28–34.
27. Flather, C. H. et al., *Minimum viable populations: Is there a "magic number" for conservation practitioners?* Trends in Ecology & Evolution, 2011. **26**(6): 307–316.
28. Goble, D. D. et al., *Conservation-reliant species.* BioScience, 2012. **62**(10): 869–873.
29. McCarthy, M. A., *Spatial population viability analysis*, in *Spatial Conservation Prioritization: Quantitative Methods and Computational Tools*, A. Moilanen, K. Wilson, and H. P. Possingham, editors. 2009, New York: Oxford University Press. 122–134.
30. Nicholson, E. et al., *A new method for conservation planning for the persistence of multiple species.* Ecology Letters, 2006. **9**: 1049–1060.
31. Pe'er, G. et al., *Toward better application of minimum area requirements in conservation planning.* Biological Conservation, 2014. **170**: 92–102.
32. Carwardine, J. et al., *Prioritizing threat management for biodiversity conservation.* Conservation Letters, 2012. **5**(3):196–204.
33. Mitchell, M., *Complexity: A Guided Tour.* 2009, New York: Oxford University Press. 368.
34. Fernandes, L. et al., *Establishing representative no-take areas in the Great Barrier Reef: Large-scale implementation of theory on marine protected areas.* Conservation Biology, 2005. **19**(6): 1733–1744.
35. Cooney, R., *The Precautionary Principle in Biodiversity Conservation and Natural Resource Management: An Issues Paper for Policy-Makers, Researchers and Practitioners.* 2004, Gland, Switzerland: IUCN. 51.
36. Smith, R. J. et al., *Let the locals lead.* Nature, 2009. **462**(7271): 280–281.
37. Game, E. T. et al., *Informed opportunism for conservation planning in the Solomon Islands.* Conservation Letters, 2011. **4**: 38–46.
38. Environmental Law Institute, *Conservation Thresholds for Land Use Planners.* 2002, Washington, DC: Environmental Law Institute.
39. Johnson, C. J., *Identifying ecological thresholds for regulating human activity: Effective conservation or wishful thinking?* Biological Conservation, 2013. **168**: 57–65.

40. Environmental Law Institute, *Conservation Thresholds for Land Use Planners*. (See reference 38.)

41. Kareiva, P., C. Groves, and M. Marvier, *The evolving linkage between conservation science and practice at The Nature Conservancy*. Journal of Applied Ecology, 2014. **51**(5): 1137–1147.

42. Douvere, F., and C. N. Ehler, *New perspectives on sea use management: Initial findings from European experience with marine spatial planning*. Journal of Environmental Management, 2009. **90**(1): 77–88.

43. USDA Forest Service, *Tongass National Forest; Land and Resource Management Plan*. 2008, Washington, DC: U.S. Department of Agriculture.

44. Fukuda-Parr, S., T. Lawson-Remer, and S. Randolph, *Measuring the Progressive Realization of Human Rights Obligations: An Index of Economic and Social Rights Fulfillment*. Economics Working Papers. 2008, http://digitalcommons.uconn.edu/econ_wpapers/200822.49.

45. Taylor, J., *Indigenous peoples and indicators of well-being: Australian perspectives on United Nations global frameworks*. Social Indicators Research, 2008. **87**(1): 111–126.

# 5

# Harnessing Knowledge: Situation Analysis, Threat Assessment, and Data

## Overview

Chapter 5 focuses on the knowledge that supports conservation plans. The chapter starts with guidance on situation analysis and threat assessment, and we describe some useful tools for conducting these studies. The outcomes of situation analyses and threat assessments are useful in guiding data assembly for a planning process and getting the balance of investment in data right. Here we cover knowledge on the three types of conservation features we identified in Chapter 3: species, ecological communities, and abiotic features. Although this chapter is far from exhaustive, we illustrate some of the data and associated products commonly used in conservation planning, and deal with major questions around their collection and use. We cover social data that can be used to improve the effectiveness of conservation actions and to support plans with human well-being objectives. Special focus is given to acquiring and using data on the costs of conservation, as this is a crucial part of planning that has generally received too little attention. We also provide a detailed guide to robustly gathering knowledge from experts and using traditional ecological knowledge—a distinctive type of expert knowledge.

## Topics

- *Situation analysis*
- *Conceptual models*
- *Threat assessment*
- *Data needs*
- *Value of information*
- *Species data*
- *Species distribution modeling*
- *Ecological communities*
- *Ecoregions*
- *Ecological integrity*
- *Abiotic data*
- *Social data*
- *Costs*
- *Discounting*
- *Net present value*
- *Expert elicitation*
- *Bias*
- *Delphi approach*
- *Traditional ecological knowledge*

## Situation Analysis and Threat Assessment

With an understanding of the context and objectives driving a conservation plan, it is time to put our knowledge and science to work analyzing the social-ecological situation in which a plan is occurring. We have combined situation analysis and threat assessment in this chapter because they both focus on understanding the nature and cause of changes in the system of interest. They are included along with data in this chapter because we see all three elements as foundational pieces of information that can then be used as part of framing conservation planning problems—the topic of Chapter 6.

### Situation Analysis

**Situation analysis** is focused on identifying the drivers and causes of changes in the objectives or conservation features of a plan. The ideal result is an explicit articulation of how socioeconomic, political, institutional, and ecological factors drive change in the systems we work in. These changes might create impacts or threats on the things we value, as well as opportunities for intervention. Other terms for this same process include *system analysis* and **conceptual models**.

A situation analysis typically constitutes a series of linked elements that have some cause-effect relationship. In this way, they should identify both proximate and ultimate drivers of change. FIGURE 5.1 is a typical situation analysis diagram used by the Conservation Measures Partnership. These diagrams usually display the conservation features that are the focus of the planning effort on the right side and the direct threats and indirect threats to these features to the left, all in different colors.

This sort of cause-effect situation mapping has traditionally been accomplished on large sheets of paper, using either markers or often different-colored "sticky" pads (FIGURE 5.2). This is a proven recipe. Increasingly, however, there are a number of free, easy-to-use software programs that perform the same function. This is one of the main uses of the Miradi software developed by the Conservation Measures Partnership. Working directly in software like Miradi has the advantage of capturing the analysis in a format that is easily shared and accessed, and it allows the situation analysis to be easily linked to other parts of the planning process, such as monitoring. The significant downside, however, is that it tends to limit participation in the actual analysis because everything has to work through the person running the software. And of course this software-based approach relies on being able to project the computer interface at a scale that a group of people can easily read and interact with.

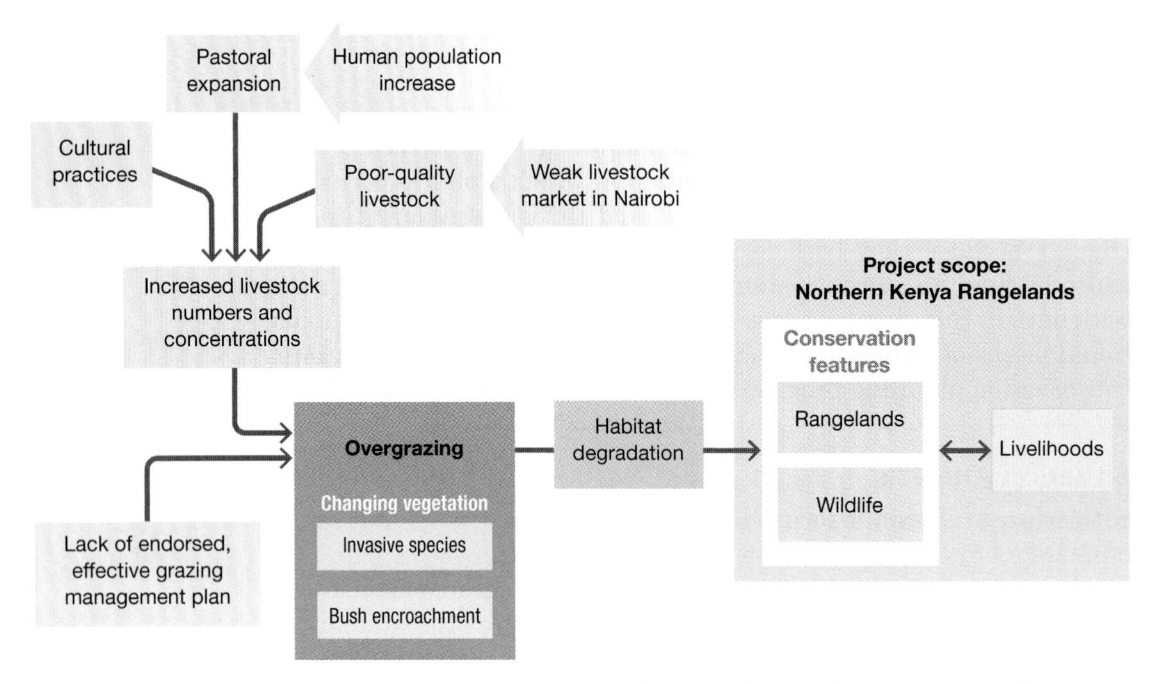


**FIGURE 5.1** Diagrammatic representation of a relatively simply situation analysis from a conservation project in the rangelands of northern Kenya. The conservation features (referred to as "Targets" in the diagram) are the rangelands and related wildlife. Direct threats to the features are noted in the red boxes, and yellow boxes highlight indirect threats.

## Threat Assessment versus Situation Analysis

For many conservation planning exercises, a situation analysis has been synonymous with assessing threats to conservation features and other values. This is understandable given that the abatement of threats to these things is what largely motivated the conservation action and plan in the first place. A full situation or system understanding is broader than just threats, especially in the complex social-ecological systems where conservation takes place. We view threat assessment as one part of a broader situation analysis.

## Threat Assessment

Understanding the processes that negatively impact our objectives is an important part of getting clarity on the situation and context in which conservation planning is taking place. Threats as they are understood in conservation are perhaps better referred to as *stressors* or *impacts* to recognize that in collaborative, multi-objective planning, referring to someone's livelihood as a threat to what you care about will not engender cooperation. Threat assessment is also important in order to construct actions that respond to relevant threats, and the knowledge gained from a threat assessment contributes to our understanding of the system and

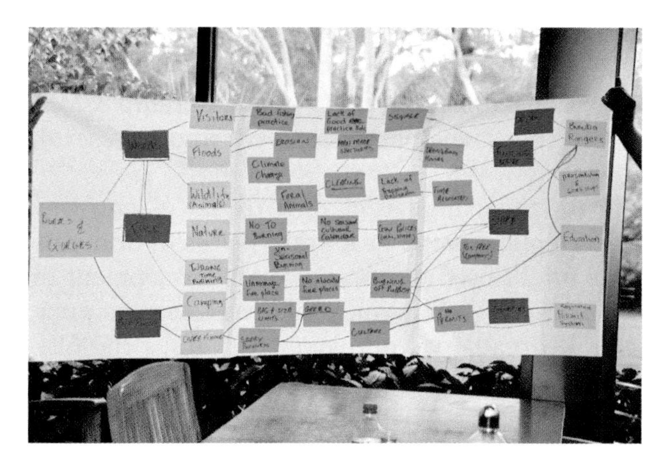

FIGURE 5.2 Using colored cards and large sheets of paper has proven to be an effective way to develop situation analyses during workshops. Left photo is from a plan developed by the Bunuba Rangers of the Aak Puul Ngantam aboriginal people in Cape York, Australia. Right photo is from a group in Peru using the MARISCO1 planning framework.

evaluation of the consequences of taking different actions. Although the purpose of a threat assessment is to gain an understanding of both absolute and relative significance of threats, it is important to distinguish between ranking threats and prioritizing threat abatement.

A threat ranking is a natural outcome of assessing a range of threats against some criteria. Ranking threats can be helpful in understanding the major influences in a system and ensuring that, when developing possible actions to achieve our objectives, we are not unintentionally missing major threats.

A highly ranked threat, however, should not automatically be considered a priority for conservation action. Whether or not a threat is a priority for action in a conservation plan depends not only on the nature of the threat but also on the actions available to address the threat. It is premature to prioritize threats without an understanding of actions and how they will affect the threat. Because threats often interact with each other, and we rarely have actions that completely eliminate threats, it can be that the most effective way to approach the objectives is by working on a number of lower-ranked threats as opposed to the highest-ranked one. We realize this will be a departure from the way many conservation planners have thought about threat assessment.

## Finding the Source

It is important to distinguish between the impact of a threat and the source or driver of that threat. Impact is expressed in terms of changes in the attributes of objectives or features. For example, the level of fragmentation might be an attribute of a habitat feature. Fragmentation is indicative of habitat condition and is the impact of a threat. The source or driver of a threat is the thing causing that change; in the case of fragmentation, it is perhaps road development. So road development is the threat. The Conservation Measures Partnership (CMP) *Open Standards*[1] refers to this dichotomy as *sources* and *stresses* of threats.

It is often possible to trace the source of a threat a long way through a system. For example, change in market prices of minerals and cheap Chinese loans might be the driver of road development in some African countries. These are sometimes referred to as the proximate or direct source of a threat (road development) versus the ultimate or indirect source of a threat (global mineral demand); see the diagram in Figure 5.1. Threat assessments generally focus on proximate or direct threats because they can typically be influenced more easily than ultimate or indirect threats.

### Threat Assessment Procedure

The conservation and ecology fields have developed a vast number of threat assessment and ranking tools for use in conservation planning. These assessment tools typically involve rating a threat against a series of criteria. Minimally, these criteria include something about the scope or extent of the threat, and something about the severity of its impact.

For example, scope might be the geographic extent of a threat (whether road development impacts all or just part of a planning region), or the extent to which it impacts all elements of a feature (e.g., even if roads are planned for just a small part of a region, they might still go across the entire population of a species). Similarly, severity might refer to the reversibility of an impact (i.e., if the threat disappears, how long it will take the feature to recover), or the sensitivity of a feature to that impact (e.g., how much smaller would we expect a species population to be in the presence of roads versus without roads).

You probably have noticed that these are essentially the same two dimensions of a vulnerability assessment—namely, exposure and sensitivity. We see little reason to distinguish between the words *vulnerability* and *threat assessment*; use whatever seems most appropriate. **Vulnerability** is commonly used when talking about climate change (see Chapter 9 on adaptation) or hazards to and potential impacts on human communities. Hazard is also used relatively synonymously with threat, and it is the term of choice for the risk assessment literature.

It can make sense to further divide scope and severity in order to get more precisely at the multiple dimensions of threat that these broad criteria cover. TABLE 5.1 illustrates a set of five criteria used by the marine conservation ecologist Kimberly Selkoe and colleagues in rating threats to the marine ecosystems of northwest Hawaii.[2] These

**TABLE 5.1** Five criteria used in rating threats to the marine ecosystems of northwest Hawaii

| Score | Scale | Frequency | Functional impact | Resistance | Recovery time | Certainty |
|---|---|---|---|---|---|---|
| 0 | No impact | Never occurs | No impact | Not applicable | No impact | Not at all certain |
| 1 | < 1 km$^2$ | Rare | < 25% of species | High resistance | < 1 year | Low certainty |
| 2 | 1–10 km$^2$ | Occasional | 25%–50% of species | Medium resistance | 1–10 years | Moderate certainty |
| 3 | 10–100 km$^2$ | Annual or regular | 50%–100% of species | Low resistance | 10–100 years | High certainty |
| 4 | 100–1000 km$^2$ | | Persistent | | > 100 years | Very certain |
| 5 | 1000–10,000 km$^2$ | | | | | |
| 6 | > 10,000 km$^2$ | | | | | |

*Source:* Data from Selkoe, K. A., B. S. Halpern, and R. J. Toonen, *Evaluating anthropogenic threats to the Northwestern Hawaiian Islands*. Aquatic Conservation: Marine and Freshwater Ecosystems, 2008. **18**(7): 1149–1165.

same criteria have gained some currency generally for characterizing threats. You can see that four of the five criteria used are aspects of scope or severity. The fifth criterion, however, is an important addition: certainty of the information. Different levels of certainty about the assessment of different threats are likely to occur because some threats are better known and better studied than others. How uncertainty is treated during a threat assessment depends on a number of value judgments and risk tolerances; Chapter 8 covers that topic in detail. At this point, it is sufficient to emphasize the importance of including some estimate of certainty in any threat assessment. Uncertainty in this case can include natural variation, knowledge gaps, or disagreement among knowledge sources.

Each threat criterion needs to be translated into a context-specific attribute. For example, the attribute used to characterize the scale of an impact needs to reflect the scale of the planning exercise and the distribution of the feature or objective in question. Ideally, these attributes will represent continuous scales in natural units, so scale might be square kilometers, with some error bounds around it. In practice, however, threat assessments are more commonly approached with bins, or categories, such as those in Table 5.1. Structuring a rating with discrete classes like this is rarely necessary but remains popular. Where possible, it is always preferable to give quantitative ratings, like expected recovery time in years, rather than purely qualitative terms, like *slow*, *moderate*, and *fast*. Such terms are inherently vague and will mean different things to different people. It can, however, be effective to give both a linguistic term, like *rarely* or *occasionally*, and then also qualify this with a quantitative value (e.g., *rarely* means "less than once per year").

Once the threat criterion and threat attribute have been established, the crux of a threat assessment is rating each threat against the objective attributes and features. To include every feature could easily lead to a massive, unwieldy assessment whose results are hard to interpret. So it generally makes sense to group conservation features and rate threats against the group. Where a very rapid indication of threats is required, it is also fine to rate threats overall against all features. In this case, the scope criteria might be an assessment of what proportion of features are impacted by that threat. Our experience has been that rapid assessments with very broad groups, or against all features, tend to give pretty similar results to more detailed assessments, suggesting that in most cases the coarser assessment will be fine.

Any good available data can be used to help inform a threat assessment, and in most cases there will be published studies on at least some of the threats. However, threat assessments typically also involve expert rating of threats. This means that the advice about expert elicitation that we present later in the chapter is highly applicable here. To emphasize the difference that a good elicitation process can mean for a threat assessment, the marine ecologist Alana Grech and colleagues found that asking experts to rank threats to sea grass globally as a group resulted in very different answers to a structured elicitation process with the same experts.[3] The structured process revealed threats that were not identified in the group exercise, and the relative rank of other threats changed substantially. Importantly, having done them both, the experts felt that the answers from the structured process were closer to the truth.

### What's Happening in the Future?

In addition to identifying current drivers, teams should also assess projected future drivers or trends and then integrate them into analyses and descriptions. For instance, in the road development example used earlier, although mining might be the current driver of an expanding road network, perhaps increasing population in a nearby city will drive development of a different type of road network in the future. This sort of forecasting is discussed in Chapter 8.

### Ranking Threats

Ranking threats is generally accomplished by giving each of the threat ratings a score and then combining the scores for all criteria to give an overall score. Where ratings are made on a consistent scale (as in Table 5.1), these scores are often simply averaged across all criteria.

As we discuss in detail in Chapter 7, this is not great practice from a multi-criteria analysis point of view—each criterion is unlikely to be exactly equal. Although in Chapter 7 we introduce ways to establish defensible weights for different criteria, how involved a threat assessment needs to be depends on how the information will be used. Because the primary use of threat assessments in a conservation plan is to ensure that we are aware of the most significant threats and consider these in developing management options, a rapid assessment without the full rigor of a multi-criteria analysis will be fine in most cases.

Threats can be ranked both overall for a system or for each conservation feature individually. FIGURE 5.3 shows the threats to northwest Hawaiian ecosystems, both overall and by ecosystem type.

## Cumulative Threats

Threats rarely affect the environment independently. The action of one threat can increase the impact of another by reducing a feature's resilience. For example, coral reefs that are subject to stress from

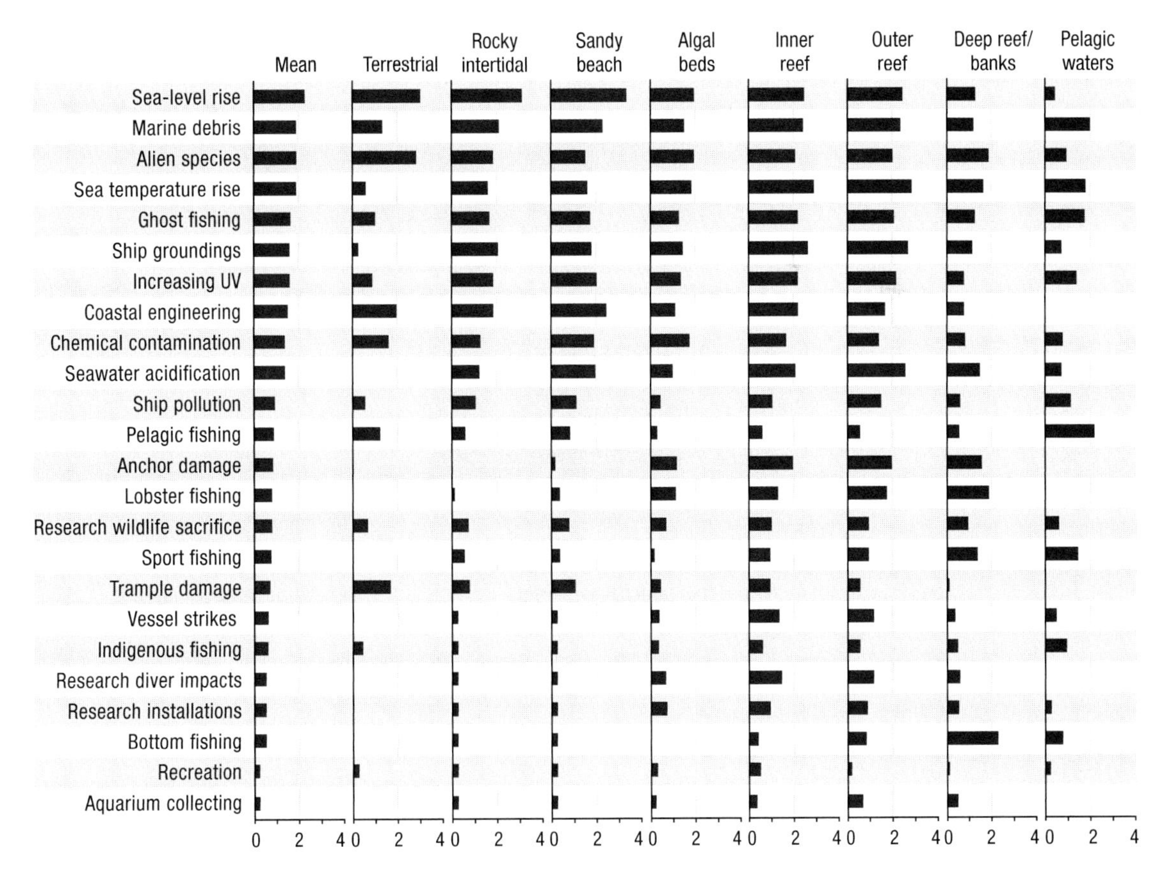


FIGURE 5.3 Assessment of threats to northwest Hawaiian ecosystems, both overall and grouped by ecosystem type. (Data from Selkoe et al., 2008.)

nutrient pollution are more susceptible to suffering from increases in water temperature associated with climate change. There is reason enough to rate threats on their own, but planners should also try to understand how threats interact to impact features or places. The impact of multiple threats is referred to as *cumulative* impact. Ideally, an assessment of cumulative threats would consider the interaction between threats that might reduce but more likely amplify the impact of the single threat. Pragmatically, however, most cumulative threat assessments involve identifying places or features exposed to, and sensitive to, multiple threats.

A cumulative threats assessment essentially involves following the same procedure described earlier for threat rating, ensuring the threat ratings are on a common scale, and then summing these either by conservation feature (FIGURE 5.4) or location (FIGURE 5.5) to give the cumulative impact. If there is spatial information about the distribution of conservation features and the intensity of different stressors, a map of cumulative impacts can be generated. For example, the freshwater conservation planner Peter Esselman and colleagues developed a cumulative threat assessment for freshwater fish habitat across the United States. They combined 15 disturbance variables in their

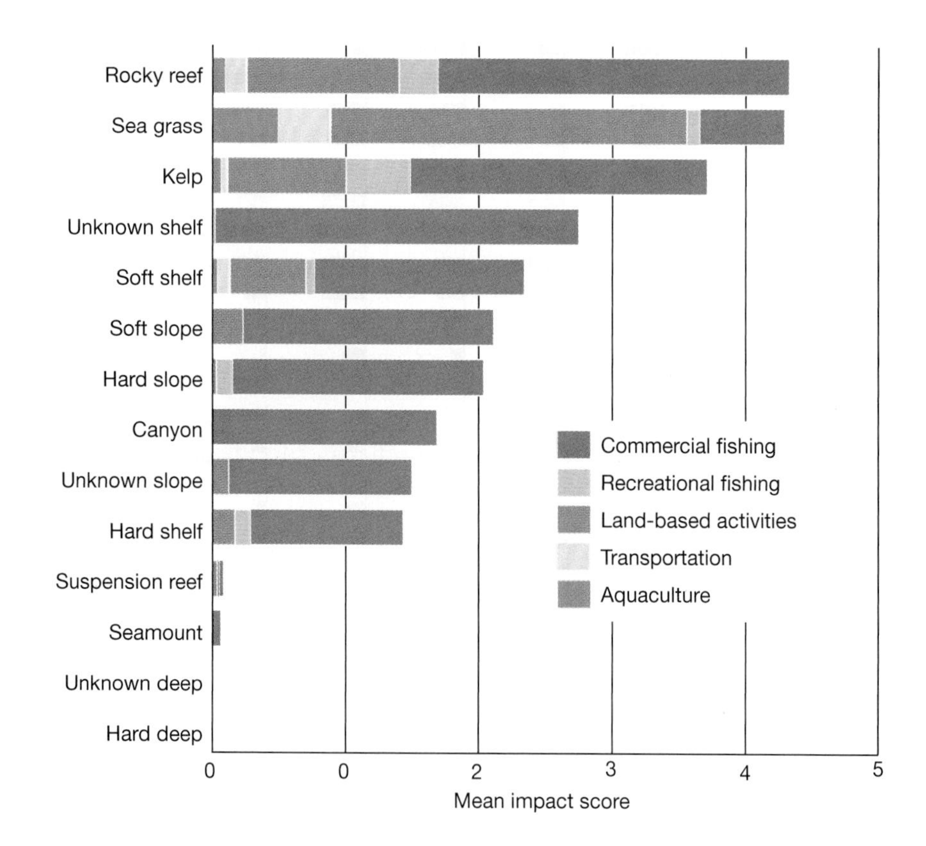


**FIGURE 5.4** Cumulative threats to different coastal and marine habitat types in British Columbia, Canada. (Data from Ban et al., 2010.)

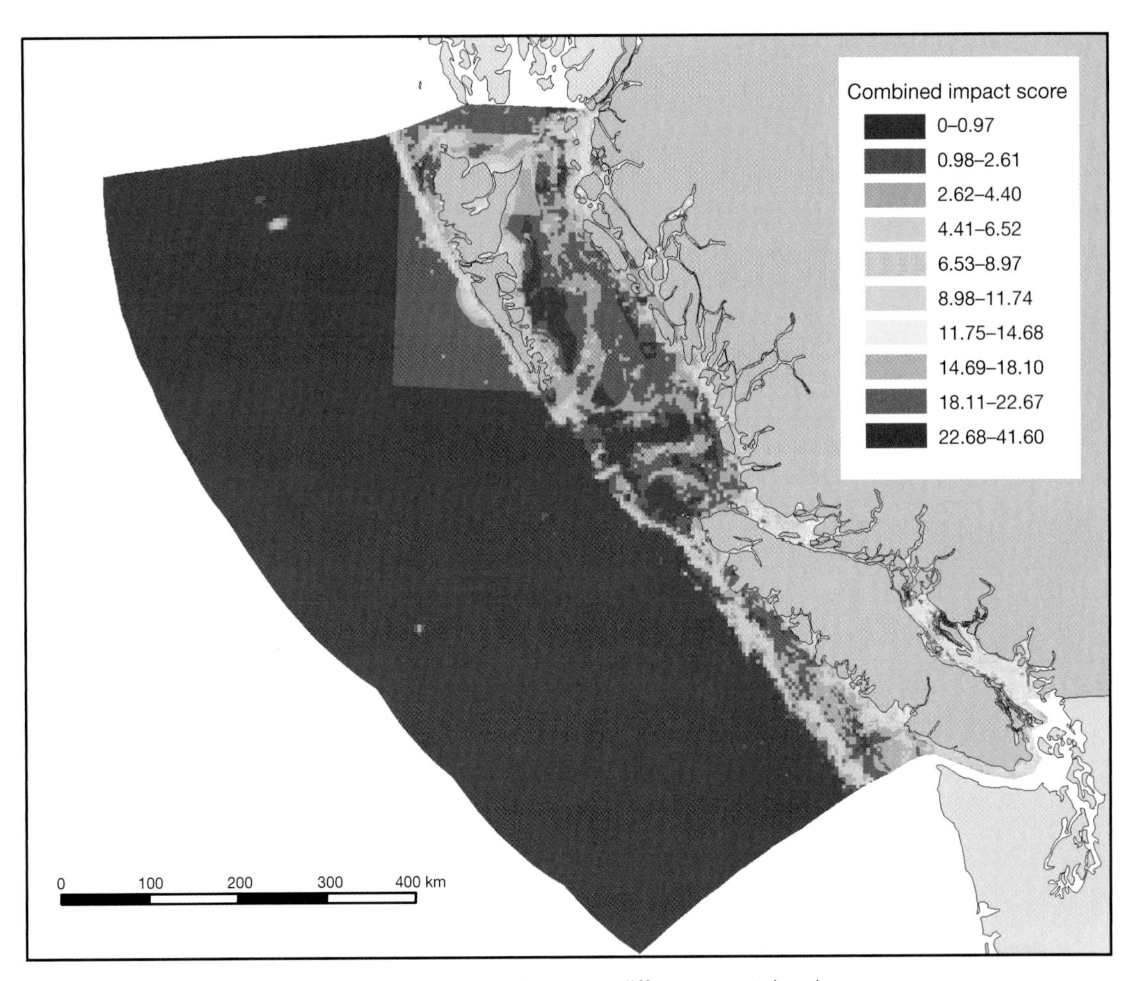


FIGURE 5.5 Spatial representation of cumulative threats to different coastal and marine habitats in British Columbia, Canada. (Adapted from Ban et al., 2010.)

threat index and used a series of linear regressions against observed fish presence to weight the impact of each different disturbance in the cumulative threat index[4] (FIGURE 5.6).

One of the most well-known efforts to investigate cumulative threats to ecological systems is the global map of human impacts on ocean ecosystems that the marine conservation biologist Benjamin Halpern and colleagues developed.[5] The same general cumulative impact method can be used for a variety of things, including identifying places with the greatest threats, conservation features with the most threats, and the most significant threats to a conservation feature.[6]

Many assumptions are involved in mapping cumulative impacts or threats this way, and they are well cataloged in a paper by two of

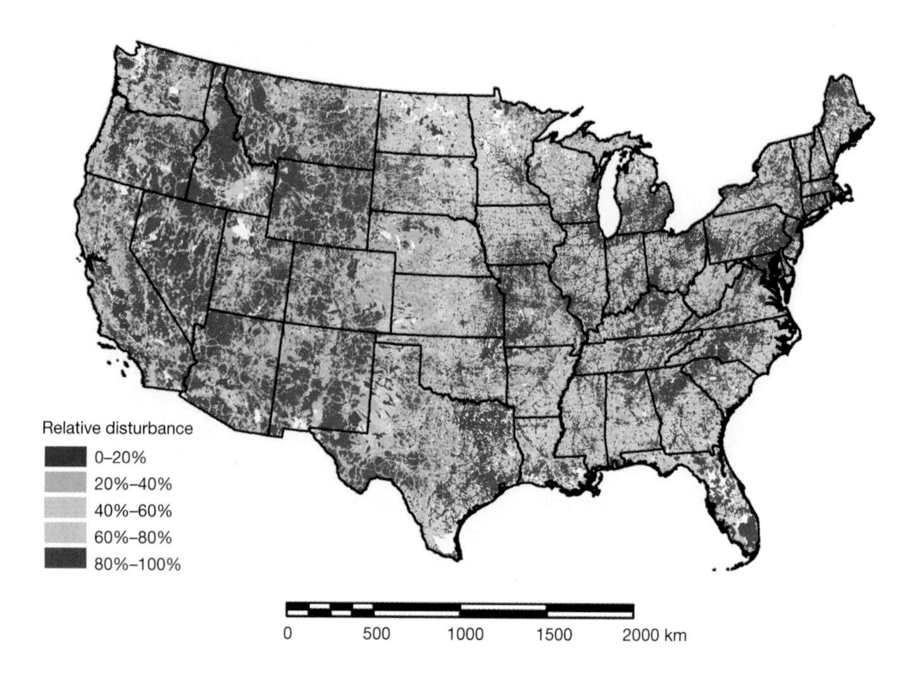


**FIGURE 5.6** Cumulative threat to river fish habitat, calculated from a weighted index of 15 types of disturbance. (Adapted from Esselman et al., 2011.)

the leaders in the field, Benjamin Halpern and Rod Fujita.[7] At the heart of these assumptions are the following three points:

1. Cumulative impacts are additive and linear.
2. Threats have the same impact in different places.
3. The data are a good approximation of the true distribution of both threats and conservation features at the scale we care about.

All three of these are almost certainly bad assumptions. The trouble is, in many cases there may be little alternative to making them as we rarely have data that would inform us about the validity of these assumptions. Despite this challenge, our view is that it remains useful for a planning process to consider cumulative impacts in this way. The remainder of this chapter is dedicated to dealing with data and data products for conservation planning.

## Data for Conservation Plans

### Data, Data Everywhere and Not a Byte to Use

Data and how they are used lie at the heart of what it means to make informed decisions. The world is awash with data. Most conservation planners have had the experience of discovering more data to support a plan than they initially anticipated. However, all the data useful for a conservation plan are rarely to be found in a single database and almost never in the format needed (e.g., extent, resolution, metric). A significant part of a conservation planner's job is being a data sleuth (**Box 5.1**). A skilled conservation planner must know where to look for data, be creative in the types of data that can support decisions, and have the ability to judge the quality of data.

For many types of data used in conservation planning, there is a vast literature, including dedicated books, for example, on species distribution models, remotely sensed vegetation data, and fisheries data. In a single chapter, in fact in an entire book, we could never hope to comprehensively treat the data useful for conservation planning. Rather, this chapter illustrates some of the data commonly used in conservation planning, and deals with major questions around its collection and use. If reading this chapter leads readers to think creatively about what data they may use in a conservation plan, even better. For instance, we have witnessed a dramatic increase in the use of citizen science data (such as bird observations[8]) being used to support conservation plans. Groups of citizen volunteers interested in assisting scientists are becoming increasingly effective at collecting and maintaining scientific data and information at different scales across the globe.[9]

In addition to discussing what might classically be considered data, we recognize that many conservation decisions will inevitably involve the use of expert opinion as well as nontraditional data (e.g., traditional ecological knowledge). Knowing how to elicit and use expert judgment robustly is, for a conservation planner, a skill at least as important as working with spatial ecological data, so we spend some time covering the key elements of working with expert data.

## How Much Data Do I Need?

Data availability is a relative concept and can be remarkably context specific. A colleague of ours from Norway once shared a story about a meeting with Norwegian lobster fishermen to discuss the establishment of some marine protected areas. Our colleague presented to the fishermen a remarkable time series of data stretching back over 100 years and illustrating significantly declining lobster numbers. One of the fishermen stood up and said, "You want to close this area to fishing and these are all the data you have?"

True: You can always have more data.

False: It is always worth getting more data.

## Linking Information Needs to the Question at Hand

Most people recognize there is some intrinsic value in data, in knowing more about the world and how it works. In this sense, all data collection and research has value, however incremental it might be. Conservation planning sometimes involves primary data collection but

## BOX 5.1 Gabon: Finding Data Where There Is None

The Republic of Gabon on the west coast of equatorial Africa is one of the most heavily forested nations in the world. The conservation biologist Michelle Lee and colleagues[a] were tasked with developing a spatial conservation plan for Gabon—a job that required at least some of the data identified in this chapter. The trouble was that Gabon's population of 1.4 million lives largely in urban areas, and for much of the country there was no existing spatial data on environments or species, nor was there any central data repository. In an instructive example of how conservation planning data can be assembled in little-charted places like central Africa, Lee and her colleagues undertook sleuthing, elicitation, and analysis to map Gabon's environmental patterns and develop a data product that could serve as a common currency for discussions about national-level planning and environmental assessment.

With no national classification of vegetation, soils, or even environments, Lee and her colleagues first built a set of land units. From the ministries responsible for mining and agriculture, they sourced what was known about geology; then, building on knowledge of how geology influences drainage and plant species composition, they identified six broad geological classes. They used a globally available digital elevation model to map topographic zones and features. Using data from Gabon's only 13 weather stations, they identified three macroscale climate zones.

To map land cover, Lee focused on distinguishing grassland, mangrove, and submontane vegetation types from the dominant lowland rainforest. These types were chosen because of the potential of these areas for industrial uses (especially grass/shrub and mangrove types), their unique species-level biodiversity (grass/shrub and submontane types), and the ecosystem services they provide (mangrove and submontane types). To map grass/shrub, they used the MODIS (satellite imagery) Vegetation Continuous Fields data set; for mangroves they digitized physical maps of the coastal zone, and submontane areas were identified based on altitude and distance from coast formulae. These different layers were then intersected to produce the land units (**FIGURE 5.1.1**).

Gabon has a wealth of endemic species, many of which are poorly known. None of the nationally listed bird or reptile fauna has enough accurately georeferenced records for predictive distribution (see the Rare Species Modeling Paradox later in this chapter) modeling. Focusing on bird and reptile species, Lee and colleagues divided the country into 25-km$^2$ planning units; and based on existing specimen or sighting records, they assigned a value of 1 to each unit with certain presence for that species. To supplement this method, they asked experts for each species to map the likely distribution of the species, assigning units a 0.7 value if they were confident of finding the species there 70% of the time and a 0.3 value if they were confident to find the species there 30% of the time. All other units were assigned a 0 for that species, indicating an assumption of the species' absence. These expert-mapped predictions attempted to

more often involves collating and synthesizing a great deal of existing data. When we've asked people what the benefit of doing a conservation plan was, both those involved and not involved frequently cite the collation of data about a region or issue.[10] Collating data can be a substantial undertaking, but one that can have value lasting well beyond the life of the plan, especially if the data are well organized and managed (**Box 5.2**).

make the most of what was known in a data-poor situation and were treated as initial hypotheses that could be verified by surveying.

Finally, to gather data on current land use in Gabon, Lee and colleagues compiled maps of existing permits for land and water uses from government ministries, agencies, and private companies. This was the first time a national database on land allocation had been compiled for Gabon.

[a] Lee, Michelle E. 2014. *Conservation and land-use planning applications in Gabon, Central Africa.* DPhil thesis. University of Oxford.

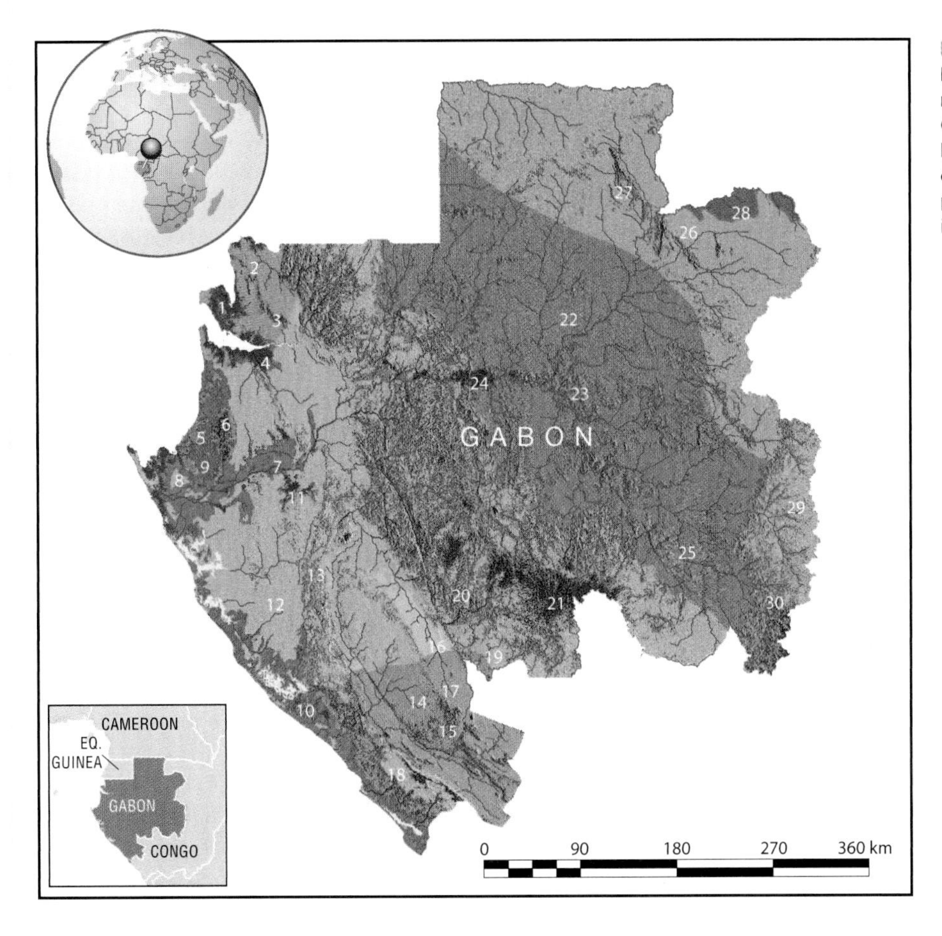


**FIGURE 5.1.1** Map of environmental units for Gabon, used as the basis for subsequent conservation planning processes in the country.

So more data are good, but are they really needed? The trouble is, there isn't a linear relationship between the amount of data we have and the quality of the decision we can make with it. Because collecting, collating, and analyzing data costs both time and money, and conservation planning is about efficient use of resources, we want to be careful to balance the desire for more data with the actual decision-making need for it.

## BOX 5.2 Data Management

Good data management is valuable for any planning process. However, it also means that data collated for the planning process can provide an independently valuable product that will likely have numerous future applications. Thinking about the life of data beyond the current conservation planning process means that conservation planners (or those responsible for data management in a planning process) need to think of themselves not just as data users but as data providers. This is clearly not the norm in conservation as it is estimated that only 1% of the data supporting published ecological studies are actually available for subsequent use.[a]

Our colleague, Jon Fisher, who has managed a lot of the data The Nature Conservancy has used and developed for conservation planning, has provided the following tips for data management in a conservation plan:

- **Understand what the data are for.** Perhaps the most critical ingredient for managing data well is understanding who will use them and how. This will drive virtually all other decisions, from what to include and exclude to editing workflows, level of complexity, format, licensing, etc. Never let technology make a decision for you (e.g., "Since Excel only allows XX fields we will have to simplify our data." There is almost always a solution that will meet the actual needs while remaining reasonably simple to set up and use. If the data are intended for wider use at the end of the project, consider format requirements and how that may influence data design (e.g., Does the spatial data have to work in ArcMap? Google Earth? Open-source GIS systems? Do end users have Excel or Access or MySQL installed and understand how to use it?). If there are subsets of the data or specific analyses you expect users to need, generate the queries in advance and make it clear to them where they are and how to use them.
- **Don't forget about metadata.** One of the most crucial aspects of data management is documenting what each data field means, and what the values in it represent (units, if using numbers that stand for something else provide a key, etc.). Where possible use field names that make this apparent, for example, "AREA_KM" can be reasonably assumed to be area in square kilometers even without metadata. Clearly specify how the data can be used and shared (e.g., "OK for any use," "for noncommercial use only"), and the appropriate attribution for it. If there are restrictions on the data, be clear about that up front.
- **Spend time on quality assurance (QA) and quality control (QC).** These tasks need to be done up front and periodically as you go. Depending on the size and complexity of your data, QA can include running a script to check for errors that you expect to crop up (and fix them if possible), or simply checking a few key properties before doing major work. If working with spatial data, be sure to at least repair geometry errors, and if possible set up a simple topology to identify errors (e.g., overlapping polygons or gaps between them that shouldn't exist) that will introduce complex errors into derivative layers. The sooner you find errors, the lower the chance you will have to redo work based on problematic data.
- **Choose simpler data structures whenever possible.** Unless working with large data sets in an environment that has to be high-performance (a web-based information system), it is generally more useful to have simpler data structures (e.g., one big table rather than a series of joined tables). This allows a user to extract or analyze data immediately without having to understand the specific data model being used. The same is true of using descriptive data values rather than symbols that require a lookup

table (e.g., "Deciduous Forest" for land cover rather than "41"); the space saved is not worth the extra work required to use the data.

- **Make a plan to share the data.** Even the best data set is not useful unless people know about it. Publishing and sharing data is often viewed as an afterthought rather than a key objective of data management. If possible, put your data on a publicly accessible website[b] with all of the relevant metadata and licensing information (if the data can't be shared publicly, you should still describe the data and who to contact to get it for approved uses). Most spatial data can currently be housed on a site like DataBasin or ArcGIS Online for free; websites change often, so check with your colleagues about where they go to look for data.

For planners who want more detailed guidance on managing data, Elizabeth Borer and colleagues provide 13 guidelines for effective data management.[c]

---

[a] Reichman, O., M. B. Jones, and M. P. Schildhauer, *Challenges and opportunities of open data in ecology.* Science (Washington), 2011. **331**(6018): 703–705.

[b] Reichman et al. (See reference a.)

[c] Borer, E. T. et al., *Some simple guidelines for effective data management.* Bulletin of the Ecological Society of America, 2009. **90**(2): 205–214.

The conservation planner Hedley Grantham and colleagues published a seminal paper on this topic.[11] They asked the following question: Would additional investments in surveys, mapping, and modeling lead to better planning decisions, given the costs of undertaking these surveys and the losses that ongoing land clearing would continue to incur while the surveys were carried out? Their results clearly showed that while investing in minimal survey data can improve a plan substantially, especially if few biodiversity data are available to begin with, there are rapidly diminishing returns in additional data (**FIGURE 5.7**).

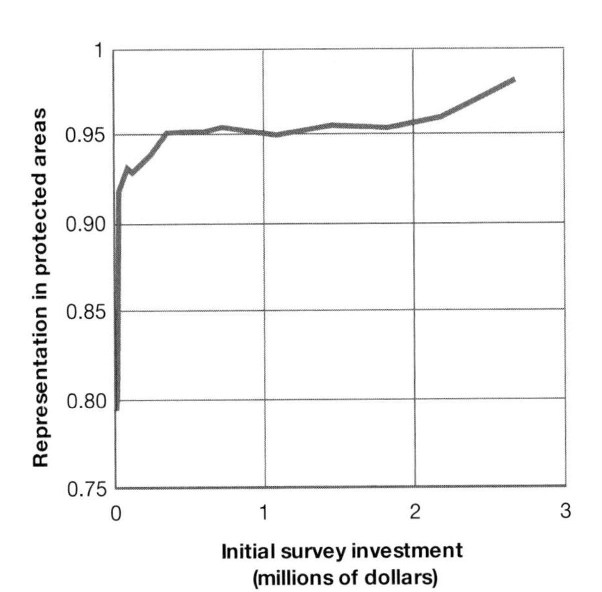


**FIGURE 5.7** Representation of Protea species in protected areas in the Fynbos biome, South Africa, as a function of investments in biological surveys. (Data from Grantham et al., 2008.)

FIGURE 5.8

Many conservation planning exercises start with an attempt to identify and pull together all available data. The cartoon in FIGURE 5.8 is probably not an unreasonable representation. Aside from the fact that this approach might involve time and energy gathering data that ends up not being used, we've also noticed that emphasizing data up front like this can put undue weight on data problems and gaps. People become concerned when a data set is incomplete or unavailable. In our experience, many of these gaps become irrelevant once it is clear how the planning process will proceed and what solution methods will be used (see Chapter 7).

In some cases we've actually seen more data act as a hindrance to conservation planning. When the Great Barrier Reef Marine Park was being rezoned, the authority responsible was under a legal obligation to consider all available data—which, given the amount of research that has happened on the Great Barrier, was an enormous amount of data. The trouble was, many of these data were unhelpful for guiding the placement of representative no-take zones; but they still required gathering, processing, and much discussion about how they could be used.

You may remember in Chapter 2, when we discussed the costs and resources involved in planning, that data availability was one of the axes—more data necessitate a larger, more costly planning process. We want to be clear that we are not against gathering additional data as part of a conservation planning process. Especially in data-poor regions, new data on biodiversity or other data can be critical to identifying important conservation areas or interventions that might be needed. For example, in the early years of WWF and TNC ecoregional plans, there was very little information on freshwater biodiversity. As this information was added to plans over time, it substantially influenced both the conservation areas that were identified as well as major conservation strategies that were implemented.

### Calculating the Value of Information

There are many uses for data, both known and as yet unknown, so the value of gathering a piece of data is difficult to determine precisely. If, however, we restrict our interpretation of *value* to the ability of data to improve the decision at hand, then it is possible to calculate what any piece of additional information is worth.

Essentially, the **value of information** is the difference between the outcome of the decision you would make with the additional information and the outcome of the decision you would make without that information. If the difference in performance between these two decisions can be monetized, then it is possible to calculate the actual financial value of the additional information. Even if the outcome of the decision cannot be easily monetized, the same logic of comparing information gain to improved outcomes can still be used. The formal term for this concept is the expected value of perfect information (EVPI).[12] Because the most common use of the EVPI concept in conservation planning is to determine where it is most important to reduce uncertainty,[13] we discuss the value of information and its calculation in detail in Chapter 8.

## Conservation Features

### Species

As we discussed in Chapter 3, species have been, and will continue to be, important elements of conservation plans. There is enormous variation, both between species and between geographies, in how much is known about a species and how many data on it are available for a conservation plan. TABLE 5.2 describes a range of data about species that can be useful in a conservation plan. We've grouped these data under three broad categories: distribution, population, and ecology or natural history. The first two categories have been staple data in systematic conservation planning, often collated under the term *biodiversity pattern.*[14]

The Italian conservation planner Carlo Rondinini and colleagues present a thorough discussion of the strengths and weaknesses of using the different types of data about species distribution for systematic conservation planning.[15] While knowledge about the natural history of species (like FIGURE 5.9) should be implicit in all conservation plans, it becomes essential when considering the likely impact of alternative actions on those species.

***Species Distribution Modeling*** To help address the bias in our knowledge of species distributions, conservation planning often relies on **species distribution models (SDMs)**. SDMs combine the known locations of species with knowledge of the environmental conditions

**TABLE 5.2** Different types of species data useful for conservation planning

| Category | Type | Notes | Possible sources |
|---|---|---|---|
| Distribution | *Point localities* | Records of known occurrence; place where the species has been found. Ideally, a point locality record will include when the species was found there (both the year and time of year that the occurrence was recorded). Many species distribution data consist of a series of known occurrences because the absence of a species is hard to confirm definitively. | • National and state agency databases<br>• International databases (e.g., IUCN Red List for mammals)<br>• Museum and herbarium records<br>• NGO data (e.g., NatureServe) |
| | *Geographic range* | A species' range is generally interpreted as the geographic boundaries of the area where a species is known to occur. It is sometimes referred to as the *extent of occurrence* and would encompass all point localities. | • National and state agency databases<br>• International databases (e.g., Red List mammals)<br>• Museum and herbarium records |
| | *Predicted distribution* | It is unlikely that all the places a species occurs have been recorded. It's also unlikely that a species can be found everywhere within its geographic range. Species distribution modeling is frequently used to predict where a species is actually likely to be present based on the suitability of the habitat (see section on species distribution modeling). | • Academic experts |
| | *Potential distribution* | The predicted distributions described earlier will generally be based on suitable habitat that we expect to be occupied by a species. Sometimes it is also important to identify locations where a species could potentially live but does not at present, such as for assisted migration or reintroduction efforts. | • Academic experts |
| Population | *Abundance* | For many species features, especially threatened and endangered ones, it is important to know not only where they live but also how many there are. The population of a species might be presented as a total population, a density, or even simply an indication of relative abundance. Just as with distribution data, it is important that data on population be identified with a particular time. In some cases abundance data are spatially explicit so that population can be used in conjunction with the distribution data described earlier. | • Wildlife and natural resource management agencies<br>• NGOs<br>• International databases (in the case of threatened and endangered species, e.g., UNEP–WCMC Red List species database) |
| Natural history/ ecology | | The list of potential data related to the natural history or ecology of a species is vast. Some of the basic pieces useful in conservation plans with species features include prey species, predators, habitat requirements, reproductive rates and requirements, and dispersal mechanisms. | • Academic experts<br>• Museum staff<br>• Traditional knowledge NatureServe databases (http://www.natureserve.org) |

FIGURE 5.9 Information about the distribution and abundance of brown bears (*Ursus arctos*) can be found in multiple public databases, such as the IUCN Red List, UNEP-WCMC, and NatureServe.

at those places in order to extrapolate where else across a landscape a species is likely to be found. The output of an SDM is generally a map showing the probability of occurrence for a species across space. In the United States, the National Gap Analysis Program managed by the U.S. Geological Survey maintains SDMs for most of the vertebrates of the United States, and these maps can be downloaded for conservation planning purposes (FIGURE 5.10). Species distribution modeling is a significant field in its own right with many thousands of useful publications; ecologists Jane Elith and John Leathwick provide an excellent introduction to SDMs.[16]

SDMs make use of a broad range of modeling techniques to explore the correlation between response (species) and predictor (environment) variables.[17] There are, however, some relatively easy to use software packages to automate species distribution modeling, such as MaxEnt.[18] Automated software can put species distribution modeling within reach of even those planners or GIS staff with no prior modeling background.

However, it is also easy to misuse SDMs, especially in the context of conservation decision making. For example, distribution modeling that leads to false positives (identifying species where they do not actually exist), a common problem when using presence-only data, can have obvious and serious consequences for a conservation plan. For planners considering using SDMs in a conservation plan, it is

FIGURE 5.10 An example of a species distribution model available online as part of the U.S. National Gap Analysis Program. This figure shows the distribution for the kit fox (*Vulpes macrotis*), one of over 1500 vertebrate species whose distributions can be extracted from the database.

important to spend the time required to get a good understanding of the assumptions embedded in the chosen SDM approach.

***The Rare Species Modeling Paradox*** Antoine Guisan, one of the world's foremost experts on species distribution modeling, describes a frustrating paradox of species modeling to support conservation planning. Predictive modeling of a species' distribution is most needed where a species is known from only a small number of records. This is frequently the case for the rare and threatened species that are an important feature of conservation plans. However, it is exactly these species for which an SDM is least useful because there are too few data points to confidently parameterize a model. Guisan and colleagues refer to this as the "rare species modeling paradox,"[19] that SDM is simultaneously the most useful and the least useful for poorly known species. As a rule of thumb, Guisan suggests that an SDM cannot be developed with confidence for any species known from fewer than five records.

***Challenges with Species Data*** A ubiquitous challenge encountered using species data for conservation planning is the variability in our knowledge both between species and within a species. We inevitably have a great deal more data on some species than others, and for any given species we are likely to have lots of data from some locations and few from others. These biases generally reflect unevenness in efforts to collect species data. We remember being struck by a map of species point data used in a conservation plan for Australia's Great Barrier Reef, showing "hotspots" of species records perfectly aligned with the location of permanent research stations on the reef. This sort of variation in knowledge is unavoidable and can present a challenge when used to make decisions about investments in different areas or species.

In some cases bias can be minimized through approaches like species distribution modeling, but there will likely remain challenging questions about what weight to place on data of different certainties. In Chapter 8, we discuss a range of techniques for dealing with uncertainty in our data and knowledge. One common uncertainty particular to species data is error in the locational data recorded along with museum accessions. Museum records are a common source of species data, especially in developing nations, but many locational records were collected before the widespread use of GPS, so it pays to be careful in checking their plausibility.

Another challenge of increasing importance when working with species data is the extent to which past information on species is representative of the present or future. The pace at which many species distributions and natural histories are changing with a rapidly changing climate is a caution on any static view of species data. In Chapter 9 we deal specifically with some of the challenges climate change poses for conservation planning for species.

Finally, one of the most significant shortcomings of species occurrence data such as point locations is that such data usually contain almost no information on the condition or quality of the population of individuals they represent. It is tempting to assume that species occurrence data represent populations of individuals that will persist over time, but that can be a tenuous assumption, especially with rare or threatened and endangered species. Although population viability data are available for only a small percentage of species, many state and provincial Natural Heritage Programs in North America maintain some level of information about the estimated viability of many at-risk plant and animal species.[20]

### Ecological Communities

***Extent and Distribution*** As introduced in Chapter 3, ecological communities are relatively distinct assemblages of co-occurring species. From a conservation planning perspective, we are primarily interested in data on the extent and distribution of ecological communities.

This requires both establishing a classification system and mapping distributions according to this system. In most cases, conservation plans will be drawing on existing classifications rather than developing new ones. Indeed, the decision to include ecological communities as conservation features (discussed in Chapter 3) tends to be driven in part by knowledge of the quality of data available to support that classification.

The classification and mapping of ecological communities typically occurs at regional, state or provincial, or national scales. These systems are also commonly referred to as ecological system or ecosystem classifications. Although some classification systems may distinguish between an ecological community and an ecological system, the difference is usually in the detail or resolution of the classification units. Regardless, they tend to serve the same function in conservation planning. We suggest looking initially to government natural resource agencies that are likely to have at least a terrestrial ecosystem classification with associated maps of distribution (such as in FIGURE 5.11), probably already in GIS format. Conveniently, a large

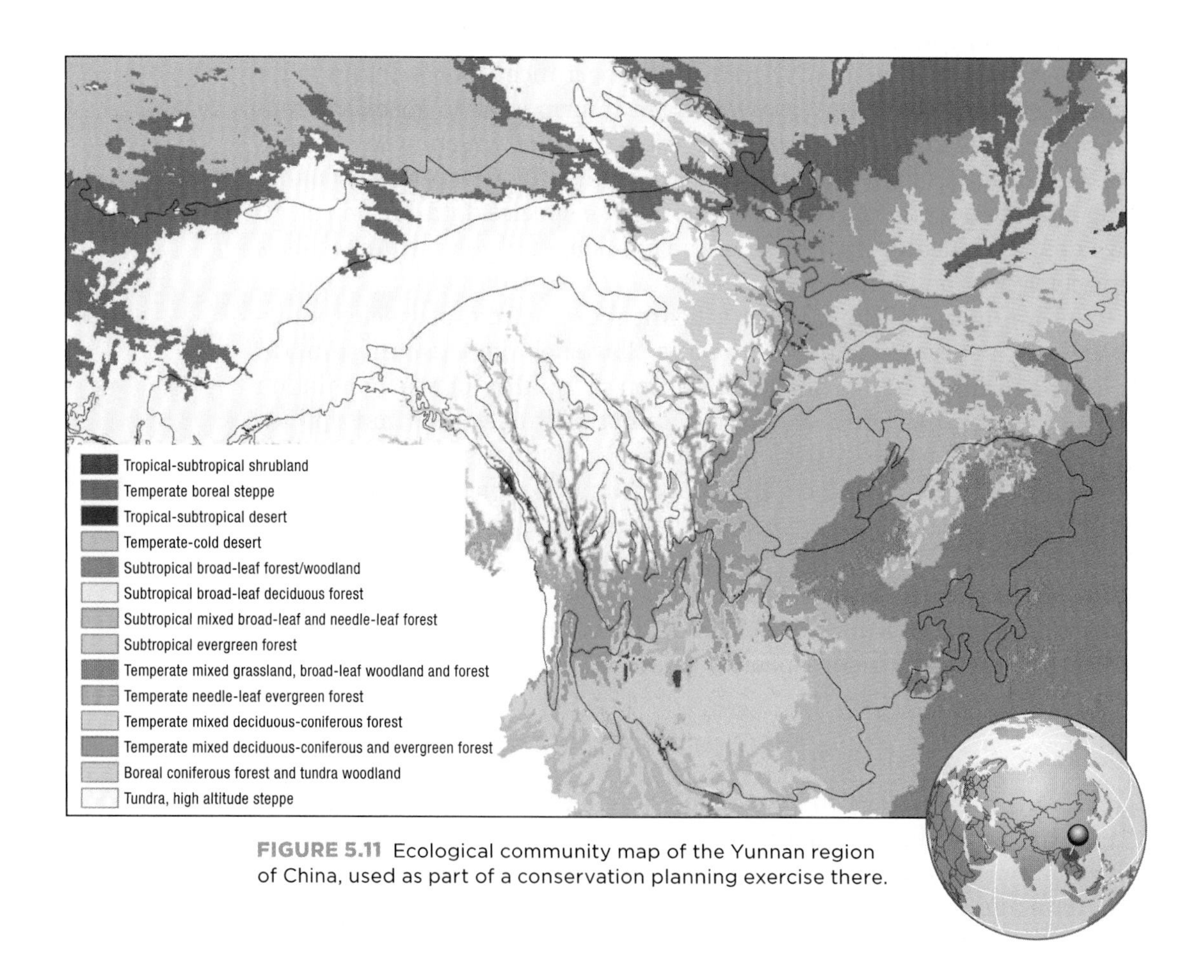


**FIGURE 5.11** Ecological community map of the Yunnan region of China, used as part of a conservation planning exercise there.

number of both global and national classifications of ecological communities are available from the ESRI ArcGIS online library, where they have been posted by other groups. Another source of community extent and distribution data is the United Nations Environment Programme-World Conservation Monitoring Centre (UNEP-WCMC).

***Ecological Integrity*** Data sets on the extent and distribution of ecological communities frequently contain little or no information about the ecological condition or integrity of these features in different places. **Ecological integrity** is a rather general term that has conventionally been linked to the perceived conservation quality of an area; all else being equal, we might prefer to conserve communities with high integrity.

The integrity of ecological communities might be influenced by things such as habitat fragmentation by roads and other infrastructure, invasive species loads, or resource extraction such as logging (e.g., FIGURE 5.12). These are either contributors to or proxies for ecological integrity; and in any region data on some of these variables will be available, even if they are from diverse sources. Unlike the classification of ecological communities, which requires detailed knowledge of flora or fauna, indices of ecological integrity can often be calculated

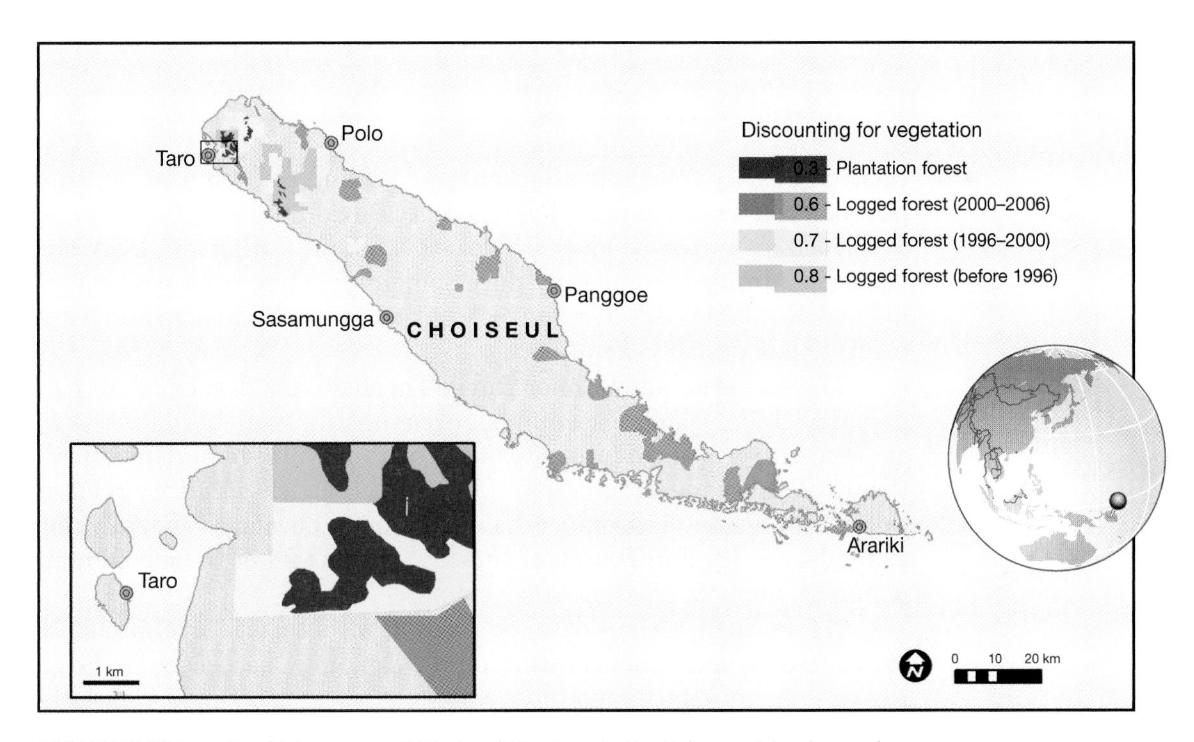


FIGURE 5.12 Logging history map of Choiseul Province in the Solomon Islands, used as a proxy for ecological conditions.

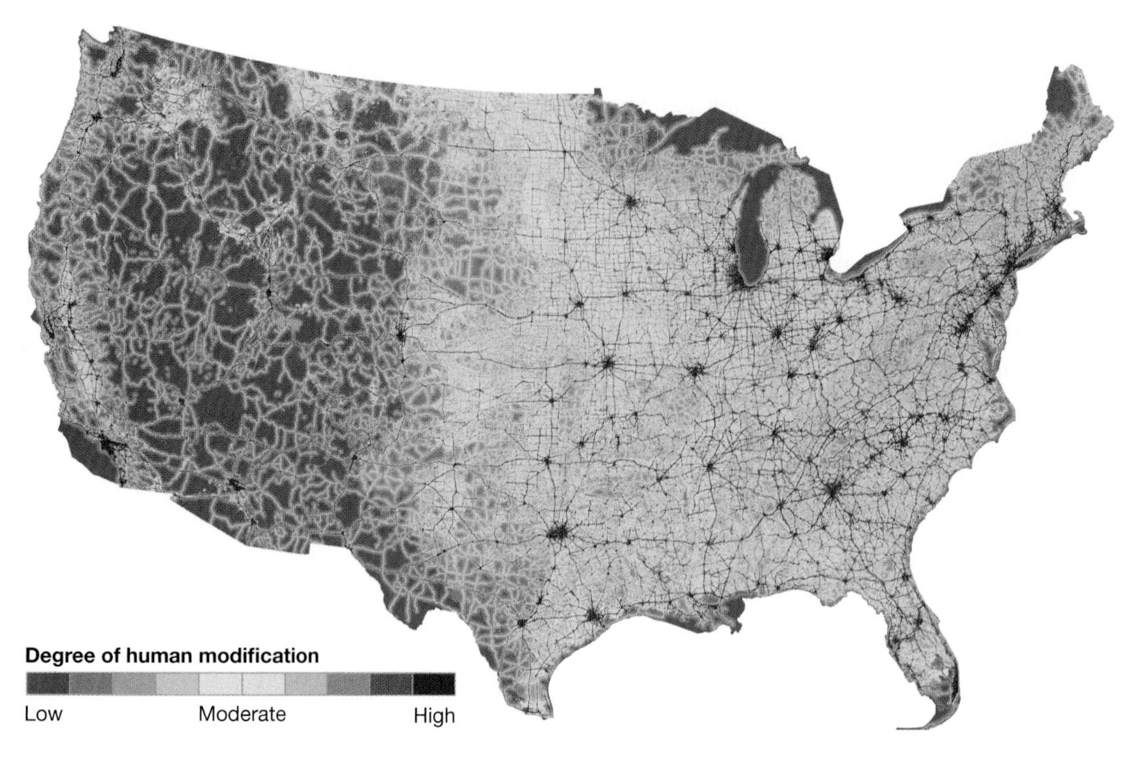


FIGURE 5.13 The degree of human modification (H) for the conterminous United States at 90-m resolution. Low levels of human activities are in green, moderate levels in yellow, and high levels in red. Major water bodies are included for reference, but water-based stressors are not included in a primary way. (Adapted from Theobald, 2013.)

relatively easily from remotely sensed data and can often be built as part of a conservation plan. For example, David Stoms and colleagues developed an integrity layer for desert lands in California, USA, for use in planning solar energy developments.[21] They used a range of publicly available GIS layers to look at the expected impacts and fragmentation of desert vegetation.

The conservation planner David Theobald has developed an elegant model that combines a number of commonly available data sets to characterize ecological integrity based on the level of human modification in a landscape (FIGURE 5.13).[22] Data on ecological integrity related to anthropogenic disturbance have also been developed globally, for example, the Human Footprint Index (FIGURE 5.14)[23] or the global map of human impacts on ocean ecosystems.[24]

While such global data on ecological integrity are probably too coarse for local conservation plans, they may be useful for large regional or national plans. As a valuable reference, David Theobald has assembled a thorough list of the commonly used, readily available spatial data sets on human modification at global and national/regional scales,[25] and provides guidance on how these data sets can

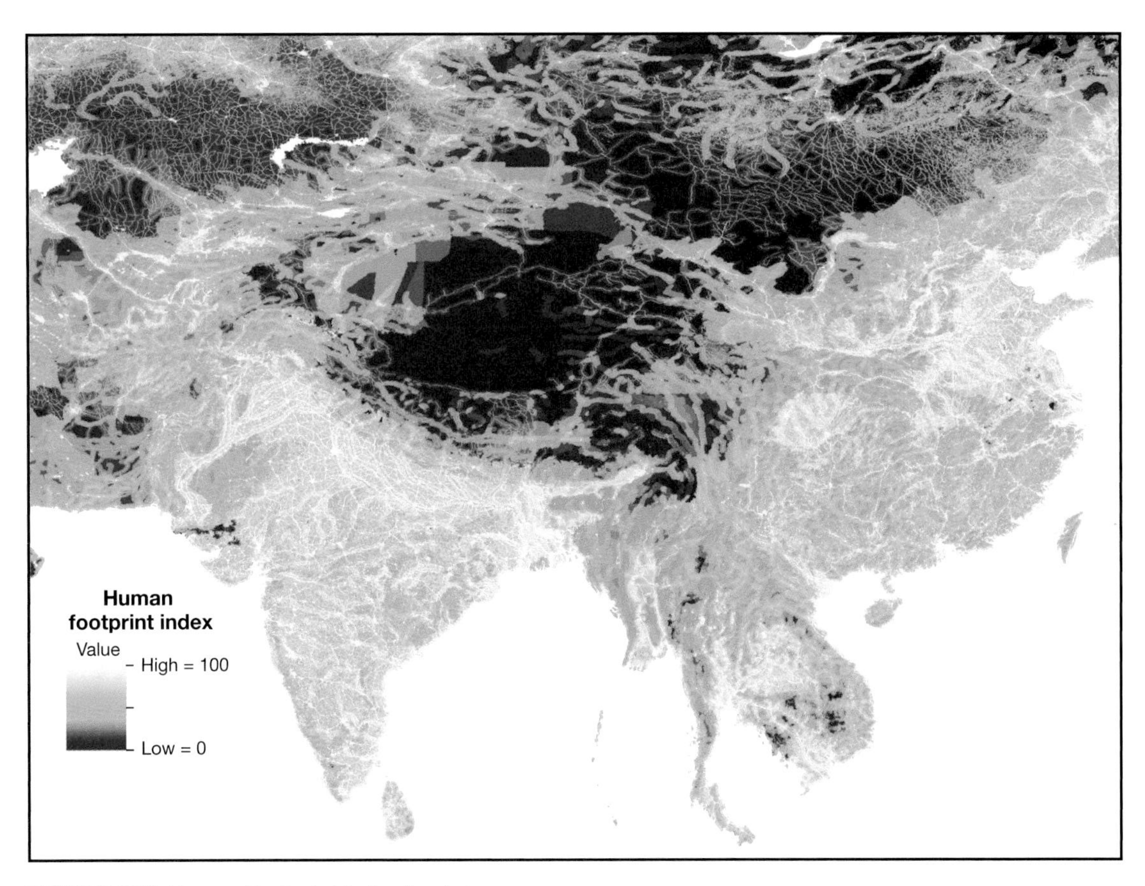


FIGURE 5.14 Human Footprint Index for Asia.

be analyzed and combined into composite indicators of the degree of human modification of the environment.

## Abiotic Feature Data

Abiotic features are those defined by geophysical properties (also referred to as environmental properties in some literature) rather than directly by the distribution of flora or fauna. Whereas conservation plans will generally draw on existing classifications of ecological communities, they will often need to develop abiotic features as part of the planning process. This is increasingly the case for conservation plans that focus on conserving the underlying environmental features of the landscape (Chapter 9). Three decisions define the task of generating conservation features based on abiotic data: which variables are used, how they are classified, and how they are assigned to features.

***Which Variables?*** Which geophysical variables should be used as the basis for abiotic features is a question for ecologists. They should be the variables believed to be most important in driving the distribution of plant and animal communities in the region. A huge number of studies explore the correlations between plant and animal distributions and abiotic variables, and understanding this correlation is at the heart of species distribution modeling. Although the broad set of variables that influence the distribution of ecological communities are consistent, their relative importance will depend on scale and local conditions in the region. TABLE 5.3 lists variables that have all been found to influence ecological communities and have been used in delineating environmental units.

Because the number of relevant variables is generally limited, it is sometimes possible to use all of them to help delineate abiotic features.

**TABLE 5.3** Abiotic variables that are commonly used to delineate environmental units because of their influence on ecological community structure

| Variable | | Common source |
|---|---|---|
| Terrestrial | Elevation | Digital elevation model (DEM) |
| | Aspect | DEM |
| | Slope | DEM |
| | Topographic position | DEM |
| | Landform | DEM |
| | Geology of bedrock | Gov't geological or mining agencies |
| | Soil/surface geology | Gov't geological or mining agencies, Global Soil Map |
| | Temperature | Local weather stations |
| | Precipitation | Local weather stations |
| | Hydrologic conditions (e.g., inundation) | DEM, vegetation types |
| | Insolation | DEM |
| Marine | Depth and bathymetry | Nautical maps |
| | Sediment type | Gov't geological or mining agencies |
| | Temperature | Ocean color satellite image |
| | Rugosity | Vessel-based acoustic sensors |
| Freshwater | Stream size | DEM |
| | Stream gradient | DEM |
| | Geology | Gov't geological or mining agencies |
| | Elevation | DEM |
| | Hydrology | Water balance models or field observations |

However, most studies that we know of or have been involved in have found that a subset of 2–4 variables has greater confidence and explains most of the variation. It is not necessary to use the same set of variables across the entire region; for example, in a conservation plan we completed for the country of Papua New Guinea, abiotic features in highland areas (> 600-m elevation) were defined by slope, geological substrate, and elevation, while in the lowlands (< 600 m) the classification was based on landform and frequency of inundation. It is also possible to reduce the overall number of variables by choosing variables that integrate others. For example, solar insolation integrates elements of latitude, aspect, and slope.[26]

***Classifying Abiotic Variables*** Each selected abiotic variable needs to be divided into classes. For example, slope might be classified by the 10 classes in TABLE 5.4. In rare cases, a particular set of classes will have obvious ecological support; for instance, a region that is flat with a single flat plateau might have only three classes of elevation. More commonly, however, variables are simply divided into largely equal intervals with a filter for ecological relevance applied. Deciding how many classes are appropriate for a variable is an exercise in striking a balance between retaining as much variability as possible and having a reasonable number of classes to work with.

**Box 5.3** describes how these selected and classified variables can be subsequently turned into a set of abiotic features.

**TABLE 5.4** Potential division of slope into classes in order to create a set of abiotic features

| Class | Slope (degrees) |
|---|---|
| 1 | 0 |
| 2 | 0–2 |
| 3 | 2–5 |
| 4 | 5–10 |
| 5 | 10–15 |
| 6 | 15–20 |
| 7 | 20–25 |
| 8 | 25–30 |
| 9 | 35–40 |
| 10 | > 35 |

*Note:* It makes no sense to have any further divisions beyond 35 degrees slope, because the total area belonging to such a class will be minuscule and will be largely bare cliff anyway.

## Ecosystem Service Data

Data on ecosystem services are often based on the same types of underlying data just discussed, particularly a combination of land use, land cover, and geology. We discuss these data in detail in Chapter 10, which focuses specifically on ecosystem services in conservation planning.

## Social Data

For much of the history of conservation planning, the term *social data* has been largely synonymous with threats to biodiversity. Data have been gathered either on direct measures of human use of natural resources (e.g., the harvest of fish or timber), or surrogate or proxy measures of human impact on the environment (e.g., human population density). These data have been used to help determine the priority of conservation action for species or habitats, either by targeting or avoiding areas with these threats. A second kind of social data frequently used in conservation plans is data on social or community support for conservation. This might be spatial data on locations where communities would or would not be supportive of

## BOX 5.3 Generating Conservation Features from Abiotic Data

**Assigning Features**

Once a relevant set of abiotic variables has been selected and classified (see main chapter text), the variables need to be combined to give a taxonomy of features that cover the complete landscape (e.g., see Figure 3.8). Although this task is impossible to do in a truly objective way, it must be done in a consistent and well-documented fashion so that it is clear how the features were produced and that the methodology and resulting product can be adequately evaluated. In general, there are two approaches to assigning a set of features from abiotic variables: spatial intersection and statistical clustering.

Spatial intersection simply involves overlaying all variables in a GIS and considering every unique combination of variables and classes as a separate feature. The upside of this approach is that it is straightforward and does not require any decisions about the total number of features. The downside is that the number of features can quickly become vast, and some of them are almost certainly purely an artifact of the data layers rather than reflecting a unique abiotic environment.

The second approach to combining variables into features is through statistical clustering or ordination. Clustering is a statistical procedure that can use any one of a large number of algorithms to assign all the data to a number of groups or clusters, the members of which are more similar to each other than they are to the other data. A good example is the *k*-means clustering algorithm, which clusters all the data into *k* number of groups by making sure that each occurrence is closer to the mean value of its group than any other group. However, the person performing the analysis needs to define what *k* is—in other words, the total number of desired features. This number will be difficult to have an intuitive feel for a priori, and even if an analyst does, defining the number based on gut feel will certainly increase the subjectivity of the result. Increasing the total number of clusters (or features) will always lead to better fit with the observed data and explain more of the variation. But there is a price to pay—the more features, the more unwieldy the data become; they are difficult to map, difficult to analyze, difficult to communicate. Fortunately, most clustering algorithms have accompanying methods for determining the best value for *k*. These can be as straightforward as plotting the number of clusters against the variation explained and looking for the point of diminishing returns.

Ultimately, it is impossible to completely remove arbitrariness from the question regarding how many features to divide the landscape into. The Australian conservation biologist Simon Ferrier,[a] who has been instrumental in developing methods to classify ecological systems for conservation planning, suggests that if in doubt, selecting the largest number of classes from those that could plausibly make sense will reduce the risk of overlooking a distinctive feature.

**Stratification**

If you are a diver and have had the good fortune to dive around islands with coral reefs, you know that the reefs on different sides of the island are very different. From the perspective of an abiotic classification, the reefs on different sides of an island often receive the same type of classifications, for example, *fringing reef slope between 5 and 15 meters deep*. But when you dive on them, you just know that those reefs are different; the corals have different growth forms, and you see different fish on them. If you were charged with conserving the marine diversity of that island, you probably would not be happy substituting a reef on one side for a reef on the other side.

The same trend of broad classifications not capturing known variation across a planning

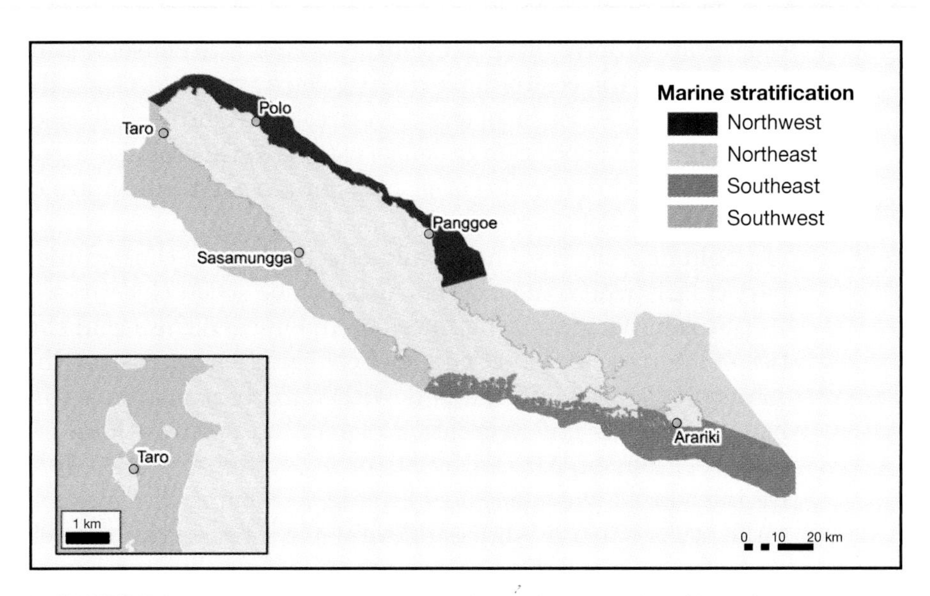


FIGURE 5.3.1 Broad divisions used to stratify an abiotic coral reef layer for a conservation plan in Choiseul Province, Solomon Islands.

region is also commonly true for community or abiotic classifications in terrestrial and freshwater environments. When establishing conservation features, planners can resolve this issue to some extent by using stratification. *Stratification* is a post hoc process whereby an existing set of features is further classified according to some biogeographic divisions. So the reef features around the island just described might now be *fringing reef slope between 5 and 15 meters deep windward side* and *fringing reef slope between 5 and 15 meters deep leeward side*, and so on. We used this exact logic to stratify the reefs of Choiseul Province in the Solomon Islands, according to exposure to different winds and ocean conditions (FIGURE 5.3.1), creating a set of features more resolved than the existing classification.

Stratification is a crude but effective way to create features that are more representative of a region's biodiversity than existing data sets. Over very large areas, such as national assessments, existing biogeographic divisions like ecoregions might serve as effective stratification units. Stratification is also a way to combine other classifications. For instance, a coarse vegetation classification might be further stratified according to an abiotic classification. No doubt, a bit of faith and a fair dose of subjectivity are involved here. Deciding whether to further stratify your features and by what divisions is a matter of deciding what will best reflect variation in natural systems across the region.

---

[a] Ferrier, S., *Mapping spatial pattern in biodiversity for regional conservation planning: Where to from here?* Systematic Biology, 2002. **51**(2): 331–363.

conservation. Or, it might be information regarding what sort of strategies would be more or less acceptable. For example, many communities in Melanesia have a strong preference for management arrangements that allow periodic harvesting for ceremonial purposes.[27]

These data on human use and impact on nature, and support for conservation, have often been incorporated into conservation plans as indicators of the cost of undertaking conservation actions. It generally costs more to work in places that are highly utilized or degraded, or where there is opposition to conservation. In this section we do not deal explicitly with either threat or cost data, both of which are covered in different parts of the chapter. Instead, this section focuses on data that can help improve the impact and effectiveness of conservation actions, particularly where the projects also include social objectives.

It is important for planners to understand the social characteristics of an area, and particularly how societies interact with the environment. People's interactions with nature are captured in data on how livelihoods and recreation depend on nature (there is clearly a close relationship here with data on ecosystem services). In Chapter 3 we introduced the concept of human well-being dimensions and how to identify which might be relevant for a conservation project. Data on these relevant dimensions serve as valuable markers of project impact, especially where human well-being is an explicit objective, but they can also provide valuable insights into the situation of a planning area. For example, high respiratory disease burdens might reflect a predominance of wood or charcoal for home cooking, something with a clear link to surrounding environments.

In most cases, some social data relevant for conservation planning will have been gathered by public agencies as part of routine data collection. Most developed countries, for example, collect a wide range of social data from household surveys, and these are usually available from a census bureau or similar such agency. In developing nations, livelihood statistics can serve as a proxy indicator for the extent of natural resource dependence in an area. A country's subnational census data usually shows types of jobs in a target area (e.g., percentage of farmers and fishers in a province/state). Demographic Health Surveys can be useful for understanding population growth, household size, and health issues.

A useful starting point for tracking down social data is to review current or recent development project documents. Most multilateral and bilateral development agencies (e.g., World Bank, Asian Development Bank, Inter-American Development Bank, USAID) post project appraisal documents online. Drawing on documentation from development projects in the region can assist conservation planners greatly in evaluating available social data as well as likely social pitfalls for projects. What social data are publicly available and easily accessible will vary enormously from place to place, even sometimes within a state or province. There will often need to be supplemental or targeted data collection, especially in developing nations.

**TABLE 5.5** contains examples of social data likely to be useful for conservation planners, along with some possible sources of these data. Social data come in both quantitative and qualitative flavors. **Box 5.4** describes the relative merits and uses of both kinds; the general consensus (and our experience) is that a mix of quantitative and qualitative data will deliver the greatest understanding of social context and trends.

For conservation plans that include **community pool resources**, the Nobel Prize–winning economist Elinor Ostrom[28] has identified a

**TABLE 5.5** Social data useful for conservation planning

| Desired information | Possible sources |
|---|---|
| Livelihoods, particularly number and value of jobs involved with natural resources | Economic and commerce statistics, industry- or sector-held data |
| Dependence on natural resources (e.g., percentage or livelihood, alternative sources of income, other available fuels, food dependency, subsistence extent, etc.) | Focus groups, household surveys |
| Use of natural areas | Focus groups, household surveys |
| Community access to natural areas | Land or water zoning and ownership, municipal settlement maps |
| Degree that natural areas or natural resources are valued by people and for what reasons, and spatial variation in this | Focus groups, formal surveys, visitation statistics, facilities |
| Gender and age differences in the previous questions | Household surveys, focus groups |
| Security and conflict, particularly over natural resources | Armed Conflict Location & Event Data Project (http://www.acleddata.com/) |
| Legal rights of communities and other stakeholders to natural resources | The IUCN/FAO/UNEP ECOLEX environmental law database (http://www.ecolex.org) |
| Natural resource governance arrangement (e.g., officially managed by the state, comanaged by the state and local communities, or managed entirely by the local communities) | |
| Social heterogeneity (e.g., ethnic, linguistic, and religious fractionalization) | Norwegian Social Science Data Services[1] (http://www.nsd.uib.no/nsd/english/) |
| Access to basic services | Census, development project reports |
| Education | Census |
| Illness and disease | Census, Demographic and Health Surveys (http://www.measuredhs.com/) |
| Social institutions and governance structures | Development project reports |

[1] Alesina, A. et al., *Fractionalization*. Journal of Economic Growth, 2003. **8**(2): 155–194.

## BOX 5.4 Quantitative versus Qualitative Social Data Collection

Social scientists once thought of quantitative data collection and qualitative data collection as fire and ice. Many argued the two were incompatible. A paradigm war ensued. Quantitative proponents argued that their approach was more objective and led to useful generalizations with predictive power. Qualitative proponents argued that all research is value bound, that it is impossible to differentiate causes and effects, and that deep, rich observational data is a better path to knowledge. Two decades of thesis and antithesis eventually led to synthesis. By the early 2000s, mixing quantitative and qualitative data collection had become the dominant paradigm in social science research.[a] Mixed-method research has been aptly called "Q-squared" because integrating quantitative and qualitative approaches results in a greater sum of knowledge than simply adding one to the other.[b]

Within conservation planning, quantitative social data are useful for inferring causality and developing evidence-based generalizations but have little to say about exactly how or why the observed impact occurred. Qualitative social data (or more accurately, social information) are useful for understanding nuances. The cheapest option for social data collection is to use qualitative information gleaned from interviews of stakeholders and experts. The more widely used option is to collect quantitative social data via a survey. The best option is to do both, thereby providing both the numbers and the stories behind the numbers.

---

[a] Johnson, R. B., and A. J. Onwuegbuzie, *Mixed methods research: A research paradigm whose time has come.* Educational Researcher, 2004. **33**(7): 14–26.

[b] Kanbur, R., and P. Shaffer, *Epistemology, normative theory and poverty analysis: Implications for Q-squared in practice.* World Development, 2007. **35**(2): 183–196.

set of 10 social variables that are directly relevant to the sustainable management of natural resources (TABLE 5.6). These serve as a useful guide to data that can significantly contribute to a conservation plan.

Where social data are largely absent, a baseline qualitative and quantitative social assessment is a useful tool for informing the detailed planning and later evaluating progress toward the fundamental objectives. Such a baseline assessment should focus on understanding local priorities and dependence on natural resources via key informant interviews and focus group discussions, and household surveys for quantitative baselines. A great deal is being written about these methods in the development literature.[29]

### Temporal and Spatial Resolution

For both social and ecological data, it is important to match the resolution of social data to the scale of the decisions at hand. For instance, if there is a decision to be made about which watersheds to work in, it would be important to have ecological data at a sub-watershed scale and

**TABLE 5.6** Ten variables that affect the likelihood of a sustainable social-ecological system (SES)

1. **Size of resource system**
"Moderate territorial size is most conducive to self-organization."

2. **Productivity of system**
"Users need to observe some scarcity before they invest in self-organization."

3. **Predictability of system dynamics**
"System dynamics need to be sufficiently predictable that users can estimate what would happen if they were to establish particular harvesting rules or no-entry territories."

4. **Resource unit mobility**
"Self-organization is less likely with mobile resource units, such as wildlife or water in an unregulated river."

5. **Number of users**
"The impact of group size on the transaction costs of self-organizing tends to be negative."

6. **Leadership**
"When some users of any type of resource system have entrepreneurial skills and are respected as local leaders as a result of prior organization for other purposes, self-organization is more likely."

7. **Norms/social capital**
"Share moral and ethical standards regarding how to behave in groups they form, and thus the norms of reciprocity, and have sufficient trust in one another to keep agreements."

8. **Knowledge of SES**
"When users share common knowledge of relevant SES attributes, how their actions affect each other, and rules used in other SESs, they will perceive lower costs of organizing."

9. **Importance of resource to users**
"Users are either dependent on the resource system for a substantial portion of their livelihoods or attach high value to the sustainability of the resource."

10. **Collective-choice rules**
"When users...have full autonomy at the collective-choice level to craft and enforce some of their own rules, they face lower transaction costs as well as lower costs in defending a resource against invasion by others."

*Source:* Data from Ostrom, E., *A general framework for analyzing sustainability of social-ecological systems.* Science, 2009. **325**(5939): 419.

social data at the district or municipal level. If two potential project sites were in different countries, we would likely not be content with national descriptions of ecological characteristics, and we should not assume that national-level social statistics are necessarily appropriate. **Box 5.5** describes some of the challenges in aligning the scale of plans with the scale of decisions and available data.

## BOX 5.5 Importance of Considering Spatial Scale in Planning

One of the most important challenges of conservation planning is a mismatch in spatial scale. This often occurs between the coarser scales at which many conservation plans are developed and the more local scales at which conservation decisions tend to be made and conservation actions implemented. Dave Theobald, a landscape ecologist and conservation planner, has faced this challenge in working with local governments in Colorado, USA, to develop and implement biodiversity conservation plans. **FIGURE 5.5.1**, taken from work by Theobald and his colleagues, illustrates the challenge of the different scales at which various ecological processes occur, at which different conservation plans are developed, and at which actions by local governments (master plan, zoning, subdivision review) tend to occur.

What are the implications for conducting planning at different spatial scales? First, many studies have shown that changes in the geographic extent of the area being considered in a plan can have dramatic effects on the number and locations of priority areas.[a] Second, changes in the grain size (the finest unit of spatial resolution that is possible with any particular data set) used in a conservation plan can also have considerable effects on a plan's priorities. In South Africa, conservation planner Mathieu Rouget developed two different systems of putative conservation areas: one based on fine-scale biodiversity data and the second based on broader-scale data.[b] Although both systems of conservation areas required similar amounts of land to meet the conservation goals of the project, the system developed with broad-scale data underestimated the

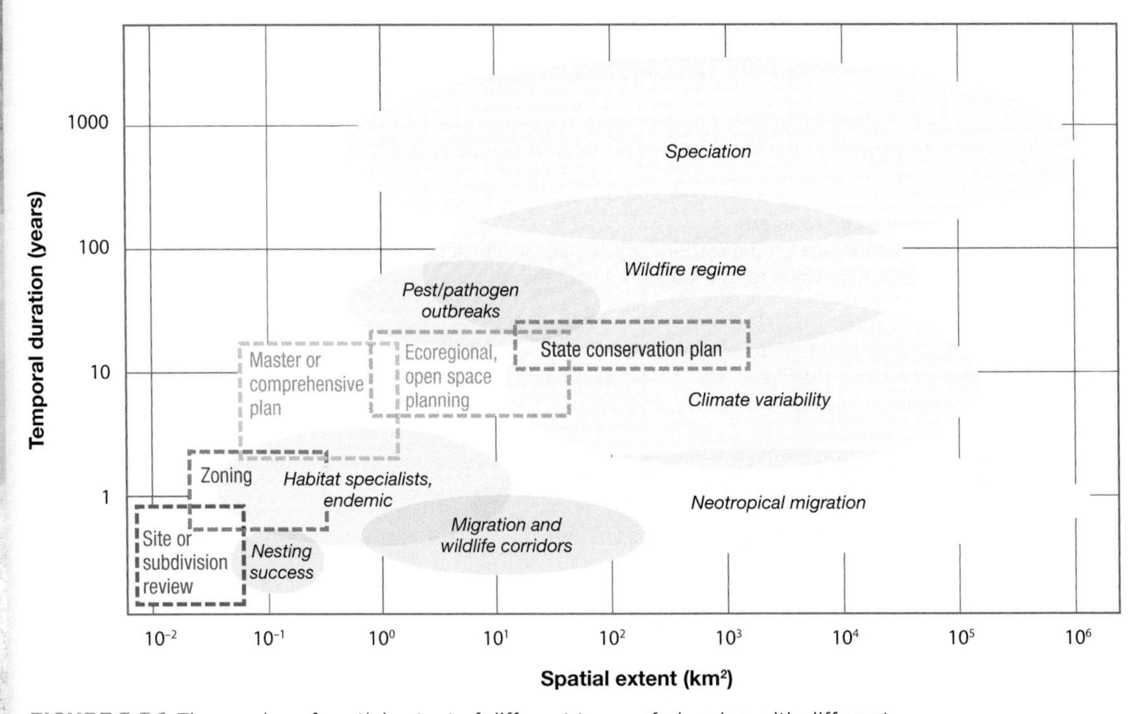


**FIGURE 5.5.1** The overlap of spatial extent of different types of planning with different ecological processes. (Adapted from Theobald et al., 2005.)

conservation value of several biodiversity features that were identified with finer-scale data and that occurred in areas that were heterogeneous and fragmented by human development. Third, several landscape-scale analyses have demonstrated that the spatial extent of a planning area is one of the variables that can influence various metrics of landscape connectivity, which in turn has implications for the design of corridors in conservation plans.[c] Finally, scale mismatches can confound the incorporation of ecological processes or social-ecological systems that may exceed the boundaries of planning units and can also result in inadequately addressing some threats that may act at a variety of scales.[d]

The effects of changes in spatial scale on conservation plans suggest that there are clear advantages and disadvantages of planning at different scales. They also raise the question of whether there is one appropriate scale for conservation planning. The simple answer is no. The appropriate scale will most often be governed by the organization or institution conducting the planning, the objectives of the project, and the data available for the plan.

For example, several international conservation organizations have developed regional or ecoregional conservation plans for areas larger than most geopolitical units such as states or provinces. Such plans have the advantage of being better able to incorporate the distributions of ecosystems and many species that are associated with particular ecosystems. On the other hand, these ecoregional scales may be a disadvantage if a government body to whom geopolitical boundaries are important is either responsible for the planning process or seen as a primary implementer.

In some situations, it may be useful to consider a set of nested conservation plans, with an umbrella plan addressing policies and strategies at a larger scale and other, more narrowly focused plans addressing finer-scale issues. At the very least, a good rule of thumb for any planning team is to work at a spatial scale larger than the unit for which they are developing a plan, to ensure that they consider what is happening in the lands and waters adjacent to the planning unit.

---

[a] Huber, P., S. Greco, and J. Thorne, *Spatial scale effects on conservation network design: Trade-offs and omissions in regional versus local scale planning.* Landscape Ecology, 2010. **25**(5): 683–695.

[b] Rouget, M., *Measuring conservation value at fine and broad scales: Implications for a diverse and fragmented region, the Agulhas Plain.* Biological Conservation, 2003. **112**(1–2): 217–232.

[c] Pascual-Hortal, L., and S. Saura, *Impact of spatial scale on the identification of critical habitat patches for the maintenance of landscape connectivity.* Landscape and Urban Planning, 2007. **83**(2–3): 176–186.

[d] Guerrero, A. M. et al., *Scale Mismatches, Conservation Planning, and the Value of Social-Network Analyses.* Conservation Biology, 2013. **27**(1): 35–44.

With regard to the scale of social data, Sheri Stephanson and Michael Mascia[30] offer the following rules of thumb for conservation planning:

- Ecoregion scale planning should employ social data at the district or subdistrict level.
- Landscape-scale planning requires subdistrict or community-level data.
- Community-scale planning will require social data at the household level.

Much of the publicly available social data collected through means such as government censuses or health and welfare statistics are aggregated at the district level, especially in rural areas, which means that finer-scale conservation planning may need to invest in supplemental social data collection to be informative. Because social data aggregated by political units typically mask a great deal of spatial heterogeneity (FIGURE 5.15), the conservation social scientist Michael Mascia warns against simply *cropping* or *clipping* social data to fit a planning region. Wherever possible, he suggests, it is far better to align social data with a planning region through disaggregation of data sets, spatially explicit modeling, or even additional targeted data collection. This means that, in general, the finer the resolution of social data, the more flexible and useful they are likely to be.

It is also important to consider the temporal resolution of the data being collected. For example, a map of ecological communities based on decades-old imagery will almost certainly be out of date. Similarly, in many countries census data are collected only once per decade, meaning that the trends reported in the data may be 10 years old. A decision usually needs to be made about whether existing data represent a reasonable current baseline.

## Cost Data

We've said it before, and we'll say it again; conservation planning is about making good decisions so that we use the resources available most efficiently. It is crucial therefore that information about the actual resources available, or required, form part of any conservation plan. The resources at play in a conservation plan can be of many sorts—dollars, staff time, expertise, equipment, and so on—but ultimately, most resources can be expressed as financial resources. Without consideration of costs, there can be no claim to efficient planning.

We dedicate a significant part of this chapter to data on the costs of conservation. We've done this because we believe that an appreciation of cost is a crucial but very frequently missing piece of conservation plans. No doubt this stems largely from the historical separation of planning as a science exercise and resource allocation as a management exercise, which we discussed in Chapter 1. If we are to remedy this and ensure conservation plans have maximum impact on the way we conserve our natural resources, then planners must engage more robustly with the resources required to implement the outcomes of plans.

There are two broad dimensions to cost data relevant for conservation planning:

1. How much does it cost to implement (including monitoring) a conservation intervention?
2. What resources are available to be committed?

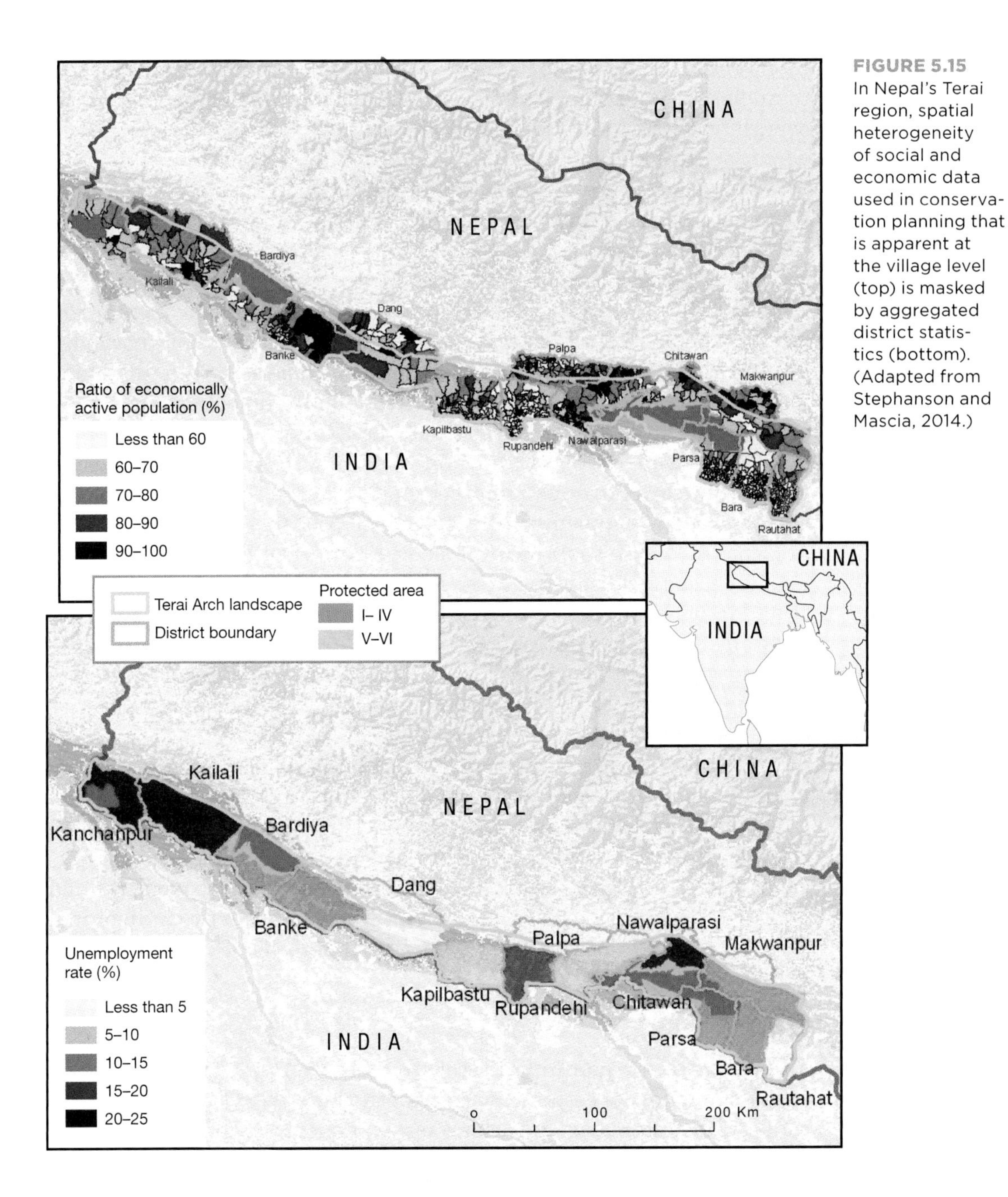


**FIGURE 5.15** In Nepal's Terai region, spatial heterogeneity of social and economic data used in conservation planning that is apparent at the village level (top) is masked by aggregated district statistics (bottom). (Adapted from Stephanson and Mascia, 2014.)

This second dimension of cost data is useful and should be included where available (especially as a constraint or cap on what is being considered), but it is not essential. Available resources are often a highly mutable quantity, sometimes depending on the outcome of the plan itself (e.g., the plan can be a tool to help raise additional funds). The first dimension of cost data—how much interventions are expected to cost—is critical. Not only does knowing the costs of interventions help make critical decisions about which interventions to implement, but those same costs will play an important role in the overall budgeting and fund-raising aspects for a conservation project, which we explore further in Chapter 11 on implementation.

It is easy to see why knowing the cost of different options is intrinsic to good decision making. The impact of an intervention might look extremely appealing, but if it costs an enormous amount, either it might be simply impossible to implement or the outcome might be worse than implementing multiple cheaper interventions. Plenty of papers are written about the inefficiency of planning in the absence of knowledge about the variation in cost between options.[31] There are equally many anecdotes about conservation projects that ended up being vastly more expensive than expected, but probably foreseeably so.

For example, when California condors were reintroduced in the United States, a huge cost, unaccounted for at the time of the decision, was the need to supplementally feed them, probably forever, due to the presence of lead in their natural foods. Similarly, in New Zealand, the long-term survival of blue ducks (*Hymenolaimus malacorhynchos*) and other native ground-dwelling birds is dependent on broad-scale and constant trapping of introduced stoats (*Mustela erminea*) to increase nesting success and productivity.[32] These are huge recurring costs to keep a species persistent in one place, and at obvious opportunity costs to other ways of using those funds.

Despite the intuitive sense of this idea, it is not necessarily easy to get an accurate understanding of how much different interventions will cost or even have cost. When we tried to determine the overall costs for different strategies within The Nature Conservancy, it was clear that the diverse sources of contribution to most conservation efforts made tracking the total cost of it extraordinarily challenging. As a field, conservation has not kept great records of the overall cost of different activities.

### Types of Cost

The conservation planning literature has given reasonable attention to costs and how to estimate them, particularly with regard to spatial conservation planning. It is clear, for example, that the costs of doing conservation vary across geographies. The conservation biologist Robin Naidoo, together with some economically minded conservation colleagues, identified five different types of costs relevant for conservation planning.[33] The focus of their work was on spatial conservation planning, so we have adapted and added to their list to make it relevant for strategies (**TABLE 5.7**).

Because actual data on costs associated with conservation are scarce, a wide range of proxies for conservation costs have been proposed and used, such as area, land value, productivity, historic catch

TABLE 5.7 Types of economic "costs" relevant for consideration in conservation plans

**Acquisition costs**

Acquisition costs are costs of acquiring rights to a resource. Typically this is a parcel of land or water, but it can include rights to a mobile resource such as a fisheries license. Acquisition of property rights can be total (i.e., the land and title are sold to a conservation agent) or partial. Partial transfers of property rights include short-term resource rental, conservation easements, and contracts between conservation agents and resource owners that exchange money for management that enhances conservation value.

**Establishment costs**

Establishment costs are one-off costs incurred at the beginning of a conservation project, such as the purchase of equipment or vehicles, or building infrastructure such as a predator fence or fish ladder.

**Management or maintenance costs**

Management or maintenance costs are those associated with management of a conservation project, such as those associated with maintaining conservation areas (e.g., maintaining fences, enforcement), monitoring a species population, or ecological stewardship such as prescribed fire and invasive species removal.

**Transaction costs**

Transaction costs are those associated with negotiating the contractual arrangements for a conservation project. These include the time and resources (e.g., legal fees) involved in negotiating and financing acquisitions, gaining the necessary agreement from relevant stakeholders, or gaining approval and certification for financial mechanisms like carbon payments. Transaction costs can be substantial, especially when staff time is taken into account.

**Damage costs**

Damage costs are those associated with damages to economic activities arising from conservation programs; for example, damages to crops and livestock from wild animals living in protected areas adjacent to human settlements can result in significant losses in income. In other cases, direct wildlife attacks might physically harm or kill humans, resulting in further economic losses.

**Opportunity costs**

Opportunity costs are profits that are forgone as a result of a conservation project—they are a measure of what could have been gained via another use of a resource had it not been part of a conservation project. In terrestrial protected areas where extractive uses are forbidden, the opportunity cost represents the profit that could be expected were the land used for extractive purposes (e.g., agriculture). From a social perspective, it is important to include opportunity costs to track the full set of consequences of conservation planning.

*Source:* Adapted from Naidoo, R. et al., *Integrating economic costs into conservation planning.* Trends in Ecology & Evolution, 2006. **21**(12): 681–687.

value, and so forth. TABLE 5.8 gives examples of different information that has been used as proxies for cost in both marine and terrestrial conservation plans, the source of these data, and the rationale behind their use. In some cases, they are actual estimates of cost (e.g., the management costs for protected areas derived by conservation biologist Andrew Balmford and colleagues), but they are still used in the fashion of proxies to reflect the cost of conservation more generally.

The examples of cost surrogates described in Table 5.8 all have compelling rationales behind their use. However, in the rare cases where surrogate values have been compared to actual costs of doing conservation, their performance has been mixed at best. The ecologist Paul Armsworth and colleagues have found that at the property level, there is a poor

**TABLE 5.8 Example, sources, and rationale for different types of cost data in conservation planning**

| Cost type | Examples of data | Source | Rationale |
|---|---|---|---|
| Management cost | Modeled annual management costs of protected areas; terrestrial,[1] marine[2] | Derived from World Bank or United Nations statistics on GNP and PPP, and nationally provided GIS layers with the location of population centers | Management costs, dependent on the country, location, and size of a protected area, are indicative of the public investment required in a conservation area. |
| | Predicted costs associated with conservation projects; threatened species recovery[3] | Elicited from experts | The actual predicted costs of successfully implementing one of the conservation options being considered. |
| Opportunity cost | Expected agricultural productivity[4] | Remotely sensed imagery | The more productive the land, the more it is likely to be valued for activities other than conservation, like farming, and the more costly it will be to displace these activities. |
| | Historical fishing revenue[5] | Provincial or national fishing agency records | Past catch trends are likely to be indicative of how important an area is to fishermen and therefore what compensation or opposition can be expected if fishing is excluded from that area. |
| Purchase cost | Current market value of land[6] | Real estate data | Market value of land is likely to be roughly indicative of how much it would cost to acquire the property for conservation. |
| Transaction cost | The number of language groups with which negotiation would need to happen[7] | Government cultural and educational data | Getting agreement on conservation management arrangements across different tribes will likely increase the transaction costs involved in establishing a conservation area. |

[1] Bode, M. et al., *Cost-effective global conservation spending is robust to taxonomic group.* Proceedings of the National Academy of Sciences of the United States of America, 2008. **105**: 6498–6501.

[2] Balmford, A. et al., *The worldwide costs of marine protected areas.* Proceedings of the National Academy of Sciences of the United States of America, 2004. **101**(26): 9694–9697.

[3] Joseph, L. N., R. F. Maloney, and H. P. Possingham, *Optimal allocation of resources among threatened species: A project prioritization protocol.* Conservation Biology, 2009. **23**: 328–338.

[4] Naidoo, R. and T. Iwamura, *Global-scale mapping of economic benefits from agricultural lands: Implications for conservation priorities.* Biological Conservation, 2007. **140**: 40–49.

[5] Klein, C. J. et al., *Spatial marine zoning for fisheries and conservation.* Frontiers in Ecology and the Environment, 2010. **8**(7): 349–353.

[6] Murdoch, W. et al., *Using return on investment to maximize conservation effectiveness in Argentine grasslands.* Proceedings of the National Academy of Sciences of the United States of America, 2011. **107**(49): 20855–20862.

[7] Lipsett-Moore, G. et al., *Interim National Terrestrial Conservation Assessment for Papua New Guinea: Protecting Biodiversity in a Changing Climate.* 2010. Brisbane, Australia: The Nature Conservancy.

relationship between remotely sensed agricultural productivity and the actual cost required to add a property to the conservation estate.[34] This is easily explained by the host of property-specific features such as topography, water access, rocks, and so on that influence property value, together with vast variation in a landowner's predisposition to biodiversity conservation and current financial position. The use of agricultural productivity as a cost surrogate is likely to be more informative at a larger scale where it might be used to prioritize between regions.

Another reason that cost surrogates might perform poorly is if they do not reflect the actual actions being undertaken. Many cost surrogates have been developed under the assumption that the principal action is akin to acquiring the land to establish and manage a conservation area. In reality, the set of actions being planned for might include policy or practice changes on private land. Without some knowledge of the conservation activities likely to be implemented in places identified as a priority, it is difficult to know what factors are likely to drive the cost.

The discussion of surrogates or proxies for conservation costs has largely been triggered by spatial conservation planning. However, increasingly we are interested in comparing diverse projects and not just a set of places. It will be the rare case when numbers on the cost of different projects are simply available in some database. Rather, getting data on the costs of projects will typically involve asking expert practitioners what different interventions are likely to cost. More and more conservation plans are doing this explicitly. In addition to following the advice later in this chapter on expert data, next we provide some guidance specifically related to estimating costs, and the sort of adjustments to these costs that might be needed to use them defensibly in a conservation plan.

## Estimating Costs

***Estimate costs in monetary units.*** Costs are naturally measured in monetary terms. Although many different things contribute to the cost of a conservation project, they can all be reduced ultimately to money. If they can't, it's a good indication that cost is not the best place for these considerations. Many aspects of conservation do not have such an easy natural unit that allows comparison across alternatives, but cost is not one of these; so don't make it more difficult than it needs to be.

***Identify the types of cost that should be considered.*** Before asking experts about the costs of interventions, it is critical to establish a classification of the costs that should be considered. Fairly generic categories will often be fine, such as staff, capital outlays, legal, equipment, travel, fund-raising, stakeholder engagement, and so forth. Clearly identifying these categories will help ensure that experts consider the same sort of costs and don't simply forget one or more elements (which is likely), and it will help ensure that the cost estimates better reflect the true cost of the intervention.

***Separate costs into time or project periods.*** It can also help sometimes to break up interventions into periods such as start-up or approval, implementation, and ongoing. This approach is useful because it is important when costs are incurred and because different interventions will have different cost profiles and different sustainability needs. Remember also not to ignore background organization costs, such as overhead and other administrative charges.

***Use history as a guide.*** Estimates can often be improved by looking at what things have cost in the past.

***Estimate ranges.*** There will usually be some degree of uncertainty in the expected cost of an intervention; this can be at least partly reflected by estimating ranges of cost (e.g., US$25,000–$40,000) rather than a single value.

***Be clear about whose costs.*** It might sound obvious, but it is critical to be clear about whose costs are being considered in the planning process. In some cases it might be relevant to consider only the costs relating to a single organization; thus, interventions likely to have multiple sources of contributing funds will appear relatively cheaper. On the other hand, if planning is occurring with partners, it might be necessary to consider all costs involved, regardless of where they are likely to arise.

***Use an "ingredients" approach to cost.*** Cost-effectiveness analysis experts point out that cost and volume (quantity) should be separated when costs are being estimated, so that the cost of something like staff time can be changed between areas to reflect legitimate differences in these costs and updated through time if need be. This means that there shouldn't just be a figure for "staff time"—the amount of time being estimated should be listed along with the estimated price per unit of staff time. This is known as an ingredients approach to cost estimation.

Similarly, the unit at which costs should be estimated will be different for different actions; it might be per hectare, per property, per year, or once off. TABLE 5.9 shows the variety of units at which the cost of different actions was estimated for a single conservation planning effort in the Kimberley region of Australia. All these estimates were obtained rapidly through a series of expert workshops.

Make the necessary adjustments when comparing costs between countries, or that occurred in the past or will occur in the future. One dollar does not buy the same amount in each country, nor is one dollar

**TABLE 5.9** Example of estimates for the cost of different conservation actions, taken from a conservation planning exercise in Australia's Kimberley region

| Action group | Objective | Activities | Costs |
|---|---|---|---|
| 1. Fire and introduced herbivore management<br>1a. Implement appropriate fire management | Increase number of clumps with 3+ year old post-fire vegetation over entire region<br>Reduce significantly the size of single fire events over entire region | Develop fire management plan for each tenure (pastoral, indigenous, DEC, and EcoFire were considered separate management models) | $0.10–0.30/ha/year depending upon land tenure and geographic location (slightly higher in northern Kimberley) |
| | | Aerial control burning in late wet season and early dry season | |
| | | Fire-scar monitoring and analysis | |
| | | On-ground burning in focal areas and around assets | $0.10–0.40/ha/year depending upon land tenure and geographic location (slightly higher in southern Kimberley) |
| | | Targeted fire suppression | $0.10/ha/year |
| | | Build relationships and capacity with landholders | $2–$2.25 million/year for each land tenure type |
| | | Education about inappropriate fire management, enforcement of regulations | |
| 1b. Manage domestic and feral cattle, plus manage other feral herbivores | Manage grazing pressure such that biodiversity is not adversely impacted, maintaining ecological function<br>Reduce impacts in sensitive areas | Pastoral tenure: define best management practice and develop a management plan | $10,000/property, once off |
| | | Pastoral tenure: extension/ education about sustainable grazing generating higher profits (and achieving successful outcomes by this) | $3,000/property, once off |
| | | Non-pastoral tenure: develop a management plan that includes landholders, raises awareness, provides employment, provides incentives to remove animals, provides options to sell or use animals, considers relationships between people and introduced animals, considers ethics | $10,000–$20,000/ group of land managers, once off |

*(continued)*

**TABLE 5.9 Example of estimates for the cost of different conservation actions, taken from a conservation planning exercise in Australia's Kimberley region (*continued*)**

| Action group | Objective | Activities | Costs |
|---|---|---|---|
| 1b. (*continued*) Manage domestic and feral cattle, plus manage other feral herbivores | (*continued*) Manage grazing pressure such that biodiversity is not adversely impacted, maintaining ecological function | All tenures: manage stock access using fences, water, fire | $300,000–$500,000/ property, once off + 10%/ year ongoing |
| | Reduce impacts in sensitive areas | All tenures: fence, muster, trap, remove feral herbivores (shooting where appropriate), using approaches shown to be effective (e.g., Judas donkey program, pig trap system used on Cape York Peninsula) | $0.15/ha/year on average (focused on preferred habitat such as along waterways) |
| 2. Weed management | Prevent invasion by new weeds and coordinate management (all regions) | Early detection and monitoring program (build on existing Northern Australia Quarantine Strategy program) | $800,000 once off + $150,000/year ongoing |
| | | Weed management strategy: workshop to liaise across groups and surveillance package, four full-time positions and support for an indigenous ranger program | $200,000 once off + $600,000/year ongoing |
| | Eradicate certain weeds of national significance (in regions specified) | Gamba grass *Andropogon gayanus*, eradicate within five years: search and spray, focus on creek lines (DL) | $500,000 once off + $10,000/year ongoing |
| | | *Mesquite Prosopis spp*: search and spray (DL, OVP) | $10,000/year ongoing |
| | | Rubber vine *Cryptostegia grandiflora*: pull out, herbicide, fire (two locations in DL and OVP) | $500,000/year ongoing |
| | | *Acacia nilotica*: search, herbicide, fire (west of Wyndham, VB) | $1,000,000 over 5 years |
| | | *Mimosa pigra*: remove one patch near Kununurra (VB) | $5,000–$10,000/year ongoing |
| | Contain and control other key weeds(in all regions, unless specified) | Grader grass *Themeda quadrivalvis*: start-up equipment, ranger costs, aerial survey (roadsides, e.g., Gibb River Road, southwest of Wyndham, VB, OVP) | $500,000 once off + $400,000/year ongoing |

TABLE 5.9 (*continued*)

| Action group | Objective | Activities | Costs |
|---|---|---|---|
| 2. (*continued*) Weed management | (*continued*) Contain and control other key weeds (in all regions, unless specified) | *Parkinsonia aculeata*: biocontrol (moth) and herbicide, eradicate in some areas, contain in others | $200,000/year |
| | | Stinking passionfruit *Passiflora foetida*: fire, biocontrol | $2,000,000 over 5 years |
| | | Neem *Azadirachta indica*: herbicide, basal spray, pull | $1,000,000 over 5 years |
| | | Bellyache bush *Jatropha gossypiifolia*, rubber bush *Calotropis procera*, butterfly pea *Clitoria ternatea*: identify realistic containment lines (catchments), map, plan, herbicide, cut and paste, good land management | $200,000 once off + $1,000,000/year ongoing |
| | | Trees like poinciana *Delonix regia*, raintree *Koelreuteria elegans* ssp. *formosana*, chinee apple *Ziziphus mauritiana*: mapping, control | $200,000 over 5 years |
| | | Buffel grass *Cenchrus ciliaris*: control around key conservation assets | $1,000,000 over 5 years |
| 3. Control of key introduced predators (cats) | Increase dingo numbers | Education to eliminate dingo baiting | $1,000,000 over 5 years |
| | | Compensation per property for animals killed by dingoes | $10,000/property/year |
| | Directly reduce cat numbers | Educate for spaying cats and controlling their access, and free spaying service (including one full-time position) | $150,000/year over entire region |
| | Improve knowledge | Research on cat ecology and treatment, including biocontrol | $12,000,000 over 5 years |

*Source:* Data from Carwardine, J. et al., *Priority Threat Management to Protect Kimberley Wildlife.* 2011, Brisbane, Australia: CSIRO Ecosystem Sciences.

today likely to buy the same as one dollar in 10 years' time. **Box 5.6** describes how costs can be adjusted to reasonably compare them across time and between countries.

## BOX 5.6 Making Cost Data Comparable

### Costs across Time: Comparing Past Costs to a Reference Year

It is often necessary to compare cost information from different time periods. For example, you might have estimates of the costs of interventions that occurred at different times over the last decade. Because the relative value of those costs changes through time (largely due to inflation), it is necessary to adjust costs so that they are comparable across years.

Several different measures of inflation can be used to adjust costs for different years, but the Consumer Price Index (CPI) is the most common. A CPI reflects the change in price to a consumer of a fixed "basket" of goods and services. Of course, this is an average change, so some things will fluctuate more significantly. For example, the cost of land related to mineral value might change far more dramatically than the CPI, but it's a good start for conservation. CPIs are available in most countries because they are typically used to index things like salaries and welfare.

To make past costs from different years comparable, first you need to pick a reference year—usually, the most recent year for which you have data. As an example, suppose we wanted to compare the cost of a project that happened in 2003 to a reference year of 2011. To do this, we would convert the 2003 cost to a 2011 cost using this formula:

Cost = cost in data year * (index value in reference year/index value in data year)

where in this case, the index value is the CPI for the relevant years, the data year is 2003, and the reference year is 2011.

Indices like CPI are often calculated more frequently than annually, so to get a single CPI value for the data and reference years, it might be necessary to take either the average CPI value for the year or the final CPI value.

### Comparing Future Costs to Present Costs

Conservation projects and interventions often have long life cycles, so planning might include interventions whose costs will be incurred annually for 20 years or more rather than all immediately following the plan. Costs that can be paid in the future are essentially cheaper. So if a conservation project was going to cost $10,000 a year to manage, in 10 years' time there is a good chance that $10,000 will not be as valuable to us as that same sum spent today (we may also have been able to use that money to generate more money).

To compare projects whose costs will be incurred over different times in the future (e.g., if one alternative involves mainly up-front costs whereas another involves annual costs), economists use the notion of discounting. Although discounting and how best to do it is a topic of substantial debate among economists, for practical purposes conservation projects can use a conventional view of discounting. This essentially means a discount rate equivalent to the opportunity cost of capital (basically the same as the returns you could expect from investing that money—often the interest rate in banks). Numerous conservation publications have assumed a discount rate of 5% as a fairly conservative estimate of capital opportunity cost.

To calculate the present value of future costs, we use this formula:

$$\textit{Present Cost} = \frac{C_t}{(1 + i)_t}$$

where $C_t$ is the cost $C$ incurred $t$ years in the future, and $i$ is the discount rate.

To calculate the full cost of a project in today's dollars, it's necessary to sum the value above across all $n$ years of the project:

$$\textit{Present Cost} = \Sigma_{t=1}^{n} \frac{C_t}{(1 + i)_t}$$

**Net Present Value**

Economists refer to future monetary values expressed in today's dollars as the net present value (NPV). The calculation of discounted costs given earlier is an example of an NPV calculation. More commonly, however, NPV is used in reference to future economic returns from actions. In the past this value hasn't been widely applicable to conservation plans because few of the benefits being considered have been monetized. This situation, however, is changing with the increase in projects that consider the value of ecosystem services, making NPV an important concept for conservation planners.

Just as money spent in the future is worth less than today, economic returns in future years are valued less than if those same returns were delivered today. Because interventions targeting ecosystem services will generally take different lengths of time to realize benefits, comparing between them requires calculating these expected values in today's money. For example, in a comparison of the ecosystem services delivered by land protection versus agricultural land uses on degraded lands in Ethiopia, the NPV of ecosystem services expected to be delivered over 30 years was compared with agricultural profits over the same time period.[a] The agricultural profits occur every year, but the ecosystem service benefits take time to accrue.

**Comparing Costs across Countries**

Conservation plans less frequently will be evaluating options in different countries, but it does happen. Comparing costs between countries also requires adjusting them. This is because the exchange rate is a poor indicator of what it actually costs to procure goods and services in a country. Anyone who has traveled overseas will appreciate this; for example, Americans traveling to Mexico generally find they can buy more for the same amount than they could in the United States, whereas if they travel to Australia they generally find that they can afford fewer things for the same amount (at least in recent years).

Costs can be compared between countries by indexing them according to purchasing power parity (PPP). PPP is a calculation of the amount of one currency needed to buy the same quantity of goods and services as one unit in a reference country—the United States. For example, conservation biologists Joselin Moore, Andrew Balmford, and colleagues used PPP to make estimates of protected area management costs comparable across countries.[b] PPP conversion factors can be found through the World Bank. A number of different ways have been proposed to compare between countries using PPP, though we do not summarize them here.

[a] Mekuria, W. et al., *Economic valuation of land restoration: The case of exclosures established on communal grazing lands in Tigray, Ethiopia.* Land Degradation & Development, 2011. **22**(3): 334–344.

[b] Moore, J. et al., *Integrating costs into conservation planning across Africa.* Biological Conservation, 2004. **117**: 343–350.

## Are Some Cost Data Better Than No Cost Data?

Available data on the cost of conservation are often poor. Not only can there be poor correlation between proxies and actual costs, but project managers don't often think in terms of those proxies, and our experience is that managers have little confidence in their relevance for conservation decisions. Just as the expert knowledge of an ecologist gives them low confidence in the abundance of birds reflecting the abundance of mammals, the expert knowledge of those who deal with the costs of conservation tells them that cost is a complex issue and proxies do a bad job of informing them. Given this, is trying to use some knowledge of costs better or worse than using no knowledge?

On the "for" side, it is often argued that the variation in costs across a region can be so substantial that even poorly correlated proxies will still be valuable in improving the efficiency of resource allocation. It is also plausible that consideration of some cost information will encourage conservation planners to seek better-informed cost information. On the "against" side, cost data that are not trusted by decision makers can undermine the influence of the plan in general. We know of many decision makers who would say that they just want the ecological information because understanding costs is the domain of project managers, not scientists.

Overall, we believe the best approach is to encourage managers and decision makers to impart their knowledge of cost and what drives it, as an integrated part of the planning process. Exactly how knowledge can most effectively and reliably be obtained from experts (like project managers) is the focus of the next section of this chapter.

## Expert Judgment

*"The plural of anecdote is data."*

**Expert judgment** is an important and ubiquitous part of conservation planning. Few conservation plans do not involve some degree of expert judgment. Using expert judgment in conservation is important for at least four reasons: (1) resources to collect the necessary primary data are limited; (2) conservation often happens in places and on topics where data are inadequate; (3) we're often interested in circumstances for which there aren't data—like what will happen in the future; and (4) often there is simply not enough time to even assemble and analyze disparate data sets, let alone collect new data.

We ask experts to give their judgment on a wide range of things: about threats, about the viability of populations, about ecosystem function, about the likely effectiveness of strategies, about the impacts on livelihoods or cultural activities of local communities of planned interventions, or the cost of different interventions. We use their answers on these topics either directly to inform our plans, or indirectly to parameterize models. In the context of conservation planning, expert judgment refers to using experts to provide their opinions about matters of fact (e.g., how much has the population of Scottish

Highland wildcats declined in the past five years?). Judgment about facts should be clearly distinguished from things that are matters of value (e.g., how many Scottish Highland wildcats should there be?). Matters of value have a firm place in conservation planning (objectives, attributes, targets, criteria weighting, etc.), but you do not need to be an expert to provide them. In conservation it can sometimes be difficult to distinguish between the two, especially for conservation scientists.[35]

Expert judgment should not be considered a substitute for empirical data. Just as there are good data and bad empirical data, there is well-collected and more reliably accurate expert judgment and there is poor expert judgment—both empirical data and expert judgment exist on a spectrum from good to bad. As the quote at the start of this section suggests, when collected rigorously, expert judgment really can be considered good data.

Getting information from experts is referred to as **elicitation**. There is an entire science known as *expert elicitation*, a subdiscipline of psychology, which is focused on getting the most accurate information possible from experts. In this section we hope to give you the knowledge to make sure you end up at the good end of the expert judgment quality spectrum.

You may wonder why you shouldn't just find the person considered the best expert and ask him or her for an opinion. Australian ecologist and risk assessment expert Mark Burgman and colleagues conducted a study that serves as a nice illustration of why this would be a poor practice.[36] In a series of conservation workshops, the researchers asked the experts present to begin by rating their own level of expertise and those of all their peers at the workshop. Overall there was a good correlation, so we can consider these people real experts by both self and peer assessment. Mark and his team then asked the experts a series of questions directly within their area of expertise, but for which it was possible to know the true answer (or close to it). For example, "Estimate the contraction of distribution (or abundance) for species X over the years 2005–2010."

The answers given by the experts in one workshop are represented by the red dots in FIGURE 5.16. How close these answers are to the truth is on the y-axis (lower values are better). As you can see, there is basically no relationship between expert status and accuracy—in fact, on average the more expert they are, the slightly less accurate they are. This same trend held across numerous workshops with different experts. This seems to be generally true—studies in a number of other fields have also found no positive relationship between "confidence" (not exactly the same as status, but closely related) and accuracy, and some studies have even found a negative relationship.

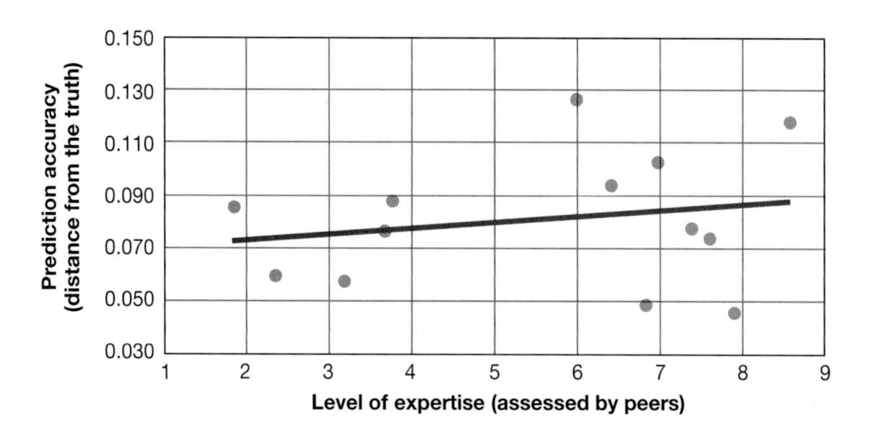


**FIGURE 5.16** The predictive accuracy of experts versus their level of expertise as assessed by peers. (Data from Burgman, McBride et al., 2011.)

## Bias—Cognitive and Motivational

Why were the experts wrong? The simple answer is that expert judgment, like all judgment, is subject to bias. When most people hear the word *bias,* they think about **motivational bias**—"of course they think tigers are the highest-priority species because they work on them!" Certainly, motivational bias is common in conservation because often the people whose judgment we are asking for do have a stake in the outcome of decisions; they may have to implement the project, they might have interest in certain species or habitats, and their judgment is colored by this involvement.

Joshua Donlan and colleagues provide a nice example of this sort of bias in an expert-based assessment of threats to sea turtle populations:[37] experts who worked on nest predation of turtle eggs consistently ranked it as the highest threat, whereas experts who worked on bycatch of turtles in fisheries routinely ranked fisheries bycatch as the highest threat.

There is, however, another kind of bias that most conservation planners are less familiar with but that is even more pervasive—**cognitive bias**. There are many kinds of cognitive bias, and in Chapter 6 we introduce them in more detail. Everyone suffers from these biases to some degree. But they become a problem when they occur among experts whose judgments we are relying on to support our decisions, especially as most people expect expert judgment to be accurate and largely free of bias. To give you an idea of how pervasive bias is, when experts are asked to provide a range of values within which they are 90% confident the truth lies—for example, what the current population of orangutans is in Borneo—these ranges contain the true value less than 50% of the time.[38] This is a substantial overconfidence bias. Once you add some group bias influence here (like groupthink or dominance), the chance of some unstructured group elicitation leading to the "true" answer is very slim.

This section is not about simply challenging expertise, but about using it better and more reliably. There are things we can do to reliably reduce the influence of bias and get better, more accurate, more robust answers from experts.

### Getting Better Judgments from Experts

So how do we get better judgments from experts? Three general approaches have been shown to work:

1. *Training.* Training experts to estimate better by repeatedly making estimates and getting feedback on their accuracy is one way to improve the reliability of expert judgment. Training is difficult in conservation because timely feedback is often challenging for the sorts of things we ask.
2. *Testing.* Testing people's expertise and using the judgments of those who demonstrate the best ability to make accurate judgment is another way to improve the use of expert judgment. There is some possibility for conservation here, and it's probably something conservation planners should do more of. One good way to "test" experts is to ask questions that will be verified in the subsequent year, for example, How much of this vegetation type do you expect to be cleared in 2012?" In future planning exercises, you can then go back to those experts who were more accurate. Testing can be a little confrontational, but if experts react badly to the idea of being tested, it's probably a good warning flag of overconfidence.
3. *Structured elicitation.* Probably the most useful approach to improve expert judgment in conservation planning is to focus on the structure of elicitation procedures. The remainder of this section deals with this topic. Structured elicitation procedures are explicit methods that anticipate and mitigate some of the most important and pervasive psychological and motivational biases.

   Many conservation planners already employ structured processes for gathering expert judgment. For example, the *Open Standards* is a clearly structured process that breaks down the question of ultimate interest (what conservation actions should we take?) into a series of discrete pieces that are more easily estimated, such as the extent of a threat. However, there is still a great deal of room for improving the structure of elicitation in most conservation plans.

### Who Should Be Considered an "Expert"?

Selection of experts is the first step in any elicitation protocol. Expertise is not just about having substantive knowledge on a subject, it's also an ability to accurately and clearly communicate judgments in a particular format (e.g., probabilities). For conservation planning purposes, the ability to be adaptive and extrapolate knowledge to new circumstances is also an important part of expertise. Not all experts are equally good at these components, but they are all important parts of being an expert.

Here are three points to consider when selecting experts:

1. *Take a broad view of expertise.* Experts are conventionally considered to have specialized knowledge not available to all, obtained through training and experience. Mark Burgman and colleagues suggest that it is important not to narrowly restrict your view of who is an expert by using sharp delineations of expertise, but rather that each claim to expertise should be examined critically.[39] In addition, the "disruption" caused by taking a broad view of expertise can be helpful in complex situations.[40]
2. *Balance the composition of your expert group.* Differences in gender, race, age, experience, background, wealth, and so on influence the way we perceive the world and the judgments we make. Sometimes these differences are relatively consistent, for example, women are less likely to be overconfident than men. This means that if, for instance, you were asking experts about the likelihood of a project being successful and for one project you asked all men and the other you asked all women, you would expect to get different answers just based on the gender of the experts you asked. The best way to control for effects like this is to ask a diverse group where possible. This is particularly important if your expert group is very small, say, 2–4 people. Intentionally "balancing" an expert group can also be useful when you suspect that motivational biases might be at play.
3. *Be transparent.* Being transparent about expert judgment in conservation planning means being clear about who are the experts involved and the process that was used to select them. You don't have to share this with the experts at the time, but this information is important for anyone who might be influenced by those judgments—there must be the opportunity for that expertise to be questioned critically by analysts, other experts, or stakeholders. We return to the issue of transparency in Chapter 6.

One of the most widely and easily used approaches for structured elicitation of judgments from experts is known as the **Delphi method**. The Delphi method was developed by the U.S. military's RAND Corporation during the Cold War and is still used extensively

## BOX 5.7 Delphi Approach to Expert Elicitation

**Step 1.** Provide instruction on the task and definitions, along with background info, studies, and so forth. This is to remove as much linguistic uncertainty as possible. Providing background information is really important because experts don't just bring knowledge, their job is also to help interpret the often fragmented signals that exist.

**Step 2.** Have everyone provide an estimate, anonymously (this is the critical part). It is ideal to ask for estimates using a four-point question technique as described in the main text of this chapter.

**Step 3.** Display the group's estimates back to all experts, and discuss any significant discrepancies. Try to retain anonymity as long as possible; however, group discussion will inevitably reveal some individual beliefs and that's okay.

**Step 4.** Give all experts a chance to revisit and revise their judgments—again anonymously.

**Step 5.** Combine the second-round estimates across experts to get a final estimate. Importantly, do not try to seek group consensus. We realize this might seem anathema to much group conservation decision making, which has emphasized consensus, but consensus is pretty much the death knell for accuracy. The estimates of experts can be combined in lots of tricky ways that consider all sorts of variation and spread, but the good news is that simply averaging across experts has proved to be the most robust approach.

by the military and intelligence community. **Box 5.7** describes the key steps of a Delphi approach.

If we go back to the expert accuracy study of Burgman and colleagues that we presented earlier, FIGURE 5.17 illustrates how, following the Delphi method—the opportunity to discuss, listen to one another, cross-examine reasoning and data, and assess their judgments—all experts were able to provide a more accurate estimate.

FIGURE 5.18 shows how, across a series of questions, the group average is nearly always more accurate and more precise than even the most accurate single expert. This is terrific news because it is easy.

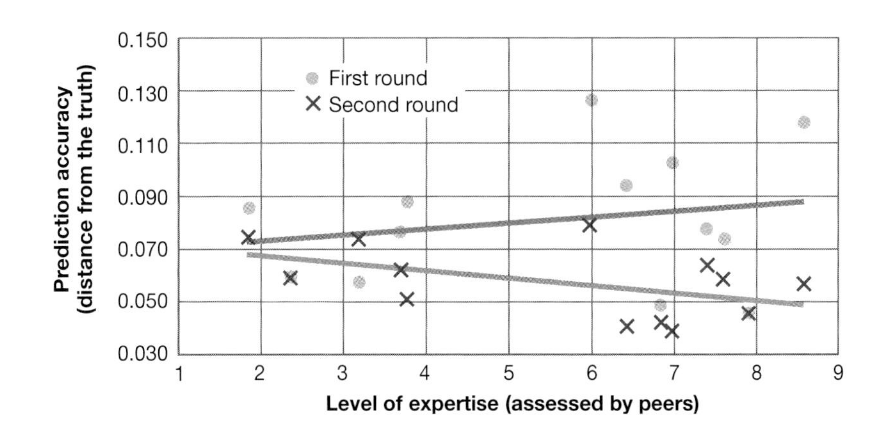


FIGURE 5.17 Change in accuracy of expert prediction between the first round of estimates (dots) and second round of estimates (crosses). Between the two rounds, experts saw and discussed the predictions of their peers. (Data from Burgman, McBride et al., 2011.)

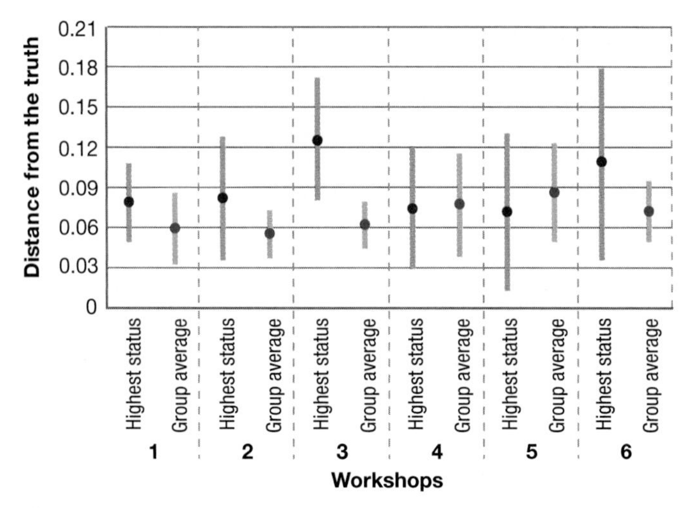


FIGURE 5.18 The accuracy of predictions averaged across a group of experts compared with the predictions of the most accurate individual experts. (Data from Burgman, McBride et al., 2011.)

However, there are two things to watch for when averaging judgments across a group of experts. The first is bimodality (FIGURE 5.19), where experts are clearly divided into two camps with different beliefs. Taking the average across all experts would likely end up with a value that no one believes is right. The second is to be careful not to lose information on the spread of estimates across experts. This information is important because if there is very little agreement using a good elicitation method, either the question has been poorly framed or there is legitimately a great deal of uncertainty in the answer. In either case, it is a flag to be careful how that estimate is used.

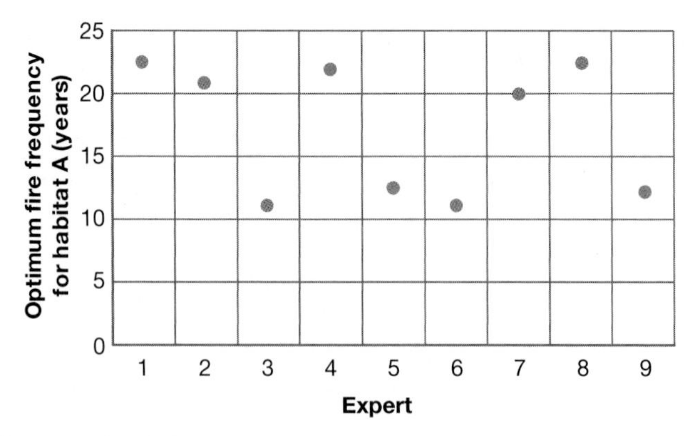


FIGURE 5.19 Example of clear bimodality in expert judgment. In this case the nine experts were divided about the optimum fire frequency for habitat A. One group believed it to be around 20 years and another put it at around 10 years.

### The Four-Point Question

Philosophers have spent a lot of time trying to work out how to get better answers from experts. A format that is currently considered likely to get closer to the truth than most others is the four-point question.[41]

1. Realistically, what do you think is the lowest the value could be?
2. Realistically, what do you think is the highest the value could be?
3. Realistically, what do you think is the most likely value is? (your best estimate)
4. How confident are you that the interval you provided contains the truth?

Why ask four questions? Because this set of questions appears to dramatically reduce overconfidence in expert judgment.[42] The first two questions help to understand the uncertainty around the estimate, and hopefully trigger experts to think about why the truth might be different from their best guess. The third question is the one that actually leads to your final estimate. And the fourth question is used to normalize the confidence intervals when displaying the data back to the group (e.g., so you can show everyone's 80% confidence interval). Without this fourth question, you may not be comparing apples with apples across experts.

Hopefully you can see that many of the types of information we may want from experts during conservation planning could be elicited using this four-point format. Here are some examples:

- Determining the natural fire interval for a habitat
- Identifying the percentage of a system influenced by invasive species
- Determining the volume of flow necessary to sustain a native fish species
- Identifying the number of individuals of a species that are killed on roads each year
- Identifying the total number of breeding pairs of a species
- Assessing what the reduction in sediment input from revegetating a piece of land might be

***Don't Ask Too Many Questions*** Advising children not to ask too many questions is a disputable idea, but it really is good practice when it comes to expert elicitation. Conservation planning often involves bringing experts to a workshop and bombarding them with questions all day. Leaders should try to limit the number of responses required to around 10. Many more than this, and the answers experts give are more likely to be influenced by previous answers rather than expertise about the particular question.

Many conservation planners who have provided input into intensive Conservation Action Planning or *Open Standards* workshops have probably even noticed this in their own answers—you start looking at what you answered for the previous question and adjust your current answer relative to that. This is called *anchoring,* and it is a form of bias. We strongly encourage you to spend the time up front by thinking carefully about exactly what you need experts to provide judgment about.

If at all possible, do not rely on the judgment of one single expert. If you absolutely have to (say, for information about a rare species in one location), then still use the four-point question. But also explicitly ask them to consider why they might be wrong, under what scenario they could be wrong, if anyone is likely to have a different opinion, and why.

Finally, keep records of estimates and how you went about eliciting them so you can refine your own process better. Drop in the odd test question. Keep track of experts who perform well and give accurate answers.

### Traditional Knowledge

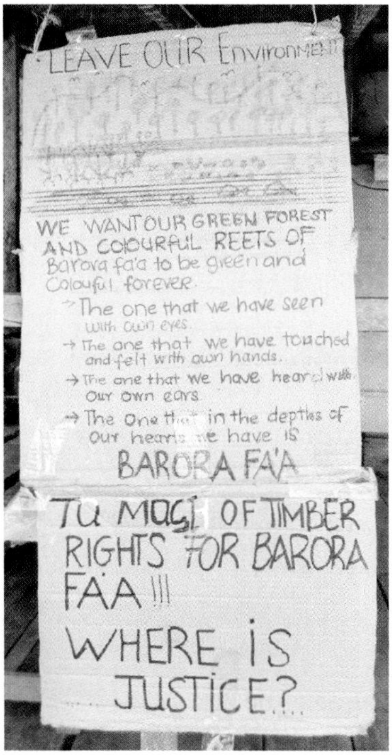

A special subset of expert knowledge is the environmental knowledge held by members of indigenous or traditional communities. This knowledge is generally referred to as **traditional ecological knowledge (TEK)** or sometimes as indigenous knowledge (because using the word traditional might falsely emphasize a static or historical perspective). TEK is characterized by practices and wisdom developed at a local scale through earning livelihoods from the environment over successive generations. Indigenous knowledge of species and environments is often embedded in traditional resource management practices and the social codes surrounding them. For many indigenous and traditional communities, the ecological knowledge base is rich and deep because the relationship between ecosystems and human well-being is more direct and fundamental than it is for most Western societies (**FIGURE 5.20**).[43] There is widespread recognition that TEK can—and often should—play a valuable role in contemporary conservation decision making. In some circumstances it might even be a required aspect of a natural resource management process. TEK is particularly important data for any kind of conservation planning done in the context of comanagement,[44] where sensitivity and consistency with traditional social codes is vital. But it can also be seen more generally as a source of useful local ecological knowledge.[45]

**FIGURE 5.20** This poster was put up in a public space by members of a community in the Solomon Islands who were concerned about logging activities degrading their natural heritage.

In an important paper on the subject, Nicolas Houde identified six dimensions or "faces" of TEK (FIGURE 5.21), and some of the key components of these faces (TABLE 5.10).[46] All of these dimensions can be relevant data for conservation plans. Nearly all conservation planners who have worked with indigenous peoples will be familiar with the first three dimensions identified by Houde. These dimensions of observation, management, and use include data on the distribution and abundance of species, the natural history

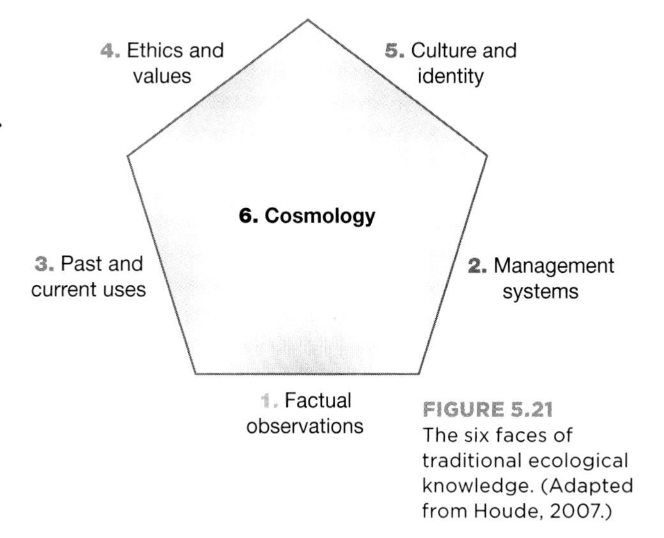


FIGURE 5.21 The six faces of traditional ecological knowledge. (Adapted from Houde, 2007.)

TABLE 5.10 Key components of the different dimensions of traditional ecological knowledge

| Face | Key components |
|---|---|
| Factual observations | Empirical observations<br>Classifications<br>Naming of places<br>Descriptions of ecosystem components<br>Understanding of interconnections<br>Spatial and population patterns<br>Ecosystem dynamics and changes |
| Management systems | Practices adapted to context<br>Methods for conservation<br>Methods for sustainable resource use<br>Methods for adapting to change<br>Appropriate and effective technologies |
| Past and current uses | Land use patterns<br>Occupancy<br>Harvest levels<br>History of the cultural group<br>Location of cultural and historical sites<br>Location of medicinal plants |
| Ethics and values | Correct attitudes to adopt |
| Culture and identity | Links life on the land, language, identity, and cultural survival |
| Cosmology | Assumptions about how things work<br>Beliefs<br>Spiritual relationship to the environment |

*Source:* Adapted from Houde, N. *The six faces of traditional ecological knowledge: Challenges and opportunities for Canadian co-management arrangements.* Ecology and Society, 2007. **12**(2): 34.

of species such as breeding patterns or relationships between species as well as ecological changes and trends in the environment, trends in catches and exploitation levels, and existing forms of management or conservation. We might expect such knowledge to be particularly extensive for species that are cultural keystone species (see Chapter 3). These dimensions of TEK are commonly collected or engaged with through the use of maps (**Box 5.8**).

## BOX 5.8 Traditional Ecological Knowledge and Maps

Traditional ecological knowledge (TEK) can be used as both a primary source of data for maps and to validate other mapped data. We have commonly used it for both purposes in the same conservation plan. For example, it is common practice to work with indigenous communities to identify the distribution of important species, important places for these species such as breeding aggregations, or culturally important locations such as places where particular resources are gathered, or where customary management arrangements are in place.

**FIGURE 5.8.1A** shows community members from Isabel Province, Solomon Islands, annotating large-scale maps with the distribution of conservation features from their TEK. **FIGURE 5.8.1B** shows what an annotated map looks like; the numbers correspond to a particular conservation feature and indicate its presence in that location. Maps like these can then be easily digitized and the data used alongside other ecological information in a conservation plan.

TEK can also play a crucial role in validating existing data on the distribution or condition of habitats and other land uses. We have often had the experience of mapping what appeared to be intact forest only to be informed by TEK that the place

(a)

(b)

**FIGURE 5.8.1** (a) Community members from Isabel Province, Solomon Islands, record the distribution of conservation features based on traditional ecological knowledge on large paper maps during a conservation planning workshop. (b) Close-up view of a distribution record.

has a long history of changes in land use and associated ecology.

Of relatively obvious but frequently overlooked importance when gathering and mapping TEK is the concordance between contemporary classifications of habitats, or even species, and traditional classifications. Depending on how the environment is used, different properties of the environment help indigenous peoples delineate the ecology of places. For example, whereas ecological classifications such as those discussed in Chapter 3 are often based on species composition, TEK may delineate habitats based on seasonal use or the occurrence of one or two resources. TEK will be more meaningful when used consistently with traditional perspectives on the environment.

Using maps is a good way to gather and check data, but it also can be a great community engagement exercise. However, translating local knowledge (traditional or not) onto a map (e.g., the extent of a vegetation type) is notoriously difficult because we do not generally see our environment with the sort of bird's-eye view typical of maps. This difficulty is compounded further in the marine environment, where spatial references to submarine habitats are generally fragmentary at best.

One approach to improving the ability to orientate knowledge on a map is to develop a 3-D map. Three-dimensional maps such as the one in **FIGURE 5.8.2,** which was developed for Manus Island in Papua New Guinea, can be created by cutting many layers of cardboard along contour lines and then using papier-mâché and model paints to complete the map. We've seen this used effectively in many locations as a highly participatory process that can engage an entire community in capturing local knowledge on a map. And they look great.

(a)

(b)

(c)

**FIGURE 5.8.2** Stages in the development of a 3-D map of Manus Island, Papua New Guinea. This was a participatory exercise to help capture and illustrate local knowledge relevant for a conservation plan. (a) & (b) Topographic variation is built up from multiple layers of cardboard cut to contour lines. (c) The finished map after being painted and annotated by community members.

Depending on how a conservation planning problem is framed and the nature of the management alternatives being considered (see Chapter 6), dimensions 4–6 of Houde's typology can also be highly relevant knowledge for a conservation plan. For example, comparisons between the suitability of different management actions might be influenced by TEK around ethics and cultural identity. More frequently, however, these dimensions play an important role in determining the objectives of a conservation plan (Chapter 3), the features used to represent these objectives (Chapter 3), and targets for these features (Chapter 4).

### Challenges of Using Traditional Ecological Knowledge in Conservation Planning

There is no shortage of debate over the use of TEK in conservation and natural resource management and about how it can most appropriately be integrated with what is generally referred to as scientific knowledge. The Australian anthropologist Marc Wohling has warned that the highly local nature of much TEK can make it a poor fit for the contemporary scale of many ecosystem disturbances.[47] He argues that it is important to be realistic about the limitations of TEK and that failing to do so can be unreasonably burdensome on indigenous peoples. Similarly, caution must be applied to ensure that TEK is not simply treated as a substitute for investment in science when indigenous stakeholders are concerned.

There is healthy debate about the extent to which the modern scientific approach should be employed to evaluate the qualities, efficacy, and parameters of TEK in the context of conservation and environmental management. In our view, it is important not to hold TEK beyond the reach of rational evaluation, but to ensure that any such evaluation occurs in concert with the owners of that knowledge.

Most conservation plans that make use of TEK will also employ scientific knowledge. There are no consistent rules for how these two types of knowledge should be combined, but Kristen Weiss and colleagues have provided a valuable summary on challenges and approaches to integrating TEK and scientific knowledge in the context of marine wildlife conservation.[48]

Any potential limitations of TEK in the context of scientific knowledge do not apply at all, however, to the use of TEK to inform the objectives, targets, or management alternatives of a conservation plan. Where there are indigenous stakeholders, we can't think of a situation where knowledge of their values, identity, and beliefs will not improve a conservation plan or the resulting conservation.

## KEY MESSAGES

- Situation analysis involves identifying the drivers and causes of changes in the systems of interest, and ideally representing them through a diagrammatic model often referred to as a conceptual model. This is a critical step in conservation planning that lays the groundwork for development of strategies and actions.
- A threat assessment is intended to understand the significance of threats for our objectives. It is important to focus on the source of threats, and to distinguish between ranking threats and prioritizing threat abatement. Ranking threats is generally accomplished by giving scores to the scope and severity of threats.
- Data and how they are used lie at the heart of what it means to make *informed decisions.*
- The use of species data in conservation planning is constantly challenged by the variability in our knowledge both between species and within a species. Species distribution modeling can help address some of the common issues.
- Ecological community data should be complemented by data on the integrity of those communities.
- Data on abiotic properties of a landscape can be extremely useful, especially when there is a paucity of ecosystem- and species-level data. However, it is impossible to entirely remove arbitrariness from the classification of abiotic variables into conservation features.
- Although social data have largely meant threat data in conservation planning, there are a wide variety of data on the social characteristics of an area that can serve numerous purposes in a conservation plan. Both quantitative and qualitative social data can be valuable and should ideally be used in combination.
- Data on how much it is likely to cost to implement a conservation action is a critical but often overlooked piece of a conservation plan. A range of techniques exist to help estimate costs accurately and adjust them for comparison with a conservation plan.
- Expert judgment is and will continue to be an important part of the data that support conservation plans. There are a range of formal elicitation methods that can substantially improve the reliability of data gathered from experts.

## References


1. Conservation Measures Partnership, *Open Standards for the Practice of Conservation, Version 2.0*. 2007. [accessed January 2, 2015]. http://cmp-openstandards.org/
2. Selkoe, K. A., B. S. Halpern, and R. J. Toonen, *Evaluating anthropogenic threats to the Northwestern Hawaiian Islands*. Aquatic Conservation: Marine and Freshwater Ecosystems, 2008. **18**(7): 1149–1165.
3. Grech, A. et al., *A comparison of threats, vulnerabilities and management approaches in global seagrass bioregions*. Environmental Research Letters, 2012. **7**(2).
4. Esselman, P. C. et al., *An index of cumulative disturbance to river fish habitats of the conterminous United States from landscape anthropogenic activities*. Ecological Restoration, 2011. **29**(1–2): 133–151.
5. Halpern, B. S. et al., *A global map of human impact on marine ecosystems*. Science, 2008. **319**: 948–952.
6. Grech et al., *A comparison of threats*. (See reference 3.)
7. Halpern, B. S., and R. Fujita, *Assumptions, challenges, and future directions in cumulative impact analysis*. Ecosphere, 2013. **4**(10): 131.
8. Sullivan, B. L. et al., *eBird: A citizen-based bird observation network in the biological sciences*. Biological Conservation, 2009. **142**(10): 2282–2292.
9. Bonney, R. et al., *Next steps for citizen science*. Science, 2014. **343**(6178): 1436–1437.
10. Bottrill, M. C. et al., *Evaluating perceived benefits of ecoregional assessments*. Conservation Biology, 2012. **26**: 851–861.
11. Grantham, H. S. et al., *Diminishing return on investment for biodiversity data in conservation planning*. Conservation Letters, 2008. **1**(4): 190–198.
12. Yokota, F., and K. M. Thompson, *Value of information literature analysis: A review of applications in health risk management*. Medical Decision Making, 2004. **24**(3): 287–298.
13. Runge, M. C., S. J. Converse, and J. E. Lyons, *Which uncertainty? Using expert elicitation and expected value of information to design an adaptive program*. Biological Conservation, 2011. **144**(4): 1214–1223.
14. Pressey, R. L., *Conservation planning and biodiversity: Assembling the best data for the job*. Conservation Biology, 2004. **18**(6): 1677–1681.
15. Rondinini, C. et al., *Tradeoffs of different types of species occurrence data for use in systematic conservation planning*. Ecology Letters, 2006. **9**(10): 1136–1145.
16. Elith, J., and J. R. Leathwick, *Species distribution models: Ecological explanation and prediction across space and time*. Annual Review of Ecology Evolution and Systematics, Vol. 40, 2009. 677–697.
17. Segurado, P., and M. B. Araujo, *An evaluation of methods for modeling species distributions*. Journal of Biogeography, 2004. **31**(10): 1555–1568.

18. Phillips, S. J., and M. Dudík, *Modeling of species distributions with Maxent: New extensions and a comprehensive evaluation*. Ecography, 2008. **31**(2): 161–175.
19. Lomba, A. et al., *Overcoming the rare species modeling paradox: A novel hierarchical framework applied to an Iberian endemic plant*. Biological Conservation, 2010. **143**(11): 2647–2657.
20. Hammerson, G. et al., *Ranking species occurrences—a generic approach*. NatureServe, 2008. Arlington, VA, USA. Also available on the web at http://www.natureserve.org/explorer/eorankguide. htm.
21. Stoms, D. M., S. L. Dashiell, and F. W. Davis, *Siting solar energy development to minimize biological impacts*. Renewable Energy, 2013. **57**: 289–298.
22. Theobald, D. M., *A general model to quantify ecological integrity for landscape assessments and U.S. application*. Landscape Ecology, 2013. **28**(10): 1859–1874.
23. Sanderson, E. W. et al., *The human footprint and the last of the wild*. BioScience, 2002. **52**(10): 891–904.
24. Halpern et al., *A global map of human impact on marine ecosystems*. (See reference 5.)
25. Theobald, D., *Landscape patterns of exurban growth in the USA from 1980 to 2020*. Ecology and Society, 2005. **10**(1): 32. http://www.ecologyandsociety .org/vol10/iss1/art32/.
26. Beier, P., and B. Brost, *Use of land facets to plan for climate change: Conserving the arenas, not the actors*. Conservation Biology, 2010. **24**: 701–710.
27. Cinner, J. E., *Designing marine reserves to reflect local socioeconomic conditions: Lessons from long-enduring customary management*. Coral Reefs, 2007. **26**: 1035–1045.
28. Ostrom, E., *A general framework for analyzing sustainability of social-ecological systems*. Science, 2009. **325**(5939): 419.
29. Khandker, S. R., G. B. Koolwal, and H. A. Samad, *Handbook on Impact Evaluation: Quantitative Methods and Practices*. 2010, Washington, DC: World Bank Publications. 262.
30. Stephanson, S., and M. B. Mascia, *Putting people on the map through an approach to integrating social data in conservation planning*. Conservation Biology, 2014. **5**: 1236–1248.
31. Carwardine, J. et al., *Avoiding costly conservation mistakes: The importance of defining actions and costs in spatial priority setting*. PLoS One, 2008. **3**: e2586.
32. Whitehead, A. L. et al., *Large scale predator control improves the productivity of a rare New Zealand riverine duck*. Biological Conservation, 2008. **141**(11): 2784–2794.
33. Naidoo, R. et al., *Integrating economic costs into conservation planning*. Trends in Ecology & Evolution, 2006. **21**(12): 681–687.

34. Armsworth, P. R. et al., *Management costs for small protected areas and economies of scale in habitat conservation*. Biological Conservation, 2010. **144**: 423–429.
35. Wilhere, G. F. et al., *Conflation of values and science: Response to Noss et al.* Conservation Biology, 2012. **26**(5): 943–944.
36. Burgman, M. A., M. McBride, et al., *Expert status and performance*. PLoS One, 2011. **6**: e22998.
37. Donlan, J. C. et al., *Using expert opinion surveys to rank threats to endangered species: A case study with sea turtles*. Conservation Biology, 2010. **24**(6): 1586–1595.
38. Speirs-Bridge, A. et al., *Reducing overconfidence in the interval judgments of experts*. Risk Analysis, 2010. **30**(3): 512–523.
39. Burgman, M. et al., *Redefining expertise and improving ecological judgment*. Conservation Letters, 2011. **4**(2): 81–87.
40. Game, E. T. et al., *Conservation in a wicked complex world; challenges and solutions*. Conservation Letters, 2013: 10.1111/conl.12050.
41. Speirs-Bridge et al., *Reducing overconfidence in the interval judgments of experts*. (See reference 38.)
42. Speirs-Bridge et al., *Reducing overconfidence*. (See reference 38.)
43. Sangha, K. et al., *Identifying links between ecosystem services and Aboriginal well-being and livelihoods in north Australia: Applying the Millennium Ecosystem Assessment framework*. Journal of Environmental Science and Engineering, 2011. **5**(7): 931–946.
44. Carlsson, L., and F. Berkes, *Co-management: Concepts and methodological implications*. Journal of Environmental Management, 2005. **75**(1): 65–76.
45. Luzar, J. B. et al., *Large-scale environmental monitoring by indigenous peoples*. Bioscience, 2011. **61**(10): 771–781.
46. Houde, N., *The six faces of traditional ecological knowledge: Challenges and opportunities for Canadian co-management arrangements*. Ecology and Society, 2007. **12**(2): 34.
47. Wohling, M., *The problem of scale in indigenous knowledge: A perspective from northern Australia*. Ecology and Society, 2009. **14**(1): 1.
48. Weiss, K., M. Hamann, and H. Marsh, *Bridging knowledges: Understanding and applying indigenous and Western scientific knowledge for marine wildlife management*. Society & Natural Resources, 2013. **26**(3): 285–302.

# 6

# Framing Conservation Planning Problems

## Overview

Previous chapters in this book have dealt with establishing objectives (Chapters 3 and 4) and developing an understanding of the social-ecological system in which the plan is embedded (Chapter 5). In this chapter we focus on the set of options that the planning process will evaluate and consider how these three elements (objectives, options, and an understanding of the system) are the core components of a well-framed conservation planning problem. We emphasize the importance of properly framing conservation planning problems to reach informed decisions. We offer guidance on developing a set of conservation options to consider, suggest that predicting the consequences of these options is among the key tasks of a conservation scientist, and identify the different ways this task can be approached. Finally, we argue that all conservation planning problems should be focused on a choice between actions, rather than species or habitats, and that all conservation planning problems should be framed as multi-objective problems.

## Topics

- *Problem framing*
- *Spatial planning*
- *Strategic planning*
- *Decision science*
- *Psychological biases*
- *Developing alternatives*
- *Estimating consequences*
- *Business as usual and counterfactuals*
- *Combining criteria*
- *Value judgments*
- *Arbitrariness*
- *Predictive models*
- *Marginal benefit*
- *Public versus private benefit*
- *Costs*
- *Theory of change*
- *Logic models*
- *Action-based planning*
- *Multi-objective planning*

## What Is a Well-Framed Problem?

Imagine you worked for a provincial wildlife agency and were asked to identify the agency's highest priorities. How would you answer this question? Chances are you would do one of two things: You might substitute this question (possibly subconsciously) for one that you can answer more easily, like what is the most threatened species in the province, what is the most economically important species in the province, what species has the agency worked on in the past, or what are the governor's favorite species? Alternatively, you might seek more clarity about what the agency's objectives and responsibilities were; what was happening ecologically, socially, economically, and politically in the province; and what sort of activities the agency could undertake. Armed with this information, you could then ask, "Given our knowledge about what is happening in the province, which of the possible activities is likely to best achieve the agency's objectives?" The latter alternative (which we suspect would be your inclination, especially if you've read the earlier chapters of this book), is what it means to *frame* or formulate a conservation planning problem. Unless problems are well framed, it is almost impossible to determine what a good solution should be (**Boxes 6.1** and **6.2**).

A *problem* can be defined as "a question to be considered, solved or answered" (from HarperCollins). Conservation planning is a collection of processes and tools designed to solve or answer questions about the best way to achieve our conservation objectives; we'll call these questions **conservation planning problems**. Conservation planning is principally concerned with answering two general types of questions: What are the best actions we can take to meet our objectives, and Where is the best place(s) to locate or deploy an action within a region of interest? As we noted in Chapter 1, these questions are sometimes referred to as the what (strategic planning) and where (spatial planning) questions. Something we hinted at early in the book is our opinion that the where question should really be viewed as a specific case of the what question. Where priority places for conservation are is dependent on what conservation activities are being considered.

As a simplistic but real example, an organization focused on revegetation will prioritize different places to work than will an organization focused on purchasing properties to avoid further loss of habitat, even though the two organizations may have the same ultimate objective of maximizing extent of protected native habitat. As such, selecting priority areas for conservation (the *where* question) involves a choice also about strategy (*what*). In many cases, like the example just mentioned, the what question may be part of the context such that the plan is focused only on solving the where problem. Our point is

### Box 6.1 Why Are Conservation Planning Problems Tough?

Conservation planning problems are generally tough problems. After all, that's why the field of conservation planning has expanded rapidly and the methods are applied to projects around the world. The alternative to deliberately and systematically tackling conservation planning problems is to simply use intuition, gut feel, and preference in an ad hoc fashion. These qualities can result in good decisions; but for at least two reasons, making good, rational, and efficient conservation decisions is cognitively difficult, and gut feel is likely to be unreliable.

First, a vast number of potential management options often cannot be systematically evaluated or even identified based on individual expert opinion. Imagine a classic conservation plan intended to determine priority places to be added to the country's national park system. Even a small country could easily have 200 potential sites. If we consider just the task of working out where to apply just this single conservation action, the number of options is given by 2 raised to the power of however many places we are considering. That means the number of possible solutions (combinations of sites) would be 2 raised to the 200th power ($2^{200}$). To give you a sense of just how many options this is to consider, the planet Earth is comprised of very roughly $2^{167}$ atoms. The obvious cognitive challenge of integrating many different pieces of data across so many options to find a good solution is no doubt one of the reasons conservation planners use sophisticated tools to solve spatial conservation problems.

On the other hand, when it comes to knowing what to do—the actions we are going to take—it's easier to believe that we might be able to rely simply on our judgment rather than formal planning or a decision-making framework. This belief exposes the second reason that good decisions are difficult; because human judgment is prey to a variety of psychological traps and biases—among them, overconfidence. We introduced the idea of these biases in Chapter 5, but in this chapter, we highlight a number of biases that often unconsciously influence the choices we make about conservation management strategies. Aside from the accountability and defensibility that comes from a rigorous planning process, these two reasons are good justification for using a formal conservation plan to guide decisions.

simply that planners should always be thinking of prioritizing places (where) in the context of particular activities (what). We address this point in more detail later in the chapter.

A third type of question that conservation plans might be asked to inform is the division of a budget across options. For example, how much should be allocated to the recovery of different threatened species? This might be thought of as a second-level question that comes beneath the what and where questions, forming the set that includes *what*, *where*, and *how much*.

When referring to problems, it's important to distinguish between a *conservation problem* and a *conservation planning problem*. Issues such as the ongoing loss of forest for expanding oil palm plantations

## Box 6.2 Common Psychological Biases That Influence Conservation Decisions

The following list outlines a small set of the many decision biases that psychologists have identified. We've tried to choose those that most obviously influence conservation decisions. These biases are not all independent, and in some of the examples we give, more than one bias is likely to be at play.

- Status quo bias: A strong preference for the current state of affairs. In conservation planning, this is likely to arise as a preference for small adjustments of existing strategies rather than large changes to an objectively better strategy.
- Anchoring: Anchoring is the tendency to give too much weight to the first piece of information offered (the "anchor"). Anchoring is what real estate agents do when they set the price of a house—try to "anchor" you around that value (so you're likely to immediately believe it is worth around that much). A simple but common manifestation of this bias in conservation is the weight people often place on the first option or action proposed.
- Sunk cost bias: Future decisions overly influenced by past spending. Objectively, only what can be achieved with future costs should influence investment decisions. In practice, however, we often find ourselves saying, "Well, we've spent all this money on this strategy already, it seems like a waste not to continue supporting it." This bias is sometimes referred to as the Concorde fallacy, referring to huge investments made in the Concorde supersonic jet by the British and French governments long after it was clear that the Concorde was economically no longer a worthwhile investment.
- Overconfidence: Overconfidence is an almost ubiquitous human condition. This trait influences conservation planning because those involved tend to be overconfident in their knowledge and intuition about systems and strategies. When experts are asked to give a range of estimates that they are 90% confident in, these ranges contain the true value less than 50% of the time.[a] Most people are overconfident, but generally women are less overconfident than men.[b]
- Base rate neglect: Tendency to ignore information on historical performance when evaluating outcomes. For example, even though we know that a substantial number of all conservation agreements will be violated, project managers are often convinced that theirs is different. It is important not to neglect information regarding past behavior, practices, and/or performance, even if that history is not directly related to the proposed action. This characterization is referred to as the *base rate*, defined as an estimate of expectation in the absence of case-specific knowledge. For example, suppose we are considering entering into a voluntary agreement around some environmental practice with a big mining corporation, and we want to predict the likely result. We might start by making a prediction of what could happen if the mining company does everything it says it will. Suppose, however, we also know that the mining company has been penalized for infringing a number of state environmental regulations each year for the past 10 years. Our prediction of success should consider this record. It is easy to believe we can overcome past failures that were due to "bad actors." We are probably all guilty of believing our case is different—but history suggests it is not, and it would be unwise to assume that it is.
- Confirming evidence: Interpreting data and information in a way that supports your beliefs. As a good example, we worked with a conservation project that commissioned a study into the ability of ecotourism to

make the project self-sustaining. Those who favored ecotourism emphasized that the study suggested sustainability was achievable within seven years. But a closer review of the study revealed that this outcome required an investment of many millions of dollars during that initial seven years, and it certainly didn't provide unequivocal support for the strategy. Skeptics of using ecotourism interpreted the same report as evidence against the strategy.

- Groupthink and dominance: Group members' striving for agreement overrides realistic appraisal. Nearly all conservation planners know about groupthink, but many are unaware of how subtle and pervasive it can be. In most cases groupthink occurs not because everyone in a group is in ecstatic agreement about an idea, but because people simply want to move forward during the often frustrating process of group decision making. To expedite the process, participants might subjugate their opinion if they feel the weight of the group is not behind it, or if they feel a dissenting opinion will slow things down. More easily detected is dominance in a group, where judgments are disproportionately influenced by a dominant member of the group. We're confident every conservation planner has seen examples of this.

We suggest that by far the most accessible introduction to these psychological traps is a book called *Smart Choices,* written by three leading decision scientists: John Hammond, Ralph Keeney, and Howard Raiffa.[c] In addition to the biases we've listed, *Smart Choices* also describes a number of other common traps, such as information availability bias. Valuably, Hammond and colleagues also offer practical suggestions for minimizing the effect of these traps and biases. Another excellent introduction is the book *The Psychology of Judgment and Decision Making,* by psychologist Scout Plous.[d]

---

[a] Speirs-Bridge, A. et al., *Reducing overconfidence in the interval judgments of experts.* Risk Analysis, 2010. **30**(3): 512–523.

[b] Lundeberg, M. A., P. W. Fox, and J. Punćcohař, *Highly confident but wrong: Gender differences and similarities in confidence judgments.* Journal of Educational Psychology, 1994. **86**(1): 114.

[c] Hammond, J. S., R. L. Keeney, and H. Raiffa, *Smart Choices: A Practical Guide to Making Better Decisions.* 1999, Boston: Harvard Business Press.

[d] Plous, S., *The Psychology of Judgment and Decision Making.* 1993, New York: McGraw-Hill.

are sometimes referred to as conservation problems. An associated conservation planning problem might be stated as "What can a government agency do, and where should it act to minimize the impact of forest loss on biodiversity while allowing for continued growth of the oil palm industry?" We have written this book largely because we believe that framing and solving conservation planning problems is an effective way to address conservation problems.

Well-formulated problems have a number of common elements. Because these elements are often associated with **decision science** and its literature (introduced in Chapter 2), they may be cast in language that is unfamiliar to many conservation planners. The concepts, however, are not difficult to grasp. From the most basic point of view, a well-framed problem requires three things:

1. Objectives with a clear **performance measure**: Objectives (as introduced in Chapter 3) are necessary so we know what we are trying to achieve, and performance measures (the attributes we discussed in Chapter 4) are necessary so we can compare the expected performance of the different options being considered. Performance is often calculated according to an objective function—literally, some formula that calculates how well an action meets the stated objectives.
2. Options: The set of alternative actions available to choose between. In the case of conservation planning, this means the different interventions being considered (something we discuss later in this chapter). Identifying a set of options can be hard, but it is an important part of framing conservation problems.
3. Consequences: Predicting what effect the implementation of an option will have on the social-ecological system, and therefore what response we might expect in the things we care about (the objectives). In conservation planning, prediction of consequences is informed by many types of data, as discussed in Chapter 5. Predictions of consequences might be formalized in statistical analyses of data or in models, or alternatively they might exist only in an expert's head.

In addition to these three basic elements, other authors who draw on decision sciences to frame problems will typically add some characterization of **constraints** (e.g., total budget, thresholds for loss; see Chapter 2), and **uncertainty** (related to our system understanding; see Chapter 8) to the list of key elements.[1]

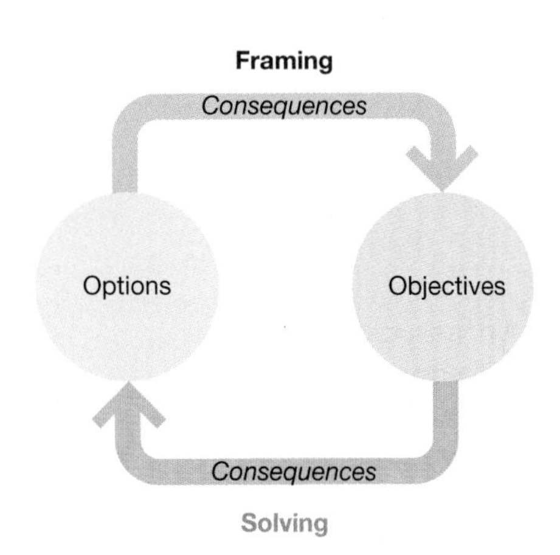


FIGURE 6.1 Framing a conservation planning problem involves identifying options and looking at their consequences for objectives, whereas solving a conservation planning problem involves evaluating these consequences to make a decision about options.

FIGURE 6.1 illustrates how framing a conservation planning problem involves identifying options and looking at their consequences for objectives, whereas solving a conservation planning problem involves evaluating these consequences to make a decision about options. Because conservation is complex business, conservation planning problems are inevitably simplifications of complex problems.[2] This is reasonable because without some simplification in problem framing, action would be paralyzed.

Understanding the context for a plan (Chapter 2) will help greatly in framing conservation planning problems in a meaningful way. A great deal of conservation planning has happened around the globe, much of it of a high quality. Unfortunately, there has also been a high prevalence of "type III errors"—finding a good answer, but to the wrong question.[3] The anecdote we gave at the beginning of Chapter 2 about the mandated planning process that didn't actually answer a question of interest to managers is a good example of a type III error. For conservation planners, awareness of type III errors should be as central to our training as type I and type II errors are for statisticians.

Of the three core elements of a conservation planning problem as just described (objectives, options, and consequences), we have already covered objectives in Chapters 3 and 4, so this chapter focuses on developing options and predicting consequences. We finish each section with rules that will improve the framing of conservation planning problems: (1) all problems should focus on a conservation action; and (2) all problems should be multi-objective.

## Options

Conservation planning is fundamentally about choice. It helps evaluate potential courses of conservation action and guides investment of resources in these efforts. The set of actions we might take to help reach our objectives—the places we might work in, the strategies we might undertake, the policies we might pursue—are the options required to frame a conservation planning problem. Developing a good set of options is really at the heart of conservation planning problems,[4] and in fact all decision problems.

Both of us have been surprised to see just how infrequently planning teams develop and evaluate options as a core part of conservation planning. Although it is a common practice and a legal requirement in many governmental management plans (e.g., the forest management plans of the U.S. Forest Service), it is a rare practice in the nongovernmental community. Conservation planners often evaluate options without realizing this is effectively what they're doing. For example, spatial plans involve the (almost too) obvious options of all the different places in a region where conservation work could occur. The final portfolio of places has been a choice between many options. In other cases planners expect a good option to be the outcome of the plan, but they do not see evaluation of multiple options as an important part of determining a "good" option. Plans are often focused on designing one promising option rather than considering a range of options. Neither of us knows how many times we have heard the rhetorical question, "Why would we consider alternative options when we already know what we're going to do?"

Conservation organizations tend to be deft at saying why something is a good idea (an important talent for sure), but we've been less diligent at articulating why a chosen option is better than other available options. As we've said before in this book, though, every good thing we do is another thing we do not do. Even where one option appears clearly favorable or inevitable for institutional reasons, we believe it is valuable to consider some alternative options (see Chapter 2 for reasons).

At the very minimum, it is good practice to consider two options and make one of them to simply do nothing. The environmental economist David Pannell calls the do-nothing option *informed inaction.*[5] The do-nothing option is important because when embedded in the planning framework described in this book, it asks teams to evaluate what would happen without the action you're considering. This **business as usual** (**BAU**) case is what helps build a **counterfactual** that emphasizes the benefit of the action (see next section on Consequences). It forces any organization or agency to ask itself a challenging question—what will happen here in the future if we don't take action?

Developing alternative options to address conservation problems should involve contextual knowledge, experience, and a dose of creativity. There is no agreed-on systematic formula for creativity. We have gathered together the following series of tips; things that we and others we've worked with have found effective in generating options as part of a conservation planning process (**Box 6.3**).

## Box 6.3 Developing Options for a Conservation Plan

The following set of options was developed for a conservation plan that one of us facilitated, following much of the guidance from the "Options" section of this chapter. The planning process was designed to identify the best conservation management and restoration options to meet the following objectives for Noosa Estuary, Australia:

- Increase fish abundance.
- Increase aquatic biodiversity.

A group of experts (drawn from diverse professional backgrounds and current roles) was asked to think broadly and freely about restoration and management options, initially as individuals and then as a group during a workshop. To limit the risk that experts would prematurely discard potential options because of perceived sociopolitical feasibility issues, the experts were explicitly instructed not to consider the sociopolitical feasibility or desirability of potential options. Although societal values must be a critical part of any decision about restoration and management of Noosa Estuary, this group of experts had not been selected for this purpose and so were not the right people to consider such values. Instead, experts were asked only to consider technical feasibility when proposing options.

A total of 14 options were developed (**TABLE 6.3.1**), and in some cases were accompanied by maps that illustrated the location of the proposed actions. These options span a diverse, and sometimes unorthodox, set of actions. The group identified the best option by formally assessing the predicted consequence of each option for the two fundamental objectives.

TABLE 6.3.1 Restoration and management options developed for Noosa Estuary

| Option | Key actions |
|---|---|
| | ***Restoration options*** |
| Restoration of sea grass | Restore sea grass habitat through improved control and redesign (e.g., swing moorings) of boat moorings. |
| Oyster reef restoration | Create oyster reefs initially in Weyba Creek and Lake Weyba. |
| Living shorelines | Replace hardened shoreline protection with structures that incorporate mangroves and oyster reefs.<br>Improve biological connectivity and extent of habitat mosaic between Noosa and Tewantin. |
| Provide habitat/hard substrate stepping-stones | Create subtidal reef structures in the main channel between Noosa and Tewantin.<br>Improve biological connectivity by increasing hard substrate and habitat mosaics. |
| Prawn restocking | Ensure restocking of prawns into the lakes and river. |
| Restoration of Kin Kin catchment | Assess current inputs and status of catchment.<br>Improve land management practices. |
| Habitat provision for raptors | Improve habitat availability for iconic raptor species by supplemental feeding.<br>Establish local education campaign and initiative. |
| | ***Management options*** |
| North Shore management/ vehicle closure | Create a “restoration zone” to restrict access.<br>Provide a buffer between recreational and commercial activities.<br>Ensure that any conservation zone includes both terrestrial dunes, beach, and nearshore areas. |
| Wake management “between the lakes” | Manage boat speed and wake.<br>Improve commercial boat design. |
| Estuary zoning (emphasis on recreational fishing) | Reduce the recreational catch.<br>Increase catch and release programs including training/education.<br>Provide support for improved fish habitat.<br>Coordinate permitting. |
| Cessation of commercial prawn trawling | Close fishing areas, particularly between the lakes.<br>Implement buyback of fishing licenses.<br>Modify fishing practices.<br>Decrease or limit catch (size or timing). |
| Better management of commercial mullet fishery | Limit catch on Noosa North Shore.<br>Provide pathways to increase product value.<br>Modify fishing practices—education. |
| Transform gillnet fishery to higher-value fishery | Transform gillnet industry to high-value line-caught industry. |
| Stormwater management | Improve the quality of water runoff flowing into the estuary through wetlands and other design features such as flow restrictors and pollution traps. |

- Make a deliberate effort to identify options: This means making space for it and doing so early in the planning process. Ideally, this would occur after identifying the objectives, but some teams have found it useful to do at the outset of a planning workshop. Doing this initial pass has the added potential benefit of highlighting preconceived preferences that may influence subsequent conversations. Most important is to recognize that a good set of options does not just appear but must be actively encouraged to appear.
- Avoid anchoring (see Box 6.2) either on actions that have been used before or those that are first proposed: This requires fighting significant psychological bias, but simply being conscious of this potential bias can help ameliorate it. In any case, it remains important to know the range of solutions that have been tried, either by paying attention to literature or by having diverse experiences in the room.
- Don't allow focus on constraints at this stage: Perceived feasibility is probably the most common constraint or reason for not proposing certain options. Many options may indeed end up being infeasible, but this is not the place to rule out strategies on this basis—let the evaluation against objectives do that. Individuals' determination of feasibility is subjective and naturally biased by their experiences. If assessment of feasibility is made a transparent and participatory process, it frequently turns out that what seems infeasible to one person is eminently possible to another. We often share the story of a senior staffer who by chance walked into a planning meeting and asked why the team hadn't considered the option of relocating a port development project. The team replied that no one considered it feasible. Thanks to his connections with the government, the senior staffer suggested this option might actually be feasible, and it eventually was. Even options that end up being infeasible might still form the nugget of successful solutions.
- Identify the bookends: What would be a great (even if unrealistic) option for each objective? What would be the worst option for each objective? You can think of great options and awful options as being positioned on either side of a range of more realistic options. These idealistic solutions can highlight vital strengths in options and often spark further options that take these strengths on board in more realistic shape.
- Follow good brainstorming advice: Much has been written on the subject of brainstorming,[6] but one piece of advice that is consistent with our experience and decision science research is to start by capturing ideas individually rather than as a group. Then group members can share their individual lists for group discussion.

- Build disruption into the process by inviting input from diverse voices, especially those outside the immediate field: This tip is likely to be particularly important when thinking about alternatives in complex systems,[7] which will be true for many if not most conservation projects. Having a diverse planning and project team as recommended in Chapter 2 will also help in developing options.
- Combine sets of actions into distinct options: As described by Robin Gregory and colleagues,[8] this process takes place during structured decision making. Alternatives can include a mix of actions, some of them on-the-ground conservation activities and others that might be focused on policy changes. It is also fine for alternative options to be at different spatial scales, for example, building a fence to keep cattle out of a particular stream versus changing fertilizer application rules for an entire state. Of course this will make direct comparison more challenging, but by no means impossible.
- Consider different levels of investment in an action or strategy as another approach in helping to develop options.

Conservation planners have different views about whether the options being considered should be fleshed out and ready for implementation, or rather be instructive of the types of options available. Which is appropriate will depend on the context and audience of the plan, but in general we see value in comparing strategies even if the fine details are not yet worked out or the option appears unrealistic. For instance, in a conservation plan for a large catchment, it might be useful to evaluate an option to remove agriculture or grazing entirely from the catchment (of course considering the potential social and economic impacts), even if the group has agreed that this option is unrealistic. There is a small chance that in doing so, you will stumble on a strategy that proves both brilliant and possible with a change of context. These extreme options provide good benchmarks for judging other options.

Regardless of your view, we believe it will generally be helpful to ensure that a range of actions are being considered. Although there is no definitive taxonomy of conservation actions, a good place to start for ideas is the typology of conservation actions that the conservation planner Nick Salafsky and colleagues have assembled.[9] Now that we have discussed examining different options for taking conservation action, we conclude with a "rule" about these actions that will improve the framing of conservation planning problems.

## Rule: Connect "Where" and "How" by Prioritizing Actions

Conservation planning has often tackled the question of place (where we should work), and the question of strategy (how we should work), in separate planning processes. As suggested throughout this book, we believe that this dichotomy is an unhelpful way to frame the problem and can diminish the relevance of a conservation plan. Although most conservation planners would be comfortable saying that the outcome of their plan is a priority set of species, habitats, or places, we argue that only *actions* can be legitimately prioritized—all options should involve some form of action.

Conservation planning is principally about informing resource allocation decisions; places, species, and habitats do not use the resources of conservation organizations and agencies—actions use resources.[10] What a conservation plan tells us is that some action associated with a location or species is a priority. Failing to acknowledge this as part of the problem framing is a recipe for inefficiency. Without being clear about actions, we cannot defensibly estimate consequences for the objectives, nor the expected cost of the option, both of which are critical elements for determining the best course of action. Conservation plans that contain lists of priority species, ecosystems, habitats, places, or other conservation features (see Chapter 3) without identifying the associated actions are an indication (although not a guarantee) that more thought needs to be directed to problem framing.

We recognize that this approach will appear unconventional to many planners. Framing conservation planning problems as a question of actions does not mean ignoring places or species; it means understanding what you are going to do at those places or for those species, why they are being selected, and effectively prioritizing combinations of places or species and actions. Quite often, conservation planners do frame problems as actions without necessarily appreciating that they are doing so. Consider the many plans conducted by (or for) government agencies for the express purpose of identifying sites for new or expanded protected areas. Protected area establishment is an action; it has clearly identifiable costs associated with it, along with factors that are likely to promote and hinder it as well as other strategies and actions that will be necessary to take in many cases in order to establish a protected area or conservation area. Most of the early conservation plans (see Chapter 1) were effectively prioritizing an action—the action of reservation. However, just as conservation has diversified its toolkit of strategies, it is important that our planning match this evolution and be explicit about the actions being considered.

One common response to our proposed framing is that while an action obviously needs to be taken for the conservation of a feature, the best action cannot always be determined a priori. In addition, many conservation plans are conducted at ecosystem or ecoregional spatial scales while more detailed data and information are often needed at finer scales of spatial resolution in order to develop some strategies and actions. We agree, but we do not view this reality as inconsistent with our framing. The subsequent or follow-up planning needed to determine the best action at a site requires resources—it is an action in and of itself. Prioritizing places or species for conducting further analysis and making subsequent decisions about resource allocation should consider factors such as these: the actions known to be available, how effective these actions are, other resources being dedicated to finding solutions for that species, habitat, or place, the resources likely to be available for that place or species, and so on.

Understanding the set of available actions, along with their costs and benefits, is more intensive and challenging than prioritizing from a list of places or species. To be clear, we are not proposing that a spatial plan for a large region now needs to include all the detail of a locally focused strategic plan. Focusing on prioritizing actions simply means being clear about the decisions the plan is supporting. We firmly believe that a conservation plan focused on identifying key areas for a region's biodiversity, even if the on-the-ground strategies in each place are unknown, can be a very sensible problem framing. For example, the plan may be intended to prioritize sites to raise awareness and push for political support for conservation in these key areas. These are both actions. By acknowledging and focusing on these actions as part of problem framing, the resulting plan is more likely to support its intended outcomes, for instance, by selecting features and communicating expected benefit in a way that resonates with the audience in question.

Viewing conservation planning as a process for simultaneously addressing the "where" and "how" questions, means that spatial planning, or identifying areas important for conservation action, becomes an aspect of strategic planning. We are already seeing a clear move in conservation planning toward action-based problem framing. **Marine spatial planning** (**MSP**) is a good example. As introduced in Chapter 2, MSP focuses on the spatial arrangement of different activities and uses of the marine environment. Each of these activities might be considered an action that contributes uniquely to achieving different environmental and social objectives and has its own distinct costs.

## Consequences

### Theory of Change and Logic Models

A **theory of change** (**TOC**) is an articulation of how an option is predicted to achieve an objective. It is an important link between the options discussed in the previous section and the prediction of consequences discussed in the following section. TOC is a term appearing with increasing frequency in conservation plans and funding proposals, especially in the nongovernmental organization (NGO) sector. Although there is no single agreed-on definition of TOC, and groups use the term in different ways, there are important common elements.

Broadly speaking, everything in this book (from determining context and identifying objectives to implementing actions) is part of constructing a TOC to guide conservation work. More specifically, a TOC is a representation of how an action or set of actions is predicted to achieve an objective. Conservation planners view a TOC as a planning tool, a process, and a communication device. It can also be a powerful fund-raising tool. Ideally, it should be all of these. Theories

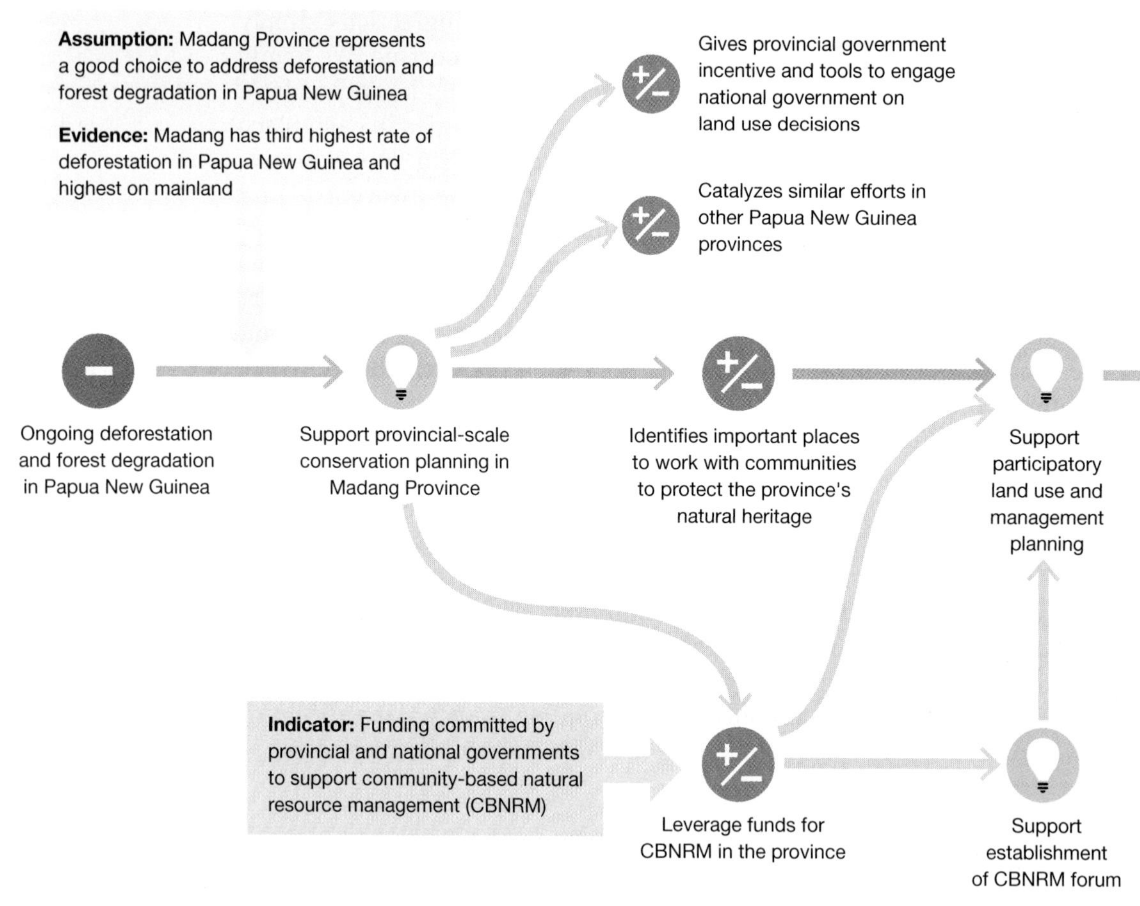


FIGURE 6.2 Graphical presentation of a theory of change for a community-based natural resource management project in Papua New Guinea. The four blue boxes are examples of text that is usually hidden but that "pops up" when readers roll their mouse over the arrows or nodes in the diagram. This pop-up approach allows a lot of information about the TOC to be presented without making the figure more crowded than it needs to be. This particular TOC was used to make decisions about where to invest science and monitoring effort given the existing strength of evidence for the theory underpinning the program.

of change can be visual or narrative summaries; the most effective TOCs contain both.

At the beginning of the book, we talked about how the aim of conservation planning is to help us get where we want to go efficiently and effectively. We have found developing a TOC to be one of the most useful exercises a planner can conduct toward this aim. A TOC could be considered a map of the route we propose taking that encompasses assumptions made about conditions we will find along the way, including road surface, number of tolls, expected traffic, amount of fuel required, risk of getting lost, and the like. FIGURE 6.2 illustrates a theory of change for a forest conservation program in Papua New Guinea.

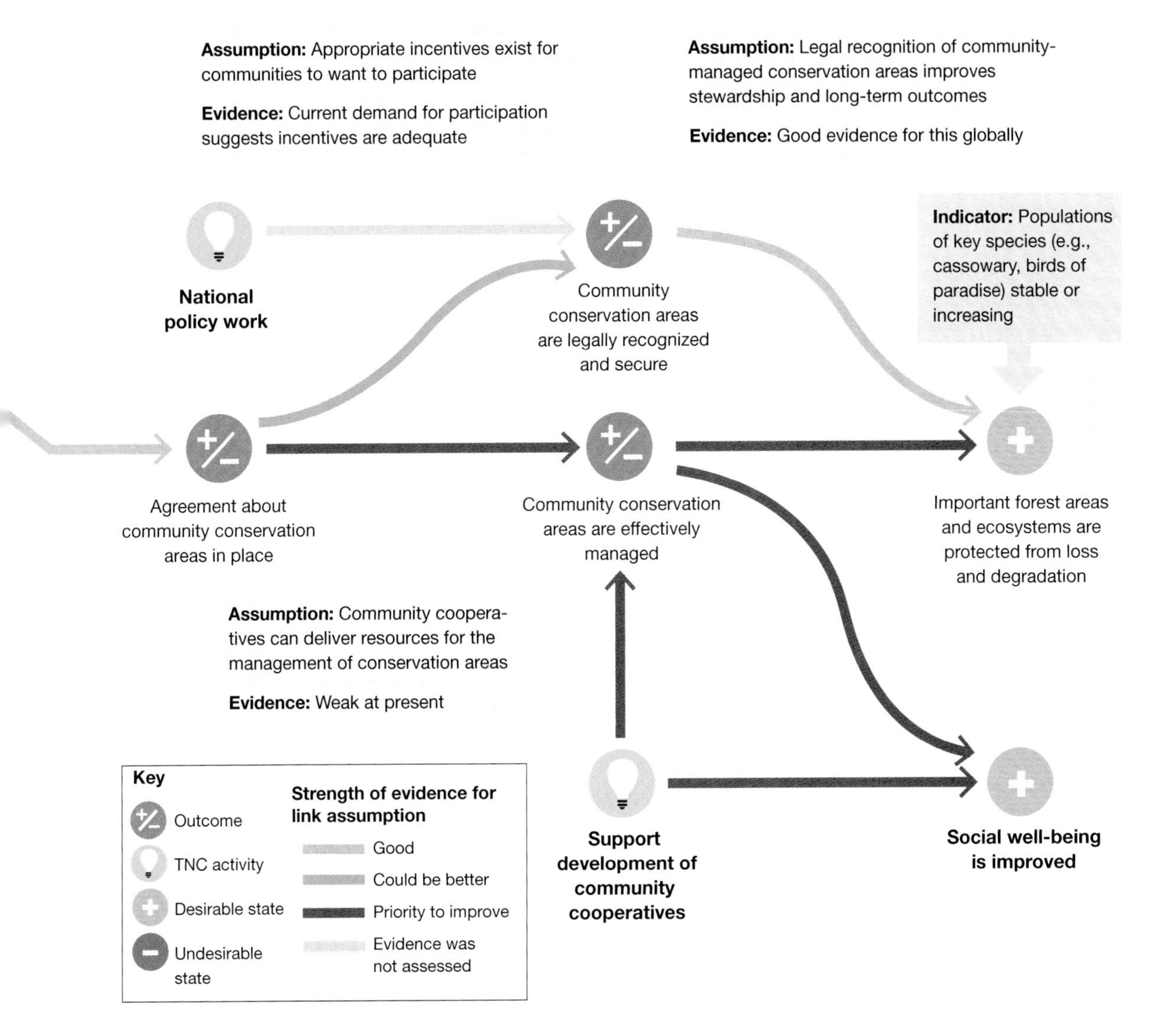

This TOC clearly articulates the assumptions of each step and also assesses the strength of evidence in support of the assumption. This consideration of underlying assumptions and context is often what distinguishes a TOC from a **logic model**. A logic model is a potential pathway of change, a logical sequence of steps from action to outcome. A TOC requires not only this logical thinking but also deeper critical reflection about the assumptions being made and what would be required to go from actions to outcomes.[11] Having made this distinction, we also think it is important not to get hung up on terminology; either term is fine provided the core concept is understood.

Another concept related to theory of change and logic models is **results chains**. For the many conservation practitioners familiar with the *Open Standards* (*OS*), Conservation Action Planning, or Miradi, the term *results chains* is everyday language.[12] In fact, there is even a useful online library of results chains developed from CAP or OS exercises (https://www.miradishare.org/).

Results chains are a series of boxes that link an action to an outcome through a series of intermediate steps (FIGURE 6.3). They can be considered a type of logic model. Conceptual models constructed as part of situation analysis, which we introduced in Chapter 5, often provide the foundation for a results chain. The software Miradi has capabilities designed specifically to help represent results chains. Although they are frequently presented as such, results chains are not a TOC. They map a (hopefully) logical sequence, but to be a TOC they must be strengthened by evidence and critical thinking about assumptions of how and why that sequence of change might come about.

A TOC fits nicely with the elements of problem framing that we've laid out in this chapter. Constructing a TOC can force planners to clearly articulate their understanding of change, which is particularly useful when it is accompanied by assumptions about how change will happen. These are the assumptions necessarily made when predicting the consequences of alternative options. We discuss the uncertainty associated with risk and assumptions in greater detail in Chapter 8.

## Predicting Consequences

To decide on what option(s) to choose, we need to know the expected consequences of the different options. For example, if a manager is considering fire management or the removal of invasive species as two strategy options for habitat conservation, we would want to know the impacts that these options are expected to have on the species or habitats we're trying to conserve, and the impact on the annual budget. Because these impacts will occur in the future, the situation requires not just estimation but a prediction or forecast. Although ultimately a guess, the prediction should be a well-informed guess that draws on the knowledge and data available about the social-ecological system.

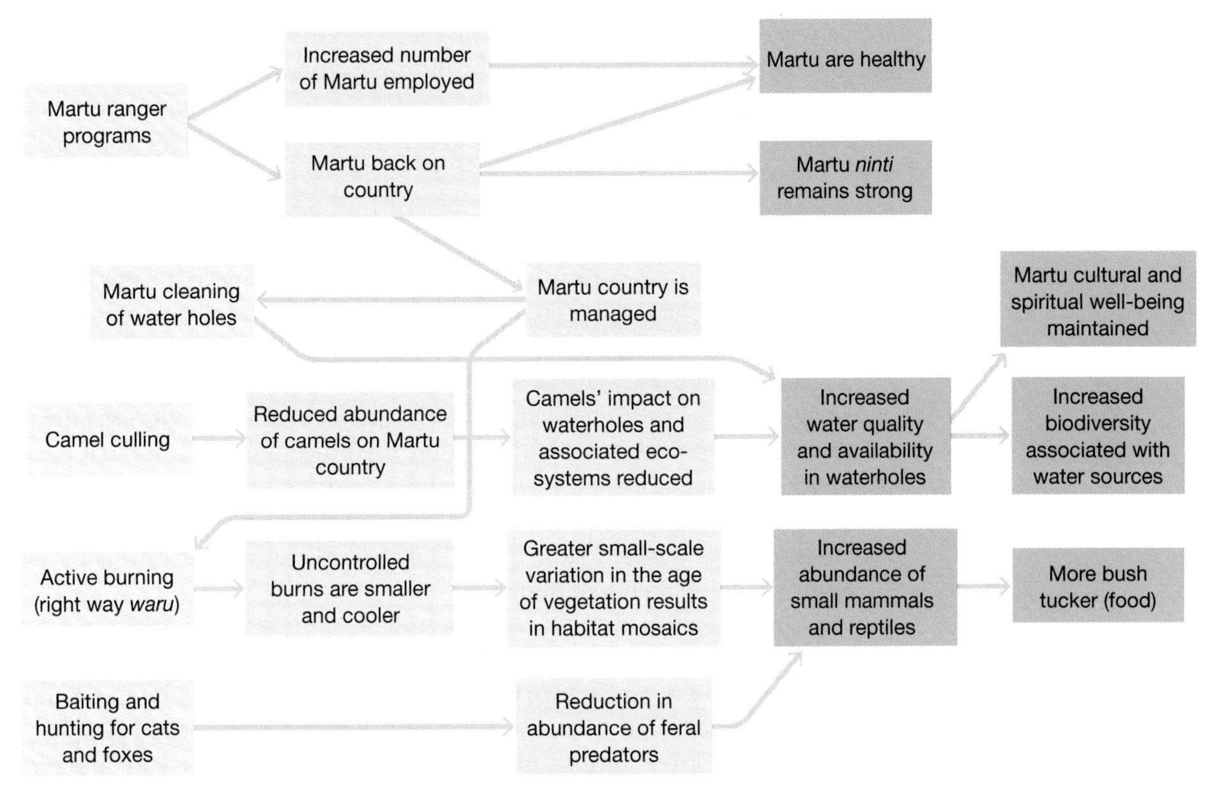


**FIGURE 6.3** Results chain for a conservation project on country of the Martu people of Western Australia. Yellow boxes indicate activities, blue boxes are intermediate outcomes, and green boxes are fundamental outcomes.

Together with objectives and options, predicting consequences is one of the three core elements of problem framing that we identified at the beginning of the chapter.

We like the term *consequences* because where there are multiple objectives, a course of action will have consequences for these objectives, and it's unlikely all of them will be perceived as benefits. A *benefit function* refers to a mathematical expression that translates how the extent of an action (or actions) affects the magnitude of the system response (achievement of objectives). While it is useful to express the relationship between actions and objectives mathematically, it is by no means necessary.

Estimating the consequences of different options is the step where science meets, and influences, our conservation decisions. Perhaps more than all others, this task should excite scientists because their knowledge and expertise are most important here. Predicting consequences is an analytical step; it is when project teams should rely on ecological scientists, economic scientists, and social scientists.

Imagine you have a forest conservation objective and want to predict the consequences of an option that involves providing local communities with gas stoves to replace the more traditional wood or charcoal stoves. Some of the knowledge that planners might need to predict the consequences include the number of households currently using wood or charcoal, the amount of wood or charcoal collected and burned, the relative importance of fuel wood collection for forest degradation compared with other things like clearing for agriculture, how much time would be saved by not collecting wood, and whether that additional time might be spent on other forest-degrading activities like hunting. It's easy to see how much thought it takes to identify the potential repercussions of a management action, especially one with multiple objectives.

Even if you rarely think of it in these terms, all of the research in conservation and related disciplines that is focused on understanding how social-ecological systems function ultimately contributes to predicting the consequences, positive and negative, of different actions. A major part of the conservation planner's job is to understand the social-ecological system that he or she is working in and interpret it through the lens of consequence.

It might sound foreign, but this task will actually be familiar to most readers. You might be surprised to learn just how much of your work involves predicting consequences. For anyone who has been involved in OS or CAP workshops, all of the time spent identifying key ecological attributes, assessing viability, ranking threats, considering feasibility, and building results chains is ultimately working toward understanding the consequence of an option on the features of interest (*targets* in OS language). Sometimes the consequence is explicitly stated as a reduction in threat or an increase in viability. At other times it is implicit because the effort to understand the most important threats is assumed to lead us to actions that will have the biggest benefit simply by working on the most important issues.

Similarly, anyone who has helped develop input files for spatial planning software such as Marxan (see Chapter 7) or conducted a systematic conservation plan has made at least implicit assessments about consequences. The act of working out what conservation features occur in a location, and therefore what conserving this location would contribute toward conserving the region's biodiversity is a statement of consequence—the consequence of taking action (or not taking action) to conserve habitat or species in that location. Sometimes the prediction of consequence is more involved than just tallying conservation features, perhaps including an assessment of habitat quality and how different levels of degradation or fragmentation influence the contribution of an area toward our objectives.

Hopefully you are starting to see what a central role predicting consequences plays in conservation planning. In the following pages, we first provide some general guidance on predicting consequences. Next, we introduce three approaches to predicting consequences. And we conclude by discussing a series of important considerations when using these approaches.

Even though conservation planners are constantly involved in estimating consequences (knowingly or unknowingly), distilling our substantial and yet incomplete knowledge of social-ecological systems into a prediction of outcomes is challenging. It is difficult work, but great benefit comes from trying to make explicit predictions about conservation outcomes (something many conservationists are reluctant to do). We know of few faster ways to learn about systems than to make predictions and get feedback on them. And the more experience you get, the better your predictive ability will be.

## General Guidance for Predicting Consequences

Here are three generally useful things to remember when predicting consequences as part of a conservation plan: (1) focus on objectives, (2) allow enough time, and (3) know the importance of counterfactuals and business as usual. In this section we briefly describe what we mean by each of these guidelines and give our rationale for highlighting them.

***Focus on Objectives*** Consequences should be predicted in terms of the established objectives. If you've identified indicators for objectives (see Chapter 4), then express the consequences in those terms. If you care about biodiversity features, ecosystem services, jobs, company profit, or whatever else, be sure to estimate the consequences for these things. This can be more difficult than it sounds because the immediate impact of options is generally some way removed from the fundamental objectives we care about (FIGURE 6.4). The temptation is often to estimate consequences only for those things our actions are directly influencing. So in the gas stove example mentioned earlier, we might estimate the number of households that would cease using wood or charcoal and the commensurate reduction in wood and charcoal use. We can probably estimate that number with a good

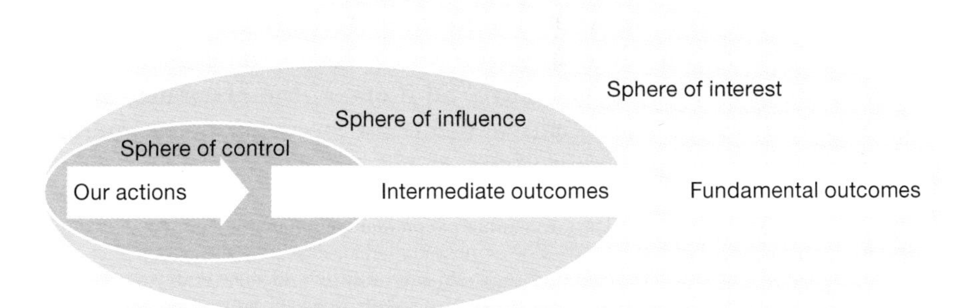


FIGURE 6.4 Predicting the consequences of options for fundamental objectives can be difficult because the immediate impact of options is often quite removed from the fundamental outcomes we care about. (Adapted from a presentation by Simon Hearn on outcome mapping.)

degree of accuracy, but what we fundamentally care about is the health of the forest.

If it is really difficult to estimate the consequences of an alternative for our fundamental objectives, then it's worth checking your logic, theory of change, and the assumptions embedded in them. Was a large leap of faith taken in there somewhere, or a significant and untested assumption? For example, we have recently witnessed a sharp increase in the number of planning teams citing the work of the Nobel Prize–winning economist Elinor Ostrom[13] in support of their theory of change.* On a theoretical level this is terrific, but on a practical level it often leaves teams with a significant gap in the mechanism between the proposed actions, such as fostering community leadership, and the impact on fundamental objectives, such as the health of coral reefs. This all makes it extremely hard to estimate the consequences of potential actions. We can't pretend it is easy, but if you can't at least make a start, then it seems reasonable to ask what confidence anyone should have in the outcomes being achieved.

Predictions of consequence often focus on the extent to which some threat is reduced. Reducing threats is important but rarely a fundamental objective. Rather, reducing a threat is one aspect of conserving the features we care about. Predicting the consequences of threat reduction for our objectives requires knowing the extent to which the threat is impacting the objective of interest and what will happen in the system when that threat is removed.

One of the most helpful prompts for keeping focused on objectives is the notion of a consequence table (**TABLE 6.1**). This is simply a table in which the options being considered form rows, the different objectives and their metrics form columns, and the consequence of each option for each objective populates the cells. You can think of completing this table as a core piece of any conservation plan. While a consequence table can itself be used as a solution method (see Chapter 7), it also serves as a great checklist for the information that needs to be gathered. It should be clear very early on which consequences will be straightforward and which will be more challenging to estimate. It is good practice to work out the current value for any attributes being used as this will provide a baseline and give an indication of how difficult the attribute is to measure. So if one of the attributes being measured is a change in the rate of deforestation, what is the current rate of deforestation in the area concerned?

---

* Elinor Ostrom's work focused on the governance of common pool natural resources—things like forests, fisheries, or grazing lands. She is most lauded for describing a set of principles or conditions by which local communites could successfully and sustainably manage local natural resources. These conditions emphasize things like self-determination of local communities and social capital. Many in conservation have said that investment in increasing properties like local social capital and participation in decision making will lead to improved governance and therefore sustainable use of natural resources.

**TABLE 6.1** An example of the predicted consequences of managing fire and feral herbivores in the Kimberley region of Australia, compared to a no-action counterfactual for a range of ecological groups. Note that the ecological groups with the lowest predicted probability of persistence are also the ones predicted to benefit the most when the counterfactual is considered. However, the counterfactual also suggests that those groups whose probability or persistence ends up being the highest actually benefit little from the proposed conservation action.

| **Bioregion: North Kimberley** | | | **Average probability of persistence with ecological function for each ecological group over 20 years** | | **Benefits of fire and herbivore management**[1] |
|---|---|---|---|---|---|
| ***Ecosystem*** | ***Ecological group*** | ***Examples*** | ***No action*** | ***Fire and herbivores*** | |
| Savanna (non-rugged) | Hollow/tree structure dependent—non-volant | Phascogale, rabbit-rat, tree rats (golden-backed, black-footed) | 0 | 0.45 | 0.45 |
| | Hollow/tree structure dependent—volant | Yellow-bellied sheath-tailed bat, cockatoos, owls | 0.9 | 1 | 0.1 |
| | Ground (surface and burrowing) dwelling—"critical weight range" mammals | Quolls, bandicoots (golden, brindled), rodents (pale field rats, Western chestnut mouse) | 0 | 0.45 | 0.45 |
| | Ground (surface and burrowing) dwelling—others | Diurnal skinks, partridge pigeons, quail, thick-knees, cisticola | 0.8 | 0.95 | 0.15 |
| | Litter dwelling | Lizards (specialist skinks, geckos) | 0.9 | 0.95 | 0.05 |
| | Granivores | Finches (Gouldian finch), pigeons, small rodents | 0.7 | 0.95 | 0.25 |
| | Insectivores | Small dasyurids, thick-heads, cisticola, ibis, fairy-wrens | 0.6 | 0.9 | 0.3 |
| | Frugivores | Emus, bowerbirds | 0.6 | 0.9 | 0.3 |
| | Nectarivores | Honeyeaters and lorikeets | 0.9 | 1 | 0.1 |
| | Herbivores | Macropods | 0.9 | 1 | 0.1 |
| | Predators | Mulga snakes, owls, dingoes, varanids, quolls | 0.6 | 0.8 | 0.2 |

[1] The differences between these persistence probabilities were calculated to determine the benefit of carrying out each action for each ecological group.

*Source:* Adapted from Carwardine, J. et al., *Priority Threat Management to Protect Kimberley Wildlife.* 2011, Brisbane, Australia: CSIRO Ecosystem Sciences.

***Allow Adequate Time*** Expect it to take time to "fill in" a consequence table. We know that many conservation planners have come to expect strategic planning to occur over a two- or three-day workshop or even shorter (this has never been the case for spatial planning because systematic conservation planning places greater emphasis on the use of spatial data). While certainly some headway and initial estimates might be made during a workshop (such as getting to the stage illustrated in Table 6.1), and sometimes a workshop will focus specifically on eliciting this sort of information from experts (as described in Chapter 5), in most cases getting good predictions of consequences will be a longer process drawing on multiple sources and approaches (see following). Remember, this is where our science expertise and knowledge most influence our conservation work—so stop and make sure you do them justice.

***Know the Counterfactuals*** Recently, a high-profile development project reported very publicly on the increase in aspects of social well-being among the communities receiving support from the project. The team's data collection was rigorous, and it appeared to be a robust trend. The problem was that the same trend was also observed across the entire country over this period. It was pointed out, equally publicly, that the outcomes being hailed as project success were in fact exactly what could have been expected were no project to have taken place! The point of this story is to emphasize the importance of the counterfactual: the value of any action should be judged not just on what the outcome looks like, but on how different the outcome would have been in its absence (see Table 6.1).

This is the same basic principle behind using a control in experimental design. Conservation problems do not typically lend themselves to rigorous experimental design, but more and more conservation initiatives are pursuing at least quasi-experimental designs.[14] The difference between a counterfactual in experimental design and in the context of estimating consequence is the need for prediction. The prediction might be informed by previous experimental work, but you can't wait to do an experiment—you need the counterfactual during conservation planning. Because we are generally talking about predicted consequences, some consider that the term *counterfactual* is not strictly accurate. Instead, we suggest evaluating the consequences of an action compared to a business as usual (BAU) scenario in which no additional actions are taken beyond what is currently in place. The BAU approach seems to help a broader audience grasp the concept—and, more importantly, the potential consequences of management actions—more quickly.

The classic example of how BAU might change our view of a conservation outcome is the case of protecting land that was unlikely to have been developed anyway, perhaps because it was too steep. The total area of land under protection might look large, but considering the total amount of intact habitat remaining, the outcome may not be that different from what would have happened without the protection. Given that the land in question is unlikely to be developed anyway, the BAU assessment might suggest that other land protection alternatives will achieve greater overall habitat protection.

Of course you can never know for sure what the future holds—as the American baseball player Yogi Berra famously said, "It's tough to make predictions, especially about the future." Still, you might expect slightly less uncertainty about a BAU prediction than about the other consequences because at a minimum you could extrapolate from current trends such as rates of decline or habitat loss, increasing storms, or growth of cities. Comparisons with BAU can also make options look better than they would initially seem. For instance, an easement in a fragmented landscape might not be predicted to result in any improvement in the current population of a carnivore, but when compared to an expected decline in its absence, just holding the line might be a pretty good outcome.

In addition to allowing assessment of the actual consequence of taking an action, a prediction about BAU outcomes also serves as the consequence for the do-nothing option that we advocated in the previous section. We've also found comparison to BAU a compelling communication tool for the strategies we decide to undertake.

### Three Approaches to Predicting Consequences

Approaches to predicting the consequences of possible actions are many. Broadly, they are on a spectrum from simply making an informed guess to devising sophisticated mathematical models (both are types of predictive models; one is simply written down in equations). A model is a simplified representation of reality. What is an appropriate approach to predicting consequences depends on what solution methods are being used (described in Chapter 7), as well as the time, resources, expertise, and data available. Whatever the method and resources, we should always be aiming for rigor and transparency.

It is beyond the scope of this book to go into the many modeling techniques that exist to help make predictions about the consequences of conservation actions. Here we introduce some general approaches to estimating consequences and also general principles that will lay the foundation for developing good predictions. The general approaches are direct estimate, criteria based, and mathematical models. This is our classification. There is no clean or established distinction between these three approaches, and they are certainly not mutually exclusive; in fact, it would be unusual if only one of these was used.

Hopefully you will take away from this section a greater appreciation of just how much of a conservation planner's job is modeling. We believe it is important for planners to recognize the significant role modeling plays, because it encourages planners to better understand the task and learn about how to do it well, that is to say, rigorously and defensibly. It is also important to remember that, by themselves, these approaches are not prioritization methods. They are ways to link options to our objectives. Prioritization, on the other hand, involves evaluating the relative desirability of these consequences. Methods for prioritizing are covered in Chapter 7.

***Direct Estimate*** Direct estimates are exactly what they sound like, a direct prediction of the consequence an action will have for a fundamental objective of a plan. This prediction will usually be based on asking experts, and it should use the expert elicitation techniques discussed in Chapter 5. Direct estimates are often a good place to start estimating consequences as they can form a rapid first pass that can usefully distinguish between options. Direct estimates, and the difficulty of making them, can also highlight areas of uncertainty, and what additional data or resources might be useful to inform estimates of consequence. TABLE 6.2 is an example of a direct estimate of consequence for a set of estuary restoration and management options, in this case simply giving the expected direction of consequence and an assessment of relative magnitude.

**TABLE 6.2** Direct estimate of consequence for a set of estuary restoration and management options. The arrow size and color indicate the expected direction and relative magnitude of change in a series of attributes due to taking the options in the left-hand column. The options listed here are a subset of those described in Box 6.3. The attributes were chosen to represent the two fundamental objectives of increasing fish abundance and increasing biodiversity.

| Option | Attributes | | | | | | | |
|---|---|---|---|---|---|---|---|---|
| | *Prawns* | *Bream* | *Whiting* | *Mullet* | *Birds* | *Yabbies* | *Crabs* | *Bio-diversity* |
| Restoration of sea grass | ↑ | ↑ | ↑ | | ↑ *# | | ↑ | ↑ |
| Oyster reef restoration | | ↑ (large) | ↑ | | ↑ * | ↓ | ↑ (large) | ↑ (large) |
| Living shorelines | | | | | ↑ * | | ↑ | ↑ |
| Provide habitat/ hard substrate stepping-stones | ↓ | ↑ | | | ↑ * | | ↑ | ↑ |

* shorebirds (e.g., migratory waders, oystercatchers, etc.); # raptors.

***Criteria Based*** Criteria-based predictions of consequence involve decomposing the estimate into a variety of factors or "criteria" that are expected to influence consequence. Consequences are independently evaluated for each criterion and subsequently combined to give an overall estimate of consequence (**Box 6.4**). This is by far the most common approach to predicting consequence for conservation planning problems. For example, there are countless prioritization schemes for habitats and species that consider criteria like rarity, threat level, and condition. These are, in essence, a criteria-based prediction of the consequence of taking conservation action on these

## Box 6.4 Combining Criteria to Estimate Conservation Benefit

To determine the consequence of an action, and therefore the benefit of taking it, estimates for individual criteria must be combined in some fashion. Although this usually involves some fairly heroic assumptions of independence and linearity of criteria, it is a useful approach for an overall evaluation of an option. Tables 6.3 and 6.4 and Figure 6.5 in this chapter illustrate common cases of combining criteria to estimate conservation benefit in a conservation plan.

Unfortunately, we have found that in the act of combining the estimates for a set of criteria, conservation planners frequently undermine their intention to be more rigorous and scientific in the way consequences are considered. Together with colleagues, one of us (ETG) has written about these "common mistakes" and how to address them.[a] Some of the key ideas are presented here. Any approach to combining criteria should aim for two qualities: to be defensible (i.e., not arbitrary) and transparent. We expand on arbitrariness and transparency next. If these two qualities are at the forefront of a planner's mind when estimating consequence, the estimates and subsequent solutions have a good chance of standing up to critical review.

### Arbitrariness

Conservation planners will find themselves tempted by and battling with arbitrariness. This occurs because we so frequently need to measure attributes and criteria using constructed scales (see Chapter 4).

The scores assigned using constructed scales are essentially arbitrary—for example, there is no objective reason to give a relatively undisturbed habitat a score of 4 rather than 5 (assuming that forest habitat in this example is assigned a value from 1—meaning of low value to conservation—to 5, meaning of high value). What these constructed scales typically represent is a set of ordinal numbers that indicate, for example, a score of 2 is better than a score of 1 and worse than a score of 3. If we restrict interpretation of such scales to simple ordinal representations between alternatives (for example, alternative X is better than alternative Y for variable *Z*), then the arbitrary nature of the numbers is not problematic. However, ordinal numbers do not convey how much better 2 is than 1; constructed ordinal scales are a problem when they're treated as a set of regular numbers to be used in some arithmetic approach to determining consequence (e.g., adding two or more variables together).

Here is an example that will be familiar to many: to help assess the viability of conservation features, The Nature Conservancy's Conservation Action Planning (CAP) process (and the conservation planning *Open Standards* software, Miradi) combines scores

*(continued)*

## Box 6.4 Combining Criteria to Estimate Conservation Benefit *(continued)*

indicating an assessment of a feature's size, condition, and landscape context using the following scale: very good, 4; good, 3.5; fair, 2.5; poor, 1. The scores for these three criteria are combined by taking the arithmetic mean. The results of the process can easily be misleading. For example, consider two habitat features, A and B, wherein habitat A receives three scores of fair, and habitat B receives two scores of good and one of poor. Based on the arithmetic mean, habitat B (8) ranks above habitat A (7.5). If the choice of scale were adjusted such that good was worth 3 rather than 3.5, habitat A (7.5) would rank above habitat B (7). As the conservation consultant Abel Wolman eloquently put it,[b] the "truth or falsity of results derived from measurements should not depend on a fortuitous choice of scale."

An easy way to check whether a prioritization result is likely to be meaningful and not arbitrary is to go back to the underlying data. In the example above, a habitat rated very good (score of 4) must be unambiguously considered four times better than a habitat rated poor (1) because this is how it is being treated when the arithmetic mean is calculated. This mistake is best avoided by estimating variables of interest on natural scales wherever possible.

Defensible approaches to combining criteria are those that are repeatable and not arbitrary. In other words, there is clear justification for the approach, and given the same underlying information, another group of people could be expected to come to a similar assessment of consequence.

### Value Judgments and Transparency

Transparency has two elements. The first is straightforward, and that is being clear about where the different data and parameters came from. Conservation planners are reasonably good at this, and this is generally what we think of when we talk about transparency. But there's a second part to transparency that conservation planners have typically paid less attention to, which is embedding value judgments in our work. We discussed this issue of value in Chapter 1 and touched on it again in Chapter 4, and it has been written about a good deal in the conservation literature.[c] We don't believe that conservation planners are being deliberately opaque when they embed value judgment, simply that this element of opaqueness is often not recognized.

Many conservation planners have an intuitive sense that criteria do not affect priority equally, linearly, or independently. A common response to this realization is to establish a set of rules for combining criteria, often presented in a look-up table (such as **TABLE 6.4.1**). Look-up tables are alluring because they are an easy way to combine criteria and because they can be developed by planners without

**TABLE 6.4.1** A look-up table illustrating how assessments of the size and context of habitat patches should be combined to determine an overall priority rank. (VG = Very good, G = Good, F = Fair, P = Poor.)

| | | Size | | | |
|---|---|---|---|---|---|
| | | *Very good* | *Good* | *Fair* | *Poor* |
| **Context** | ***Very good*** | VG | VG | G | G |
| | ***Good*** | VG | G | F | F |
| | ***Fair*** | G | F | P | P |
| | ***Poor*** | F | F | P | P |

formal training in modeling. However, this ease of creation can lead to error.

Rules and look-up tables reflect the values, beliefs, assumptions, biases, and risk tolerances of their creators. For example, Table 6.4.1 (from a prioritization process that we were involved with) shows how assessments of the size and context of habitat patches should be combined to determine overall priority rank. In this case, a fair for size and a fair for landscape context result in an overall score of poor. Logically, this rating might be expected to yield an overall score of fair; but the rules in the table could reflect the planner's belief that an interaction occurring between these variables further reduces the conservation importance of the habitat patch at low scores. Alternatively, the rating might reflect the planner's assessment that scoring either size or context of a patch as fair might be possible, but giving both factors a score of fair would be unrealistic. It's difficult to be sure.

Similarly, a poor for size and a very good for context results in an overall score of good, whereas a poor for context and a very good for size results in an overall score of fair. Again, it's unclear whether this rating means the conservation planner believes context should have more effect on priority than size, addressing context is more feasible than addressing size, or why this assessment of effect is limited to this combination of scores. Interpreting these values and judgments becomes a daunting prospect when the look-up tables contain three or more variables.

The principal issue here is not the funny math, but that these beliefs are not transparent and therefore not open to critique. Conservation plans are intended to reduce bias and promote objectivity or at least to be explicit about assumptions, bias, and their effects so that assumptions and the resulting priorities can be effectively contested by interested parties. Many involved in decision-making philosophy consider contestability the formative property of a defensible prioritization process.[d] Rather than promoting transparency, planning methods that include, for example, look-up tables and combinatorial rules actually obfuscate the reasons behind the prioritizations by burying a series of value judgments and assumptions beneath a numerical veneer. Such a use of numbers simply formalizes unacknowledged bias and endows the process with a false sense of credibility.

Instead of using this sort of flawed prioritization process, it would be better to acknowledge that priorities are based on individual intuition, bias and all. Donors and the public can then judge whether they are comfortable with their resources being prioritized this way. Our experience has been that when value judgments and intuition are made transparent, they are indeed likely to be challenged, but the resulting discussion ultimately strengthens the planning process.

A prioritization conducted by The Nature Conservancy to help make decisions about establishing new conservation projects in Africa is a good illustration of how this mistake can be addressed.[e] Staff involved in the plan believed that, from The Nature Conservancy's perspective, the relative conservation priority of each country was affected by the distinctiveness of the biodiversity, extent of land clearing, level of fragmentation of the remaining habitat, extent of the existing protected area network, and quality of governance in the country. To make value judgments about these variables transparent, one of us (ETG) asked employees involved to sketch functions that reflected their belief about how each variable related to conservation priority (**FIGURE 6.4.1**).

Having a strong preference for conserving less-fragmented habitat, for example, is perfectly legitimate, but this preference should be clearly distinguished from a scientific assessment of the effects of habitat fragmentation on conservation outcomes.[f] These sketched functions were then turned into mathematical expressions and used as part of a return-on-investment prioritization process. Although drawing heavily on the experience, opinions,

## Box 6.4 Combining Criteria to Estimate Conservation Benefit *(continued)*

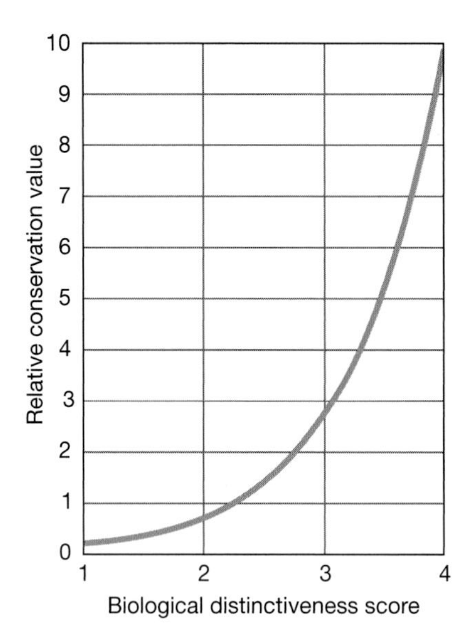


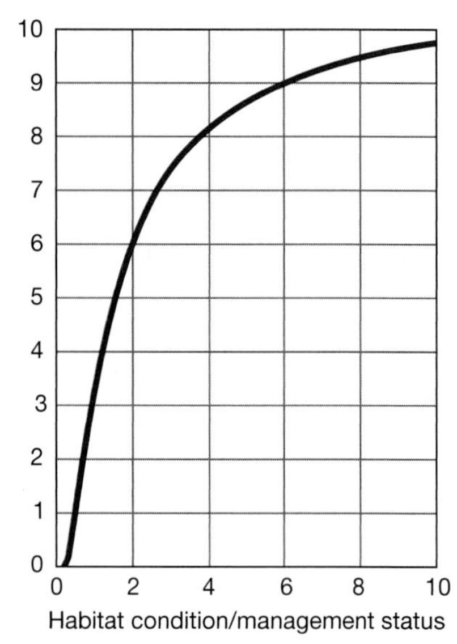


**FIGURE 6.4.1** Subjective functions linking habitat condition and biological distinctiveness to conservation value. These functions were sketched by participants in a conservation planning exercise to make value judgments associated with estimating conservation benefit more transparent. (Data from Tear et al., 2014.)

and values of the employees involved, the prioritization did so in an explicit fashion that made it possible to identify and contest these beliefs.

---

[a] Game, E. T., P. Kareiva, and H. P. Possingham, *Six common mistakes in conservation priority-setting.* Conservation Biology, 2013. **27**: 480–485.

[b] Wolman, A.G., *Measurement and meaningfulness in conservation science.* Conservation Biology, 2006. **20**: 1626–1634.

[c] Wilhere, G. F. et al., *Conflation of values and science: Response to Noss et al.* Conservation Biology, 2012. **26**(5): 943–944.

[d] Burgman, M. A., *Risks and Decisions for Conservation and Environmental Management.* 2005, Cambridge: Cambridge University Press.

[e] Tear, T. H. et al., *A return-on-investment framework to identify conservation priorities in Africa.* Biological Conservation, 2014. **173**: 42–52.

[f] Wilhere et al., Conflation of values and science (see reference c); Failing, L., and R. Gregory, *Ten common mistakes in designing biodiversity indicators for forest policy.* Journal of Environmental Management, 2003. **68**: 121–132.

habitats or species. Criteria like rarity and threat level are proxy measures for the overall consequence for biodiversity; they rely implicitly on the counterfactual that under a business-as-usual scenario, rare and threatened features are likely to be lost, thereby diminishing biodiversity. However, because the process of decomposing criteria is not done explicitly, many planners and participants do not recognize that they are actually predicting consequence through this process. The criteria-based approach is most obviously formalized in

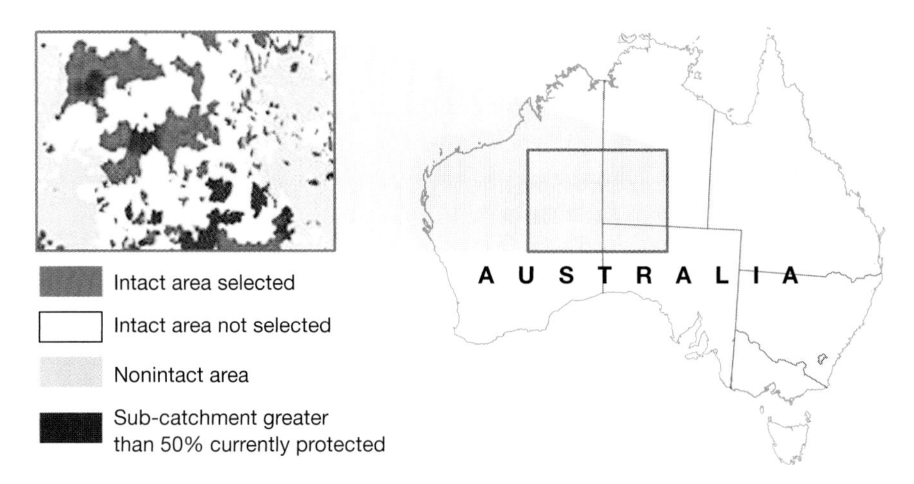


FIGURE 6.5 A familiar but often unrecognized example of criteria being combined to estimate the benefit of a conservation action. This is a solution from a spatial prioritization for Australia where sub-catchments are being used as planning units. The benefit of taking conservation action in one of the sub-catchments is estimated by combining criteria for the bird species present, the vegetation type, and the wilderness quality of the planning unit. (Adapted from Klein, C. J. et al., *Spatial conservation prioritization inclusive of wilderness quality: A case study of Australia's biodiversity.* Biological Conservation, 2009. **142**(7): 1282–1290.)

Multi-Criteria Decision Analysis (MCDA)—discussed in detail in Chapter 7—but plays an important and often unrecognized role in all solution methods (e.g., FIGURE 6.5 and TABLES 6.3 and 6.4).

Table 6.3 shows this process in action as part of a Conservation Action Plan, the assumption being that conservation action will help ameliorate the threats so the consequence of a strategy to address a threat is related to how significant that threat is to begin with. Table 6.4 shows an example from Canada, where the consequence for plant biodiversity of taking action for a particular rare plant species in Saskatchewan province is estimated by the combination of three criteria: (1) whether it is present in other Canadian provinces; (2) the number of ecoregions within the province where it occurs; and (3) the level of threat to the ecoregions where it occurs. Figure 6.5 illustrates criteria being combined within spatial prioritization using the software Marxan (see Chapter 7 for details about Marxan). Given the percentage of conservation plans that use one of these two (or closely related) methods, it is easy to see just how widely the criteria approach to benefit is employed.

How are criteria different from the objectives and attributes that we've already introduced? The practical distinction between the two is that objectives are what we are trying to achieve, whereas criteria reflect factors that in some way influence the expected consequence of an option. For example, common criteria might relate to the alignment of a strategy with an institution's interest and expertise. Alignment

**TABLE 6.3** A familiar but often unrecognized example of a series of criteria being combined to help estimate the benefit of a conservation action. This table is from a Conservation Action Plan for the Meli Mountain region of China. In this case, the perceived severity of a threat to the criteria shown here is combined to produce an overall threat rating that is directly related to the benefit of working on that threat.

| Project-specific threats (common taxonomy*) | Alpine mosaic (alpine meadow, alpine areas, and alpine-subalpine shrub/alpine-rhododendron shrub) | Cold–temperate coniferous forest (spruce-fir forest) | High-gradient stream systems | Low-latitude glaciers | Sclerophyllous evergreen broad-leaved forests | Temperate coniferous–broad-leaved midland forests | Ungulates | Overall threat rank |
|---|---|---|---|---|---|---|---|---|
| Fuelwood collection (gathering terrestrial plants) | – | Very high | High | – | Very high | High | – | Very high |
| Unplanned tourism and recreation areas (tourism and recreation areas) | High | Medium | High | Very high | – | – | Medium | High |
| Construction timber collection (logging and wood harvesting) | – | Very high | – | – | High | High | – | High |
| Farming activities (work and other activities) | – | – | High | – | Very high | – | High | High |
| NTP collection (gathering terrestrial plants) | Very high | High | – | – | Medium | Medium | Medium | High |
| Excessive ranging (livestock farming and ranching) | Very high | Medium | – | – | Medium | High | – | High |
| Global warming (temperature extremes | High | – | – | Very high | – | – | – | High |
| Subsistence/illegal timber collection (logging and wood harvesting) | – | Very high | – | – | – | Medium | – | High |
| Forest fire (fire and fire suppression) | – | High | – | – | High | High | – | High |
| Poaching (hunting and collecting terrestrial animals) | – | – | – | – | – | – | Very high | High |
| Road construction (roads and railroads) | – | – | High | – | – | High | Medium | High |

**TABLE 6.4** Example of a criteria-based approach to estimate the consequence of conservation action for different plant species in Saskatchewan, Canada. For each plant species (effectively options for conservation action), the scores for a series of criteria are used to indicate which is most important for conserving plant biodiversity (effectively the biodiversity consequence of working on that species).

| Criterion | Description of components | |
|---|---|---|
| Regional responsibility (RR) | Species exists in no other Canadian province or territory (occurs only in Saskatchewan): 5.0 score | |
| | Species exists in 1–2 other Canadian provinces or territories: 4.0 score | |
| | Species exists in 3–5 other Canadian provinces or territories: 3.0 score | |
| | Species exists in 5–7 other Canadian provinces or territories: 2.0 score | |
| | Species exists in 7 or more other Canadian provinces or territories: 1.0 score | |
| Local rarity (LR) | SRANK 1: 2.5 score | Species exists in 1 ecoregion: 2.5 score |
| | SRANK 1.5: 2.0 score | Species exists in 2 ecoregions: 2 score |
| | SRANK 2: 1.5 score | Species exists in 3 ecoregions: 1.5 score |
| | SRANK 2.5–3: 1 score | Species exists in 4–5 ecoregions: 1.0 score |
| | SRANK 3.5–5 or not assessed: 0.5 score | Species exists in 6–9 ecoregions/data deficient: 0.5 score |
| Habitat vulnerability (HV) | Species exists only in highly threatened ecoregion(s) (> 27 threats): 2.5 score | Threats exist and are listed: 2.5 score |
| | Species exists primarily in ecoregion(s) faced with high (> 27 threats): 2.0 score | Threats exist but are not listed: 2.0 score |
| | Species exists in ecoregion(s) faced with high (> 27) and/or moderate (13–15) and/or low (3–8) threats: 2.0 score | Threats are expected but cannot be confirmed: 1.5 score |
| | Species exists only in ecoregion(s) faced with low (3–8) threats: 1.0 score | No known threats, likely to occur in the future: 1.0 score |
| | Species exists only in ecoregion(s) faced with low (3–8) anthropogenic threats: 0.5 score | No known threats, unlikely to occur in the future: 0.5 score |

*Source:* Adapted from Kricsfalusy, V. V., and N. Trevisan, *Prioritizing regionally rare plant species for conservation using herbarium data.* Biodiversity and Conservation, 2014. **23**(1): 39–61.

of strategy with institutional expertise is unlikely to be an objective in its own right, but a well-aligned strategy is more likely to be effectively supported, so we might expect that this will boost the expected impact of the strategy. In this way, criteria can reflect institutional concerns that are not fundamental objectives but simply desirable properties of a chosen option.

Criteria can also represent more fundamental things about a social-ecological system. For example, in estimating the benefit of protecting different forest patches in Papua New Guinea, one of the criteria considered was the logging history of the area. This criterion was used to modify the predicted biodiversity benefit of protecting that forest patch so that more recently logged areas were predicted to deliver a smaller benefit than those with longer times since being logged. When it comes to biodiversity, the extent and severity of threats faced by the features of interest are frequently used as criteria. These are rarely referred to as criteria so are not usually recognized as such; however, they are factors that influence the consequence of a particular option for our objectives—which is the definition of *criteria* we're using here.

Criteria need to either directly or ultimately reflect our objectives and chosen attributes. They may be used to ensure that the consequences we are considering reflect the true intent of the objectives, so the benefit toward a broad objective like biodiversity conservation might include criteria such as taxonomic distinctiveness, rarity, cultural or financial value of species, and condition, level of fragmentation, or the percentage remaining for habitats. Lists of criteria, however, should never be used as a substitute for clear identification of objectives.

***Mathematical Models*** The third general approach to predicting the consequence of an option is the use of mathematical models. It is a gray line between what we're calling mathematical models and the criteria-based approaches just discussed. As the ecologist and decision scientist James Nichols elegantly describes,[15] all three of these approaches (direct estimate, criteria based, and mathematical models) depend ultimately on conceptual models that are simply a set of ideas about how the world, or a subset of the world, works.

A mathematical model is created by translating these ideas into equations. In a general sense, predicting consequence is the basic function of all mathematical models.[16] Mathematical models can be a powerful tool to help estimate consequence; however, we fully acknowledge that for many conservation plans, mathematical models are either unavailable or the effort required to build them is too great to be worth it. At the same time, mathematical models are more commonly available than is frequently recognized. Many models useful for conservation planning have been built by academics or research agency staff to explore specific pieces of a system's dynamics. Such models often require only minor adjustments to provide information immediately useful for predictions of consequence.

For example, the Soil and Water Assessment Tool (SWAT)[17] is a set of hydrological models that can be used to estimate the consequence of changes in land use and land cover on environmental flows.[18] We might use such a model to judge the consequences of various actions that would reduce or mitigate land use change that could have negative consequences for water flows and/or water quality. It is also common for models to provide information on some objectives and not others, so these model outputs may need to be subsequently combined with a criteria-based approach. All mathematical models are intended to reflect empirical relationships, but necessarily require simplifying and abstracting these relationships.

TABLE 6.5 is a cursory introduction to some of the types of mathematical models that we have seen usefully applied to predicting consequence as part of a conservation plan. Our hope is to briefly introduce conservation planners to the different types of models to consider using, and to know generally what modelers are talking about when they use the terms described. The models identified in Table 6.5 are all predictive models; a term that reflects how they are used rather than a particular mathematical approach. These models are not discrete classes; for example, a model might be a stochastic model and a simulation model.

### Important Considerations When Predicting Consequences

In the previous section we introduced the different approaches to estimating the consequences of alternatives. We now cover some important considerations when doing so: estimating costs, handling threat and vulnerability, using attributes and surrogate measures of consequence, weighing the relationship between different levels of investment in an option and the predicted consequence, and determining marginal benefit and diminishing returns. We have chosen these topics as aspects of consequence prediction that can be tricky and because we are often asked questions about them.

***Cost*** Taking action costs money, or at least resources and effort that themselves cost money. Cost can thus be thought of as a consequence of a chosen option—in other words, "If we choose this option, it will cost us this much." Of course, our evaluation of different alternatives will be informed by the resources available and where they might come from; but from the viewpoint of problem framing, we have found it useful to think about costs as a consequence.

In Chapter 5 on data, we introduced the broad notion of cost in conservation planning and some of the data that might be used to understand it. How cost is used will depend on the chosen approach to solving the conservation planning problem (Chapter 7). We call it out specifically here as a reminder of what an important piece of the conservation planning problem it is. It is worth remembering that the best people to estimate costs are not necessarily the same experts involved in predicting other consequences. Knowing the cost of different options typically requires relevant project management expertise.

***Threat and Vulnerability*** A great deal of conservation action is motivated by threats to natural systems. It is therefore not surprising

**TABLE 6.5** Mathematical modeling approaches useful for predicting the consequence of options

| Approach | Description and example |
| --- | --- |
| Statistical | Statistical models describe how combinations of variables relate to another variable (a response variable, which is the thing we're interested in predicting). Statistical models are built by seeing how different combinations of variables explain empirical data. Statistical models are used extensively in ecology and in social sciences. There are many types of statistical models and many books written on their use in conservation environmental and social sciences; most readers will at least be familiar with regression-based models. |
| Stochastic | In a stochastic model, the probability of an outcome is determined by predictable relationships between variables and also by incorporating an estimate of randomness for the input variables. Stochasticity is often used where uncertainty exists in the parameters for a model. For example, Kerrie Wilson and colleagues represent land conversion using a stochastic model in determining which parts of Indonesia to invest in.[a] |
| Deterministic | In a deterministic model there is no randomness, and so the outcome is determined entirely by the starting values for the model variables. If you know where you start from, you will know where you end up. They are used lots in engineering but less so in conservation because stochasticity is often an important influence on outcomes. However, deterministic models can still be useful as representation of the "mean" outcome, and examples in conservation include metapopulation models used for predicting the consequence of losing or conserving a particular patch or population.[b] |
| Simulation | Rather than a type of model, simulation is really the application of a model with an uncertain (stochastic) element to it. Simulation usually involves repeatedly forward-projecting a model to get an idea about either average response patterns or changes through time. Simulation is often used to explore the impact of different fisheries or marine management actions.[c] |
| Matrix | A type of model based on matrix algebra. Particularly common for modeling populations and the outcomes of demographic changes in a population (say, through harvest). Different age (or stage) classes in a population are represented by different elements in the matrix. For example, the conservation ecologist Jonathon Rhodes and colleagues use a matrix model to look at the expected population of koalas under a range of recovery actions.[d] |
| Spatial | Any of the above models, but specifically where the spatial relationship between variables is important in determining the outcomes and the spatial arrangement of variables (say, land use types) is an outcome of interest. Spatial models are often used to predict future changes in land use such as urban or agricultural expansion. |
| Bayesian | Type of statistical model that allows for updating belief in the likelihood of particular outcomes based on additional observation data. Useful for prediction because a wide variety of evidence can be used.[e] |

| Approach | Description and example |
|---|---|
| Network | A model that is structured around the relationship between a series of connected nodes, most often used in conservation to represent landscape or metapopulation connectivity.[f] |
| Population viability analysis | A particular type of stochastic single-species model, generally analyzed using simulations, used to estimate probabilities of a population going extinct over a given time frame.[g] |
| Fuzzy cognitive models | A type of qualitative model based on a set of linked variables (or nodes) and the degree of influence (positive or negative) that variables have on each other. Useful in conservation because they can be developed and parameterized from cognitive mental models of a system (or for a theory of change) and can therefore be built in a highly participatory fashion.[h] |

[a] Wilson, K. A. et al., *Prioritizing global conservation efforts.* Nature, 2006. **440**(7082): 337–340.

[b] For example, Cabeza, M. et al., *Conservation planning with insects at three different spatial scales.* Ecography, 2010. **33**(1): 54–63.

[c] For example, Okey, T. A. et al., *A trophic model of a Galápagos subtidal rocky reef for evaluating fisheries and conservation strategies.* Ecological Modeling, 2004. **172**(2): 383–401; and Brown, C. et al., *Effects of climate-driven primary production change on marine food webs: Implications for fisheries and conservation.* Global Change Biology, 2010. **16**(4): 1194–1212.

[d] Rhodes, J. R. et al., *Using integrated population modeling to quantify the implications of multiple threatening processes for a rapidly declining population.* Biological Conservation, 2011. **144**(3): 1081–1088.

[e] For example, Carroll, C. et al., *Hierarchical Bayesian spatial models for multispecies conservation planning and monitoring.* Conservation Biology, 2010. **24**(6): 1538–1548.

[f] For example, Treml, E. A., and P. N. Halpin, *Marine population connectivity identifies ecological neighbors for conservation planning in the Coral Triangle.* Conservation Letters, 2012. **5**(6): 441–449.

[g] For example, Sebastián-González, E. et al., *Linking cost efficiency evaluation with population viability analysis to prioritize wetland bird conservation actions.* Biological Conservation, 2011. **144**(9): 2354–2361.

[h] For example, Nyaki, A. et al., *Local-scale dynamics and local drivers of bushmeat trade.* Conservation Biology, 2014. **28**(5): 1403–1414.

that decisions about conservation investments can be strongly influenced by the level of threat to particular features or places. Actions are often taken to diminish threats or protect things that are vulnerable. As we discussed in Chapter 5, the concepts of threat and vulnerability are closely linked, and vulnerability is the sensitivity of a feature to a particular threat.

Threat and vulnerability often play a central role in framing conservation planning problems—just think about how many planning efforts start with a threat or vulnerability assessment and base subsequent decisions on these. In our experience, evaluating threat or vulnerability in the absence of consequences may be undesirable for a couple of reasons.

First, threat is often used as a criterion without consideration of how an action would actually affect it. In other words, problems are framed such that the more threatened a feature is, the higher its priority. This is not a well-framed planning problem, because decisions about priority should be determined not just by the level of threat but also by what difference the available actions will make. Threatened species lists or Red List status are classic examples of threat without assessment of consequence—they offer useful information, but not a well-framed conservation resource allocation problem. Something that is highly threatened but for which potential conservation actions are unlikely to make a difference not only wastes money, but diverts resources from other potentially successful actions. If an action is unable to mitigate the threatening process, the continued presence of that threat may ultimately spell failure for efforts focused on a particular conservation feature. In such cases, our fundamental objectives of conserving a habitat or species might be best advanced by prioritizing the features or places that are least threatened. **Box 6.5** provides an example of this tricky framing issue in the context of threats to coral reefs.

Furthermore, considering threats in the absence of consequences of taking particular actions may also result in substituting a threat for something more fundamental (like persistence) without carefully thinking about how the two are linked. Even when we ask, "How much will our action diminish this threat?"—which is a big improvement over just asking, "How big is the threat?"—the removal of a threat is not perfectly correlated with the consequence for conservation features, because other unaddressed threats might remain. For example, removing the threat of habitat loss in an area of tropical forest may not be well correlated with the persistence of mammal populations in those forests if hunting is also going on.

Both of these issues can be avoided by addressing threats and vulnerability in the context of assessing the consequence of options. Likewise, focusing on actions as options rather than as the features themselves will help ensure that threat and vulnerability are framed in the most rational and defensible manner.

***Attributes and Surrogacy*** In Chapter 4 we introduced the idea of attributes that we would use to measure change in our objectives. These attributes should form the currency for predicting consequences. For example, the column heads in Table 6.2 represent attributes used to predict the consequence of potential management options for Noosa Estuary.

We raise this point here because, although the consequences we care about are consequences for our objectives, it is often necessary to rely on surrogate attributes when assessing these consequences. Remember from Chapter 3 that surrogates are what we use when we are unable to measure exactly what we care about. Many assessments of consequence draw heavily on surrogates (see Chapter 3); for example, the ecologist Joshua Goldstein and colleagues estimate the consequence of different restoration actions in Hawaii by looking at their

## Box 6.5 Should We Protect the Most Vulnerable or the Least Vulnerable?

Does being vulnerable make something a high conservation priority or a low priority? Although vulnerability assessments are an almost routine part of conservation planning, how these assessments should rationally influence conservation priority has received almost no explicit attention. On one hand, it seems obvious that higher vulnerability should equal higher priority for conservation; those vulnerable features are the most in need of the boost in survival prognosis offered by conservation action. On the other hand, however, it could also be argued that highly vulnerable features are bad choices for conservation because they have lower potential for successful conservation. We have seen it interpreted both ways, even within the same organization.

One of us investigated this question specifically, in the context of coral reef conservation.[a] Coral reefs are threatened by stressors such as bleaching (FIGURE 6.5.1) and cyclones, which are largely beyond the control of local conservation but can strongly influence the fate of reef habitat even inside protected areas. However, protecting a coral reef from local stressors such as fishing and pollution can greatly improve its chances of recovering from the disturbance caused by bleaching or cyclones, and reefs vulnerable to these disturbances are likely to benefit the most from protection.

So, should areas of high disturbance vulnerability be places to target or avoid? The answer depends on how we measure the attributes of success, and on what is happening on the reefs that don't get protection. If success is measured simply by the health of reefs inside conservation areas, then high vulnerability should mean low conservation priority. If on the other hand we are measuring success by the health of the larger reef systems (including protected and unprotected reefs), then the answer is more complicated. Essentially, if the cumulative stressors are low enough that reefs without protection continue to function adequately, priority should be given to the most vulnerable reefs. If, however, unprotected reefs are expected to have a gloomy fate, then it is best to consolidate the health of the least vulnerable reefs—so high vulnerability equals low priority. This latter case seems increasingly likely to dominate given the predicted increase in disturbance frequency on reefs due to a changing climate.

The point of this example is to illustrate that it's not enough to simply assume a relationship between vulnerability and conservation priority. The consequence of each alternative needs to be assessed. If this evaluation of consequence against objectives is done carefully, and problems are structured well, the way things like vulnerability are treated should be easily evident.

[a] Game, E. T. et al., *Should we protect the strong or the weak? Risk, resilience and the selection of marine protected areas.* Conservation Biology, 2008. **22**: 1619–1629.

FIGURE 6.5.1 Coral reef that is "bleached" as a result of high temperatures.

expected impact on native birds, even though the ultimate objective is to restore functioning native ecosystems.[19] Using surrogates like this is an important part of pragmatic conservation planning.

***Dose and Response*** The discussion of consequence so far has largely been about clear cases where an action is taken or not; for example, if we protect this patch of forest, the viability of the orangutan population will increase by some amount. There will, however, be plenty of cases (if not all cases) for which different amounts of effort can be invested in an action with different consequences expected depending on the amount of investment. For example, there is a relationship between the number of forest patches we protect and the population of orangutans that this protection can help deliver. Thus, predicting consequences can be thought of as equivalent to defining a dose-response model[20] in medicine. Treating consequence in this manner is obviously more complicated than predicting the consequence for a single level of investment, but it is also more realistic and opens up the possibility of using a wider range of the solution methods described in Chapter 7.

Estimating consequence in this dose-response fashion is typically accomplished in one of two ways. The most straightforward is simply to identify a set of potential investment alternatives, treat each of these as options being considered, and predict the consequences for each. For example, Anna Roberts and colleagues looked at reduction in phosphorus input into the Gippsland Lakes of Victoria, Australia, under a series of different funding options (TABLE 6.6). The second way to establish a dose-response estimate of benefit is to have a formal model where one of the terms or variables is the amount of resources invested. For example, Rob Alkemade and colleagues have developed a model that links rangeland biodiversity to cattle production.[21] This can be used as a dose-response model if we assume that conservation actions would require a reduction in cattle production, which would have both a clear cost (the dose) and a clear benefit to biodiversity (the response).

***Marginal Benefit and Diminishing Returns*** The benefit of a conservation action depends partly on what conservation has already occurred. For example, an action to create additional conservation areas in a region that is already well protected might be considered of less benefit than conserving the same area in a region with very little existing protection. This concept is known as **diminishing returns**, or in economics as *marginal benefit theory*. To rephrase the conservation area example, marginal benefit theory means that increasing overall protection of a habitat from 5% to 10% would be considered of greater benefit for biodiversity than if the same percentage increase took the habitat from 50% to 55% protected.

TABLE 6.6 **14 different types of investment in land management to reduce phosphorus (P) inputs into the Gippsland Lakes, Australia. These different levels of investment can be considered like different "doses" from which the consequences of different objectives can be predicted. The predicted consequences of these options are illustrated in Chapter 7 (see Table 7.5).**

| Scenario | Description |
| --- | --- |
| 1 | 40% P reduction by 2030, based on the 10-year average load to the lakes |
| 2 | 30% P reduction by 2030, based on the 10-year average load to the lakes |
| 3 | 20% P reduction by 2030, based on the 10-year average load to the lakes |
| 4 | 10% P reduction by 2030, based on the 10-year average load to the lakes |
| 5 | $2 million/year for 5 years, followed by funding to maintain works |
| 6 | $5 million/year for 5 years, followed by funding to maintain works |
| 7 | $10 million/year for 5 years, followed by funding to maintain works |
| 8 | $2 million/year for 5 years, followed by no ongoing funding |
| 9 | $5 million/year for 5 years, followed by no ongoing funding |
| 10 | $10 million/year for 5 years, followed by no ongoing funding |
| 11 | Payments for all current recommended practices at current rates for all industries, and including management of riparian areas for rivers and smaller streams |
| 12 | As for scenario 11, but excluding riparian management of smaller streams |
| 13 | Current incentive rates for irrigated-dairy current recommended practices, full enforcement of effluent management, no riparian management |
| 14 | Enforcement of farm effluent management only |

*Source:* Data from Roberts, A. M. et al., *Agricultural land management strategies to reduce phosphorus loads in the Gippsland Lakes, Australia.* Agricultural Systems, 2012. **106**(1): 11–22.

Marginal benefit is an important concept in predicting consequence because it reflects how an expected change in an attribute actually works toward the objectives. Marginal benefit also acknowledges that the consequence of an action is likely to change through time as we do more. Take, for example, a conservation project aimed at increasing the viability of a ground-breeding bird. Suppose the likelihood of extinction over 20 years starts off at 80%; but after a few years of intensive conservation efforts to remove invasive pests that threaten the bird, the likelihood of extinction has been reduced to 20%. Further investment, although still increasing the viability of the population, does so to a smaller and smaller extent each year. This effect is known as diminishing returns. Diminishing returns are an important concept when using return-on-investment solution methods (see Chapter 7), and they are frequently implemented through the use of a species-area curve (see Chapter 4).

## Rule: All Plans Should Consider the Consequences for Multiple Objectives

We finished the section on options with a rule about focusing on actions, and we conclude this section on predicting consequences with our second rule for the chapter: all plans should consider the consequences for multiple objectives.

Almost everyone involved in conservation recognizes that trade-offs between objectives are inevitable when considering how to use our land, sea, and freshwater resources. Land is needed to grow food and fiber; water is needed for drinking and agriculture. A trade-off is made when an option is associated with a positive consequence for one objective but a negative consequence for another. Trade-offs exist because multiple objectives are involved and they are not perfectly correlated, so the same action will not deliver the best outcome for all of them. Conservation of nature must ultimately be balanced with achievement of sometimes competing social and economic objectives.

A conservation plan can be deemed multi-objective in one of two ways. It might involve a conservation organization or agency trying to achieve multiple objectives, for instance, biodiversity and human well-being objectives. Alternatively, the plan might involve a number of actors planning together but each aiming to achieve their own objectives, for instance, conservation, fisheries, shipping, mining, and development groups participating in a marine spatial planning exercise. In both cases, multiple objectives inevitably mean trade-offs:[22] for example, priority areas for biodiversity are not always going to deliver the most ecosystem services; and conservation strategies will sometimes limit economic opportunities.

It is important to note that in situations where a number of groups with different interests are conducting planning together, it cannot be assumed that a conservation organization would adopt the objectives of other sectors (in fact, we think it is important not to blur the line about who is representing which objectives). Rather, knowing the different objectives enables a plan to elucidate compromise solutions between different stakeholders, or trade-offs between incommensurable objectives, and help people transparently reduce conflict where possible. Framing conservation planning problems as multi-objective is a strong platform for engagement with a wide range of stakeholders.

It is our view that all conservation plans are likely to be multi-objective to some extent. This does not mean that useful insights cannot be gained by framing problems with a single objective, but that to be effective in supporting decisions, these insights need to be placed in the context of multiple objectives.

In Chapter 7 we focus on tools that can help solve conservation planning problems that are framed in multi-objective terms and focused on a range of conservation actions as different possible options (our two rules for this chapter). We hope that from this chapter it is clear what a crucial role problem framing plays in conservation planning. In Chapter 7, our attention is on turning these well-framed problems into great solutions.

## KEY MESSAGES

- Unless conservation planning problems are well framed, they are unlikely to be useful in informing decisions. Clear problem framing is important because conservation problems are complex, and a range of psychological biases can detrimentally influence our judgment about the best solutions.
- According to decision science, a well-framed problem has three core elements: objectives, options, and prediction of consequences.
- A good set of alternative options is a critical component of good planning, but the task of developing good options is frequently neglected. Developing good conservation options requires deliberate, focused effort.
- Predicting the consequences of different actions, including doing nothing, is a key task of a conservation scientist. There are many ways to approach this task, but it often involves some form of model development.
- All conservation plans should be framed as a choice between potential actions. Although it has been common practice, the priorities arising from a conservation plan should not be species or habitats, but rather actions associated with conserving these features. Under this paradigm, spatial conservation planning can be viewed as one aspect of strategic planning.
- All conservation problems are to some extent multi-objective problems, and they should be framed as such.

## References

1. Possingham, H. P. et al., *Making smart conservation decisions,* in *Conservation Biology: Research Priorities for the Next Decade,* M. E. Soule and G. H. Orians, editors. 2001, Washington, DC: Island Press.
2. Game, E. T. et al., *Conservation in a wicked complex world; challenges and solutions.* Conservation Letters, 2013: 10.1111/conl.12050.
3. Schwartz, S., and K. M. Carpenter, *The right answer for the wrong question: Consequences of type III error for public health research.* American Journal of Public Health, 1999. **89**(8): 1175–1180.

4. Sutherland, W. J. et al., *Solution scanning as a key policy tool: Identifying management interventions to help maintain and enhance regulating ecosystem services.* Ecology and Society, 2014. **19**(2): 3.
5. Pannell, D. J., *Public benefits, private benefits, and policy mechanism choice for land-use change for environmental benefits.* Land Economics, 2008. **84**(2): 225–240.
6. Diehl, M., and W. Stroebe, *Productivity loss in brainstorming groups: Toward the solution of a riddle.* Journal of Personality and Social Psychology, 1987. **53**(3): 497; and Brown, V. R., and P. B. Paulus, *Making group brainstorming more effective: Recommendations from an associative memory perspective.* Current Directions in Psychological Science, 2002. **11**(6): 208–212.
7. Game et al., *Conservation in a wicked complex world.* (See reference 2.)
8. Gregory, R. et al., *Structured Decision Making: A Practical Guide to Environmental Management Choices.* 2012, Oxford, UK: Wiley-Blackwell. 299.
9. Salafsky, N. et al., *A standard lexicon for biodiversity conservation: Unified classifications of threats and actions.* Conservation Biology, 2008. **22**(4): 897–911.
10. Game, E. T., P. Kareiva, and H. P. Possingham, *Six common mistakes in conservation priority-setting.* Conservation Biology, 2013. **27**: 480–485.
11. Vogel, I., *Review of the Use of "Theory of Change" in International Development.* 2012, London: UK Department for International Development (DFID).
12. Margoluis, R. et al., *Results chains: A tool for conservation action design, management, and evaluation.* Ecology and Society, 2013. **18**(3): 22.
13. Ostrom, E., *Governing the Commons: The Evolution of Institutions for Collective Action.* 1990, Cambridge: Cambridge University Press.
14. Mascia, M. B., C. Claus, and R. Naidoo, *Impacts of marine protected areas on fishing communities.* Conservation Biology, 2010. **24**(5): 1424–1429.
15. Nichols, J., *Using models in the conduct of science and management of natural resources.* In *Modeling in Natural Resource Management: Development, Interpretation and Application,* T. M. Shenk and A. B. Franklin, editors. 2001, Washington, DC: Island Press. 11–34.
16. Nichols, *Using models in the conduct of science.* (See reference 15.)
17. Gassman, P. W. et al., *The Soil and Water Assessment Tool: Historical Development, Applications, and Future Research Directions.* 2007, Ames: Center for Agricultural and Rural Development, Iowa State University.
18. Baker, T. J., and S. N. Miller, *Using the Soil and Water Assessment Tool (SWAT) to assess land use impact on water resources in an East African watershed.* Journal of Hydrology, 2013. **486**(0): 100–111.
19. Goldstein, J. H., L. Pejchar, and G. C. Daily, *Using return-on-investment to guide restoration: A case study from Hawaii.* Conservation Letters, 2008. **1**(5): 236–243.

20. Hajkowicz, S., *The evolution of Australia's natural resource management programs: Towards improved targeting and evaluation of investments.* Land Use Policy, 2009. **26**(2): 471–478.

21. Alkemade, R. et al., *Assessing the impacts of livestock production on biodiversity in rangeland ecosystems.* Proceedings of the National Academy of Sciences, 2013. **110**(52): 20900–20905.

22. McShane, T. O. et al., *Hard choices: Making trade-offs between biodiversity conservation and human well-being.* Biological Conservation, 2011. **144**(3): 966–972; and Bradford, J. B., and A. W. D'Amato, *Recognizing trade-offs in multi-objective land management.* Frontiers in Ecology and the Environment, 2012. **10**(4): 210–216.

# 7

# Solving Conservation Planning Problems: Methods and Tools

## Overview

Here we introduce a set of methods and tools for solving the conservation planning problems we framed in Chapter 6. *Solved* in this case means identifying the options that will best advance our objectives. Solution approaches we describe span a range of sophistication from consequence tables to mathematical optimization. Within the conservation planning literature, solving spatial conservation planning problems (spatial prioritization) has been one of the most well-treated subjects, and although we cover some general principles of spatial prioritization, our focus is firmly on how spatial prioritization can solve contemporary conservation planning problems that include considerations like multiple actions and connectivity. We also cover Multi-Criteria Decision Analysis (MCDA) and return-on-investment analysis, and then describe ways to present and navigate trade-offs between options. We finish with a discussion about choosing solution approaches and using them in combination with each other.

## Topics

- *Spatial prioritization*
- *Marxan with Zones*
- *Zonation*
- *Connectivity*
- *Zoning*
- *Consequence tables*
- *Dominance*
- *Multi-Criteria Decision Analysis*
- *Weighting*
- *Analytic Hierarchy Process*
- *Swing weighting*
- *Return on investment*
- *Optimization*
- *Heuristics*
- *Rules of thumb*
- *Trade-offs*
- *Pareto optimal*
- *Efficiency frontier*

## What Does It Mean to "Solve" a Conservation Planning Problem?

This chapter introduces a set of methods and tools that we and other conservation planners have found helpful in solving the sort of conservation planning problems we framed in Chapter 6 (e.g., determining the best action to increase fish abundance in an estuary). "Solving" conservation planning problems is a key step in making informed decisions, but we are not suggesting that conservation planning problems have one right answer. By *solved,* we simply mean critically evaluating the information available in order to identify the options that will best advance our objectives.

When people hear the term *conservation planning,* they usually think about the sort of methods and tools contained in this chapter. Methods and tools for solving conservation planning problems are the part of planning that has received the greatest attention in the scientific literature, often under the term *decision support.* And so they should; these methods and tools allow us to take our knowledge and say, "This is the best course of action." These methods and tools and the analyses associated with them enable us to highlight the superiority of particular options (strategies and actions) by identifying why and how an option is better than other alternatives.

We have endeavored here to cover a range of the most useful and frequently used tools in a conservation planner's toolbox. We describe them roughly in order of sophistication: consequence tables, Multi-Criteria Decision Analysis (MCDA), return on investment (ROI) analysis, spatial prioritization, optimization, and trade-off analysis. The classification of methods and tools is our own. Few planners will be expert at all of these, but it's valuable to know that a range of solution methods exist.

The set of methods and tools we cover is neither exhaustive nor mutually exclusive. Most conservation plans are likely to draw on more than one of the methods here; at the end of this chapter we include a section on how these approaches can be used in combination with each other.

## Generalized Consequence Table Concept

Solving conservation planning problems does not necessarily mean using sophisticated quantitative tools; it means using an appropriate tool for the job. In many cases, good solutions can be found without fancy software or analysis, simply by looking carefully at the estimated consequences of the options being considered. The starting point for this comparison should be a consequence table, as described

**TABLE 7.1** Example of a consequence table. The top row identifies a set of five fundamental objectives and one strategic objective (see far right column) for a rangeland conservation project. The second row is the set of indicators used to assess consequences for these objectives. In the left-hand column are four possible strategy options. The numbers in the table are predictions about the consequence of each strategy option for each of the objectives.

| | **Increase security** | **Increase household income** | **Maximize rangeland health** | **Maximize wildlife populations** | **Maximize financial sustain-ability** | **Maximize alignment with orga-nizational expertise** |
|---|---|---|---|---|---|---|
| ***Strategy options*** | ***Number of cattle rustling incidences per year*** | ***Number of house-holds receiving conservancy-related income*** | ***Percentage of perennial grass cover*** | ***Population size of key species*** | ***Index of financial sustain-ability*** | ***Index of alignment*** |
| Livestock program | 60 | 30% | 30% | 1000–1500 | $500K | Low |
| Range management | 50 | 80% | 60% | 2100–3000 | $150K | High |
| Increase scouts | 30 | 10% | 20% | 2000–2200 | $200K | Low |
| Promoting tourism | 50 | 40% | 20% | 700–1100 | $100K | Medium |

in Chapter 6. A consequence table shows the options being considered down one side, the objectives across the top, and the estimated consequence of each option on each objective in the cells (**TABLE 7.1**). Because such tables often contain a lot of information, some tricks can be used to help zero in on the better alternatives.

## Dominance

Start by removing all options that are dominated. An option is dominated if there is another option that performs at least as well against all objectives and better for at least one objective. If such a situation exists, there is no rational reason for choosing the dominated option, so it can be disregarded. For example, in the consequence table shown in Table 7.1, the "Livestock program" option is dominated by "Range management" because this option performs at least as well or better on all objectives.

Drawing a line through an option may attract the ire of someone on a planning team. If this happens, it is important to probe why they might disagree with the decision—most likely it will reveal either an additional objective that was not considered in the consequence table, or a personal bias. The former might require going back and adding another objective; the latter simply highlights that planning is doing what it should.

Planners who use consequence tables effectively will often speak of *practical dominance*.[1] These are cases where one option is not strictly dominated but performs slightly better than an option on one or more objectives and is strongly outperformed by that option on at least one objective. From a practical point of view, options that are very close in performance on an objective might reasonably be considered equivalent for that objective, especially given the likely uncertainty in the estimates of consequence. Removing options that are practically dominated will often allow planners to greatly narrow a set of options.

The notion of removing dominated and practically dominated options is a useful and straightforward planning tool because it does not require quantitative estimates of consequence, weighted objectives or criteria, or independence between objectives or criteria. In all cases, planners should at least attempt to look for dominated options in a consequence table. Identifying non-dominated options will often narrow the options enough to make a good decision.

Establishing a consequence table is also the perfect platform from which to use the next two types of methods we cover: Multi-Criteria Decision Analysis and return on investment.

## Multi-Criteria Decision Analysis (MCDA)

**Multi-Criteria Decision Analysis** (**MCDA**) is a broad term that encompasses a range of methods for incorporating multiple objectives into the evaluation of alternative options. Equivalent terms include *multi-criteria decision making* (MCDM) and *multi-criteria analysis* (MCA). As we discussed in Chapter 6, most conservation planners do not immediately equate criteria and objectives; however, a classically formulated MCDA would consider each item in an objective hierarchy (see Chapter 3) as criteria.

Criteria should include all those considerations by which the performance of an action should be evaluated (but lists of criteria are no replacement for clearly identified objectives). If we use the example illustrated in the section on consequence tables (see Table 7.1), we could use MCDA to solve this problem by considering the items of the top row (security, income, rangeland health, wildlife populations, cost, alignment with organizational expertise) as the means by which we will assess the strategic options in the left-hand column.

MCDA is used extensively in business management, and although they are frequently not identified as such, a great many conservation prioritization schemes (e.g., the *Open Standards*) are essentially MCDAs. MCDA is most commonly used for prioritizing actions, but it can also be applied to spatial conservation decisions. Alexander

Moffett and Sahotra Sarkar provide a good review of the potential application of MCDA approaches to the design of conservation area networks.[2]

The basic elements of an MCDA are outlined in **Box 7.1**. The essential premise is that a *weight* reflective of relative importance is assigned to each criterion. This weight can be either the perceived influence of that criterion over the outcome or the relative preference for outcomes that satisfy that criterion (e.g., if criteria include fish abundance and tourism access, a particular stakeholder might view one as more important than another). Criterion weights are then combined with the performance values of options for criteria in order to calculate an overall score by which options can be compared. In the language of MCDA, this score is known as an option's **multi-attribute utility**. Although there are many ways that criterion weights and performance scores can be combined to calculate multi-attribute utility, by far the most common is known as a linear additive benefit function or model:

$$U_j = \Sigma_{i=1}^{n} W_i V_{ij}$$

where $U_j$ is the overall benefit or utility of option $j$, $W_i$ is the weight given to criterion $i$, and $V_{ij}$ is the value of option $j$ for criterion $i$. In simple terms this means that for each option, the performance score for criterion 1 is multiplied by the weight given to criterion 1; the same calculation is made for each additional criterion, and then these are all added together. The result is a consequence score for each option.

As mentioned in Chapter 6, calculating consequence this way is never a true and complete depiction of reality (as with all models), but it is a simple way to provide good insight into the relative outcomes of different options. Using a linear additive utility model requires making the substantial assumption that the scores for criteria are independent.

## Box 7.1 Basic Elements of Multi-Criteria Decision Analysis (MCDA)

1. Establish a set of criteria against which the desirability of options is to be judged.
2. Assign weights to criteria. Weights reflect the importance of criteria in determining the outcome. Weights might be assigned individually or as consensus among a group.
3. Combine an assessment of the performance of each option for each criterion with the weight for that criterion.
4. Aggregate scores for each option across all criteria to give an overall assessment of performance or utility.

*Source:* Adapted from M. A. Burgman, *Risks and Decisions for Conservation and Environmental Management*. 2005, Cambridge, UK: Cambridge University Press.

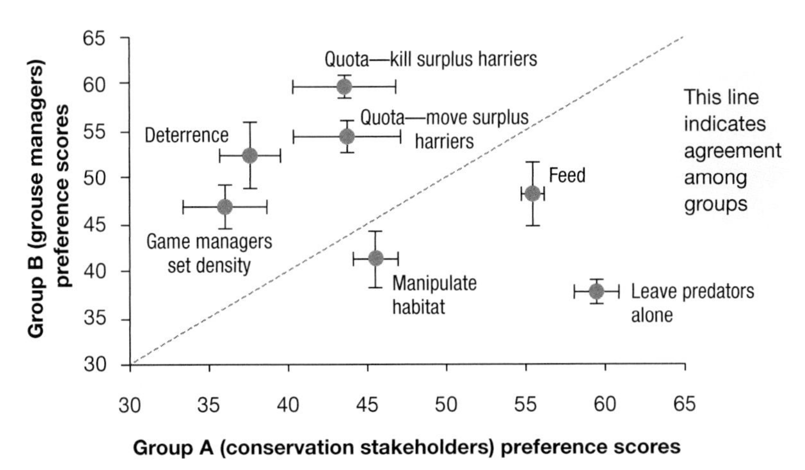


FIGURE 7.1 Graph illustrating the result of a Multi-Criteria Decision Analysis (MCDA) conducted by conservation biologist Stephen Redpath and colleagues to help select options for the management of harriers (*Circus cynaeus*) in the Scottish moorlands. Seven management options are evaluated, and the plot shows the overall score and rank as given by conservation stakeholders (x-axis) and grouse managers (y-axis). Grouse (*Lagopus lagopus scoticus*) are an important game bird in Scotland, and there is conflict between conservationists and grouse managers because harriers prey on grouse. The error bars around each option illustrate the level of agreement within each stakeholder group based on individual differences in criterion weighting. Data from Davies, A. L., R. Bryce, and S. M. Redpath, *Use of Multicriteria Decision Analysis to Address Conservation Conflicts*. Conservation Biology, 2013. **27**(5): p. 936–944.

This means that the score an option receives for any criterion is completely unaffected by the scores for other criteria. However, a range of other functions are available for aggregating criteria scores and weights that do not require such strict assumptions. For example, due to the lack of independence among criteria, MCDA is used frequently in the health field to produce quality-of-life assessments that are based on multiplicative utility functions (e.g., the Health Utilities Index[3]).

Ultimately, MCDA gives you a rank order of the options under consideration. FIGURE 7.1 illustrates the result of an MCDA conducted by conservation biologist Stephen Redpath and colleagues to help select options for the management of harriers (*Circus cynaeus*) in the Scottish moorlands.[4] **Box 7.2** describes how MCDA can be used to solve planning problems that involve identifying the best option to take in particular places, as we advocated in Chapter 6.

## Normalizing Data

As we saw in Chapter 4, indicators—and hence criteria—can be measured on very different scales. To calculate utility using a linear additive function, the performance values of options for different criteria must be measured on the same scale; otherwise, those criteria measured on scales with larger numbers will be unintentionally weighted

## Box 7.2 Example of Spatial MCDA

### Where Can We Most Effectively Invest in Revegetation?

The West Hume area of Southern New South Wales (NSW), Australia, is predominantly mixed-cropping agricultural land and lies within an important area for both salinity and biodiversity management. Although revegetation can assist with ameliorating both of these issues, the most effective places to revegetate to achieve biodiversity objectives are likely to be different from those that are the best to address issues of land salinity.

To help prioritize locations to invest in revegetation that balances multiple objectives, the Murray Catchment Management Authority (CMA) collaborated with Australia's Commonwealth Scientific and Industrial Research Organization (CSIRO) to take a spatially explicit multi-criteria approach to solving the conservation planning problem.[a] They used the free software, Multi-Criteria Analysis Shell for Spatial Decision Support (MCAS-S), developed by the Australian Bureau of Agricultural and Resource Economics and Sciences. MCAS-S combines multi-criteria analysis (MCA) with geographic information systems (GIS) to provide helpful decision support tools that can be used even by planners without much experience in MCDA or spatial analyses.

Essentially, the MCAS-S tool helps weight a series of preferences for where, in this case, revegetation should take place to meet different objectives. To do this, it employs the AHP weighting process described in this chapter. **TABLE 7.2.1** illustrates these preferences and their respective weights for the West Hume catchment case. It then combines these weights with a series

**TABLE 7.2.1** Weightings for a series of criteria about priority areas for restoration to improve biodiversity and dry land salinity

| Guidelines | | Weighting | (rank: 1 = most important) |
|---|---|---|---|
| *Biodiversity guidelines (rules)* | | | |
| 1. | Revegetate for geographical dispersal | 0.0974 | (6) |
| 2. | Revegetate biophysically heterogeneous areas | 0.0541 | (9) |
| 3. | Revegetate rare broad vegetation types | 0.1038 | (5) |
| 4. | Revegetate areas with rare species | 0.0935 | (7) |
| 5. | Revegetate in areas with dense patch distribution | 0.1502 | (1) |
| 6. | Revegetate close to large patches | 0.1302 | (4) |
| 7. | Revegetate close to streams | 0.1492 | (2) |
| 8. | Revegetate enclosed areas | 0.0731 | (8) |
| 9. | Revegetate to form corridors | 0.1305 | (3) |
| 10. | Revegetate land with low production potential | 0.0179 | (10) |
| **Total** | | **1.000** | |
| *Salinity guidelines (rules)* | | | |
| 11. | Revegetate areas: responsive groundwater flow systems | 0.1678 | (1) |
| 12. | Protect high-value water resources | 0.0939 | (6) |
| 13. | Protect high-value biodiversity assets | 0.1039 | (5) |
| 14. | Protect high-value built assets | 0.0730 | (7) |
| 15. | Revegetate soils with high salt stores | 0.1262 | (3) |
| 16. | Revegetate high-recharge-potential areas | 0.1551 | (2) |
| 17. | Revegetate in areas with high rainfall | 0.0683 | (8) |
| 18. | Revegetate away from saline discharge zones | 0.0641 | (9) |
| 19. | Revegetate low-value agricultural land | 0.1180 | (4) |
| 20. | Revegetate areas with high forest production potential | 0.0299 | (10) |
| **Total** | | **1.000** | |

of spatial data layers to provide a straightforward and highly participatory approach to prioritizing conservation action. **FIGURE 7.2.1** illustrates the MCAS-S interface showing a series of spatial layers combined through MCDA to identify the soils most at risk in Northern Territory, Australia.

[a] Lesslie, R. G. et al., *The application of a simple spatial multi-criteria analysis shell to natural resource management decision making*, in *Landscape Analysis and Visualisation*. 2008, Berlin: Springer-Verlag. 73–95.

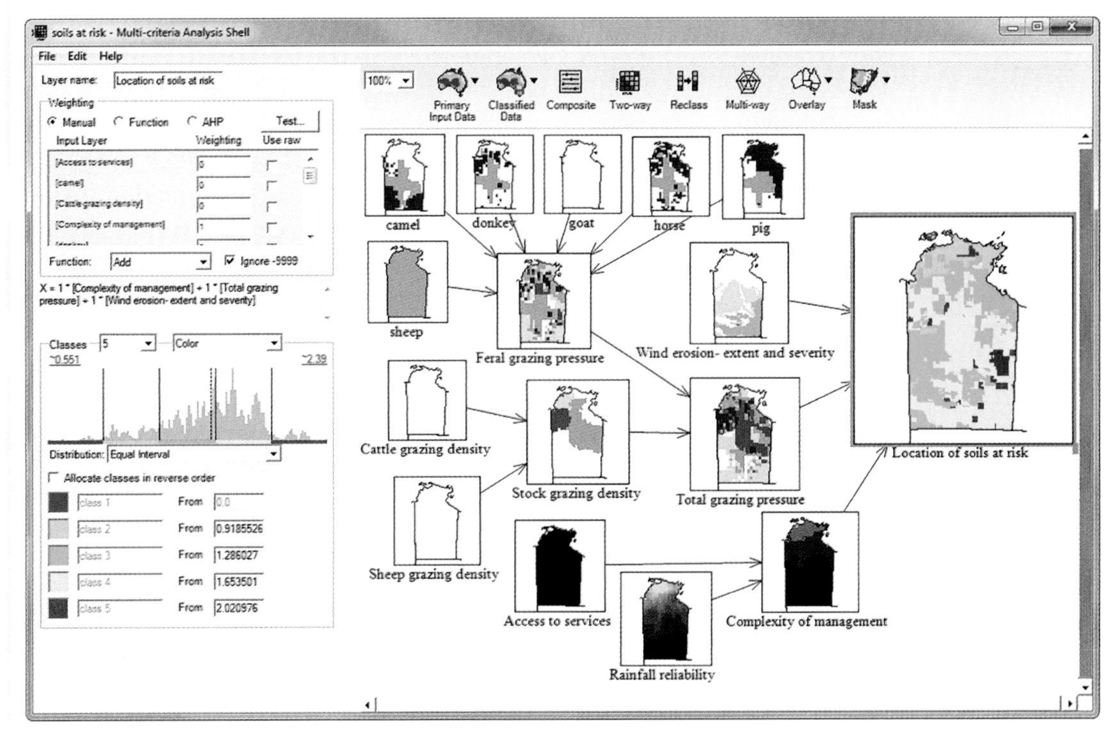


**FIGURE 7.2.1** MCAS-S graphical interface showing a series of spatial layers combined through MCDA to identify the soils most at risk in Northern Territory, Australia.

more heavily.[5] The process of converting data to the same scale is known as *normalizing*.

There are different procedures for normalizing data, but one that is considered robust for use in MCDA[6] is given by the formula

$$V'_{ij} = \frac{V_{ij} - min[V_i]}{max[V_i] - min[V_i]}$$

where $V_{ij}$ is the performance of option $j$ for criterion $i$, and $min[V_i]$ and $max[V_i]$ are the minimum and maximum possible values for criterion $i$. The prime mark after $V$ on the left-hand side of the equation is standard notation that indicates the parameter is a transformation of another parameter.

### Weighting Criteria

It should be clear to all planners that criteria in a set are unlikely to have equal importance in determining outcomes. Weighting is intended to help ensure that the calculation of utility actually reflects the perceived importance of the different criteria. The weighting given to criteria can significantly change the outcome of an MCDA, so the systematic and transparent weighting of criteria is the true substance of an MCDA.

An important and underappreciated point about weights is that they are meaningful only with reference to the observed range of outcomes for each criterion. In other words, the importance of a criterion in influencing a decision depends both on its inherent importance to the objectives and on how well or poorly the options under consideration perform for that criterion. For example, a stakeholder group might value recreation opportunities over biodiversity conservation, but if the options being considered vary little and all are satisfactory in their consequences on recreation opportunities, then it does not make sense to assign the recreation criterion a relatively greater weight. Giving greater weights to criteria that do not vary much has the effect of making options seem more similar in utility value, which is not informative for decision making. This means that it is premature to weight criteria before knowing the expected performance of options against these criteria (hence the importance of some form of consequence table).

Weights of criteria in an MCDA embed the value judgments of those doing the weighting. As such, weights should not be seen as objective assessments of the relationships between criteria, as we might expect when constructing a formal model of consequence as discussed in Chapter 6.

There are many ways to go about weighting criteria in an MCDA; here we describe three of them.

#### Direct Weighting

The simplest approach to weighting criteria is to assign weights directly to each criterion following discussion about its relative importance. This approach is rapid but assumes that some group agreement can be reached about the relative importance of different criteria. Because direct weights are not the result of a structured elicitation, they are likely to embed value judgments and biases in an opaque fashion. However, they may be appropriate in cases where repeated experience has provided experts with tested knowledge about the importance of different criteria (e.g., the sort of organizational conditions that will mean a project is adequately supported).

One straightforward approach to direct weighting is to ask participants to give each criterion a score out of 100 (with 100 being the most important). Because the sum of weights for all criteria should equal 1 if an additive utility model is being used, this can be easily accomplished by dividing each weight by the sum of all weights.

## Analytic Hierarchy Process

A frequently used and sophisticated way to establish criterion weights in MCDA is through the use of the **Analytic Hierarchy Process** (**AHP**). The method was developed by the mathematician Thomas Saaty, and a great deal has been written about it, including Saaty's original papers.[7]

AHP involves pairwise comparisons between criteria, scoring each pair on a scale of 1 (criteria are of equal importance) to 9 (one criterion is of extreme importance compared with the other), as outlined in TABLE 7.2. Of course, it is also important to know which of the two criteria are of more or less importance, so it is often helpful to think

TABLE 7.2 Scoring system for pairwise comparisons between criteria in the Analytic Hierarchy Process. Each pair of criteria is scored on a scale of 1 (criteria are of equal importance) to 9 (one criterion is of extreme importance compared with the other).

| Intensity of importance | Definition | Explanation |
|---|---|---|
| 1 | Equal importance | Two activities contribute equally to the objective. |
| 2 | Weak or slight | |
| 3 | Moderate importance | Experience and judgment slightly favor one activity over another. |
| 4 | Moderate plus | |
| 5 | Strong importance | Experience and judgment strongly favor one activity over another. |
| 6 | Strong plus | |
| 7 | Very strong or demonstrated importance | An activity is favored very strongly over another; its dominance is demonstrated in practice. |
| 8 | Very, very strong | |
| 9 | Extreme importance | The evidence favoring one activity over another is of the highest possible order of affirmation. |
| Reciprocals of above | If activity *i* has one of the above nonzero numbers assigned to it when compared with activity *j*, then *j* has the reciprocal value when compared with *i*. | This is a reasonable assumption. |
| 1.1–1.9 | If the activities are very close . . . | It may be difficult to assign the best value. When compared with other contrasting activities, the size of the small numbers would not be too noticeable, yet they can still indicate the relative importance of the activities. |

of the 9-point scale as a line drawn between two criteria (FIGURE 7.2). If there is a hierarchy of objectives (and criteria), then AHP involves pairwise comparisons of all objectives at the same level and in the same cluster based on their importance to the common "parent" objective immediately above. The pairwise comparisons are then used to calculate weights based on Saaty's math.

Pairwise comparison is a nicely structured way to elicit weights that can help avoid a number of common psychological traps as described in Chapter 6. It is also quite intensive; for example, an MCDA process with 7 criteria would require making at least 21 pairwise judgments (depending on how the problem is structured). The number of judgments being asked of those involved could be a source of inconsistency in these judgments related to violations of preference logic. That is, if someone believes criterion A to be more important than criterion B, and criterion B to be more important than criterion C, then logically, criterion A must be more important than criterion C. However, it is surprising how often violations of this logic creep into participants' judgments during an AHP process. Checking for inconsistencies like these is an important part of using AHP.

Many software packages can help with an MCDA by performing the AHP and sometimes embedding the AHP in a more complete MCDA. The programs range from highly sophisticated and automated software, often created for profit and used by big companies, to basic, free software typically originating from academic institutions. It is also possible to conduct an AHP in Microsoft Excel. You can easily find any of these software programs by typing "Analytic Hierarchy Process software" into an Internet search engine. In our experience, the more expensive software is generally more user-friendly and contains automated methods for identifying the sort of inconsistencies flagged earlier, and they present easy-to-interpret and visually appealing results. Free software generally requires more effort from the user in preparing data and interpreting analyses.

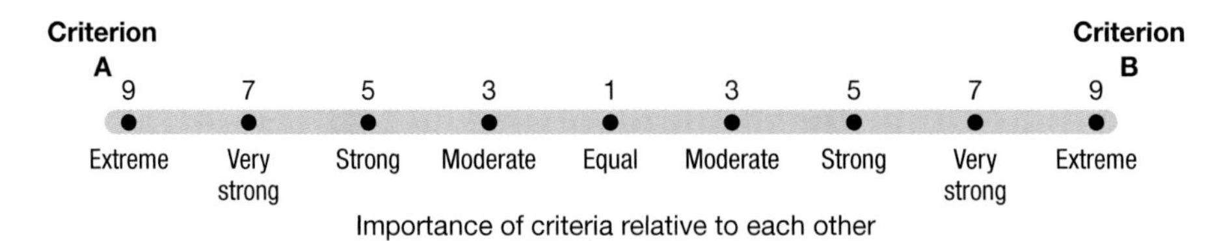


FIGURE 7.2 Graphical representation of the 9-point scale used to score the relative importance of a pair of criteria in the Analytic Hierarchy Process (AHP). AHP is a method for deriving the weights assigned to criteria as part of a Multi-Criteria Decision Analysis (MCDA). The position on the horizontal line indicates how important one criterion is relative to the others. So a score of 1 indicates that criteria A and B (in this case) are of equal importance. A score of 5 on the right-hand side would indicate that criterion B is of very strong importance compared with criterion A.

### Swing Weighting

Swing weighting is a straightforward but robust approach to weighting criteria. It was introduced to us by the decision scientist Mike Runge. Essentially, the process involves imagining a series of hypothetical options or outcomes for which all criteria have the worst score except for one, which takes the best score. An additional scenario is also included as a baseline, which has the worst score across all criteria. This means if there are seven criteria there should be eight hypothetical outcomes. Participants are then asked to rank each of these outcomes from most desirable to least, and give each a score. The most desirable outcome should be scored 100, and the baseline scenario (obviously the least desirable) receives a score of 0. All other scenarios are scored relative to these two values, and two or more scenarios can receive the same score (we have seen some planners insist that criteria have a *hard rank,* i.e., that no two can be equally important—but this practice only serves to discard useful information and obscure true relationships among criteria). The weight given to each criterion is then simply the score for the scenario in which that criterion is maximized, divided by the sum of the scores given to all scenarios.

This technique is known as swing weighting because the weights reflect the criteria that participants would most want to swing to the best plausible level of achievement. The great strength of swing weighting is that it helps those involved recognize that the range of outcomes for a given criterion is critical in deciding how much importance to place on it. Swing weighting also has the advantage of being faster and less prone to inconsistencies than AHP because judgments are made across all criteria rather than pairwise. However, it cannot easily handle complex hierarchies in the way AHP can. Swing weighting is easily done in Microsoft Excel; TABLE 7.3 shows how a swing weighting spreadsheet can be constructed.

### What to Do with Weights Once You Have Them

After using one of the techniques just described to elicit weights for criteria from individual stakeholders or experts, planners often find themselves wondering what to do with this series of weights. Finding a solution using MCDA will generally require adopting a single set of criterion weights in order to calculate utility and rank the alternatives.

The most obvious approach is to average the weights for each criterion across all participants. This may be a reasonable approach but should never be undertaken without also examining the differences in the weights given by participants. An important step is to highlight where there are significant differences in the weights between individuals, and discuss why this might be so. For example, in assessing criteria impacting an endangered species' recovery potential, the

**TABLE 7.3** Example of a swing weighting table. Five hypothetical locations for a watershed conservation project (in this case, different cities) are being compared based on the five criteria given along the top (they are intended to represent the objective of maximizing the amount of avoided sedimentation). Five hypothetical options are needed, so that each can take the maximum possible value for one of the five criteria (highlighted in yellow). These values are the maximum actually observed in the real set of cities being considered. All other criteria in the hypothetical option take the lowest value observed for that criterion among the real options. The baseline scenario takes the lowest value for all criteria. Stakeholders then score the relative desirability of each of these hypothetical options (scores are at far right) and use them to calculate the weight for that criterion (bottom row).

| | Avoided sedimentation | | | | | |
|---|---|---|---|---|---|---|
| | **Does the city use suface water?** Proportion of surface water | **Is there natural land cover in areas important for erosion control?** Proportion of natural land cover in high slope areas | **Is the area prone to erosion, given soils and climate?** Proportion of watershed classed as medium, high, or very high erosion risk | **Is there an urgency to act?** Decline over time in natural land cover on high slope areas (2005–2009) | **Could land protection of high slope areas help?** Proportion of slope not protected | **Score** |
| Baseline | 0.13 | 0.00 | 0.00 | 0.00 | 0.00 | 0 |
| City 1 | 1.00 | 0.00 | 0.00 | 0.000 | 0.00 | 100 |
| City 2 | 0.13 | 0.88 | 0.00 | 0.000 | 0.00 | 40 |
| City 3 | 0.13 | 0.00 | 1.00 | 0.000 | 0.00 | 20 |
| City 4 | 0.13 | 0.00 | 0.00 | 0.100 | 0.00 | 10 |
| City 5 | 0.13 | 0.00 | 0.00 | 0.000 | 1.00 | 5 |
| Weight | 0.57 | 0.23 | 0.11 | 0.06 | 0.03 | |

Australian government conducted an MCDA. One participant weighted high cadmium as far more influential than the other participants. When this difference was identified and queried, it turned out that this individual was highly knowledgeable about cadmium poisoning and was able to share information with the group that altered other participants' assessments of the importance of this criterion.

## Sensitivity Analysis

It is good practice to explore the sensitivity of MCDA results to changes in these weights. Sensitivity analysis can be useful to flag criteria with a major influence over outcomes, so that the accuracy of consequence assessments for these criteria can be checked and improved. Sensitivity analysis can also help identify alternatives that rank highly regardless of the importance assigned to different criteria—we would have high confidence in choosing such robust alternatives.

In its simplest form, **sensitivity analysis** involves changing the weights for one criterion while holding the others constant (known as single-dimension sensitivity analysis). There are numerous other multidimensional approaches to sensitivity analysis, where the weights for multiple criteria are changed at the same time and the results compared.[8] However, undertaking robust sensitivity analysis is often more technically demanding than the MCDA itself. Some MCDA software packages have automated sensitivity analyses that are very helpful. While we do not want to discourage anyone from undertaking sensitivity analysis, we believe it should be considered a necessity only for very significant and high-risk decisions.

## Return on Investment (ROI)

In the field of conservation, **return on investment** (commonly referred to simply as **ROI**) has come to be a rather general term for prioritization schemes that explicitly consider the cost of the options being considered. These schemes are in contrast to prioritization schemes based principally on the distribution of biodiversity, or biodiversity and threats to it.

In economic parlance, a return on investment is just that: the ratio of the amount of money you receive relative to the amount you need to invest in some endeavor (e.g., fund into a share portfolio, or new equipment to improve the output of a business). The comparison between different options based on this ratio between investment and return is known as a *cost-benefit analysis.*

Conservation ROI analysis belongs to a general class of economic analysis known as *cost-effectiveness analysis.* Economists consider something to be a cost-effectiveness analysis rather than a cost-benefit analysis when the outcome or return side of the equation is not monetized (expressed as a dollar value). This tends to be the case in conservation where the return might relate to the number of species saved or habitat protected (the exception to this might be prioritizations involving ecosystem services).[9] Cost-effectiveness analysis was largely pioneered and is most widely used in health sciences, particularly when looking at the effectiveness of different interventions where the desired metric might be human lives saved. Examples include deciding between detecting and treating cases of tuberculosis or applying different levels of vaccination coverage.[10] The common notion behind ROI or cost-effectiveness is that the expected return or outcome from a conservation action should be balanced against the cost of achieving that outcome.

Although they are not always recognized as such, systematic conservation plans[11] are often a sort of ROI analysis. This is particularly so when they are implemented using optimization software programs such as Marxan[12] or Zonation[13] (described in the section titled "Spatial

Prioritization" later in this chapter), which have algorithms that work to minimize the cost of the set of places required to meet established conservation targets.[14] If the cost data used with such software reflect the relative cost of taking a conservation action in particular places, then this is essentially performing an ROI analysis.[15]

ROI analysis can also guide strategic decisions about investment between different options. For example, Adam Barlow and colleagues[16] compare the cost-effectiveness of 18 different options to reduce human–tiger conflict in Bangladesh. Possible options include killing problem tigers, collaring tigers, different kinds of fences and deterrents, and education programs; effectiveness of each option is judged by the number of expected human and tiger lives saved per dollar spent.

In our experience, there is generally enough uncertainty around the costs of different options that minor differences in the relative ROI of alternatives should be interpreted with caution, especially as cost-effectiveness is not the only consideration in selecting an action. In some ways the most relevant information contained in an ROI analysis is at the bottom end—those options that appear to deliver a poor ROI. The transparency of ROI analyses should at least cause decision makers to carefully justify the selection of options that have relatively poor cost-effectiveness.

The cost-effectiveness, *CE,* of taking action *i* in place *j* can be given by the general equation

$$CE_{ij} = \frac{B_{ij} \cdot Pr_{ij}}{C_{ij}}$$

where $B_{ij}$ is the benefit of taking action *i* in place *j*, $C_{ij}$ is the cost of taking that action, and $Pr_{ij}$ is the probability that if taken, the action will deliver the expected benefit. This last term is not strictly necessary but is good practice, easy to do, and increasingly expected. Again, it is worth emphasizing that neither the benefit nor the cost portions of this equation need to be explicitly stated in financial terms.

## Ranking

The most straightforward ROI solution to a conservation planning problem is to simply rank each option by its cost-effectiveness. This is essentially answering the question, "What is the most cost-effective option in which to invest the next dollar?"

Joshua Goldstein and colleagues describe the use of an ROI analysis to rank six alternative approaches to the reforestation of montane pastureland in Hawaii.[17] Each restoration approach represents a different set of land use transitions. Based on an average 200-hectare parcel in their study area, the authors model conservation return as the expected increase in mean native bird and plant understory density under each of the six alternative restoration approaches. Required investment was modeled as the net present value (NPV) of expected restoration management over 50 years. To calculate the ROI of each restoration approach, the authors simply divided the expected increase in native bird and understory plant density by the expected cost of that restoration transition pathway (FIGURE 7.3).

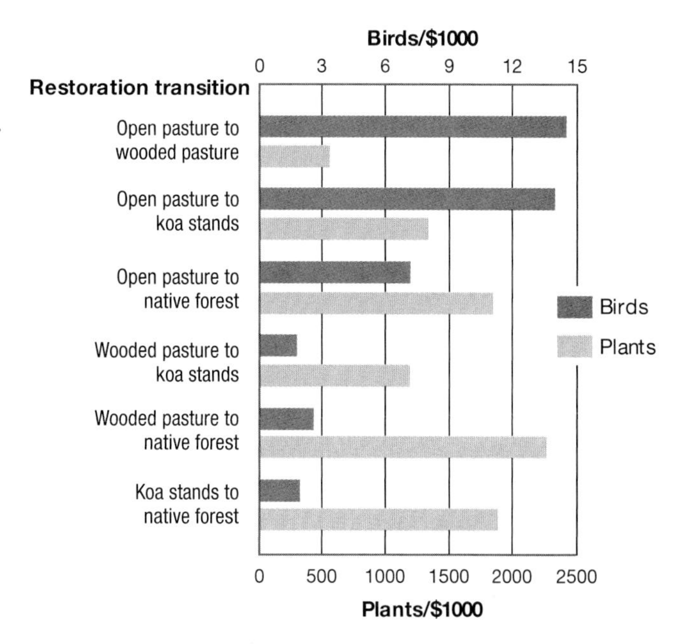


FIGURE 7.3 Return-on-investment ratios for a range of restoration options for montane pastureland in Hawaii. Each restoration approach represents a different set of land use transitions: open pasture, wooded pasture, *Acacia koa* ("koa") dominated regeneration stand, and native forest. The return or benefit is given for both birds and understory plants. (Data from Goldstein et al., 2008.)

## ROI with Actions and Locations

The example just mentioned uses ROI analysis either to prioritize actions (restoration approaches) without specific reference to the location where they will occur. ROI analysis can also be used to solve problems that require considering both action and location simultaneously, essentially answering the question, "What should we do where?"

The reefs of the Coral Triangle in the western Pacific Ocean are threatened by unsustainable fishing practices as well as by runoff containing sediment, nutrients, and other organic pollutants. The threat from fishing can be ameliorated by establishing marine protected areas, but mitigating the threat posed from runoff requires terrestrial conservation actions in coastal catchments.

The conservation planner Carissa Klein and colleagues describe an ROI analysis to prioritize conservation investment between these two activities across the 16 marine ecoregions of the Coral Triangle.[18] To model conservation return, the authors use a formula developed by marine conservation biologist Benjamin Halpern and colleagues to estimate the overall level of threat from both marine and terrestrial sources being experienced by reefs in each ecoregion.[19] Assuming that these threats decline linearly as the proportion of protected ecoregion increases, conservation return per action per ecoregion (or *ecoaction* as the authors call it) is defined as the expected reduction in overall threat to reefs from protecting an area of either reef or land in that ecoregion. The cost of investing in each possible ecoaction is calculated as the sum of the annual management and opportunity costs (see Chapter 5 for detail on what these costs mean) associated with marine and terrestrial protected areas in those locations. Protected area management costs were calculated based on models developed by Andrew Balmford and colleagues and Joslin Moore and colleagues and subsequently generalized by Michael Bode and colleagues,[20] while opportunity costs were based on expected agricultural profitability in terrestrial areas and expected catch value in marine areas.[21] Each action in each ecoregion was then ranked according to its expected ROI (FIGURE 7.4).

This ranking provides two useful pieces of information for prioritization. First, it provides an ROI comparison of marine versus terrestrial action (for marine conservation) within each ecoregion. Second, it offers a comparison of the expected ROI (both marine and terrestrial) between ecoregions across the Coral Triangle. Except for one ecoregion (North Philippines), marine protection always has a higher ROI than terrestrial protection within an ecoregion. On the other hand, when ROI ranks are compared across all ecoregions of the Coral Tri-

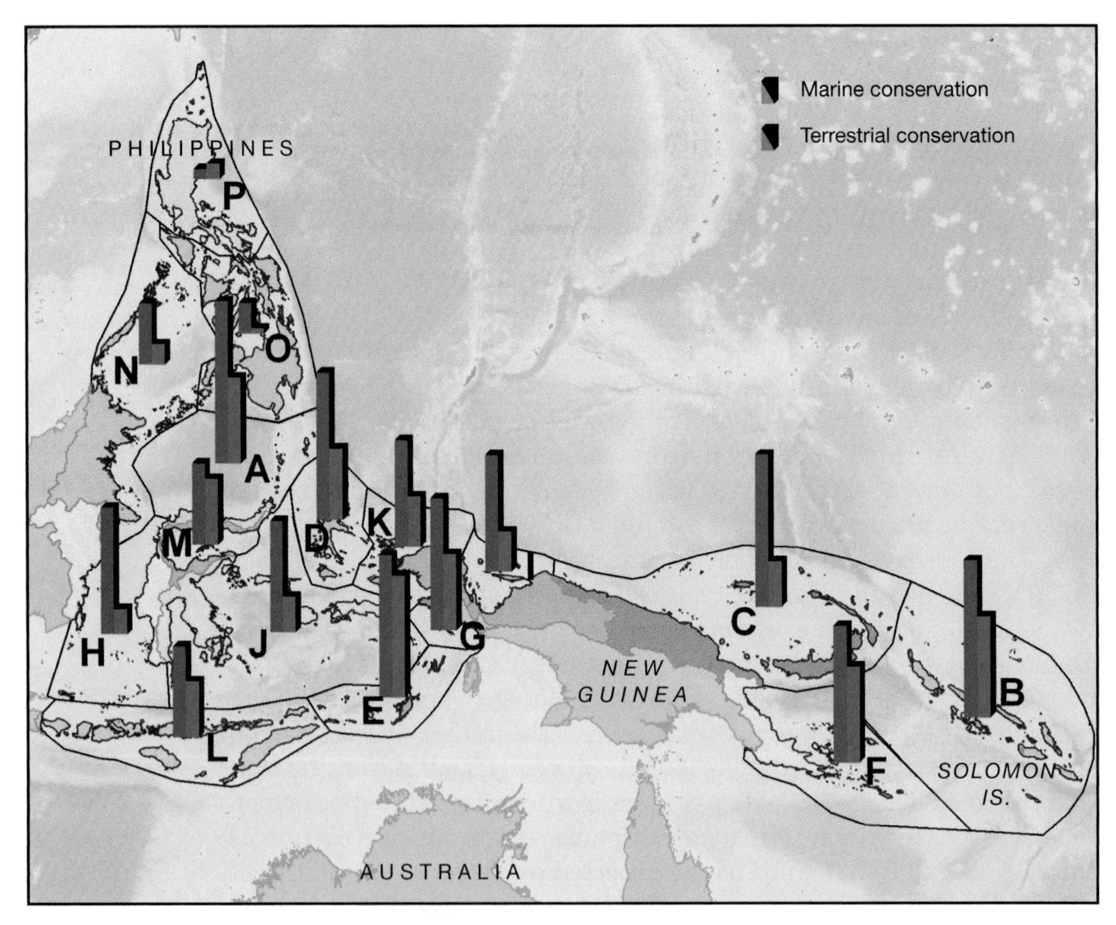


FIGURE 7.4 Relative priority for investment in marine and terrestrial conservation actions for coral reef conservation in different ecoregions across the Coral Triangle (bar height indicates relative rank). Letters labeling ecoregions indicate rank order for marine conservation. (Adapted from Klein et al., 2010.)

angle, terrestrial action in some ecoregions often has a higher ROI than marine action in others.

## INFFER

One of the few automated tools we know of for supporting ROI analyses in conservation and environmental management is INFFER (Investment Framework for Environmental Resources, http://www.inffer.org). Developed by Australian environmental economist David Pannell and colleagues, INFFER strikes a balance between rigor and ease of application. As a framework, INFFER embeds ROI into a comprehensive workflow that guides users through the information needed for a rigorous ROI analysis and then automates the integration of this information into an ROI calculation.

Much of this analysis is achieved through an online project assessment form. INFFER is based on identifying environmental *assets* (equivalent to the conservation features introduced in Chapter 3) and projects aimed at conserving these assets. The output is a return-on-investment (benefit-to-cost) ratio for each project that can be used to rank alternative projects and allocate resources accordingly. **TABLE 7.4** contains the information considered in the INFFER ROI

**TABLE 7.4** Information included in the INFFER return-on-investment analysis

| |
|---|
| Identification of the environmental asset(s) to be protected or enhanced, including spatial location and extent |
| The significance or value of each environmental asset, relative to others |
| The proportion of the asset's value that would be protected or improved as a result of the project (if it were completely successful) |
| The threats that are affecting or are likely to affect the environmental asset |
| Specific, measurable, time-bound goals for each asset |
| Works and actions that are proposed to be undertaken to achieve the goals |
| The time lag between undertaking the project and the generation of benefits |
| The future degree of environmental damage with and without the proposed works and actions |
| The risk of technical failure of the project |
| Positive and negative spin-offs from the project (e.g., impacts on other environmental assets) |
| The likely extent of adoption by private landholders of the works and actions that would be required to achieve the stated goals |
| The risk that, despite new public investment, private landholders will adopt new works and actions that would further degrade the environmental asset |
| Legal approvals required to undertake the works and actions |
| The policy mechanisms/delivery mechanisms to be used to encourage and facilitate uptake of the required works and actions |
| Sociopolitical risks |
| Short-term costs of undertaking the project |
| Annual maintenance costs required to ensure ongoing benefits |
| The risk of not obtaining those essential maintenance costs, such that project benefits are lost |

*Source:* Adapted from Pannell, D. J. et al., *Integrated assessment of public investment in land-use change to protect environmental assets in Australia*. Land Use Policy, 2012. **209**(2): 377–387.

**TABLE 7.5** Return-on-investment (expressed as "Benefit : cost ratio") assessment of several potential management options to reduce phosphorus inputs in the Gippsland Lakes, Australia. In addition to the reduction in phosphorus (P) and the total cost, the benefit : cost ratio was based on consideration of factors such as the likelihood of an option being successfully implemented.

| Scenario | % P reduction | Cost[1] | Benefit : cost ratio |
|---|---|---|---|
| 1. 40% P reduction | 40 | 994 | 0.02 |
| 2. 30% P reduction | 30 | 223 | 0.2 |
| 3. 20% P reduction | 20 | 80 | 1.0 |
| 4. 10% P reduction | 10 | 16 | 2.6 |
| 5. $2M/year for 5 years, with ongoing annual costs | 9 | 23 | 2.0 |
| 6. $5M/year for 5 years, with ongoing annual costs | 18 | 114 | 0.6 |
| 7. $10M/year, for 5 years, ongoing annual costs | 22 | 142 | 0.6 |
| 8. $2M/year for 5 years, no ongoing funding | 7 | 10 | 4.4 |
| 9. $5M/year for 5 years, no ongoing funding | 7 | 25 | 1.7 |
| 10. $10M/year for 5 years, no ongoing funding | 10 | 49 | 0.9 |
| 11. Current payments including rivers and smaller streams | 17 | 192 | 0.4 |
| 12. As for 11, minus smaller streams | 13 | 30 | 1.8 |
| 13. Current payments irrigated dairy + effluent enforcement, no riparian management for rivers or smaller streams | 9 | 25 | 1.7 |
| 14. Effluent enforcement | 6 | 16 | 2.8 |

[1] Expected present value (A$ million) of costs over 25 years.

*Source*: Adapted from Roberts, A. M. et al., *Agricultural land management strategies to reduce phosphorus loads in the Gippsland Lakes, Australia*. Agricultural Systems, 2012. **106**(1): 11–22.

calculation. It is important to consider this information in any ROI analysis, even if the INFFER framework is not being used.

Anna Roberts and colleagues in the Australian state of Victoria used the INFFER framework to evaluate the ROI of a series of catchment management options for reducing phosphorus inputs in a lake system. This output is summarized in **TABLE 7.5**. As these results indicate, some of the options can deliver a rapid but relatively small reduction in phosphorus (P) loads quite cost-effectively, but investing further in P reduction yields steeply diminishing returns.

## Spatial Prioritization

For many conservation planners, using spatial prioritization tools like Marxan to help select the most important places for conservation action is essentially synonymous with conservation planning. Systematic conservation planning (SCP),[22] the most recognized approach to conservation planning, has focused predominantly on prioritizing places to conserve, and spatial prioritizations have featured prominently in the conservation literature. **Spatial prioritization** is, by a good margin, the most well-treated subject in conservation planning, as we observed in the opening chapter of this book when discussing questions that conservation planning helps answer. This is not surprising given there is something distinctly visceral about places. The organization we work for has a long history of planning just around place.

We suspect that spatial prioritization problems have featured in the literature more prominently than strategic planning has, and academics have more enthusiastically embraced them. This likely happens because spatial prioritization problems are easily bounded and are often supported by readily available GIS data. The set of options being considered is just the list of possible sites or planning units, whereas it is more difficult to cleanly bound strategic planning problems.

Due to the ready availability of excellent literature on spatial prioritization (including entire books),[23] we do not dedicate much space to general background on spatial prioritization. **Box 7.3** briefly describes the history of solving spatial prioritization problems and includes some of the most useful references and examples.

However, spatial prioritization is also an area of rapid advances. In this chapter we focus principally on those advances we feel are most significant for solving contemporary conservation planning problems: consideration of multiple actions and costs, consideration of connectivity, and the ability to easily interact with multiple scenarios. Because the approach to solving spatial prioritization problems is so often defined by the choice of software, we briefly describe a couple of the most widely used and useful planning software packages, Marxan with Zones, and Zonation, both of which embody the advances we mentioned earlier. We do, however, open this section with some of the general principles of spatial prioritization.

## General Principles

There is an enormous body of detailed and practical advice for solving conservation planning problems that involve spatial prioritization. Much of this advice can be found under the topic of Systematic Conservation Planning (SCP) and takes the shape of books, journal articles, and best practice guidelines. Together with Madeleine Bottrill and Robert Pressey, one of us has identified a set of core elements for solving spatial conservation planning problems that are relevant

## Box 7.3 A Two-Minute History of Spatial Prioritization

Spatial prioritizations have predominantly focused on the problem of choosing places to establish conservation areas. Beyond interest and opportunism, the challenge of determining which places represent priorities for conservation from a planning point of view was initially approached through the use of heuristic rules (i.e., rules of thumb). This rule-of-thumb approach usually just involved listing each potential site and the species of interest that lived there, ranking them according to the heuristic rule, and working down the list as time or resources allowed. Classic examples include the richness heuristic, where priority is given to the site with the largest number of species; the rarity heuristic, where priority is given to the site with the rarest species; and the greedy heuristic, where priority is given to the site that has the greatest number of species not conserved within existing conservation area. There are many variations and improvements on these, and although they can deliver quick and often good solutions, they also have a number of shortcomings that could prove costly for conservation.[a]

Beginning in the 1980s with the work of Australian ecologist Jamie Kirkpatrick,[b] conservation planners began using more sophisticated optimization algorithms to help solve spatial prioritization problems. This move to optimization was made possible because of the clear formulation of spatial prioritization problems, often referred to as the *reserve*

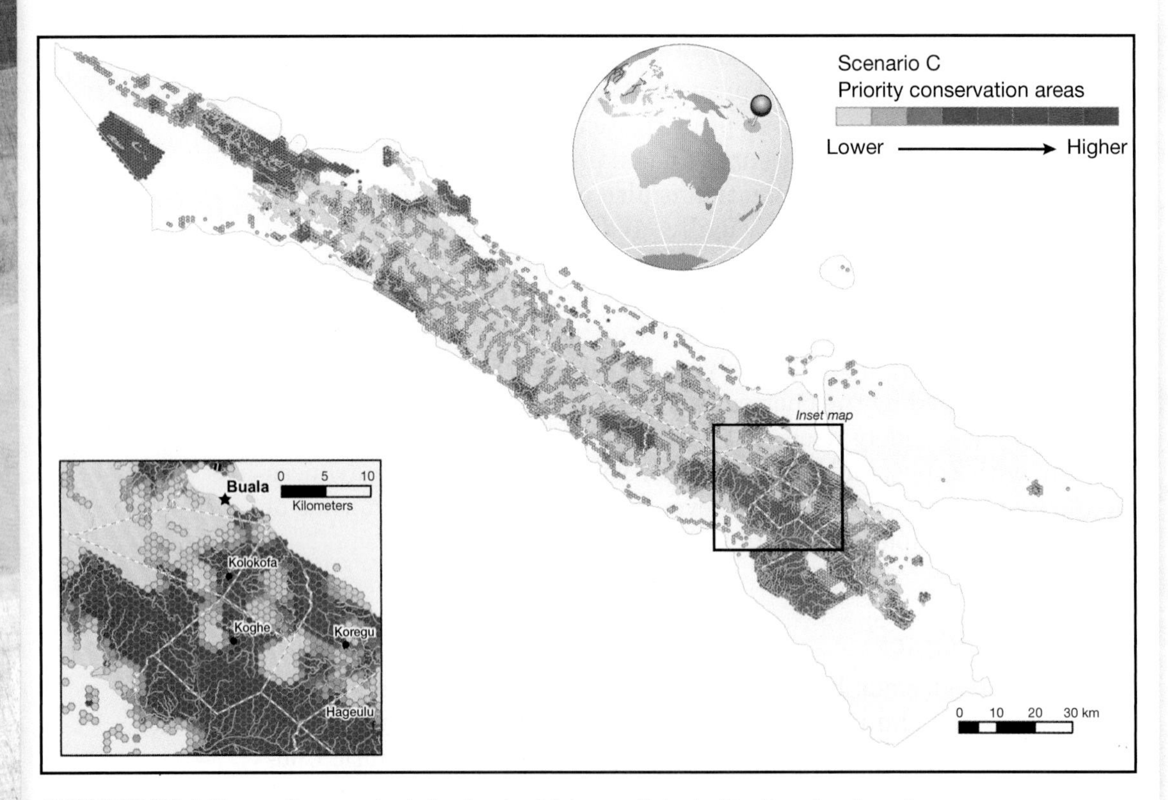


FIGURE 7.3.1 A Marxan "summed solution" output for a spatial prioritization developed for Isabel Province, Solomon Islands.

*selection problem.* This problem essentially involves selecting a set of sites for conservation that adequately captures the desired representation of biodiversity, and does so at a cost that is socially and politically acceptable. In a classic paper,[c] Jeffery Camm and colleagues describe alternative formulations of the reserve selection problem.

Many early examples of the reserve selection problem were solved using integer programming methods (see later in this chapter) that required the scientists involved to be proficient in computer programming. However, as the size of the areas being planned for and the amount of data increased, more sophisticated optimization (or approximation) methods were needed. At the same time, there was a rapid increase in the number of spatial prioritization problems being tackled—and by conservation scientists and project managers without computer programming skills. To meet both these needs, automated software tools were developed to help with spatial prioritization.

Although a large number of automated spatial prioritization tools have been developed, arguably the three most widely used and longest lasting have been Marxan (**FIGURE 7.3.1**), developed by Hugh Possingham and Ian Ball; C-Plan, developed by Bob Pressey and Matt Watts; and Zonation, developed by Finnish conservation biologist Atte Moilanen. Each of these software programs approaches the problem slightly differently, and substantial documentation about all of them is available on the Web.

There is little doubt that these tools and their solutions have influenced conservation decisions around the globe. They have been extensively adopted by government agencies and nongovernmental organizations, and they have played a significant role in high-profile conservation outcomes such as the rezoning of the Great Barrier Reef Marine Park in Australia.

---

[a] Pressey, R., H. Possingham, and J. Day, *Effectiveness of alternative heuristic algorithms for identifying indicative minimum requirements for conservation reserves.* Biological Conservation, 1997. **80**(2): 207–219.

[b] Kirkpatrick, J. B., *An iterative method for establishing priorities for the selection of nature reserves: An example from Tasmania.* Biological Conservation, 1983. **25**: 127–134.

[c] Camm, J. D. et al., *A note on optimal algorithms for reserve site selection.* Biological Conservation, 1996. **78**(3): 353–355.

for all spatial prioritization efforts (**TABLE 7.6**). Similarly, the intimate relationship between the development of spatial conservation prioritization and SCP means that many of the fundamental principles of SCP are generally applicable to spatial prioritization. The Finnish conservation planners Aija Kukkala and Atte Moilanen have reviewed the core concepts of spatial prioritization and their treatment in the peer-reviewed literature, including concepts like representation, complementarity, threat, and irreplaceability.[24] Although their work is concerned with planning from an academic point of view, it is a useful summary of the concepts underpinning much of the development in spatial prioritization methods.

**TABLE 7.6 Three core elements of solving spatial conservation planning problems**

**Commitments.** Commitments apply to areas that form the non-negotiable infrastructure of a conservation plan. These are areas already managed for the purpose of nature conservation and that contribute to meeting the targets for conservation features. They include things like existing protected areas, public lands designated to protect natural values such as forest reserves or research areas, private areas managed for conservation, and areas under traditional local management customs that protect natural values such as taboo areas. Identifying these existing spatial commitments is the intention of the process of *gap analysis* (see Chapter 1). It is politically and psychologically important to acknowledge the contributions made by existing conservation areas and other conservation activities, and can ensure that a resulting spatial prioritization is as complementary and efficient as possible.

**Configuration.** Configuration refers to the spatial arrangement and juxtaposition in the landscape or seascape of conservation areas and other conservation mechanisms. Understanding configuration requirements is critical to effective spatial prioritization. Many considerations can influence configuration requirements; two of the most common are compliance and connectivity. Configuration can influence subsequent compliance with spatial management by making it easy for people to know when they are in a conservation area, for example, through boundaries that follow natural features (e.g., ridge lines or rivers), or other recognized demarcations (e.g., lines of latitude or longitude particularly in marine systems, or political boundaries such as state or provincial boundaries). The need for conservation efforts to be configured in a way that helps sustain the connectivity of this ecological process has been given a great deal of attention in spatial prioritization, and we discuss it in this chapter.

**Preferences.** Preferences are characteristics of areas that help planners resolve choices between alternate possible sites for conservation. They reflect a more nuanced view of the value of different areas for conservation than is captured simply through presence or absence of conservation features. For example, two areas of the same habitat might not be considered of equal value because they are exposed to different levels of anthropogenic disturbance—we may have a *preference* for conserving the less disturbed area.

## Prioritizing Locations for Multiple Actions

Historically, most spatial conservation planning has been framed as a binary question: either places are a priority or they are not—a very black/white, in/out view. This simplification serves conservation planning very well when the principal purpose of such prioritizations is to identify sites to create new conservation areas. However, as conservation projects have become more ambitious, human well-being and ecosystem service delivery objectives are often included (see Chapter 10), and the role of the matrix of lands and waters outside conservation areas is better understood and appreciated.[25] The "in versus out" approach to spatial planning involves too much simplification for navigating complex strategies and actions that often transcend these conservation area boundaries, and doing so would likely miss many opportunities for efficient and effective solutions.

The key change in spatial prioritization in response to the changing nature of conservation decisions has been the development of methods that allow for prioritization of locations for multiple actions, all with differing consequences and costs for conservation objectives. This change means that the places identified as priorities are the best places to undertake a particular, previously identified action. A related outcome of this change is that whereas conservation planners were once concerned primarily with ecologically intact landscapes and watersheds, the field has now moved to attaching some level of conservation value to most areas across a planning region except those that have been heavily developed or converted (FIGURE 7.5). We hope it is easy to see how this is a perfect fit with the action-based problem framing we introduced in Chapter 6.

The multiple-action evolution of spatial planning arose principally through a desire to allocate areas to different *zones* within a larger conservation area. Zoning is a relatively specific concept, but the idea of a zone is easily generalized to the spatially restricted application of any action. Explicitly recognizing multiple actions as part of spatial planning has the potential to greatly improve the ability of conservation plans to adequately meet multiple objectives. The pro-

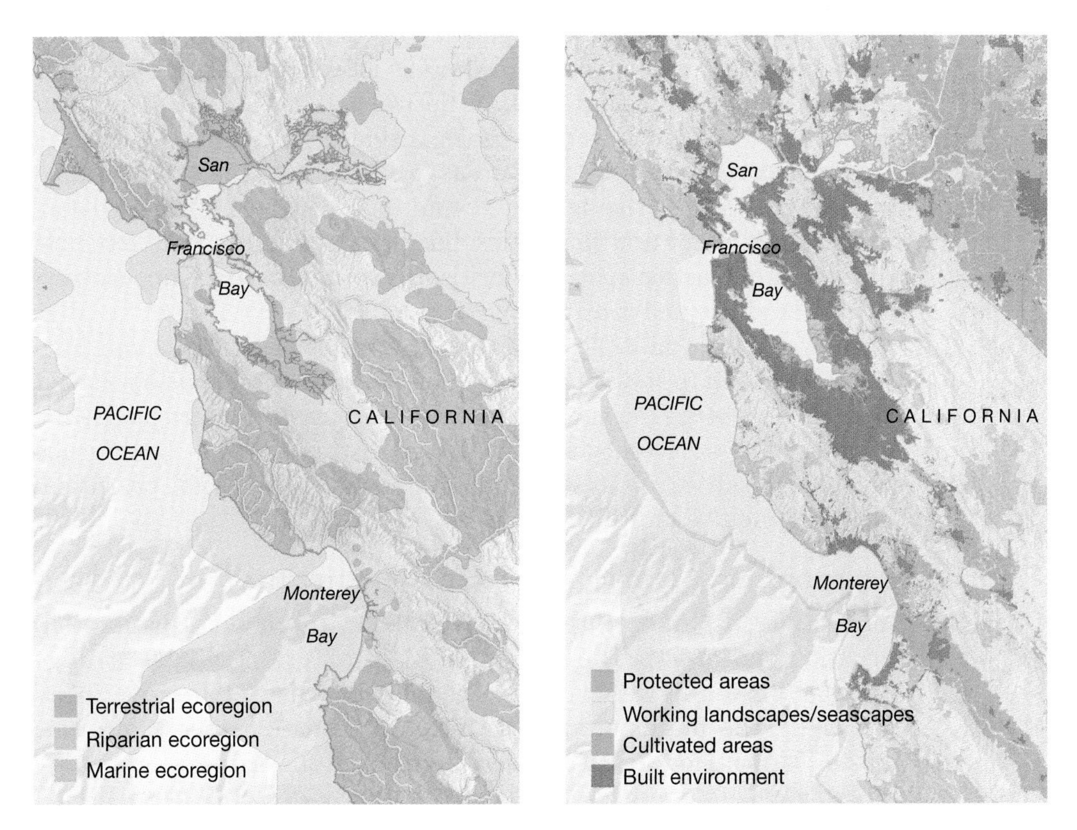


FIGURE 7.5 Conservation's changing view of landscapes. The image on the left shows just two types of land in central California, USA: intact habitat and everything else. The image on the right shows four types of land use: conservation, multiple use, agricultural, and urban. Many conservation plans are now concerned with planning for all these land use types.

cess of marine spatial planning (MSP) introduced in Chapter 2 is a good example of this evolution of spatial prioritization. This evolution also helps position the results of conservation planning to be more easily integrated with land use planning that happens at local areas of government throughout the world (see Chapter 10).

Consider the simplified but not uncommon problem of identifying ways to meet conservation targets in an area of tropical forest while at the same time trying to improve the economic prosperity of the region through expansion of oil palm plantations and the continuation of logging and agricultural activities of local communities. Conservation, oil palm, logging, and agriculture are all different actions. Each action has different consequences for conservation and economic prosperity, and it comes at a different cost for users of that landscape. For example, selective logging of a forest, while not as good for conservation as a fully protected forest, is likely to allow a substantial amount of biodiversity to persist. It is also likely to have lower opportunity and management costs than a national park because some management costs can be passed to the logging company. Oil palm plantations have the least biodiversity benefit but are important for regional jobs and economies and so have a very high opportunity cost if good areas for oil palm are set aside for strict conservation. It might also be desirable to avoid having oil palm plantations directly adjacent to conservation areas. A good solution to this problem requires allocating areas to all of these land use types in a way that maximizes the delivery of multiple objectives. Kerrie Wilson, with Erik Meijaard and colleagues, has provided an excellent example of how multiple-action spatial prioritization can be used to support complicated land use and conservation decisions.[26]

Considering multiple actions in spatial prioritizations can be a useful approach to solving conservation planning problems because it provides a rigorous way to integrate conservation planning with the planning efforts of other sectors. Later in this section, we describe some of the software tools available to assist with spatial prioritizations that include multiple objectives.

### Incorporating Connectivity in Spatial Prioritization

Connectivity is an important consideration in solving spatial conservation planning problems, especially in an era of climate change (Chapter 9). While we do not see connectivity as an objective in its own right, the movement of organisms or resources between areas can be essential for population persistence and the ecological integrity and function of many ecosystems.[27] As such, connectivity patterns can strongly influence the expected consequence of an alternative. This result will be increasingly true as conservation in general trends from a focus on the protection of "static" features to a focus on landscapes and seascapes that support the persistence and functioning of ecosystems and their biodiversity. Connectivity has also proved to be a concept that resonates with decision makers and the public, and in that sense has become an organizing principle for a lot of conservation efforts—consider big connectivity projects such as Yellowstone to Yukon in North America, and the Mesoamerican Biological Corridor of Central America. Connectivity will frequently be an important part of finding a good solution to spatial conservation planning problems.

There is a clear link between connectivity and an emphasis on planning for multiple actions. Connectivity between conservation-focused areas can often be achieved through sensible management or "softening" of the matrix of lands and waters between conservation areas. In fact, thinking about connectivity has the benefit of encouraging a whole-landscape or whole-system view of conservation planning. A lot of recent conservation planning research effort has focused on the connectivity of habitats and species through time given the expected changes in their locations as a consequence of a changing climate. We deal specifically with this topic in Chapter 9.

Getting good data on the connectivity of organisms in a landscape can be challenging. However, an increasing number of models and modeling tools are available to help (see Chapter 5). For example, in North America, the Connectivity Analysis Toolkit has been widely used in conservation planning exercises.[28] In this chapter we assume that some of this information is available and focus on several examples of how connectivity information can be integrated into widely applied approaches to solving spatial conservation planning problems. These different approaches cover a spectrum of sophistication; the most appropriate solution method for a given planning process will depend heavily on the availability of relevant connectivity data.

A straightforward approach to connectivity is that used by the conservation planner Robert Smith and colleagues to design new conservation areas in the Maputaland region of Mozambique, Swaziland, and South Africa.[29] After using Marxan to identify a set of core conservation areas for their species and habitat conservation features, Smith and colleagues manually identified a set of important linkages between these areas that were considered by experts to be important for the movement of large vertebrates (FIGURE 7.6). They then locked in these areas together with the core areas and re-ran the analysis to ensure that habitat being protected in these corridor areas was considered toward target achievement and that the remaining conservation areas were as complementary as possible. This sort of manual, expert-based solution approach is easy to implement and well suited to projects aiming to emphasize large landscape linkages.

In cases where substantial data on connectivity patterns is available, it makes sense to treat connectivity in a more sophisticated fashion. Ideal connectivity data include both the direction and

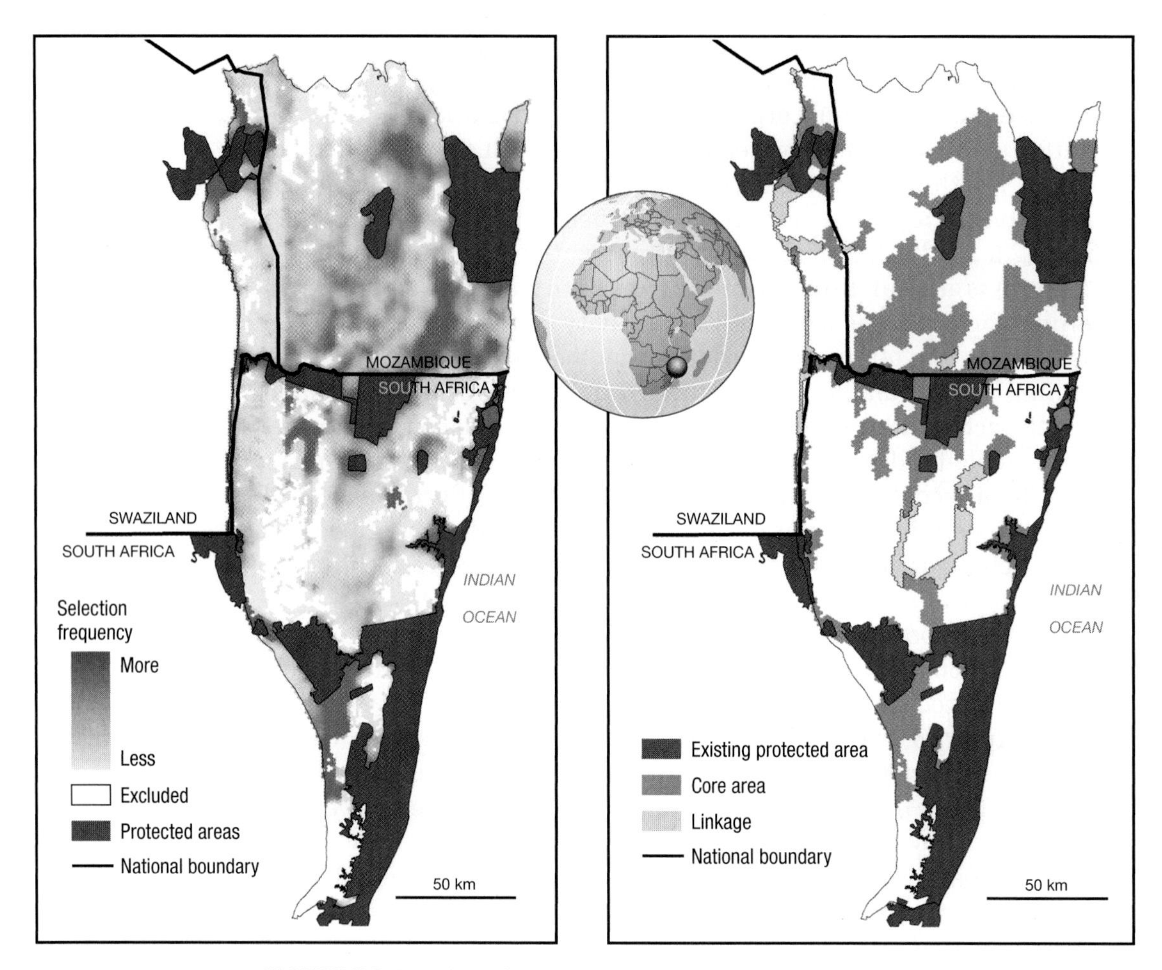


**FIGURE 7.6** Ensuring adequate connectivity in spatial prioritization. The figure on the left hand shows a prioritization output for the Maputaland region of Mozambique, Swaziland, and South Africa, and the one on the right shows how the core areas identified by this prioritization were supplemented with "linkage" areas to ensure adequate connectivity. (Adapted from Smith et al., 2008.)

strength of connectivity processes between all parts of the landscape under consideration. Because many spatial prioritizations are based on planning units, connectivity can often be summarized by a table that describes the strength of connectivity between different planning units (**TABLE 7.7**).

A connectivity table allows planners to use knowledge about asymmetric connectivity patterns, which is particularly important for features like river systems because much of the connectivity flows in a single direction. We might, for example, consider some conservation actions in the upper reaches of a river system even if we expected heavy degradation of the lower reaches, but the reverse is unlikely to be true: we would not want to focus conservation effort on the lower reaches if, for example, it did not also protect the upper reaches that a particular target fish species depends on for reproduction.

Both Marxan and Zonation software can use connectivity matrices as inputs that allow connectivity to be a consideration in spatial prioritization. They do this essentially by including a connectivity component in the objective function to penalize solutions that do not capture important connectivity patterns. The process of finding good solutions to spatial prioritizations that include connectivity considerations has been described well by Maria Beger and colleagues, who also provide an example of planning that considers the spatial pattern of larval dispersal on coral reefs.[30]

A good example of addressing connectivity in spatial prioritization is provided by the freshwater conservation planners Virgilio Hermoso and Simon Linke.[31] They illustrate how multiple types of freshwater connectivity (FIGURE 7.7) can be integrated into a spatial prioritization, including longitudinal connectivity along river networks as well as lateral connectivity between rivers and wetlands, which is important for many aquatic and terrestrial species.

TABLE 7.7 Example of a planning unit connectivity table. The table contains information on the strength of connectivity between different planning units. The connectivity considered is asymmetrical, such that the strength of connection from planning unit 1 to 4, for example, is different from the strength of connection going from planning unit 4 to 1. This may occur, for instance, in freshwater systems where the connectivity from upstream to downstream planning units is stronger than from the reverse.

| From planning unit # | To planning unit # | Connectivity strength |
|---|---|---|
| 1 | 4 | 0.1285 |
| 1 | 5 | 0.157503 |
| 1 | 8 | 0.093988 |
| 1 | 10 | 0.182587 |
| 1 | 11 | 0.119218 |
| 4 | 1 | 0.092371 |
| 4 | 8 | 0.162792 |
| 4 | 10 | 0.31625 |
| 4 | 11 | 0.15391 |
| 4 | 23 | 0.11925 |
| 10 | 1 | 0.19456 |
| 10 | 8 | 0.52397 |
| 10 | 11 | 0.26658 |
| 10 | 12 | 0.206547 |
| 10 | 15 | 0.340125 |
| 10 | 23 | 0.204071 |

## Connectivity across Realms

Conservationists have long recognized that terrestrial, freshwater, and marine realms are closely connected and what happens in one can influence the other. One of the most easily recognizable connections is the impact that coastal terrestrial and freshwater systems have on nearshore marine environments (FIGURE 7.8). Although conservation plans often cover multiple realms (e.g., terrestrial and freshwater features, or marine and terrestrial features), explicit consideration of the impact that one has on another has been less common,[32] and challenging. Fortunately, there has been recent progress on spatial prioritization approaches that incorporate the important process of connectivity between realms.

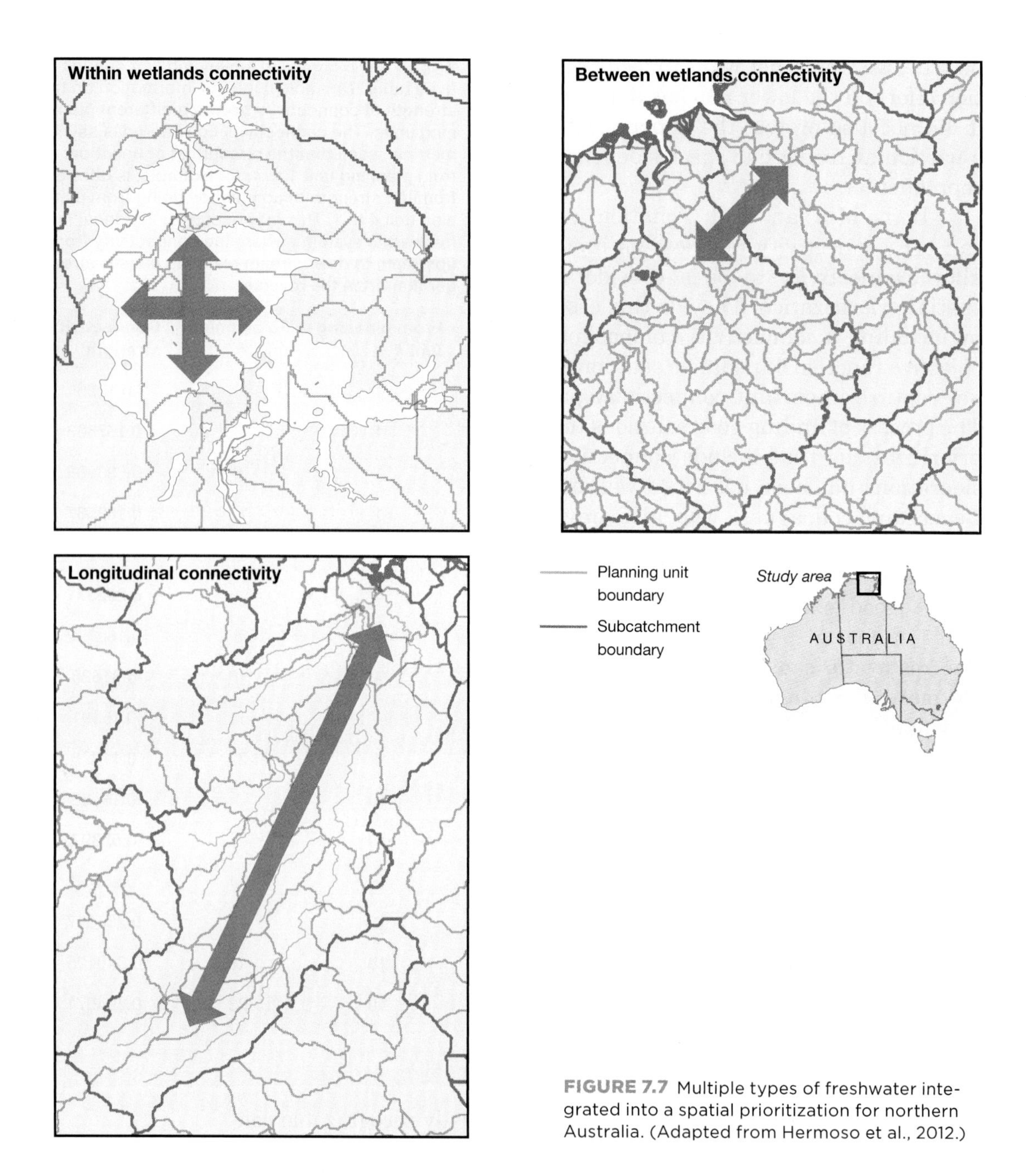


FIGURE 7.7 Multiple types of freshwater integrated into a spatial prioritization for northern Australia. (Adapted from Hermoso et al., 2012.)

Together with colleagues, marine conservation biologists Carissa Klein and Stacy Jupiter have led innovative work to investigate how terrestrial conservation efforts can be prioritized so that forest conservation delivers significant benefits to adjacent coral reefs.[33] For all forest areas on Fiji's three largest islands, Klein and Jupiter estimated the consequence of forest conservation efforts for their contribution toward

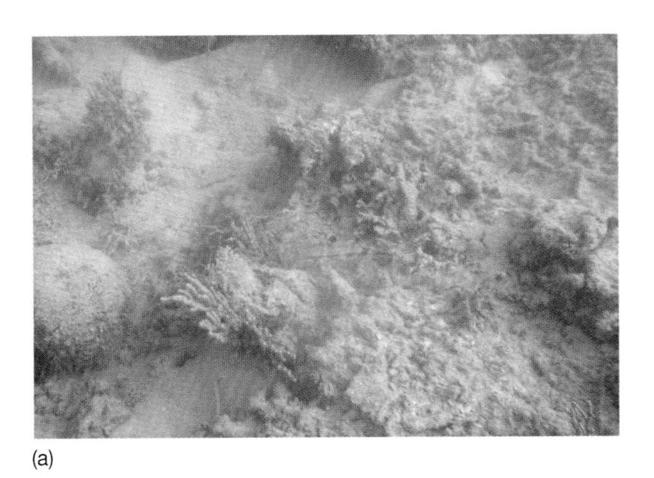

(a)

(b)

FIGURE 7.8 (a) Coral reef covered in sediment. (b) A major source of sediment. A logging pond and access road on the coast of Isabel Island, Solomon Islands.

terrestrial conservation targets and also for what they contribute to coral reef health by minimizing terrestrial pollution flows onto those reefs. The benefit to coral reefs included an assessment of the amount of terrestrial pollution that would reach the reefs if the forest were cleared, and also of the relative importance of this pollution for the health of reefs given other threats such as fishing. This cross-realm connectivity benefit function was then incorporated into a Marxan analysis that has been used by Fiji's Protected Area Committee to modify existing priority boundaries to better meet terrestrial conservation goals while improving coral reef condition through runoff prevention.

## Software

We have included a brief discussion of Marxan with Zones and Zonation. These are by no means the only software programs that can solve these problems, but we have chosen them here because they are both widely used, free, well supported, and continually being updated to meet new spatial prioritization challenges.

### Marxan with Zones

Globally, **Marxan** has been the most widely used spatial prioritization software.[34] The core functionality of Marxan is to identify a set of potential conservation areas that meet the target set for each conservation feature while minimizing the cost of doing so. This formulation resulted in spatial prioritizations that contained essentially two types of areas: conservation areas, and everything else. Marxan with Zones is the evolution of the Marxan family of software to meet the challenges described earlier of planning for multiple actions.

Marxan with Zones solves target-based spatial planning problems. Targets (discussed in Chapter 4) can be for specific features in specific zones, for example, the amount of board feet available in a logging zone or the amount of coral reef in a no-take zone. Or, where multiple zones might contribute toward meeting the target for a conservation feature, the target could be for the whole landscape. A number of good examples of its use in multiple-action spatial planning are available in the literature.[35]

***Inputs*** Like Marxan before it, Marxan with Zones spatial prioritization is based on a set of *planning units.* As well as summarizing the occurrence of conservation features and other relevant activities (like fishing) within each planning unit, Marxan with Zones requires some additional inputs that are characteristic of planning for multiple actions.

First, it requires specifying the set of zones to be planned for. A zone can be thought of as a type of management action or arrangement that will take place. For example, a forest landscape might have zones such as plantation forest, production forest with different degrees of cutting intensity, recreation forest, nature preserve, and unallocated areas. A zone can be a formal designation of land or sea management as defined by legislation, or it might simply be a proposed activity, like restoration.

Second, Marxan with Zones requires specifying the compatibility of conservation features and other activities (like logging or fishing) with each of the zones. This is the key step in estimating the consequences (from Chapter 6) of different possible management arrangements for conservation features and other activities. TABLE 7.8 is an example of a compatibility table built for a marine spatial planning exercise that used Marxan with Zones.

If we consider the conservation feature "sea grass" in Table 7.8, we can see that in a tourism zone, it is likely to be somewhat degraded but not completely removed and will therefore still have some conservation value. But if sea grass is in an industrial zone, it likely will retain very little conservation value and will not contribute much toward meeting conservation targets. On the other hand, because sand mining is incompatible with management of marine reserves, it would be excluded from that zone, and anyplace allocated as a reserve would affect sand mining potential; the same would be true for the tourism zone. In contrast, sand mining and fishing may be compatible to some extent, such that a fishing zone would not totally exclude sand mining value from an area and vice versa. This process is the same as that of discounting (discussed in Chapter 5), where a hectare of feature X in zone Y may be worth less than a full hectare if we expect the activity to have a negative impact on the habitat or biodiversity there.

Putting numbers on the compatibility of conservation features with different zones requires integrating both the social values

**TABLE 7.8** Compatibility table built for a marine spatial planning exercise that used Marxan with Zones. This table illustrates the compatibility of different features with different zones and the activities taking place within them.

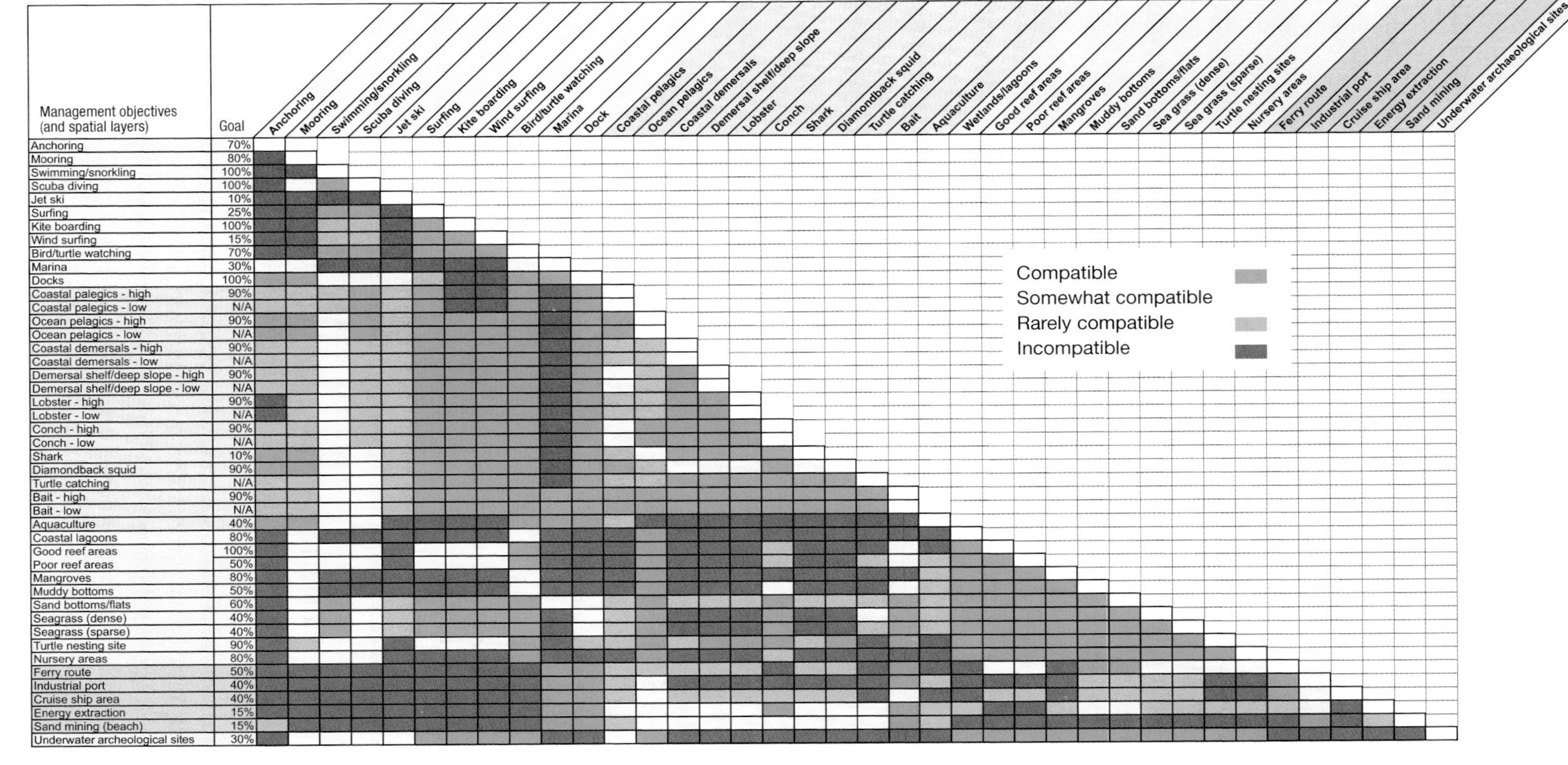


| Management objectives (and spatial layers) | Goal |
|---|---|
| Anchoring | 70% |
| Mooring | 80% |
| Swimming/snorkling | 100% |
| Scuba diving | 100% |
| Jet ski | 10% |
| Surfing | 25% |
| Kite boarding | 100% |
| Wind surfing | 15% |
| Bird/turtle watching | 70% |
| Marina | 30% |
| Docks | 100% |
| Coastal palegics - high | 90% |
| Coastal palegics - low | N/A |
| Ocean pelagics - high | 90% |
| Ocean pelagics - low | N/A |
| Coastal demersals - high | 90% |
| Coastal demersals - low | N/A |
| Demersal shelf/deep slope - high | 90% |
| Demersal shelf/deep slope - low | N/A |
| Lobster - high | 90% |
| Lobster - low | N/A |
| Conch - high | 90% |
| Conch - low | N/A |
| Shark | 10% |
| Diamondback squid | 90% |
| Turtle catching | N/A |
| Bait - high | 90% |
| Bait - low | N/A |
| Aquaculture | 40% |
| Coastal lagoons | 80% |
| Good reef areas | 100% |
| Poor reef areas | 50% |
| Mangroves | 80% |
| Muddy bottoms | 50% |
| Sand bottoms/flats | 60% |
| Seagrass (dense) | 40% |
| Seagrass (sparse) | 40% |
| Turtle nesting site | 90% |
| Nursery areas | 80% |
| Ferry route | 50% |
| Industrial port | 40% |
| Cruise ship area | 40% |
| Energy extraction | 15% |
| Sand mining (beach) | 15% |
| Underwater archeological sites | 30% |

*Source*: Adapted from Agostini et al., 2015.

ascribed to different resources and the available science. This will necessarily involve a good degree of subjectivity that is certain to make some participants wary, an appropriate response given the challenges we are aware of when estimating thresholds related to human activity impacts on conservation features.[36] It is, however, precisely what natural resource managers are asked to do all the time, just in a more explicit and transparent fashion.

Third, although not necessary, Marxan with Zones allows consideration of more complex representation of costs than has been possible in other spatial prioritization software. Rather than a single measure of cost, Marxan with Zones retains information about the costs to each different actor or stakeholder in the landscape of assigning a particular planning unit to a particular zone. For example, allocating an area with very productive soils to strict conservation will likely come at a higher cost to the region's agriculture than an area with poor or degraded soils.

Fourth, to allocate a series of management actions or zones across a region requires understanding the desirability of different zone adjacencies. For example, is it okay for plantation forest to be directly adjacent to a nature reserve, or would we prefer having some production forest between the two? This concept is the rationale behind the well-established protected areas notion of core and buffer zones; the difference here is that the relationships between a wide variety of zones need to be considered. Marxan with Zones allows users to specify the relative desirability of the different spatial relationships between zones.

***Outputs*** Marxan with Zones uses a sophisticated simulated annealing algorithm to deliver two main types of solutions to spatial prioritization problems. Simulated annealing is like the process of annealing glass—as the glass cools, fewer and fewer adjustments are possible; Marxan is initially very flexible in the solutions it explores, but as the process proceeds and it homes in on a good solution, it makes fewer adjustments.

The first output from Marxan with Zones is an actual allocation of zones across a planning region that is most efficient in terms of meeting targets and minimizing costs (FIGURE 7.9). The second outputs are solutions that indicate the importance of different areas for different zones (FIGURE 7.10). Between these two outputs there is a great deal of flexibility in the sort of problems that can be tackled. For instance, instead of an actual spatial solution, Marxan with Zones could be used to inform the general balance of allocation to different zones that would meet multiple objectives (conservation, production, jobs), such as 15% protected area, 20% oil palm, and the remaining in reduced-impact logging. These numbers could then form the basis of policy advocacy.

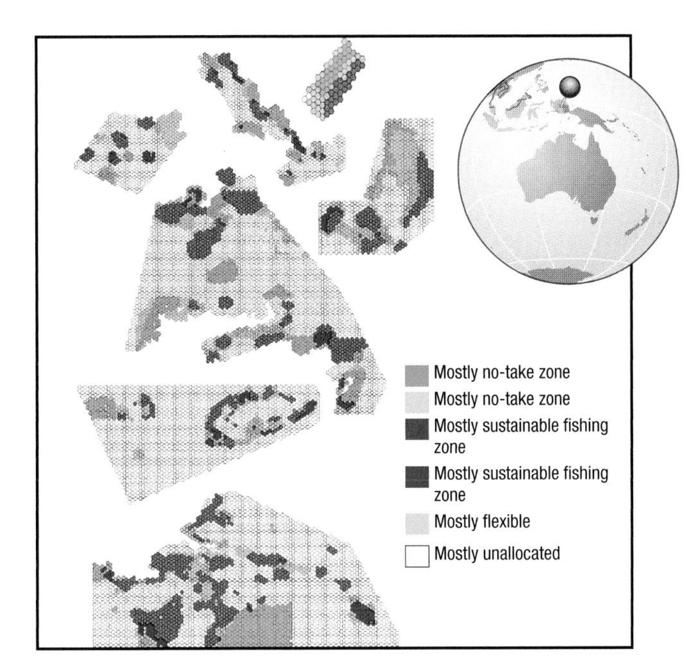


FIGURE 7.9 Marxan with Zones output; the allocation of marine areas to different zones in Raja Ampat, Indonesia. (Adapted from Grantham, H. S. et al., *A comparison of zoning analyses to inform the planning of a marine protected area network in Raja Ampat, Indonesia.* Marine Policy, 2013. **38**: 184–194.)

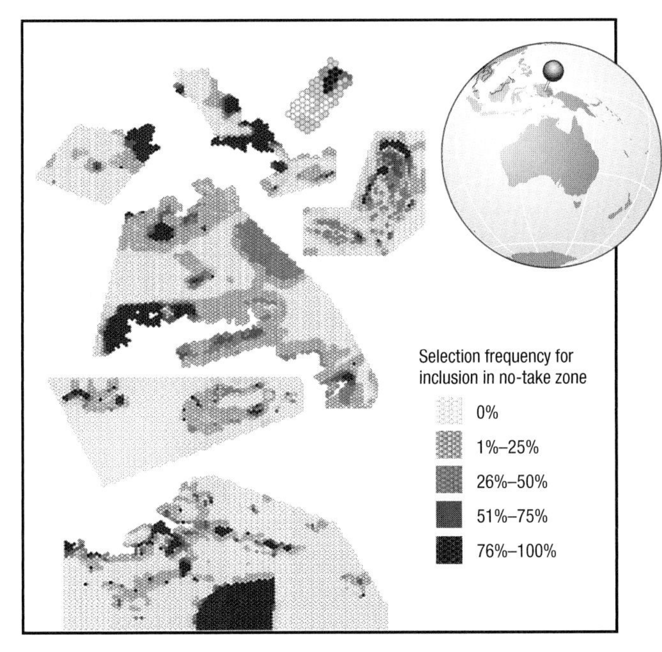


FIGURE 7.10 Marxan with Zones output; the relative importance of different areas in Raja Ampat, Indonesia, for inclusion in the "no-take" zone, as indicated by selection frequency. (Adapted from Grantham et al., 2013.)

## Zonation

The second software tool we address is **Zonation.**[37] The core functionality of Zonation is the hierarchical ranking or prioritization of a landscape based on conservation value (FIGURE 7.11). This tool solves spatial prioritization problems by identifying the most important and cost-effective areas to include in core conservation zones. Unlike the

**FIGURE 7.11**
Zonation output; the hierarchical ranking or prioritization of landscapes in Australia based on conservation value. Areas in red represent the top 2% of the country based on the conservation importance ranking (calculated using Zonation).

Marxan family of software, Zonation is not based on achieving a set of targets (Chapter 4), but rather on giving a picture of the relative conservation value of the entire landscape (although targets can be used). Zonation analyses are also generally not based on planning units but rather on the raw distribution of conservation features and costs, which in practice means its solutions can achieve a finer resolution than that of solution methods based on planning units. Zonation has been used widely in terrestrial, freshwater, and marine settings, and it is supported by good documentation and many published papers.

Like Marxan, Zonation was designed to consider the distribution of biodiversity conservation features in a region and identify a complementary and efficient set of conservation areas, or to evaluate the impact of potential development plans. Its functionality has since been extended to include consideration of connectivity, a sophisticated treatment of uncertainty in input data, and the constraints imposed by planning across multiple administrative regions.[38] This latter functionality is particularly useful in cases where legislation dictates that a certain amount of a conservation action must happen on each property or region. For example, the conservation planner Jim Thompson worked with colleagues in Brazil to identify priority areas for restoration so that each landowner could be in compliance with a forest code requiring 80% of his or her land to be forested (**FIGURE 7.12**).

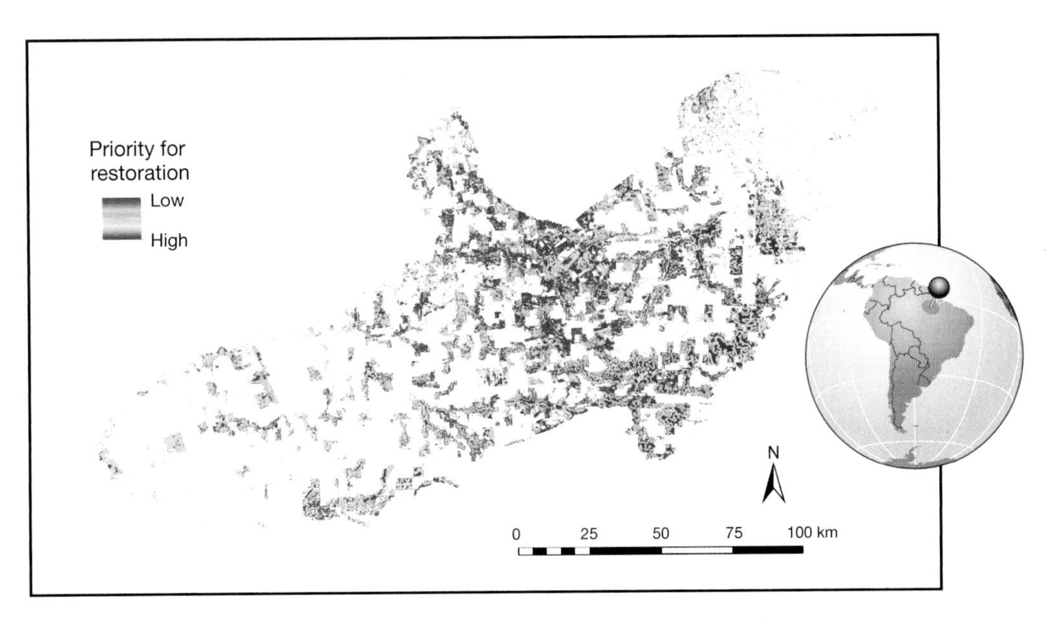


FIGURE 7.12 Zonation output illustrating priority areas for restoration in the Brazilian Amazon district of Paragominas, so that each landowner could be in compliance with a forest code requiring 80% of his or her land to be forested. Red areas represent those locations having the highest priority for restoration activity and blue areas the lowest priority. Only those places where restoration is possible are displayed with a color.

The focus of Zonation is firmly on prioritizing the conservation of biodiversity features. Despite its name, Zonation does not have the multiple-zone functionality of Marxan with Zones. However, it can still be used to identify priority sites for different conservation actions and the balance of these actions across a region, provided that only one action is possible in each location. For example, Atte Moilanen (the developer of Zonation) and colleagues[39] use Zonation to prioritize management actions across three different land use types in New Zealand: dairy farming, dry-stock farming, and plantation forest (FIGURE 7.13). To do so, they estimate the expected change in current biodiversity conditions from a set of actions on each landscape, and combine this estimate with the costs of each management action (TABLE 7.9), and the distribution of biodiversity to develop a landscape prioritization of management actions.

One of the main reasons planners use spatial prioritization software like Marxan and Zonation is that they allow planners to make use of complex optimization procedures that would otherwise require expertise in mathematics and computer programming. Mathematical optimization can be a powerful tool for helping to solve a range of conservation planning problems; we explore this process further in the next section.

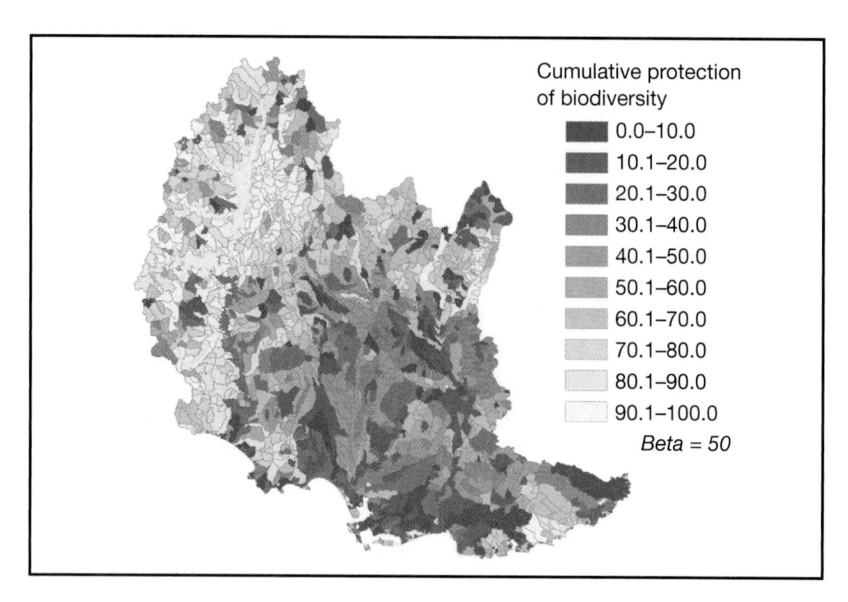


**FIGURE 7.13** Zonation output showing priority areas for management actions across three different land use types in New Zealand: dairy farming, dry-stock farming, and plantation forest. (Table 7.9 describes the types of management that are assumed to be undertaken for each land use type.) Here, the highest-priority areas are in dark green, and it is assumed that the appropriate management action is taken for the particular land use occurring in that place. The color divisions on the scale represent the cumulative protection of the region's biodiversity that would be achieved by undertaking the proposed management actions over the corresponding area of the figure. (Adapted from Moilanen et al., 2011.)

**TABLE 7.9 Management actions for three different land use types in New Zealand, along with their costs and expected conservation impact**

| Land use | Action | Cost | Combined impact |
|---|---|---|---|
| Dairying | Three-wire electric fencing | \$5/m fencing costs, or \$1,000/100 m of stream length | 60% reduction in diffuse nitrogen inputs |
| | Riparian planting 5 m either side of stream, including three years maintenance | \$20,500/ha, or \$2,050/100 m of stream length | Effective increase in native catchment cover of 33% |
| | Establish artificial wetland on tile drain discharges | \$550/ha applied across 1% of those parts of the planning unit in dairy production | Effective increase in native catchment cover of 33% |
| Dry-stock | Five-wire fencing | \$18/m fencing costs, or \$3,600/100 m of stream length | 30% reduction in diffuse nitrogen inputs |
| | Riparian planting 5 m either side of stream, including three years maintenance | \$20,500/ha, or \$2,050/100 m of stream length | Effective increase in native catchment cover of 33% |
| Plantation forest | 5 m set back from both sides of streams for all operations | Assumed opportunity cost of \$5,000/ha or \$500/100 m of stream length | 30% reduction in diffuse nitrogen inputs |
| | | | Effective increase in native catchment cover of 33% |

*Source*: Adapted from Moilanen, Leathwick, and Quinn, 2011.

## Optimization

The term *optimizing* is often used in a general sense to imply "doing as well as possible." In that sense, all of the methods described so far are a form of optimization. However, in this section we focus on mathematical **optimization** and the stricter definition this implies: identifying the option that will deliver the best possible outcome for an objective (typically by maximizing or minimizing a **mathematical function** that describes the objective). For example, in communicating about a plan, a conservation planner might say something like, "We optimize for biodiversity and connectivity." Strictly speaking, this is not possible because only one objective can be optimized at a time. Biodiversity and connectivity might, however, be combined into a single mathematical function—but then the **utility function** is optimized, and not the elements it contains.

Mathematical optimization is a highly quantitative process that can be powerful and compelling. To be able to optimize a solution is sort of the holy grail of planning—knowing you have identified the most efficient solution possible. However appealing the idea may be, there are also good reasons that optimization in the mathematical sense may not be necessary or desirable for solving conservation planning problems. These include, in no particular order:

- *Where can you go from optimal?* Few solutions to conservation planning problems can be implemented without negotiation, and few decision makers will be satisfied without at least seeing some alternatives. Identifying the optimal solution is arguably less important than finding a range of potentially good solutions.
- *Mirage of optimality.* In nearly all cases there will be significant uncertainty in the data used during conservation planning. Given this uncertainty, including uncertainty in the strategies (solutions in this parlance) that are being applied, it can be somewhat misleading to talk about an optimal solution, especially if it performs only slightly better than the alternatives. Think of this result as a bit like overlapping error bars. The optimality you are reporting might simply be a mirage due to data uncertainty.
- *Precision versus speed.* Finding optimal solutions can be computationally intensive, and perhaps more importantly, often requires substantial time to set up the problems and work with folks who are mathematically proficient enough to solve them. This means that compared to other tools in the planner's toolkit, optimization can take longer to propose some alternatives. Where time is critical, this might be a factor.

- *Problem framing in conservation is difficult.* There are likely to be many nuances that simply cannot be adequately captured in the mathematical framing of a problem. This does not mean the results are not informative, but it does mean that strict optimality might not be the most important outcome.
- *Look to the bottom, not the top.* Decision makers have many reasons for choosing solutions that do not appear optimal. Some are legitimate, and some are simply the result of bias. Decision makers are typically risk averse, so while they may not enthusiastically take advice about "optimal" solutions, their ears are likely to prick up at the mention of bad solutions that should be avoided. The reputational risk posed by choosing a solution that was predicted to perform poorly can be more compelling than the importance of choosing the alternative predicted to perform optimally.

Having said this, conservation planning is about good resource allocation, and optimization is a natural tool for this. We believe that optimization still has a strong role to play in conservation planning, for example, to identify land use allocations that maximize the conservation benefit for a particular economic output. Here we briefly introduce a few of the more common optimization methods employed in conservation planning. This section is by no means a user guide to optimization methods. We include this introduction principally to give planners at least some familiarity with these terms, so they can use this familiarity to track down more detailed guidance on optimization methods as needed. Much of the spatial prioritization discussed earlier in this chapter is based on optimization algorithms. And while familiarity with the algorithms is not a prerequisite of their use, it does greatly help to understand how they find solutions and the assumptions that are being made in doing so.

## Optimization Methods

Mathematical optimization requires that conservation planning problems be expressed in mathematic equations. In practice, this means that there is a mathematical function that can characterize the consequence of an option as discussed in Chapter 6 and tell us how one option performs compared with another. Occasionally the optimal solution can be found simply through clever calculus; however, in general, optimization in conservation planning problems works through a process of searching through potential options to find the best. Different optimization algorithms are basically different processes for searching through potential solutions. As we discussed in Chapter 6,

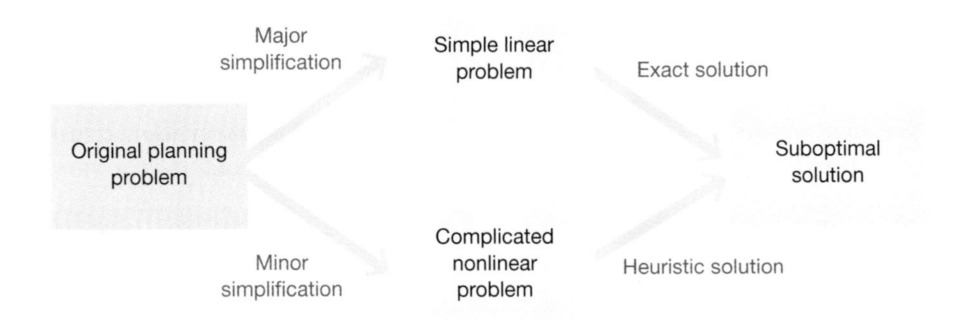


FIGURE 7.14 Two paths to a suboptimal solution. (Adapted from Moilanen and Ball, 2009.)

the number of potential solutions to conservation planning problems can be vast.

The optimization algorithms used most in conservation planning fall into two classes: exact algorithms and heuristic algorithms. Exact algorithms, as the name suggests, guarantee finding the absolutely optimal solution. Heuristic algorithms, on the other hand, are not guaranteed to find the optimal solution but are efficient ways to find solutions that should be reasonably close to optimal.

Finding exact optimal solutions is computationally very intensive; to be tractable on personal computers, the process requires some fairly restrictive assumptions about the structure of the problem (see the linear programming description that follows). In most cases, this means that planners must choose between either simplifying the problem in order to use exact optimization, or using a more realistic problem but accepting an approximation of the optimum solution. In practice this means that both exact and heuristic optimization approaches are finding good but not strictly optimal solutions, just through different pathways. The conservation planning software developers Atte Moilanen and Ian Ball summarize this nicely in FIGURE 7.14.[40] Both Marxan and Zonation employ clever heuristic algorithms so that they can find solutions to problems that are far too big for exact optimization.

### Exact Optimization

The most common type of exact optimization used in conservation planning is linear optimization or linear programming (computer scientists refer to optimization simply as programming). Linear programming can be used to find exact solutions to even very large problems, but it requires the assumption (among others) that the function describing benefit or consequence is linear. Put simply, a linear function means that the benefit of taking two actions should be simply the sum of taking each action on its own.

If you plotted investment in a conservation action on one axis and the benefit or consequence for an objective on the other axis, a linear function would be a straight line. Of course most conservation planning problems do not have benefit functions like this, but it is a very common and useful simplification. A particular type of linear optimization known as *integer programming* has also been used widely in conservation planning. Integer programming requires the additional assumption that the variables being selected are integers (e.g., 0, 1, 2, 3, etc.). Integer programming has been used a great deal in reserve selection, where the parcel is either 0 (not selected) or 1 (selected), thereby satisfying the integer requirement. It has been used widely in forestry and in other land use planning where parcels are allocated to a discrete set of potential land uses. Much has been written about the use of linear programming and integer programming in conservation planning.

### Heuristic Optimization

As many conservation planning problems cannot be sensibly simplified into linear equations, it would take too long to find exact solutions (think of having to run a computer for years to get one answer). Envisioning a typical regional conservation plan makes this task easy to appreciate. There may be dozens of species that are treated as conservation features, each with different targets. There may be other types of features as well, such as ecosystems or ecological processes. There may also be social objectives with accompanying features that are part of the plan. In addition, there may be several layers of "cost" data that could include land monetary values or data on the ecological integrity or viability of different features. All told, there could be scores of variables influencing the selection of conservation areas and nearly countless possible solutions to this "problem." Heuristic optimization algorithms basically use clever tricks to avoid searching all potential solutions and still find a good one.

In essence, **heuristic algorithms** take a solution, make a small change to it, and if it results in a better solution, they accept it; if it is worse, they reject it. By doing a large number of these small tests, the algorithms hopefully converge on something close to the optimal solution. Because they do not test all possible solutions, the solution that a heuristic algorithm settles on can be different depending on what solution it starts with. This situation is typically explained by analogy with hill climbing—imagine you were trying to reach the top of a mountain range but could only take steps upward; as soon as you got to the top of one hill, you would be stuck as you could not go down in order to reach the base of a higher peak (FIGURE 7.15).

To help ensure the solution is as close to optimal as possible, heuristic algorithms typically involve a large number of tests (iterations), which are repeated multiple times with different starting solutions (runs—or in our metaphor, different routes up the mountain) to try to ensure that as much of the solution space is covered as possible. For example, the simulated annealing algorithm used in Marxan is a heuristic optimization, and users will be familiar with defining how many iterations and how many repeated runs it should undertake. To help search more solutions and avoid the hill-climber's dilemma, most heuristic algorithms also introduce some randomness or stochasticity into the solutions that are tested. In addition to simulated annealing, other useful heuristic optimization algorithms include stepwise or iterative local search heuristics, and a collection of algorithms known as genetic algorithms (search algorithms that mimic the process of natural selection). Moilanen and Ball provide an excellent introduction to heuristic algorithms.[41]

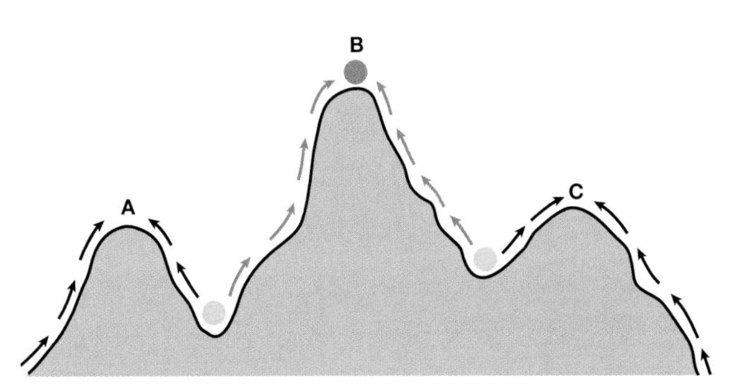


FIGURE 7.15. The hill-climber's dilemma. Imagine a climber who is trying to walk to the highest point in this set of hills (B) but is told that she can only go up and not down. If she starts at the edge of the hills, she can climb to the top of peaks A or C but will get stuck there because she can't go down the valley and then go up to point B. If she starts in the valleys between the hills, she can reach the highest point, assuming she goes in the right direction initially. This analogy illustrates a problem faced by heuristic optimization algorithms; because there is no guarantee of finding the highest (optimum) solution from every starting point, it's important to repeat the process multiple times from different starting points to ensure the best chance of finding the true optimum (the highest peak in the climber's example).

## Optimization Software

Mathematical optimization is not easy and generally requires some expertise in programming and mathematics. Obviously, popular spatial prioritization software like Marxan and Zonation provide a user-friendly way for non-programmers to use rigorous optimization algorithms. However, a wide range of commonly used software packages, including R, Matlab, SAS, and even Microsoft Excel, have automated optimization routines that can be easily adapted for planning problems. These still require formulating the problems mathematically.

One example of a relatively automated optimization program designed for conservation planning and adaptable to a wide range of problems is RobOff (**Box 7.4**). The example in Box 7.4 also nicely illustrates how optimization can be used to solve problems framed as combined spatial and strategic planning problems, as we advocated in Chapter 6.

## Box 7.4 RobOff

Developed by Atte Moilanen and his team, RobOff is a freely available software tool. It was designed to help conservation planners who are unfamiliar with programming to use optimization in solving strategic conservation planning problems. RobOff enables planners

**INPUTS**

**ENVIRONMENTS**

Vegetation type

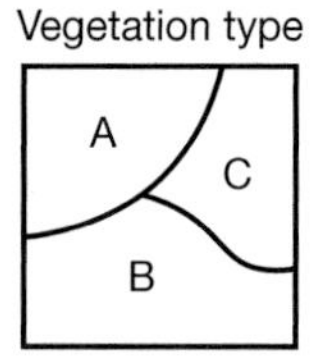


Condition

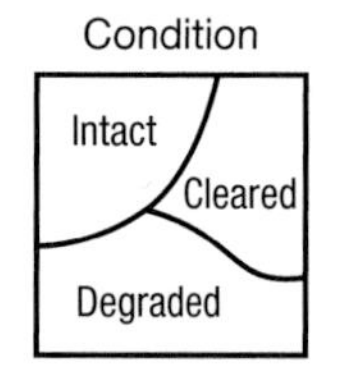


Protection

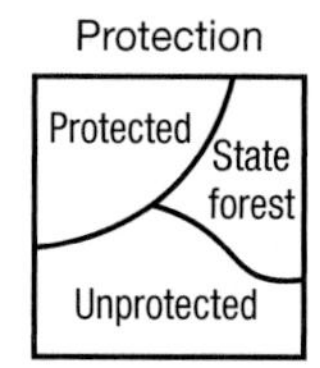


Land use } Threat

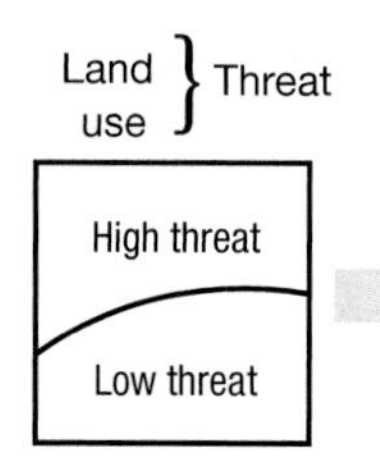


**Environmental map**

where each pixel has a unique code based on input layers

e.g., ENV 1: forest intact—on unprotected land—low threat

**FEATURES**
- Sp 1
- Sp 2
- . . .
- Sp 504

**ACTIONS**
- Do nothing
- Protection
- Active restoration plus protection

**COST**

Cost of performing a given action in a given environment (incl. land acquisition, transaction and management cost)

**RESPONSES** Define response curves of all guilds to all actions in all environments where they occur

**ENV 1**

Sp 1
Sp 2
Sp 3
. . .

Do nothing

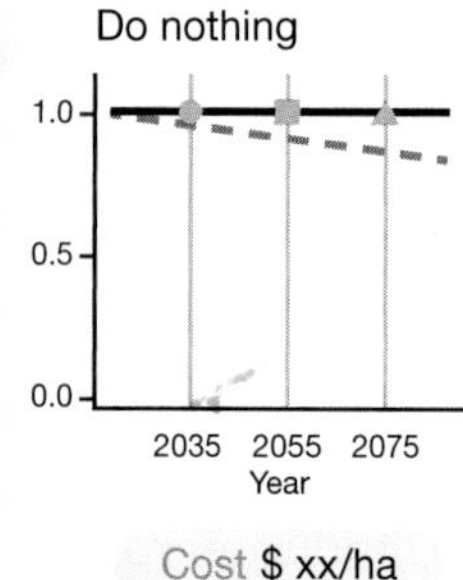


Cost $ xx/ha

Protection

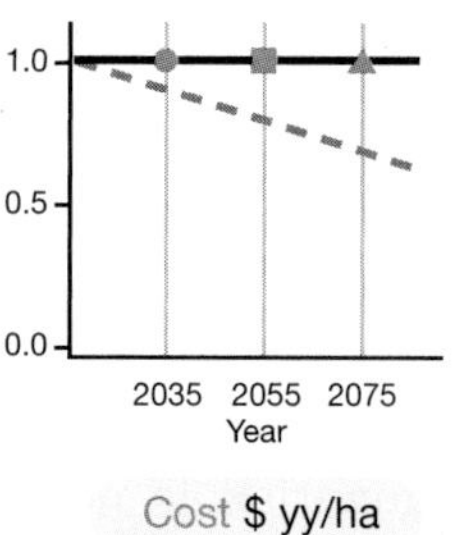


Cost $ yy/ha

Restoration

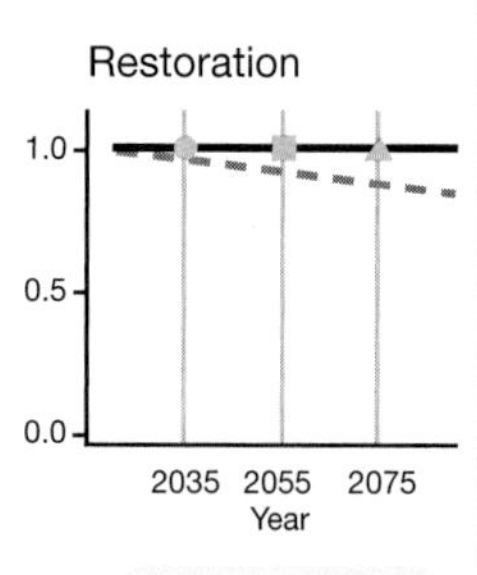


Cost $ zz/ha

**ENV 2**

Sp 4
Sp 5
Sp 6
. . .

Do nothing

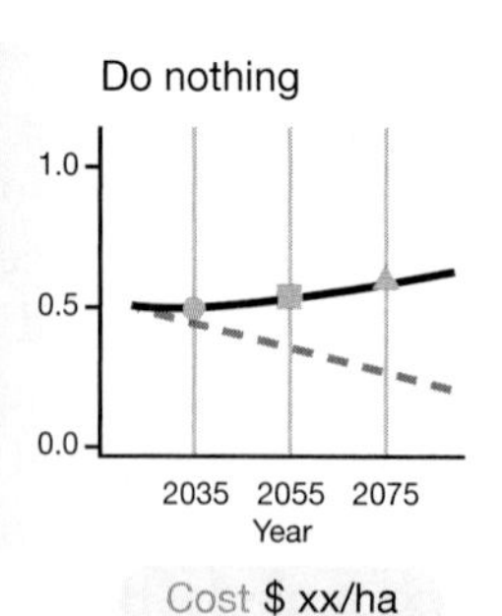


Cost $ xx/ha

Protection

1.0
0.5
0.0
2035 2055 2075
Year

Cost $ yy/ha

Restoration

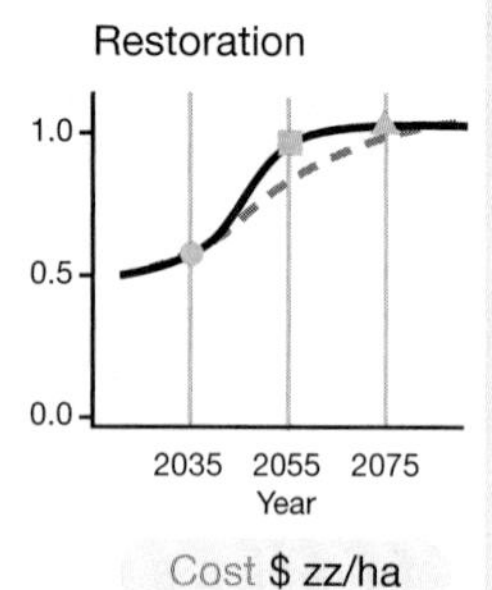


Cost $ zz/ha

**FIGURE 7.4.1.** An overview of the RobOff process. For each environment being considered, a series of available actions are defined, along with their respective costs and the expected response of species living in those habitats.

to use both exact (exhaustive search) and heuristic (genetic) algorithms to solve the general problem of determining the best actions to take in different environments to conserve a set of conservation features. The "environments" in RobOff might be different types of habitat, but they could equally be different types of land tenure—for instance, private land versus public land. The framework allows users to specify how each conservation feature responds to each of the possible actions, and which actions are possible in different environments. The solutions are a recipe for how much should be spent on each type of action in each different environment. These solutions serve as a sort of optimal investment portfolio.

As with all optimizations, RobOff users need to provide mathematical functions that describe the consequence of each action for each feature being considered. Efficient optimization also depends on providing the cost of taking each action in each different environment. Despite this challenge, RobOff makes powerful optimization analysis accessible to a wide range of practitioners.

Together with the World Wildlife Fund (WWF), the conservation planner Heini Kujala and colleagues used RobOff to guide a large conservation planning effort in Australia, aimed at working out the priority set of

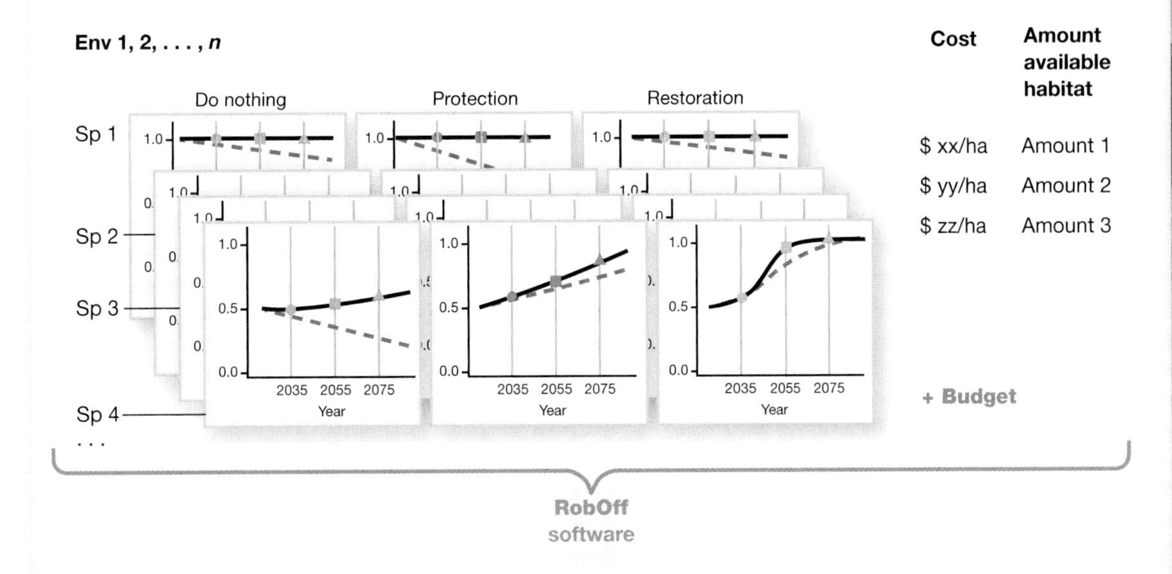


**Optimal allocation of resources** (i.e., best set of actions for each environment given the total budget)

ENV 1

| | |
|---|---|
| - Do nothing | 5% |
| - Protection | 80% |
| - Active restoration plus protection | 0% |

ENV 2

| | |
|---|---|
| - Do nothing | 93% |
| - Protection | 0% |
| - Active restoration plus protection | 7% |

FIGURE 7.4.2. How the inputs described in Figure 7.4.1 are combined in RobOff. The resulting outputs show which actions should be taken in which habitats and the level of investment in each action.

(continued)

**Box 7.4 RobOff** *(continued)*

actions that would best conserve Australia's biodiversity in a changing climate. They were interested in allocating resources across three potential actions: protect land, protect land and undertake active revegetation, and do nothing. The different environments being considered were a combination of habitat type and the condition of the vegetation in them (e.g., rainforest-intact, grassland-degraded, etc.). Their benefit functions described how different vegetation types and threatened species are likely to respond to the different actions being considered.

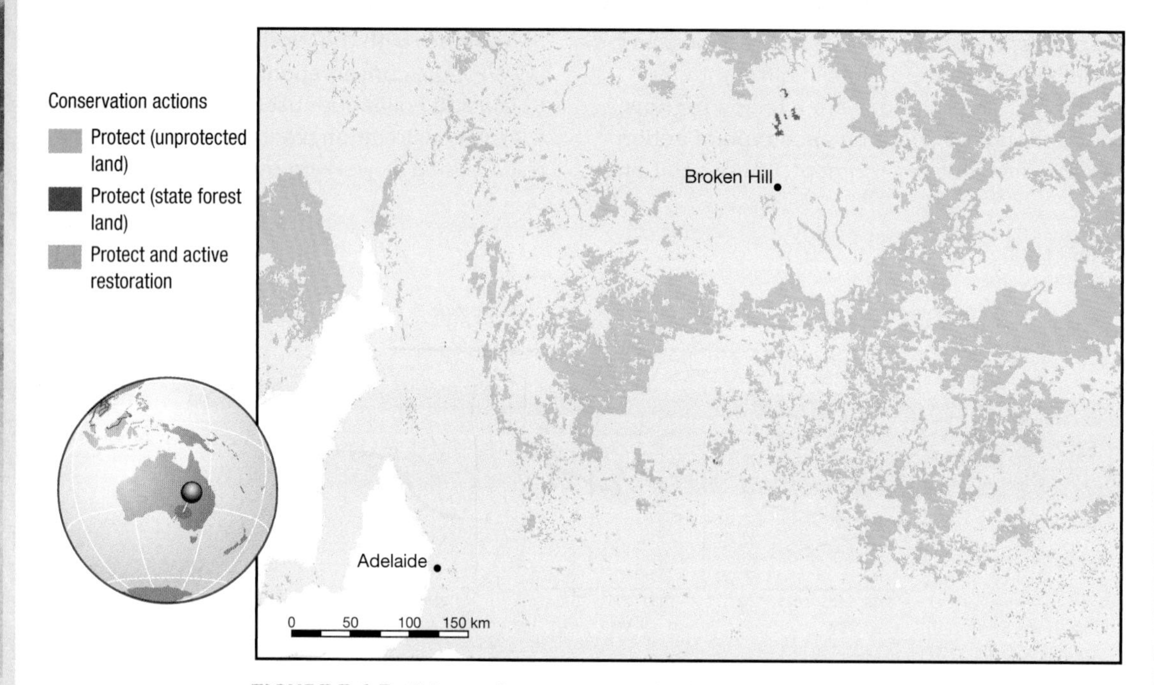


FIGURE 7.4.3. This graph represents RobOff output showing the priority areas for three different conservation actions in the state of South Australia, Australia.

## Rules of Thumb

Because optimization requires expertise that most conservation planners and natural resource managers do not have, many conservation planning researchers strive to come up with simple applied **rules of thumb** that do a reasonable job of approximating the solutions found through optimization. Rules of thumb are technically heuristic solution methods. How well a rule of thumb performs will vary from problem to problem, but a rule of thumb that has been tested against optimal solutions is probably better than one that has not.

The conservation planner Kerrie Wilson and colleagues have found in a number of cases that the optimal allocation of resources can be easily approximated by working in the places where the biggest short-term benefit is expected.[42] Similarly, together with colleagues, one of us developed a rule of thumb to approximate the optimal solution to the problem of whether to direct protection to the most vulnerable or least vulnerable coral reefs. The rule of thumb suggested that if unprotected reefs are likely to be in a satisfactory state, protection should go to the most vulnerable reefs; but if unprotected reefs are likely to be degraded, then protection should go to the least vulnerable reefs.[43]

Developing a rule of thumb for solving a conservation planning problem is a good way of having an influence beyond the plan in question. It also can help reduce the resources that need to be spent on planning for similar strategies in other places.

## Presenting and Navigating Trade-Offs

So far in this chapter, we have presented five distinct methods for solving conservation planning problems. Trade-off analysis, on the other hand, is a more general approach to evaluating alternatives when there are multiple objectives. Some form of trade-off is likely to be evaluated as part of all the solution methods described in this chapter and is, in fact, at the heart of conservation planning.

A trade-off exists when achievement of one objective comes at the expense of the achievement of another objective; that is, objectives are competing. For example, the amount of biodiversity conserved in a landscape is likely to reduce the amount of food that can be produced. If a conservation planning problem includes competing objectives, there will inevitably be trade-offs. In these cases the best solution cannot be decided by a priori construction of a multi-attribute objective function, but rather by exploring the trade-offs involved in actual solutions. For instance, in the food versus biodiversity example just mentioned, the weight stakeholders give to biodiversity objectives relative to food production objectives will be influenced by the consequence of biodiversity conservation for food production; most stakeholders would find a solution that reduced food production to below what is needed to meet local demands an unpalatable solution.

**Trade-offs** are naturally presented as consequential relationships between things we care about. For instance, when planning the allocation of a landscape to different activities, there is likely to be a trade-off between food production (the amount of food able to be produced) and biodiversity conservation (the number of species conserved). The more land we use to grow food, the fewer are the species likely to survive; and similarly, the more land we dedicate to conservation, the less food we can produce from that landscape (assuming the same level of productivity). By identifying different combinations

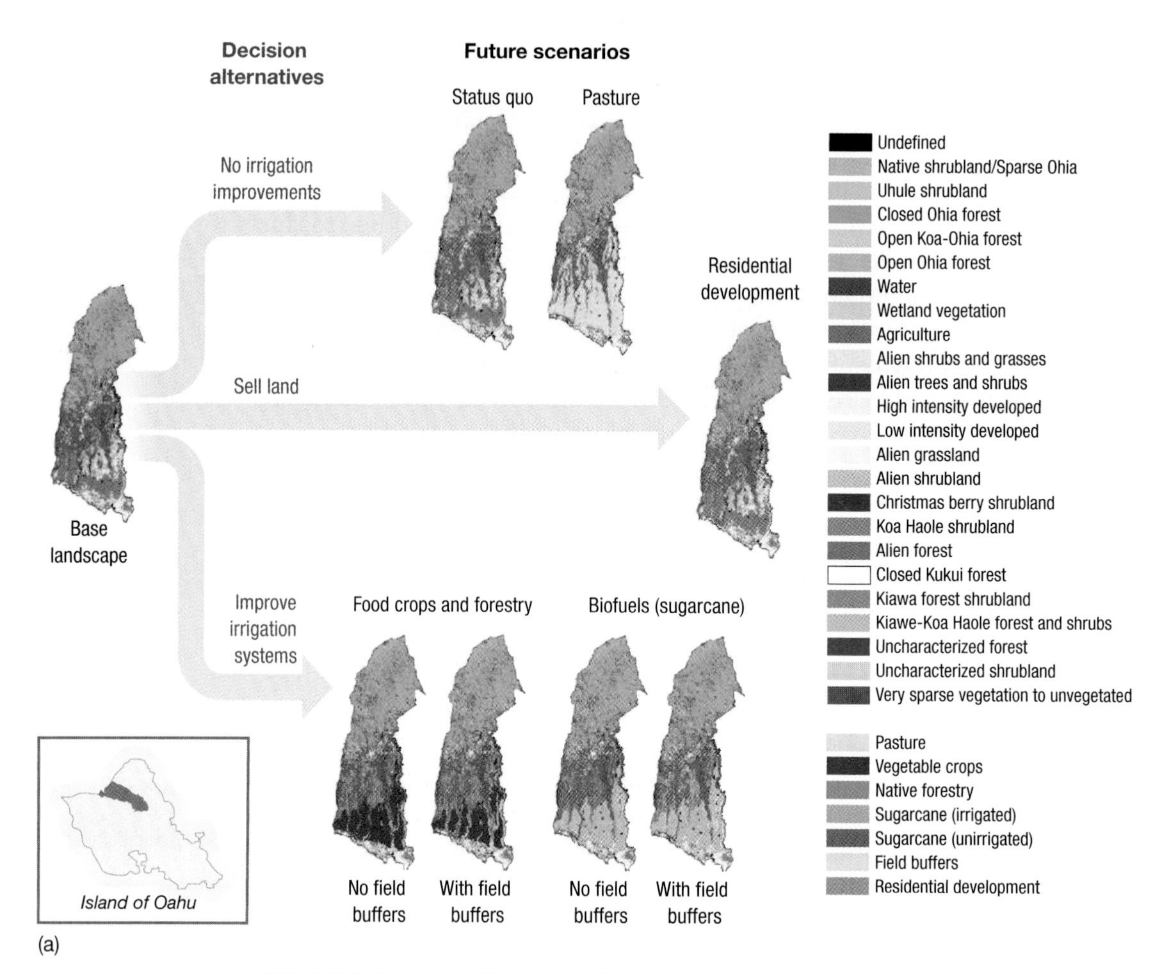


**FIGURE 7.16** Trade-offs between financial gain, carbon storage, and water quality for three different land use development alternatives on the North Shore of Oahu, Hawaii, USA. Part (a) shows what these different development scenarios look like, and part (b) shows how this would affect the three properties of interest. (Adapted from Goldstein, J. H. et al., *Integrating ecosystem-service tradeoffs into land-use decisions.* Proceedings of the National Academy of Sciences USA, 2012. **109**(19): 7565–7570.)

and extent of land use placement, and plotting the expected food production and biodiversity conserved for each, we are able to see the consequence of improving one of these objectives in terms of loss for the other (**FIGURE 7.16**). This requires characterizing the biodiversity and food production expected from different land use combinations (predicting consequences as described in Chapter 6). Emerging ecosystem service valuation and mapping tools like InVEST (http://www.naturalcapitalproject.org) are designed to help provide these predictions (see Chapter 10).

Robin Gregory and colleagues have identified four goals with respect to evaluating trade-offs in natural resource management that we think provide a nice summary and are worth repeating here.[44]

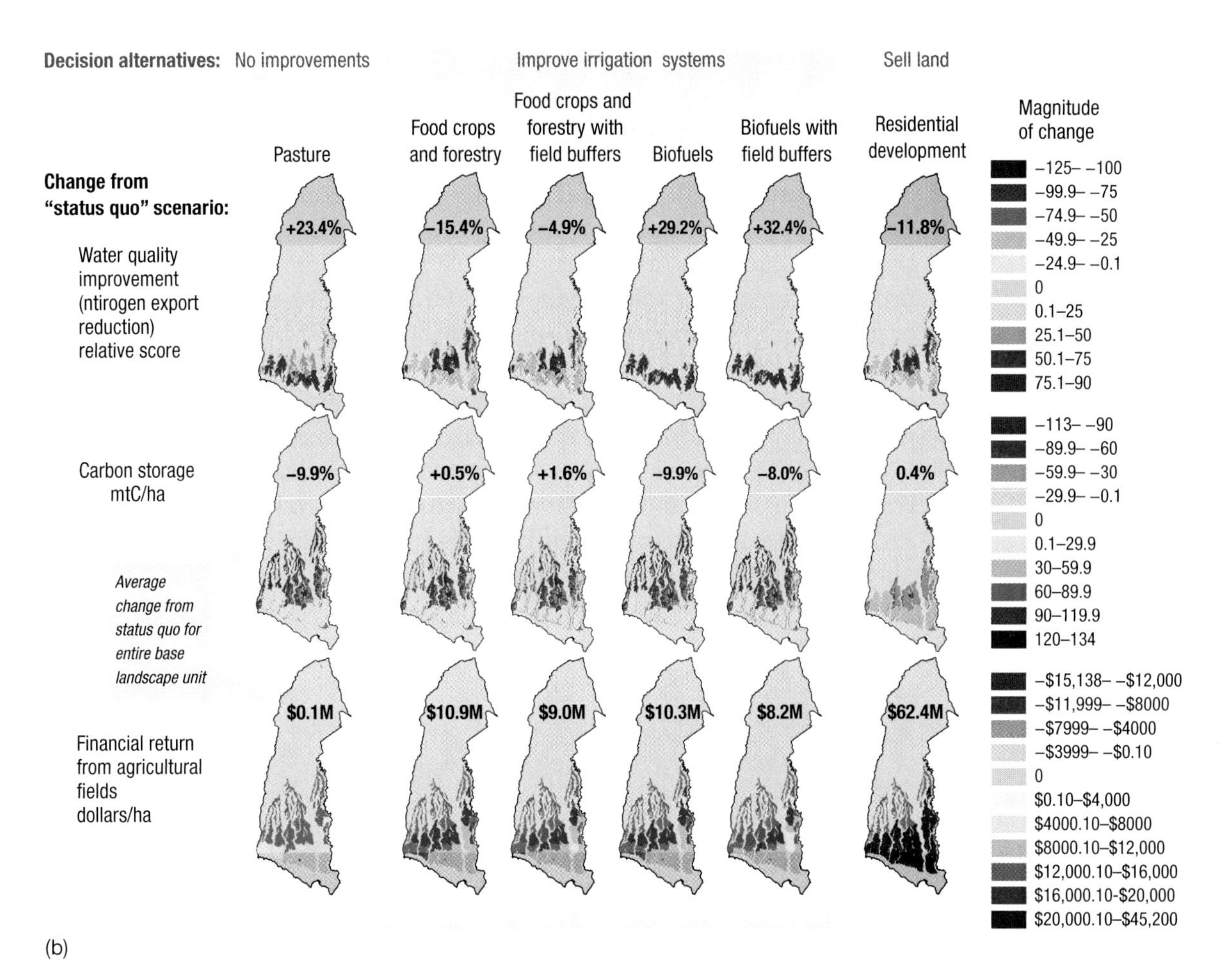


(b)

FIGURE 7.16 (continued)

1. To avoid unnecessary trade-offs by developing alternatives that can deliver on multiple objectives wherever possible
2. To expose unavoidable trade-offs and promote constructive deliberation about them
3. To make trade-offs explicitly and transparently informed by a good understanding of consequences and their significance
4. To create a basis for communicating the rationale for a decision

You have probably noticed that so far we've talked about trade-offs between only two objectives, and yet it is clear that most conservation plans have more objectives than this. Our emphasis on trade-offs between two objectives is really just a matter of pragmatism. While it is still reasonably straightforward to visualize trade-offs between three objectives, doing so between four would require looking at the trade-off plot in 3-D (which, believe it or not, we have witnessed).

Three-dimensional visualization is probably the future of graphical display, but fortunately there are many conservation planning problems that, while not two-dimensional, can be usefully summarized in terms of two dimensions (like the one in **FIGURE 7.17**).

All analytical tools ultimately involve simplification of highly complex social-ecological systems, and we believe that illustrating trade-offs can be a useful conservation planning tool, especially for problems involving two major considerations—say, food and carbon, or carbon and biodiversity. Illustrating trade-offs between a couple of dimensions can be a powerful advocacy tool, both for planning itself and for particular management practices. Simplifying the problems means they can often tell a compelling and easily understood story (e.g., one that shows improved production logging can help reduce trade-offs between biodiversity protection and production revenue).

The use of trade-off visualizations to help inform decisions is probably most well developed in economics. Economists use trade-off anal-

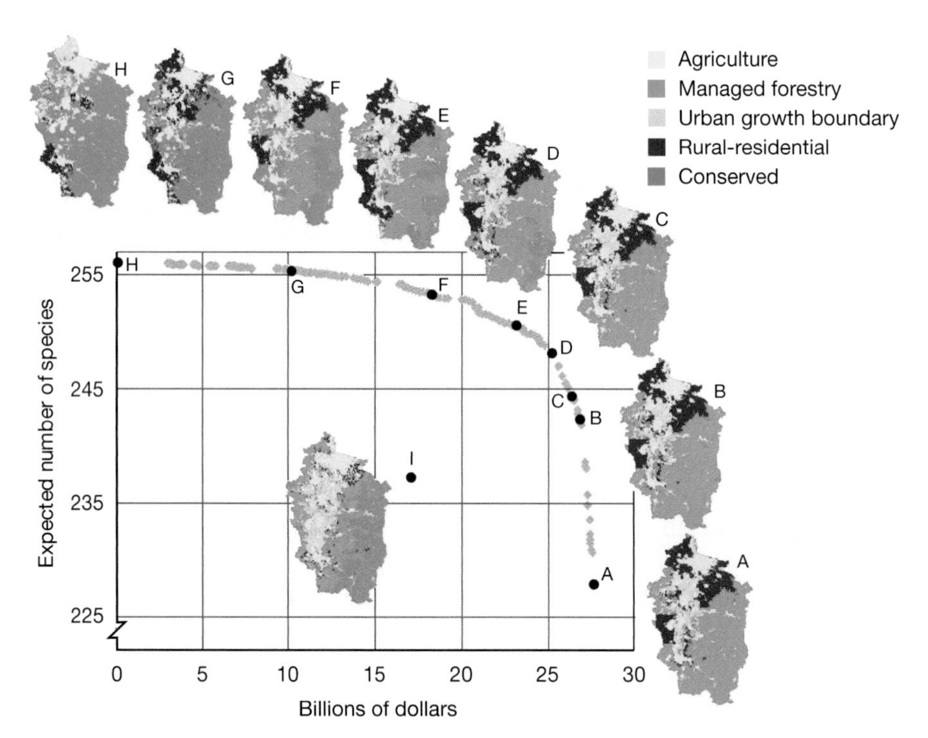


**FIGURE 7.17** Different land use patterns on the Pareto frontier for species conservation and economic gain in the Willamette Basin, Oregon, USA. Scenarios A–H are all Pareto efficient in that neither economic gain nor species conservation can be improved without a loss in the other objective. These scenarios thus reflect necessary trade-offs between economic gain and conservation gain. In contrast, scenario I shows the current land use pattern, which is highly inefficient because—as scenarios B–F illustrate—it is possible to find alternate land use arrangements that make simultaneous gains in both economic returns and conservation. (Data from Polasky, S. et al., *Where to put things? Spatial land management to sustain biodiversity and economic returns.* Biological Conservation, 2008. **141**: 1505–1524.)

ysis to check that alternatives are as efficient as possible, or in their language, **Pareto optimal**. An alternative is Pareto optimal when no improvement in one objective can be made without simultaneously diminishing the achievement of at least one other objective. Think of them as non-dominated solutions using the language we introduced earlier in this chapter. Solutions that are Pareto optimal are considered to be on the **efficiency frontier** (see Figure 7.17). Alternatives that are not Pareto optimal are inefficient because they involve unnecessarily sacrificing the achievement of one or more objectives, whereas alternatives along the efficiency frontier represent trade-offs that cannot be avoided (as in Figure 7.17).

Constructing efficiency frontiers allows planners to see the extent to which solutions (either current or proposed) are inefficient (FIGURE 7.18). This is certainly one of the most useful applications of trade-off analysis because it points to where things can be improved and makes an argument for doing so. In practice, nearly all current situations and most proposed alternatives will be inefficient to some extent because some existing decisions about actions will already have been

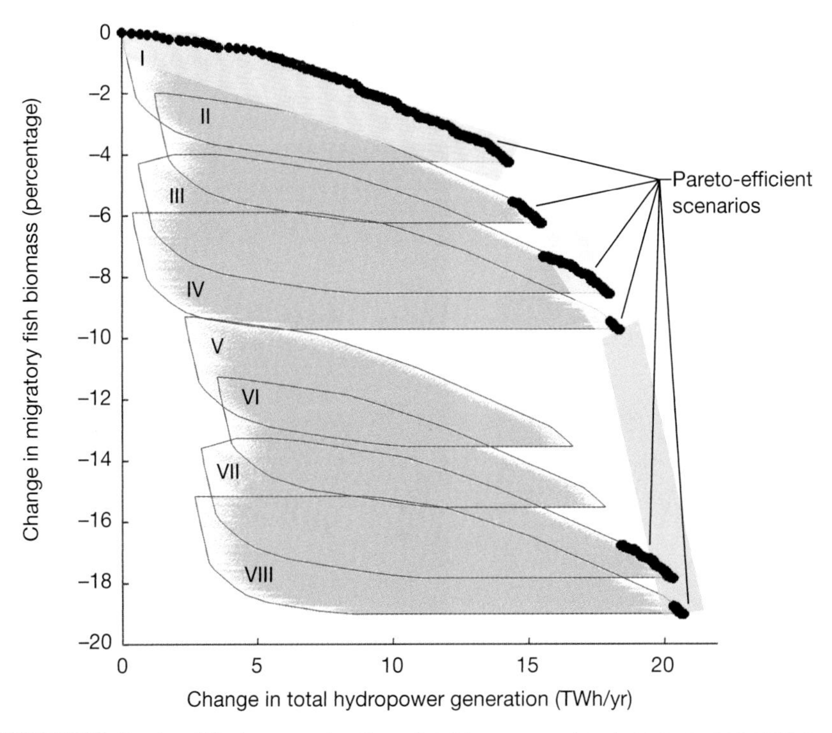


FIGURE 7.18 Trade-off between migratory fish biomass and hydropower generation for different dam configuration options on the Mekong River. Chart areas I–VIII represent dams on different sections of the Mekong. Only those options in black are Pareto efficient, because they achieve the most fish biomass possible for a given hydropower generation. (Data from Ziv, G. et al., *Trading-off fish biodiversity, food security, and hydropower in the Mekong River Basin.* Proceedings of the National Academy of Sciences USA, 2012. **109**(15): 5609–5614.)

made without considering efficiency—the slate is almost never clean. So in the case of Figure 7.17, many factors other than economic gain and biodiversity gain influence land use arrangements in the Willamette Basin, such as where people have historically settled and which lands are managed by different government agencies. Similarly, in Figure 7.18 political influences like the need to create jobs in certain areas may mean that dams on the Mekong are built away from the efficiency frontier.

Once again, efficiency frontiers are typically constructed between two objectives, and occasionally between three objectives, which would give an efficiency surface. Efficiency frontiers are easy to construct in theory but often difficult in practice. The standard approach is to take one objective and, across its entire range of potential values, find the best you can do on another objective. This process is typically accomplished through optimization. Doing this optimization can be a serious computational task in its own right.

A useful and more easily accomplished starting point is simply to plot the consequence for the two or three objectives of each of the alternatives being considered. Or, develop a hypothetical set of alternatives that favor one or the other objective to varying degrees. For example, The Nature Conservancy and the University of Tennessee collabo-

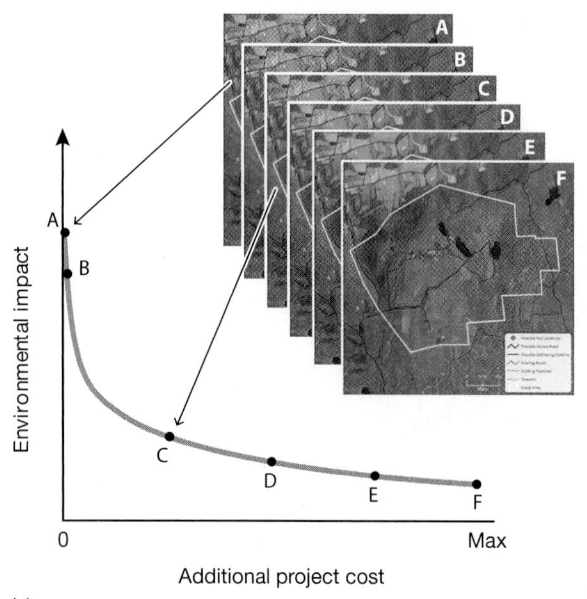

FIGURE 7.19 (a) Different gas infrastructure layout options along the Pareto frontier for project cost and environmental impact. The layouts and their relative costs and benefits were determined using EnSitu, a shale gas infrastructure siting tool for the Appalachian Mountains, developed by Austin Milt from the University of Tennessee in Knoxville and Tamara Gagnolet from The Nature Conservancy. (b) Shale gas infrastructure in Pennsylvania, USA.

rated to develop a GIS tool that helps propose a series of alternative infrastructure patterns for shale gas development and then sites these along an efficiency frontier of project cost and environmental impact (FIGURE 7.19).

Arguably a more difficult task than illustrating trade-offs is moving from illustration to deciding on the best option. In theory (and if we were truly passive, objective observers), all solutions located on the efficiency frontier are equally good. Of course, in practice this is where the influence of other objectives comes into play (either consciously or unconsciously). For example, decision makers often demonstrate reluctance to rezone inefficient land uses because of the political battle it might entail. There are, however, ways to look at trade-offs to identify points that might be desirable. Trade-off curves often exhibit points where the rate of decrease in one objective increases rapidly as the other objective increases. These are referred to as points of inflection (FIGURE 7.20). Such points are not always obvious, and there is no guarantee that these represent a desirable solution. However, they are often valuable to identify because they represent the point at which losses for each objective are minimized. With increasing distance from an inflection point, small improvements in one objective represent significant losses in another objective, and decision makers increasingly have to favor one objective over another.

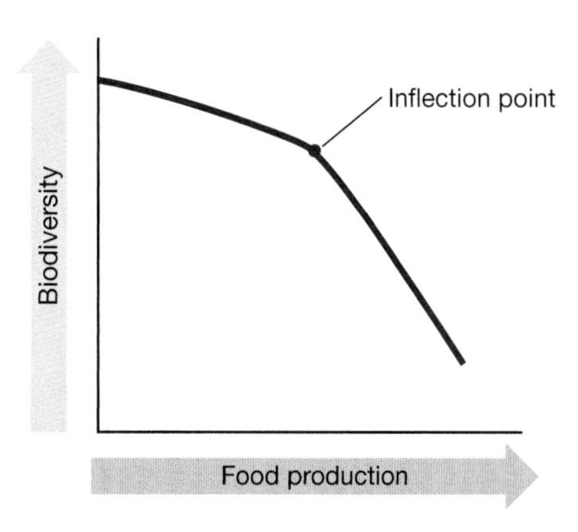


FIGURE 7.20 Inflection point in a hypothetical trade-off curve between food production and biodiversity.

## Trade-Off Caution

The conservation policy scientist Paul Hirsch elegantly warns that just because something can be framed as a trade-off, that does not mean it can indeed be traded off.[45] Hirsch and colleagues suggest that, frequently, particular stakeholders might feel that things such as cultural heritage or the right to make a living off the land cannot be traded off at all.

Another way to think about this is that stakeholders' assessment of consequence for a particular objective may have a clear threshold rather than being a continuous function. For example, cultural heritage can either be protected or not—there is no partial delivery of this objective. Wilderness advocates often hold similar attitudes. Great care must be taken to avoid trade-off analysis that appears callous because in many cases the trade-off affects people, not just some generalized commodity. Similarly, from a political and communication point of view, formal trade-off analysis can present a challenge because it involves acknowledging the possibility that an action or policy may have negative consequences for values that people care about.

## Choosing the Best Methods and Tools

The methods and tools described in this chapter are not mutually exclusive. In many cases, it makes sense to use them in concert. Consequence tables provide the foundation for MCDA and ROI. Optimization is often used to find good options within a trade-off analysis or to allocate resources as part of an ROI analysis. Spatial solutions might incorporate an ROI element, and then a small number of promising solutions might be subjected to a trade-off analysis or MCDA. Evaluating trade-offs is often an important aspect of spatial prioritization, and various trade-offs are likely to be analyzed as part of all the solution methods described here.

The ability to recognize the best methods and tools to solve the problem at hand is part of the skill that a good conservation planner brings to the job. All conservation planning problems could be approached with more than one type of method or tool; however, the way the problem is framed will obviously make some solution methods more useful than others. Looking at the way others have solved similar problems, especially if you have access to the peer-reviewed literature, is a good start. In discussions on planning context and team composition in Chapter 2 and situation analysis in Chapter 5, we identified some factors that should also influence the selection of a solution method. Foremost on the internal side is the expertise available to the planning team. Of course, not all (probably few) conservation planners will have skills and experience in the range of methods and tools identified in this chapter. Nonetheless, partnerships between agencies, conservation organizations, and academic institutions will often result in highly effective planning teams that have the skills to use a variety of methods and an understanding of realistic options and strategies that can be acted on with the various stakeholders involved.

The amount of time available for the planning process will also likely have some bearing on the choice of solution method. However, application of the solution method is rarely the most time-consuming step in a planning process. For example, there is a temptation to see something like a consequence table as a rapid option in time-pressed situations. It certainly can be a short process, especially if the consequences are largely based on expert judgment, but it can equally require hours of building or adapting models to help understand the likely consequences of different actions.

On the external side, stakeholders will influence the appropriate choice of solution method. Importantly, and contrary to many observers' expectations, we have found that the participation of less formally educated stakeholders does not necessarily mean that less sophisticated planning tools must be used. In fact, sometimes the opposite is true.

One of us was involved in a planning process in the Republic of Micronesia, where it was clear from the problem framing that a very straightforward solution method, akin to a consequence table, would be sufficient to guide the decision at hand. However, the communities involved were disappointed that more sophisticated spatial prioritization was not being done because they had heard about its use in other places and equated it with taking the planning and engagement with them seriously. Conversely, we have both been involved with plans in developed countries where highly educated stakeholders have reacted negatively to the use of decision support software, which some stakeholders consider "black box" solutions until they become more familiar with the methodology.

Communication about how a chosen method or tool is helping to integrate information and find good solutions is critical. We discuss this subject further in Chapter 11.

The significance of the decision being informed should also influence the choice of solution method. Major decisions about land, sea, or freshwater use require a commensurate investment in ensuring the optimal solution. Suppose the plan involved zoning acceptable uses across an entire province or state; this would affect many people's livelihoods and likely have a long-term conservation impact one way or another. In this situation, it will almost certainly be beneficial to use the best possible tools for such a plan and to carefully assess risk and the impact of uncertainty, which are the topics of the next chapter.

## KEY MESSAGES

- "Solving" conservation planning problems means critically evaluating the information available in order to identify the options that will best advance our objectives. It is a key step in making informed decisions.
- A range of methods and tools are available to help solve the sort of conservation planning problems framed in Chapter 6. These tools include consequence tables, Multi-Criteria Decision Analysis (MCDA), return on investment (ROI) analysis, spatial prioritization, and mathematical optimization.
- Trade-offs are likely to be evaluated as part of all the solution methods described in this chapter. There are tools that can help present these trade-offs and find good solutions despite them.
- The methods and tools used to solve conservation planning problems are not mutually exclusive. In many cases, it makes sense to use them in concert.

- Selecting the most appropriate conservation tools will depend on several factors. These include the stakeholders involved, available expertise, level of investment in the planning process, decisions that will be affected, and data available for use with a particular tool.

## References


1. Gregory, R. et al., *Structured Decision Making: A Practical Guide to Environmental Management Choices*. 2012, Oxford, UK: Wiley-Blackwell. 299.
2. Moffett, A., and S. Sarkar, *Incorporating multiple criteria into the design of conservation area networks: A minireview with recommendations*. Diversity and Distributions, 2006. **12**: 125–137.
3. Feeny, D. et al., *Multiattribute and single-attribute utility functions for the Health Utilities Index Mark 3 System*. Medical Care, 2002. **40**(2): 113–128.
4. Redpath, S. et al., *Using decision modeling with stakeholders to reduce human–wildlife conflict: A raptor–grouse case study*. Conservation Biology, 2004. **18**(2): 350–359.
5. Steele, K. et al., *Uses and misuses of multicriteria decision analysis (MCDA) in environmental decision making*. Risk Analysis, 2009. **29**(1): 26–33.
6. Margules, C. R., and S. Sarkar, *Systematic Conservation Planning*. 2007, Cambridge, UK: Cambridge University Press. 304.
7. Saaty, T. L., *How to make a decision—the analytic hierarchy process*. European Journal of Operational Research, 1990. **48**(1): 9–26; and Saaty, T. L., *Axiomatic foundation of the analytic hierarchy process*. Management Science, 1986. **32**(7): 841–855.
8. Butler, J., J. Jia, and J. Dyer, *Simulation techniques for the sensitivity analysis of multi-criteria decision models*. European Journal of Operational Research, 1997. **103**(3): 531–546.
9. Kareiva, P. et al., editors. *Natural Capital: Theory and Practice of Mapping Ecosystem Services*. 2011, Oxford, UK: Oxford University Press. 365.
10. Edejer, T. T. et al., editors. *Making Choices in Health: WHO Guide to Cost-Effectiveness Analysis*. 2005, Geneva: World Health Organization.
11. Margules, C. R., and R. L. Pressey, *Systematic conservation planning*. Nature, 2000. **405**(6783): 243–253.
12. Ball, I. R., and H. Possingham, *MARXAN (V1.8.2): Marine Reserve Design Using Spatially Explicit Annealing, 2000*.
13. Moilanen, A., and H. Kujala, *Zonation Spatial Conservation Planning Framework and Software V. 2.0, User Manual*. 2008, Helsinki: University of Helsinki.
14. Game, E. T., and H. S. Grantham, *Marxan User Manual; for Marxan Version 1.8.10*. 2008, Brisbane: University of Queensland and Pacific Marine Analysis and Research Association.

15. For example, Carwardine, J. et al., *Avoiding costly conservation mistakes: The importance of defining actions and costs in spatial priority setting.* PLoS ONE, 2008. **3**: e2586.
16. Barlow, A. C. D. et al., *Use of an action-selection framework for human-carnivore conflict in the Bangladesh Sundarbans.* Conservation Biology, 2011. **24**(5): 1338–1347.
17. Goldstein, J. H., L. Pejchar, and G. C. Daily, *Using return-on-investment to guide restoration: A case study from Hawaii.* Conservation Letters, 2008. **1**(5): 236–243.
18. Klein, C. J. et al., *Prioritizing land and sea conservation investments to protect coral reefs.* PLoS ONE, 2010. **5**(8): e12431.
19. Halpern, B. S. et al., *A global map of human impact on marine ecosystems.* Science, 2008. **319**: 948–952.
20. Balmford, A. et al., *The worldwide costs of marine protected areas.* Proceedings of the National Academy of Sciences USA, 2004. **101**(26): 9694–9697; Moore, J. et al., *Integrating costs into conservation planning across Africa.* Biological Conservation, 2004. **117**: 343–350; and Bode, M. et al., *Cost-effective global conservation spending is robust to taxonomic group.* Proceedings of the National Academy of Sciences USA, 2008. **105**: 6498–6501.
21. Naidoo, R., and T. Iwamura, *Global-scale mapping of economic benefits from agricultural lands: Implications for conservation priorities.* Biological Conservation, 2007. **140**: 40–49.
22. Margules and Pressey, *Systematic conservation planning.* (See reference 11.)
23. Moilanen, A., K. A. Wilson, and H. P. Possingham, *Spatial Conservation Prioritization: Quantitative Methods and Computational Tools.* 2009, New York: Oxford University Press. 328.
24. Kukkala, A. S., and A. Moilanen, *Core concepts of spatial prioritization in systematic conservation planning.* Biological Reviews, 2013. **88**(2): 443–464.
25. Franklin, J. F., and D. B. Lindenmayer, *Importance of matrix habitats in maintaining biological diversity.* Proceedings of the National Academy of Sciences USA, 2009. **106**(2): 349–350.
26. Wilson, K. et al., *Conserving biodiversity in production landscapes.* Ecological Applications, 2010. **20**: 1721–1732.
27. Hilty, J. A., W. Z. Lidicker Jr., and A. M. Merenlender, *Corridor Ecology: The Science and Practice of Linking Landscapes for Biodiversity Conservation.* 2006, Washington, DC: Island Press; and Theobald, D. M., *Exploring the functional connectivity of landscapes using landscape networks,* in *Connectivity Conservation,* K. R. Crooks and M. Sanjayan, editors. 2006, Cambridge, UK: Cambridge University Press. 416–443.
28. Carroll, C., B. H. McRae, and A. Brookes, *Use of linkage mapping and centrality analysis across habitat gradients to conserve connectivity of gray wolf populations in western North America.* Conservation Biology, 2012. **26**(1): 78–87.

29. Smith, R. J. et al., *Designing a transfrontier conservation landscape for the Maputaland centre of endemism using biodiversity, economic and threat data.* Biological Conservation, 2008. **141**(8): 2127–2138.

30. Beger, M. et al., *Incorporating asymmetric connectivity into spatial decision making for conservation.* Conservation Letters, 2010. **3**: 359–368.

31. Hermoso, V., M. J. Kennard, and S. Linke, *Integrating multidirectional connectivity requirements in systematic conservation planning for freshwater systems.* Diversity and Distributions, 2012. **3**(5): 448–458.

32. Tallis, H., Z. Ferdana, and E. Gray, *Linking terrestrial and marine conservation planning and threats analysis.* Conservation Biology, 2008. **22**(1): 120–130.

33. Klein, C. J. et al., *Evaluating the influence of candidate terrestrial protected areas on coral reef condition in Fiji.* Marine Policy, 2014. **44**: 360–365.

34. Ball, I. R., H. P. Possingham, and M. E. Watts, *Marxan and relatives: Software for spatial conservation prioritization,* in *Spatial Conservation Prioritization: Quantitative Methods & Computational Tools,* A. Moilanen, K. A. Wilson, and H. P. Possingham, editors. 2009, Oxford, UK: Oxford University Press. 185–210.

35. Klein, C. J. et al., *Spatial marine zoning for fisheries and conservation.* Frontiers in Ecology and the Environment, 2010. **8**(7): 349–353; Venter, O. et al., *Using systematic conservation planning to minimize REDD+ conflict with agriculture and logging in the tropics.* Conservation Letters, 2013. **6**(2): 116–124; and Agostini, V. N. et al., *Marine Zoning in Saint Kitts and Nevis: A Path towards Sustainable Management of Marine Resources.* 2010, Washington, DC: The Nature Conservancy.

36. Johnson, C. J., *Identifying ecological thresholds for regulating human activity: Effective conservation or wishful thinking?* Biological Conservation, 2013. **168**(0): 57–65.

37. Moilanen, A. et al., *Spatial Conservation Planning Framework and Software: ZONATION User Manual 3.1.* Metapopulation Research Group, University of Helsinki, FI. [accessed November 11, 2013]. cbig.it.helsinki.fi/software/zonation/].

38. Moilanen, A., and A. Arponen, *Administrative regions in conservation: Balancing local priorities with regional to global preferences in spatial planning.* Biological Conservation, 2011. **144**(5): 1719–1725.

39. Moilanen, A., J. R. Leathwick, and J. M. Quinn, *Spatial prioritization of conservation management.* Conservation Letters, 2011. **4**(5): 383–393.

40. Moilanen, A., and I. Ball, *Heuristic and approximate optimization methods for spatial conservation prioritization.* Spatial Conservation Prioritization, 2009: 58–69.

41. Moilanen and Ball, *Heuristic and approximate optimization methods.* (See reference 41.)

42. Wilson, K. A. et al., *Prioritizing global conservation efforts.* Nature, 2006. **440**(7082): 337–340; and Wilson, K. A. et al., *Conserving biodiversity efficiently: What to do, where, and when.* PloS Biology, 2007. **5**(9): 1850–1861.

43. Game, E. T. et al., *Should we protect the strong or the weak? Risk, resilience and the selection of marine protected areas.* Conservation Biology, 2008. **22**: 1619–1629.
44. Gregory et al., *Structured Decision Making.* (See reference 1.)
45. Hirsch, P. D. et al., *Acknowledging conservation trade-offs and embracing complexity.* Conservation Biology, 2011. **25**(2): 259–264.

# 8

# Uncertainty and Risk

## Overview

The world is uncertain. Conservation planning is uncertain. There is likely to be uncertainty both in our current knowledge and in our knowledge of how things will change. Where this uncertainty may result in negative outcomes, there is also risk. In this chapter we describe different types of uncertainty and the importance of recognizing and acknowledging uncertainty in the planning process. We describe three general ways to respond to uncertainty in a conservation plan: minimizing it, compensating for it, or finding solutions that are as robust as possible to the uncertainty that is present. We introduce scenario analysis as a tool for exploring the impact of uncertain futures. We conclude with a section on risk assessment within conservation planning, with guidance on identifying, assessing, and prioritizing the management of risks.

## Topics

- *Uncertainty*
- *Epistemic uncertainty*
- *Linguistic uncertainty*
- *Probability*
- *Likelihood of success*
- *Interval judgments*
- *Transparency*
- *Value of information*
- *Minimizing uncertainty*
- *Robustness*
- *Minimax*
- *Sensitivity analysis*
- *Info-Gap*
- *Monte Carlo simulation*
- *Scenarios*
- *Scenario analysis*
- *Risk assessment*
- *Risk ranking*
- *Pre-mortem*

## Uncertainty

Uncertainty is a pervasive part of conservation planning. There is obvious uncertainty in the data inputs to conservation plans: uncertainty about whether habitats or species are where we think they are, how much of them there is, the condition they are in. There is uncertainty in the cost of doing conservation work, and uncertainty in the willingness of individuals, communities, or organizations to participate. There is uncertainty in how these things change over time and about what will happen in the future. For instance, will the habitat or species still be there? This forward-looking uncertainty gets a lot of attention because climate change is a major source of it, but it also includes uncertainty about future policies and funding. There is uncertainty in how social-ecological systems function,[1] which means uncertainty in our theories of change and the models we build to describe the behavior of the systems we work in. There is even likely to be uncertainty in the way conservation planning problems are framed and the choices we make in doing so: uncertainty in the meaning of our objectives, uncertainty in the features and targets we choose to represent these objectives, or even the way we've characterized the planning context. Each of these uncertainties propagates through a conservation plan and leads to uncertainty in their outputs.

We don't believe the pervasiveness of uncertainty makes conservation planning any less important or relevant; in fact, we think the sense-making nature of conservation planning is even more critical to good decisions in an uncertain world. It does, however, make it imperative that conservation planners be able to recognize uncertainty, be explicit about the presence of uncertainty when working with stakeholders, manage it appropriately, and understand what it means for the decisions they're informing. Uncertainty in our data and knowledge does not necessarily equate to uncertain decisions—highly uncertain inputs can still lead to very robust decisions if the uncertainty would not alter the best course of action. Additionally, even a highly uncertain decision can still be an informed one.

For each of our chapters, it would be possible to write an accompanying chapter discussing the uncertainty in just that chapter. We haven't done that. Instead, this chapter introduces different types and sources of uncertainty and some of the ways to respond to uncertainty in a conservation plan. We focus particularly on how to respond to uncertainty when solving conservation planning problems; that is, the uncertainty that most directly affects our choice of alternative strategies. We illustrate these responses with some common manifestations of uncertainty in conservation planning.

## Kinds of Uncertainty

There are multiple ways to group and classify the uncertainties in conservation plans. In exploring the consequences of uncertainty on different conservation planning approaches, the conservation planner William Langford and colleagues[2] identify four useful classes of uncertainty: (1) uncertainty in the inputs (e.g., how reliable are the data being used?), (2) uncertainty in the planning process (e.g., do the chosen objectives reflect the true objectives of stakeholders?), (3) uncertainty about future changes in the ecological and social systems in question, and (4) uncertainty in the ability of the chosen method to find the best solution (e.g., how are scores for different criteria being combined?), as described in Chapter 6.

What these four classes do not characterize is how the uncertainty actually arises—what the source of the uncertainty is. The ecologist Helen Regan and colleagues have developed a well-established, general taxonomy of uncertainty.[3] They divide the sources of uncertainty into two types: **epistemic uncertainty** and **linguistic uncertainty**. In their words,

> Epistemic uncertainty is uncertainty associated with knowledge of the state of a system and it includes uncertainty due to limitations of measurement devices, insufficient data, extrapolations and interpolations, and variability over time or space. Linguistic uncertainty, on the other hand, arises because much of our natural language, including a great deal of our scientific vocabulary, is under-specific, ambiguous, vague, context dependent, or exhibits theoretical indeterminacies.

In **TABLE 8.1** we list the most relevant sources of uncertainty that Regan and colleagues identify under their two broad types, along with examples of where they can occur in conservation planning.

## Acknowledging Uncertainty

From our perspective, the most important and also probably the easiest task relating to uncertainty in conservation planning is simply acknowledging its presence. The very act of recognizing uncertainty will start you and others thinking about how it might be affecting the outcomes of a plan. Admitting its existence is also vital to the credibility of a plan. Plans that ignore or obfuscate uncertainty risk undermining not only the decisions being supported by that plan but also the process of conservation planning and decision making more generally. One of us has argued that not acknowledging uncertainty and the concomitant risk of not achieving the objectives is one of the most common mistakes made in conservation planning.[4] Ways to acknowledge and communicate uncertainty when presenting the outputs of

**TABLE 8.1** Types of uncertainty common in conservation planning

| Uncertainty | |
|---|---|
| ***Epistemic uncertainty*** | ***Example*** |
| Measurement error | Uncertainty associated with the accuracy of remotely sensed land cover data, or estimates in population size |
| Natural variation | Uncertainty in population trends because of natural inter-annual variation in population (e.g., seabird populations can fluctuate substantially between years based on weather conditions) |
| Randomness and stochasticity | Impact of unpredictable political events (e.g., a rapid change of government) on conservation funding |
| Model uncertainty | Uncertainty in our theories of change for ecological or social systems |
| Subjective judgment | Expert-based assessments of the condition or threat to conservation features |
| ***Linguistic uncertainty*** | ***Example*** |
| Vagueness | Uncertainty in the precise meaning of many words used in the objectives of plans (e.g., resilient, sustainable, sufficient, healthy, persistent, natural, viable, or threatened) |
| Context dependence | Uncertainty as to how threatened a species or habitat is if it is highly threatened in one place but not in another |
| Underspecificity | Uncertainty because statements are too general (e.g., "sea level is predicted to rise by 10 cm" is underspecific because its impact depends on the time period of this rise and also whether the coastline is rising or subsiding) |

a conservation plan (such as using saturation and texture on maps instead of hard lines and colors) are topics we visit in greater detail in Chapter 11 on communication.

Following are four ways to acknowledge uncertainty in a conservation plan. They are not mutually exclusive, nor are they all relevant to every plan.

***Use probabilities.*** The natural unit of uncertainty is **probability**. Many of the epistemic uncertainties identified in Table 8.1 can be represented as probabilities. Probabilities can be powerful and flexible tools, provided a few basic rules are adhered to. If you switched off when probability theory was being taught, or perhaps were never taught it, we encourage you to take a primer in the basics of probability.

***Use intervals.*** A second useful way to acknowledge uncertainty is to use intervals or ranges instead of single values. In conservation,

FIGURE 8.1 The endangered Hirola antelope (*Beatragus hunteri*), restricted to a small coastal region of Kenya.

intervals are most commonly associated with estimates of populations. For instance, the population of the Hirola antelope (*Beatragus hunteri*, an endangered antelope species for the coastal region of Kenya; FIGURE 8.1) is estimated at between 1100 and 1900 individuals. As discussed in Chapter 5, intervals are a particularly useful way to acknowledge uncertainty when eliciting judgments from experts. Intervals will generally have a likelihood attached to them—say, 90%—but unless this likelihood has been estimated explicitly, it is not always known.

***Estimate the likelihood of success.*** Not all conservation strategies are equally likely to be successful. Uncertainty around the success of a strategy should modify our assessment of the expected outcome. For example, in calculating the expected return on investment for threatened-species recovery projects in New Zealand, the conservation decision scientist Liana Joseph and colleagues asked experts to estimate the probability that projects would be successful at recovering the species.[5] This estimate includes both the project's probability of being implemented effectively and, if implemented, its probability of being successful. A theory of change or results chain is likely to contain many points of uncertainty because each linkage in the chain contains assumptions. However, in practice it is far more expedient and pragmatic to give a single estimate for the **likelihood of success**. In most instances these estimates will be a subjective probability elicited from experts. Experts often find it easier to make a subjective estimate of success likelihood on a linguistic scale (such as in FIGURE 8.2), and then convert it to a set of probabilities.

***Be transparent about risks.*** All conservation strategies have a chance of failure. Uncertainties that might lead to failure are risks. In the final section of this chapter, we detail how risks can be identified and prioritized. We also discuss how doing so can improve conservation outcomes and be an important communication tool.

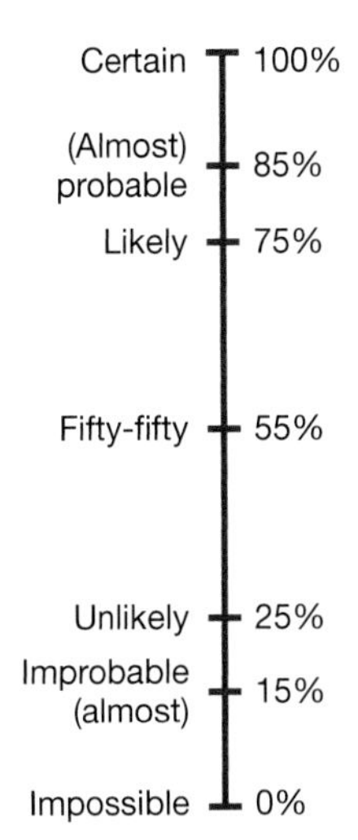


FIGURE 8.2 Experts often find it easier to make a subjective estimate of success likelihood on a linguistic scale and later convert it to a set of probabilities.

## Responding to Uncertainty

Now that we've acknowledged how much uncertainty a conservation plan can involve, what can we do to improve our decisions in the face of this uncertainty? As noted earlier, this section focuses particularly on responding to uncertainty during the stage of solving conservation planning problems and selecting the places and strategies to work on. We've grouped potential responses to uncertainty into three general types: (1) minimize uncertainty, (2) compensate for the uncertainty, and (3) live with the uncertainty (by finding a solution that is most robust to it).

We describe these three types next, along with some of the tools for applying them in a conservation plan. Then we'll dive deeper to examine the broad and frequent application of scenario analysis and risk assessment methods across the field of conservation planning. The tools we chose to include in this section are those we see as most usefully and practically applied in a conservation plan, given common constraints on time and expertise.

### Minimize Uncertainty

Some (but not all) uncertainty can be reduced. Natural variation (the fluctuations inherent in physical and biological processes; see Table 8.1) cannot generally be reduced, whereas uncertainty arising from the lack of knowledge can be. The trouble is, doing so nearly always has a cost. Thus, it is usually necessary to prioritize which uncertainty is most worth investing resources to minimize.

A deliberate attempt should always be made to reduce linguistic uncertainty, and Jan Carey and Mark Burgman have nicely illustrated some approaches.[6] In this section, our attention is on ways to reduce epistemic uncertainty. As discussed earlier in the chapter, epistemic uncertainty will lie mainly in the data and models used as inputs to a conservation plan, and in our understanding of system function, perhaps articulated through a theory of change.

There are some obvious ways to reduce uncertainty, for instance, by collecting more data, asking more people, or obtaining more recent or accurate data by using techniques such as ground-truthing. The question we need to ask ourselves concerning collecting more or better data should be: is it worth it? The answer depends not only on the level of uncertainty and the cost of reducing it but also on how much the uncertainty might affect the decisions being made. In other words, how much better would our decision making be if there were no uncertainty? This is the crux of a concept known as the *expected value of perfect information,* which we elaborate in **Box 8.1**.

Near the end of this chapter we go into detail about risk assessment in conservation planning. In many ways, risk assessment as we

## **Box 8.1** Value of Information

Not all uncertainty matters. The significance of uncertainty in a conservation plan is determined not so much by the level of uncertainty but rather by what a conservation program would do differently if that uncertainty didn't exist. This is the cornerstone of a concept known as the *expected value of perfect information* (EVPI). In its most straightforward form, the EVPI measures the improvement in expected outcome if all uncertainty was resolved before making a decision. An improvement might occur if we would undertake a different conservation strategy with perfect knowledge than we would with uncertain knowledge. This potentially improved (best case) outcome can then be weighed against the resources (money and time) that would be required to remove that uncertainty.

Formal analysis of the value of information is a well-established tool in decision science and is used extensively in design and evaluation of clinical trials.[a] It has only recently started to gain traction in environmental and conservation decision making. The conservation decision scientists Joslin Moore and Mike Runge provide an elegant example of using EVPI in the evaluation of strategies to control invasive willow (*Salix cinerea*) in Australia's threatened alpine bog ecosystems.[b] Identifying the best strategy to control willow is difficult because knowledge of the population dynamics of willow in the region, and particularly its interaction with fire, is highly uncertain. After analyzing the EVPI, Moore and Runge were able to conclude that the outcomes of invasive species treatment activities were limited by the overall budget and that resolving uncertainty in willow population dynamics made little difference to the expected performance unless the budget was greatly increased. This is valuable information for conservation managers who are often faced with scientists emphasizing the importance of resolving uncertainty in our knowledge.

One challenge of the EVPI concept is that the quantitative information and analytic expertise required for its calculation has restricted its application in conservation. However, we have found it useful to conduct even a rapid, subjective and qualitative assessment of how new knowledge might alter a conservation decision. This can often be accomplished as simply as asking a manager, decision maker, or scientist, "What piece of information would make you change your mind about the best strategy to take?" This can be an effective way to engage in a discussion about the relative importance of uncertainty, and it rapidly highlights places where research effort might need to be invested.

---

[a] Yokota, F., and K. M. Thompson, *Value of information literature analysis: A review of applications in health risk management.* Medical Decision Making, 2004. **24**(3): 287–298.

[b] Moore, J. L., and M. C. Runge, *Combining structured decision making and value-of-information analyses to identify robust management strategies.* Conservation Biology, 2012. **26**(5): 810–820.

describe it is also a tool to help prioritize which uncertainties are most important to minimize or at least understand.

Ironically, it is also possible to minimize uncertainty by having fewer data—or more correctly, by using fewer data. There will generally be different levels of confidence in the data for different conservation features. For instance, we might know the distribution of coral reef habitat quite accurately for an archipelago but have only uncertain knowledge of the true distribution of sea grass within the same archipelago.

In a plan we were recently involved in, there was substantial uncertainty in the accuracy of our knowledge of lakeshore beaches around western Lake Erie, USA, due to the age of the data and the combination of contributing sources. Because these data were strongly influencing the outputs, in this case restoration priorities, it was decided not to use those data and to exclude beaches as a conservation feature. The result was a more certain but less comprehensive plan. It will not always be possible to simply discard features with poor data, but conservation planning should always involve some review of whether there is enough confidence in a data layer to use it.

Reducing uncertainty by gathering more data does not mean that planning, decision making, or taking action need to be put on hold. This is the essence of adaptive management—learning while doing. In many cases, we can more quickly resolve uncertainty—say, about the impact of a change in the fire regime of an ecosystem of interest—by undertaking conservation actions. We deal with adaptive management in detail in Chapter 11, so it is sufficient here to recognize that experimentation and adaptive management can be important tools in minimizing uncertainty.

Closely related to adaptive management is the idea of being dynamic in our planning and decision making. Much uncertainty arises simply because we project the outcomes of conservation actions into the future—a highly uncertain future. The outcomes of a plan intended to span the next 10 years will likely be more uncertain than those of a plan dealing with conservation activities over the next 2 years. Therefore, another way to reduce uncertainty in a plan is to shorten the time period it covers. This may necessitate revisiting the plan more frequently and budgeting resources accordingly.

### Compensate

Imagine a conservation plan designed to inform about priority places to conserve a state or district's biodiversity by protecting examples of all the major habitat types in that state. The biodiversity conservation outcome of protecting a set of sites proposed in the conservation plan might be doubtful because of uncertainty in our mapping of current habitat extent, or whether that habitat will exist in the same place in the future with changes in climate. The outcome might also be obscure because of uncertainty in whether a location will remain available for conservation (i.e., not converted to an incompatible land use), or whether conservation in each place will be effective at conserving the habitat there. Where there is uncertainty in any of these variables, there is a chance that the total amount of each habitat that ultimately ends up being protected will actually be less than what we believe the chosen solution is delivering.

The most straightforward way to address this kind of uncertainty is to compensate for it by factoring in additional conservation action to offset the losses that are expected given uncertainty in our current knowledge and future condition. In the example from the previous paragraph, this would mean protecting a greater proportion of each habitat type to offset the difference between the target protection and the expected outcome given uncertainty. Gary Allison and colleagues describe this as an "insurance factor"[7] and demonstrate how the technique can be used to determine, for example, the additional amount of marine habitat that needs to be protected to compensate for the expected impact of oil spills and hurricanes, both highly uncertain events that can affect even protected habitats.

Vivitskaia Tulloch and colleagues have demonstrated how to implement a similar approach with the conservation planning software in Marxan, in their case to deal with uncertainty in the accuracy of habitat data for coral reefs in Fiji.[8] A neat feature of their method—other than that it can be done in freely available software—is allowing planners to set a level of confidence that they want to compensate to; for example, "We want to be 90% sure that we will meet our target of 17% protection for habitat Y."

Compensation is a very precautionary approach to uncertainty, and while it might be prudent in many cases, it can also face the possible criticism of inefficient resource use. Uncertainty is rarely uniformly distributed, so it may be possible to undertake less conservation but in places and ways with greater certainty around the outcome. This is the key to the uncertainty response we describe next—finding robust solutions.

### Live with It—Finding Robust Solutions

Conservation planning often happens under tight deadlines, limited budgets, and the constraints of a preestablished planning context (see Chapter 2). Combined, these things mean that it is frequently not possible to reduce uncertainty by gathering more data or to compensate for it by increasing the amount of conservation that happens. In such cases, the response to uncertainty should be to seek **robust solutions** that are as sound as possible. We use robust in this case to mean solutions that would perform well despite potential changes in the planning problem due to uncertainty.

It is easy to imagine how an alternative that does well under one set of inputs might be expected to do poorly if these inputs changed. For example, for a planned gold mine that would threaten a globally significant salmon run, the best strategy under current conditions might be for the government to compensate the companies involved for all or part of lost revenue if the project did not go forward. However, if the price of gold on the world market were to increase even 1%, this strategy might no longer be feasible, requiring an alternative strategy of establishing mitigation efforts or placing more stringent environmental regulations in place for mine operations. Because there is substantial uncertainty about the price of gold even in the near future, the government's decision to compensate the company versus an alternative one of requiring mitigation efforts that are riskier for the salmon but more robust to the uncertainty in gold prices is a critically important trade-off to consider. This example illustrates what can make robust solutions difficult to determine; there is often a potential trade-off between expected performance and robustness. As such, what makes a robust solution is not simply an analytical question, but is also determined by the risk tolerance of planners and decision makers.

A wide range of methods and tools have been developed to help identify robust solutions to planning problems, particularly in engineering, and many of these can be useful for conservation planning. Techniques like sensitivity analysis or scenario analysis will be familiar to most readers, whereas other methods like Monte Carlo modeling or Information-Gap analysis might be less so. All the methods for finding robust solutions take the same general approach: testing a range of uncertain values and seeing how this affects the expected performance of the alternatives being considered. Some approaches prescribe bounds on a set of possible conditions and look for the most robust solution within these limits. Others anchor on a "most likely" best guess for the uncertain items, and then gradually expand the search space around the best guess, looking for the worst case at each level. The major difference between methods is in how the uncertainty is represented, which in turn depends a lot on how much we know about the uncertainty.

As funny as the concept sounds, we can sometimes estimate how much uncertainty there is. Take, for example, modeled species distributions (discussed in Chapter 5). These models do not tell us that a species lives in one place and not another, but instead they indicate how likely it is that a species lives in a place (FIGURE 8.3). We may assume that the level of uncertainty we have in the species being in

**FIGURE 8.3** This species distribution model indicates presence likelihoods for the desert tortoise (*Gopherus agassizii*) in the Mojave and parts of the Sonoran Deserts in the U.S. Southwest. (Adapted from Nussear, K. et al., *Modeling habitat of the desert tortoise [Gopherus agassizii] in the Mojave and parts of the Sonoran deserts of California, Nevada, Utah, and Arizona.* U.S. Geological Survey open-file report, 2009. **1102**: 18.)

any given place is directly related to the likelihood of occurrence as predicted by the model; the less likely, the more uncertain (this is not considering uncertainty in the original species observation data and climate data used to build the species distribution model, sometimes referred to as second-order uncertainty).

Quite a lot of work has been done on responding to this type of uncertainty as part of spatial or systematic conservation plans. A good example comes from the Pantanal wetlands of Brazil, where conservation planner Reinaldo Lourival and colleagues looked at uncertainty in the presence of different successional vegetation types due to uncertainty in the flooding pattern.[9] To establish likelihoods of each vegetation type being in each planning unit, they simulated floods and vegetation changes for 50 years into the future. They then used these likelihoods as probabilities within the software Marxan, which was able to find solutions that were likely to meet conservation targets despite uncertainty in the future distribution of vegetation types. This same approach is widely applicable to many spatial planning problems with uncertainty.

In cases where we don't really know how much uncertainty there is, finding robust solutions depends on testing a range of possible values. This testing can be done in a qualitative way, for instance, by discussing how changing our assumptions about how a strategy or action might work could alter the expected direction of an indicator. More typically, however, this response to uncertainty is associated with quantitative methods.

In TABLE 8.2 we describe three approaches to finding robust solutions (sensitivity analysis, Monte Carlo simulation, and Info-Gap analysis) that have been used in the context of conservation planning. Although scenario analysis also belongs in this list, we see it as such a broadly useful approach for conservation planners that we address it in detail in the next section of this chapter. With the possible exception of sensitivity analysis, these are highly quantitative approaches requiring programming skills that we certainly don't expect all conservation planners to possess. They are, however, valuable approaches to be aware of, especially for conservation plans that use formal models.

***Minimax and Other Simple Rules*** You may have noticed that while the approaches just described explore the robustness of alternatives to uncertainty, they do not actually identify which alternative is the best choice. As mentioned at the beginning of the section, the best choice will depend on what performance is being sacrificed in pursuit of robustness and what the risk tolerance of planners and decision makers is. However, some simple rules (or heuristics) can help guide a choice once the performance across a range of uncertainties has been tested.

A rule we have found to be consistently useful in determining the most robust solution is known as ***Minimax***, or in other words, minimizing the maximum loss in the worst-case scenario. The crux of the Minimax strategy is choosing the alternative with the least worst outcome across the uncertainties that have been explored (FIGURE 8.4). A nice thing about rules like Minimax is that they apply equally to qualitative scenario analysis (see next section) and to quantitative

**TABLE 8.2** Approaches to testing the robustness of conservation planning solutions to uncertainty

| Approach | Description and example |
|---|---|
| Sensitivity analysis | More a general class of procedure than a specific approach. Sensitivity analysis involves changing an uncertain parameter according to some rule and looking at the impact this has on the solution. This change might simply be a slight variation of a parameter in either direction, or it could be an entirely different value based on a different set of assumptions. For example, any conservation plan that uses the expected cost of an alternative in solving a conservation problem will have made a series of assumptions relating to the chosen cost. It is good practice in such cases to change the cost based on a different but still plausible assumption and see how this affects the performance of different solutions. In the case of costs, it would be a valuable sensitivity analysis to determine the best solutions if cost were ignored; this gives an indication of how important cost is in determining the best solution and thus how important uncertainty in this value might be. |
| Monte Carlo simulation | Monte Carlo simulation involves randomly assigning uncertain parameters a new value according to a preestablished probability distribution, and doing this repeatedly (say, hundreds of times), to build a picture of the impact of uncertainty in that parameter. Unlike sensitivity analysis, Monte Carlo simulations allow testing over the full range of an uncertain value. This approach is commonly used to simulate the effect of uncertainty in the input to models, especially GIS models, hence its frequent relevance to conservation planning. The geographers John Gallo and Michael Goodchild used Monte Carlo simulations to explore the impact of landowners' willingness to sell on conservation priorities in California.[a] |
| Info-Gap analysis | Info-Gap (or information gap) is an engineering technique used in cases of severe uncertainty, where very little is known about a parameter in question.[b] The basic Info-Gap asks which alternative can tolerate the greatest uncertainty and still meet our minimum performance requirement. Ben Halpern and colleagues have used Info-Gap to help work out the most robust spacing for marine protected areas given severe uncertainty about the distance of larval dispersal.[c] Info-Gap is the most quantitatively challenging of the approaches described here, but the conservation planning software Zonation includes a function allowing users to employ Info-Gap methods to explore robust solutions.[d] |

[a] Gallo, J., and M. Goodchild, *Mapping uncertainty in conservation assessment as a means toward improved conservation planning and implementation.* Society & Natural Resources, 2012. **25**(1): 22–36.

[b] Ben-Haim, Y., *Information-Gap Theory: Decisions under Severe Uncertainty.* 2001, London: Academic Press.

[c] Halpern, B. S. et al., *Accounting for uncertainty in marine reserve design.* Ecology Letters, 2006. **9**(1): 2–11.

[d] Pouzols, F. M., M. A. Burgman, and A. Moilanen, *Methods for allocation of habitat management, maintenance, restoration, and offsetting when conservation actions have uncertain consequences.* Biological Conservation, 2012. **153**: 41–50).

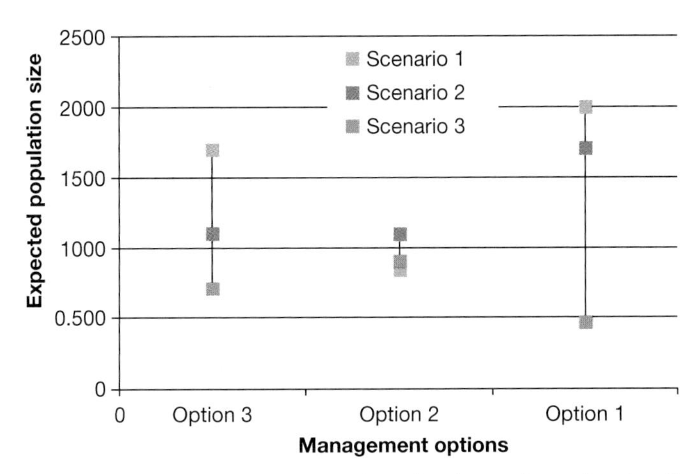


FIGURE 8.4 Example of a Minimax strategy for selecting options, showing the expected consequence of three different management options under three scenarios. The consequence in this case is the expected population size of an endangered species. In this example, option 2 is the Minimax choice: although it is expected to deliver a lower or equal population size than the other two options under two of the scenarios, across all scenarios it is the option with the least worst expected outcome.

modeling exercises. Minimax is also very easy to communicate and, in our experience, resonates well with decision makers. A risk-averse decision maker may want to choose a Minimax strategy to limit exposure to negative outcomes.

In some cases a clear minimum performance may be required of an alternative, as in projects that offset impacts somewhere else. In these cases, a more appropriate decision rule is to choose the alternative expected to deliver this performance over the greatest range of uncertain values. This method is sometimes referred to as robust satisficing.

As the environmental economist Steve Polasky and colleagues have argued,[10] decision rules like Minimax are quite different approaches to decision making than the traditional tactic of selecting alternatives based on the highest expected utility (an economic phrase referring essentially to performance against objectives). But these rules are perhaps better suited to the uncertainties inherent in complex social-ecological systems, especially in an era of global change.

## Scenarios and Scenario Analysis

Imagine that an effective global carbon market is in place and all nations have agreed to strict emissions caps that keep the price of $CO_2$ offsets at US$20 a ton or above. This policy guarantees a vast and sustainable flow of revenue to landholders, communities, and governments who protect forests and peat bogs, virtually eliminating deforestation globally.

This scenario paints an optimistic picture. But scenarios like this are not intended to be predictions, nor are they a choice that someone in charge of a conservation project could make—rather, they are learning tools. What would it mean for conservation if this scenario became reality? What would it mean for other industries? For the economies of developing nations? Analyzing scenarios can help us understand and prepare for the inevitably uncertain future.

### Scenarios versus Alternatives

The term **scenario analysis**, or **scenario**, is widely used in planning and can mean various things. Frequently, things that are more like options or alternatives being evaluated are referred to as scenarios. Here we try to distinguish between scenarios and options.

An **option** or alternative is a choice that can be made, whereas a scenario is a constructed manifestation of the future in which the chosen alternative must exist. We acknowledge that the line between the two can be pretty blurry. For example, if we investigated the consequence of changing an entire district to a single land use type—say, native forestry plantation—to see what the carbon sequestration and biodiversity outcomes would be, is this a scenario or an alternative? It could be either: it provides a useful learning tool because we get a baseline for comparison and can explore the consequence of a future in which this happens; and however unlikely, it might be an option being considered by policy makers. All alternatives have uncertainty associated with their outcomes because we cannot perfectly predict the future; scenarios describe some of this uncertainty through intentional manipulations of imagined futures along the axes of uncertainty.

The landscape and systems ecologist Erin Bohensky and colleagues[11] suggest that scenarios are best suited to exploring situations

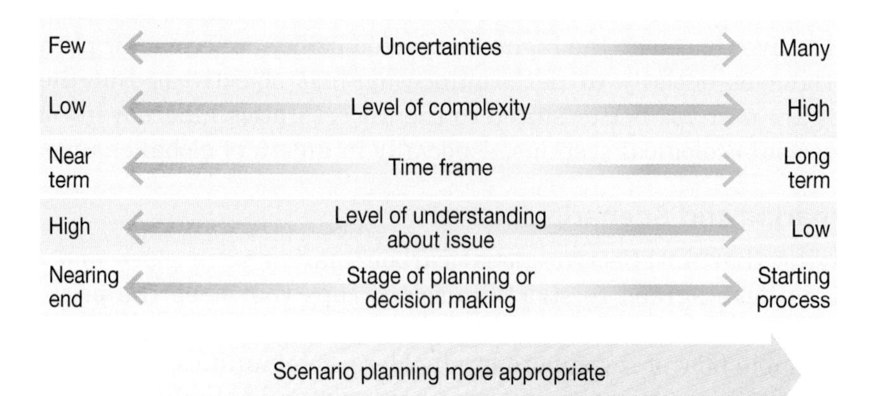


**FIGURE 8.5** Factors that make scenario analysis an appropriate choice for dealing with uncertainty in conservation planning. (Adapted from Rowland et al., 2014.)

where uncertainty is high and controllability is low. For example, climate change and global governance are largely beyond the control of conservation decision makers, even in a large region. In these situations, scenarios can help to illuminate the consequences of these global drivers of change and to formulate robust local responses. We think that this "controllability" test is a good way to distinguish between scenarios and alternatives. Similarly, in a detailed guide to scenario planning for natural resource management,[12] Erika Rowland and colleagues graphically present some of the factors that make scenario analysis an appropriate choice for dealing with uncertainty in conservation planning (FIGURE 8.5).

### Scenario Analysis

The initial premise of scenario analysis is to develop a small set of possible future scenarios that describe how some of the main uncertainties—demographic trends, policies, markets, budgets, degree of climate change (Chapter 9), or stakeholder support—might behave. Some of the most globally recognizable scenarios are the different emissions scenarios developed by the Intergovernmental Panel on Climate Change (IPCC) to explore uncertainties in national policies and social behavior and the impact on climate change (FIGURE 8.6).

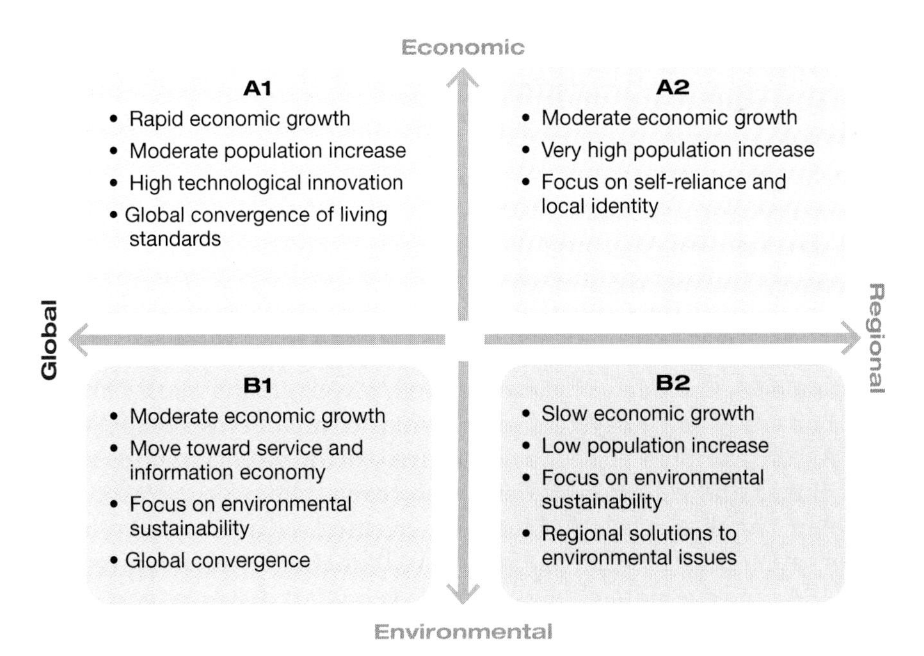


FIGURE 8.6 The four scenarios developed by the Intergovernmental Panel on Climate Change (IPCC) to describe alternative storylines about divergent yet plausible future global development pathways. (Adapted from Nakicenovic, N., and R. Swart, *Emissions Scenarios: A Special Report of Working Group III of the Intergovernmental Panel on Climate Change.* 2000, Cambridge, UK: Cambridge University Press.)

The set of alternatives or options being considered are then imagined to be occurring against the backdrop of these different scenarios with the aim of learning about how they would be expected to perform. For example, we might assess the performance of coastal adaptation options under each of the different IPCC scenarios or under a series of sea-level-rise models based on different assumptions. Sometimes scenarios might yield parameters that can be used in a formal predictive model (like the extent of sea-level rise expected under each of the IPCC scenarios), but scenario analysis can equally be accomplished simply by describing how we think an option might perform under a set of future conditions. The aim, as with sensitivity analysis, is to see how robust an option is to a range of possible futures. This information can then be used to refine alternatives to make them more robust, or discuss the broader conditions needed for success of the option.

As a straightforward example, in evaluating strategies for conservation of urban ecosystems in and around Stockholm, Ulla Mörtberg and colleagues[13] explored the consequence of two possible urban growth scenarios. The first scenario, which they termed "Compact," included urban growth policies that emphasized energy efficiency and minimized transport distances and costs. The second scenario, termed "Urban Nature," included urban growth policies that emphasized protection of urban green spaces and planned but distributed development.

Mörtberg and colleagues speculated that the first scenario would likely result in the loss of urban biodiversity and ecosystem values, which then would diminish the engagement and support of urban communities with conservation. Some of the consequences may be offset, however, by reduced loss of biodiversity in the peri-urban area because of a reduction in urban sprawl. The second scenario, they suggested, maintained urban biodiversity and ecosystems, but required greater ongoing budgets for transport infrastructure and energy availability because of growth in low-density housing, some of which was likely to come at the expense of peri-urban areas. With these two scenarios, the researchers highlighted the consequence for urban biodiversity projects under different policy environments, and made obvious the need to seek greater clarity around the biodiversity and ecosystem service value of urban ecosystems.

As an example of scenario analysis on a more regional scale, Joan Baker and colleagues[14] developed three scenarios (current land use plan trend, a more development-oriented scenario, and a more conservation-biased scenario) for the Willamette River Basin in Oregon, USA, and evaluated how well wildlife habitat, fisheries and invertebrates, and water availability/use performed under each scenario.

Exactly how alternatives are analyzed under different scenarios will depend on the method used to evaluate the consequences of those alternatives, as discussed in Chapter 6. One of the most straightforward approaches we have seen is to combine scenario analysis with

Multi-Criteria Decision Analysis (MCDA, discussed in Chapter 7). Michelle Hamilton and colleagues[15] provide some good examples of how these two tools can be combined to evaluate the robustness of alternatives to a set of scenarios. In essence, the process involves looking at how the importance of different criteria would change under different scenarios. **Box 8.2** describes a scenario analysis used to evaluate energy management options for military installations.

## Box 8.2 Example of Scenario Analysis for Different Energy Management Alternatives

Deciding on appropriate investments in energy infrastructure is challenging because the adequacy of any solution depends on a range of highly uncertain factors, including technology, geopolitics, socio-demographics, and environmental regulation. These deep uncertainties make energy decisions a good candidate for employing scenario analysis.

Michelle Hamilton and colleagues[a] provide an elegant example of how scenario analysis can be combined with Multi-Criteria Decision Analysis (MCDA; see Chapter 7) to support decisions around energy security for major military and industrial installations in the United States.

1. A series of criteria was identified (**TABLE 8.2.1**) that reflected the energy security

**TABLE 8.2.1.** Criteria for evaluating energy development options

- Reduce operation and management costs
- Increase self-sufficiency
- Reduce energy consumption
- Reduce foreign energy
- Reduce carbon emissions
- Increase energy efficiency
- Increase renewable energy
- Reduce foreign rare earths
- Reduce cyber vulnerability
- Increase technology innovation

**TABLE 8.2.2.** Alternative energy security options for major military and industrial installations being considered

- Microgrids with integrated control
- Solar cogeneration with battery backup
- Geothermal with cognitive building management
- Microturbines and fuel cells with cogeneration
- Innovative HVAC and dynamic windows
- Electric fleet vehicles with optimization

**TABLE 8.2.3.** Scenarios characterizing uncertainties likely to influence energy security considerations

| Scenario | Combinations of emergent conditions and major uncertainties |
|---|---|
| Green movement | Future energy incentives and regulations; green political movement; carbon tax/legislation |
| National security | Conflict in oil-producing countries; conflict in countries with significant rare earth resources; cyber-attacks to energy infrastructure |
| Islanding | Deterioration and vulnerability of private utility infrastructure |
| Technology innovation | Lack of private investment in R&D; increased pressure for the government to foster innovation and provide proof of principle |
| Economic | Slow economic development; cutbacks in government |

*(continued)*

## Box 8.2 Example of Scenario Analysis for Different Energy Management Alternatives *(continued)*

objectives of decision makers and would be used to evaluate alternative energy development options. Six alternative energy system options were under consideration (**TABLE 8.2.2**).

2. MCDA was then used to evaluate the performance of each alternative against each criterion, weight the importance of each criterion, and combine these two to give an overall score for each alternative. This score is referred to as the baseline.
3. Five future scenarios were developed that characterized a range of deep but important uncertainties and emergent conditions (**TABLE 8.2.3**).
4. An expert-based assessment was used to determine whether each scenario was likely to lead to an increase or a decrease in the relative importance of the evaluation criteria (**TABLE 8.2.4**). This assessment was based on linguistic terms like *major increase* or *minor increase*, which were subsequently used to adjust the criteria weights in a standard way.

**TABLE 8.2.4.** The importance (weight) of each criterion is updated for each scenario

| | | Scenarios | | | | |
|---|---|---|---|---|---|---|
| **Criteria** | **Baseline** | ***Green Movement*** | ***National Security*** | ***Islanding*** | ***Technology Innovation*** | ***Economic*** |
| Reduce O&M costs | Baseline relevance of criteria (no change) | Minor increase | Minor increase | Minor increase | | Minor increase |
| Increase self-sufficiency | | | | Major increase | | |
| Reduce energy consumption | | Major increase | | | | Major increase |
| Replace foreign energy | | | Major increase | | | |
| Reduce carbon emissions | | Major increase | | | | |
| Increase energy efficiency | | Major increase | | | | Major increase |
| Increase renewable energy | | Major increase | | | | |
| Reduce foreign rare earths | | | Major increase | | | |
| Reduce cyber vulnerability | | | Major increase | | | |
| Increase technology innovation | | | | | Major increase | |

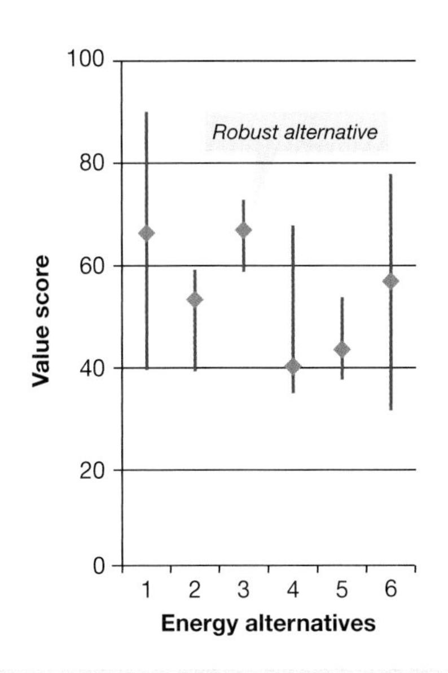


5. By recalculating an overall score for each alternative under each scenario and then plotting these scores (FIGURE 8.2.1), it was possible to determine the alternative that was most robust to the uncertainty captured in the scenarios (in this case, geothermal with cognitive building management).

[a] Hamilton, M. C. et al., *Case studies of scenario analysis for adaptive management of natural resource and infrastructure systems.* Environment Systems & Decisions, 2012: 1–15.

**FIGURE 8.2.1.** Value scores for the six energy alternatives under five scenarios. (The diamond under each alternative represents the baseline score; the bar extension from the diamond represents the range of scores of the alternatives.)

### How Do You Design Scenarios?

Designing a small set of useful scenarios is the first, and usually the most challenging, step in scenario analysis. Many approaches have been used to develop scenarios, and because scenarios are learning tools, those involved often find the process of developing scenarios as useful as analyzing them. Regardless of whether you are developing scenarios with a group of stakeholders or a few experts, the common starting point is to identify the key drivers of change in the landscape of interest. *Drivers* are the range of factors that we believe will be most influential in shaping what the future looks like.

Emily McKenzie and colleagues[16] from the Natural Capital Project have put together a useful handbook on scenario analysis in the context of planning for ecosystem services. Although they use the term *scenarios* more broadly than we do, our definition corresponds to what they call *exploratory scenarios.* In their guide they list a set of common drivers for scenarios (TABLE 8.3), which is an excellent starting point for most conservation plans.

A generalized approach to scenario development would include the following process:

1. Identify the three to five most important drivers of change.
2. For each of these drivers, identify possible future trends with a small number of categories (e.g., remain the same, small increase, big increase), or bifurcating decisions (e.g., implement a policy or not).

**TABLE 8.3** Common drivers of change relevant for building scenarios when conservation planning

| Category | Drivers |
|---|---|
| Social and demographic | Population growth or decline<br>Migration<br>Cultural values<br>Awareness<br>Poverty<br>Diet patterns<br>Education<br>Religious values |
| Technological | Technological innovation<br>Technology choice |
| Economic | Economic growth<br>Trade patterns and barriers<br>Commodity prices<br>Demand and consumption patterns<br>Income and income distribution<br>Market development |
| Environmental | Climate change<br>Air and water pollution<br>Introduction of invasive non-native species |
| Political | Macroeconomic policy<br>Other policies (e.g., subsidies, incentives, taxes)<br>Land use plans, zoning, and management<br>Governance and corruption<br>Property rights and land tenure |

*Source:* Data from McKenzie et al., 2012.

3. Create a framework by grouping these trends along two to three axes of uncertainty.
4. Develop a set of coherent storylines (a narrative about what may happen in the future) that draw on the possible trends and cover as much of the space in the framework as possible.

**TABLE 8.4** shows a set of drivers of social and environmental change in the Gariep River Basin of South Africa, and how they were grouped into four scenarios. These scenarios were then used to explore uncertainty in outcomes for ecosystem services and human well-being in the basin.[17]

There is a good deal of debate in the scenario analysis community about how plausible each scenario and its storyline need to be. Part of the point of scenario analysis is explicitly *not* to focus on what is likely;[18] extreme, low-probability scenarios can still be very useful

TABLE 8.4 Key bifurcations in drivers of change used to distinguish scenarios for ecosystem services in South Africa

| Driver | Market forces | Policy reform | Fortress world | Local resources |
|---|---|---|---|---|
| ***Political, economic, and social environment*** | | | | |
| National governance | | | | |
| Structures | + | ++ | – | – |
| Civil society | – | + | – | + |
| National economic growth | ++ | + | – | – |
| Distribution of wealth | – | + | – | – |
| National social and environmental policy | – | + | – | – |
| HIV management | + | ++ | – | – |
| ***Demographic trends*** | | | | |
| Birth rate | Medium | Low | High | High |
| Mortality rate | Medium | Low | High | High |
| Urbanization | Increasing | Increasing | Increasing | Constant |

Symbols: ++, exceptionally strong; +, strong; –,weak or nonexistent
*Source:* Data from Bohensky et al., 2006.

in scenario analysis. It is critical, however, for scenarios to be coherent; a scenario must encapsulate a logically consistent story about the future world. For instance, if a scenario for the Willamette River example given earlier emphasized more development but increased both suburban areas and farmland, it would not be coherent as development in this region generally occurs at the expense of farmland.

Scenarios can be challenging and time-consuming to develop, especially as they typically reside in the domain of macro trends that are outside the control and experience of most conservation planners. For this reason, it can be helpful to use a set of existing scenarios of large-scale change as a starting point. In addition to the IPCC scenarios already mentioned, another set of scenarios frequently employed for planning related to the environment are those developed as part of the Millennium Ecosystem Assessment[19] (FIGURE 8.7).

A challenge with global scenarios like those of the Millennium Ecosystem Assessment is ensuring their relevance at the scale of planning, which is typically much smaller than the entire world. This means translating global trends into national and subnational trends. A good example of this process is provided by the geographers Michael Hill and Rhonda Olson, who used scenario analysis to explore

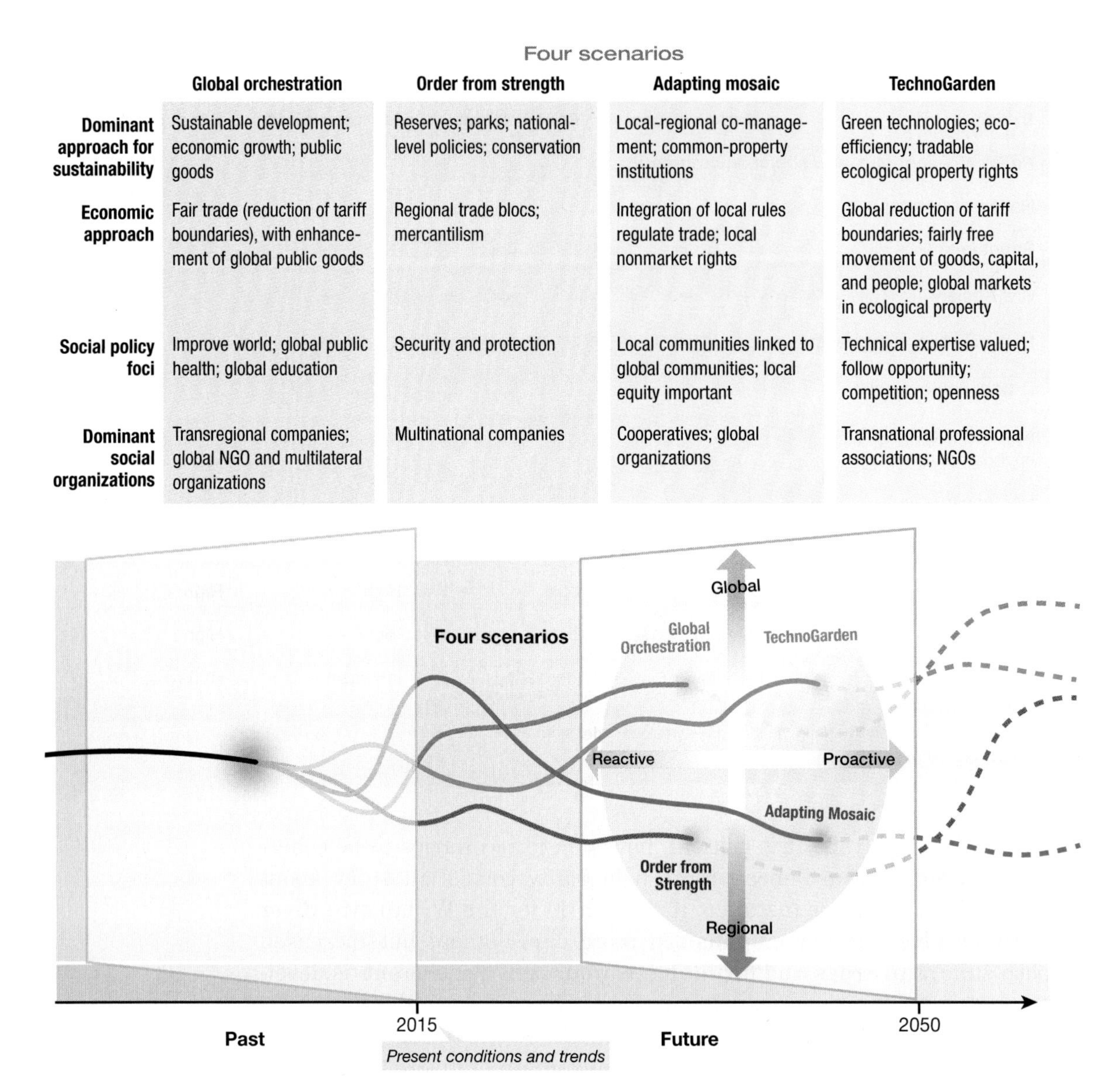

Four scenarios

| | Global orchestration | Order from strength | Adapting mosaic | TechnoGarden |
|---|---|---|---|---|
| **Dominant approach for sustainability** | Sustainable development; economic growth; public goods | Reserves; parks; national-level policies; conservation | Local-regional co-management; common-property institutions | Green technologies; eco-efficiency; tradable ecological property rights |
| **Economic approach** | Fair trade (reduction of tariff boundaries), with enhancement of global public goods | Regional trade blocs; mercantilism | Integration of local rules regulate trade; local nonmarket rights | Global reduction of tariff boundaries; fairly free movement of goods, capital, and people; global markets in ecological property |
| **Social policy foci** | Improve world; global public health; global education | Security and protection | Local communities linked to global communities; local equity important | Technical expertise valued; follow opportunity; competition; openness |
| **Dominant social organizations** | Transregional companies; global NGO and multilateral organizations | Multinational companies | Cooperatives; global organizations | Transnational professional associations; NGOs |



FIGURE 8.7 The four scenarios of global development constructed as part of the Millennium Ecosystem Assessment (MEA). (Adapted from Millennium Ecosystem Assessment, *Ecosystems and Human Well-Being: Synthesis.* 2005, Washington, DC: Island Press.

possible future outcomes of policy options for agriculture, energy, and biodiversity in North Dakota, USA.[20] Hill and Olson developed a set of scenarios and accompanying storylines by tracing different pathways through four sets of well-established global scenarios, translating these to a national effect, and then describing the effect of these national trends on three issues of interest in North Dakota (FIGURE 8.8). One interesting finding from their exploration of these scenarios was that climate change had the biggest influence on out-

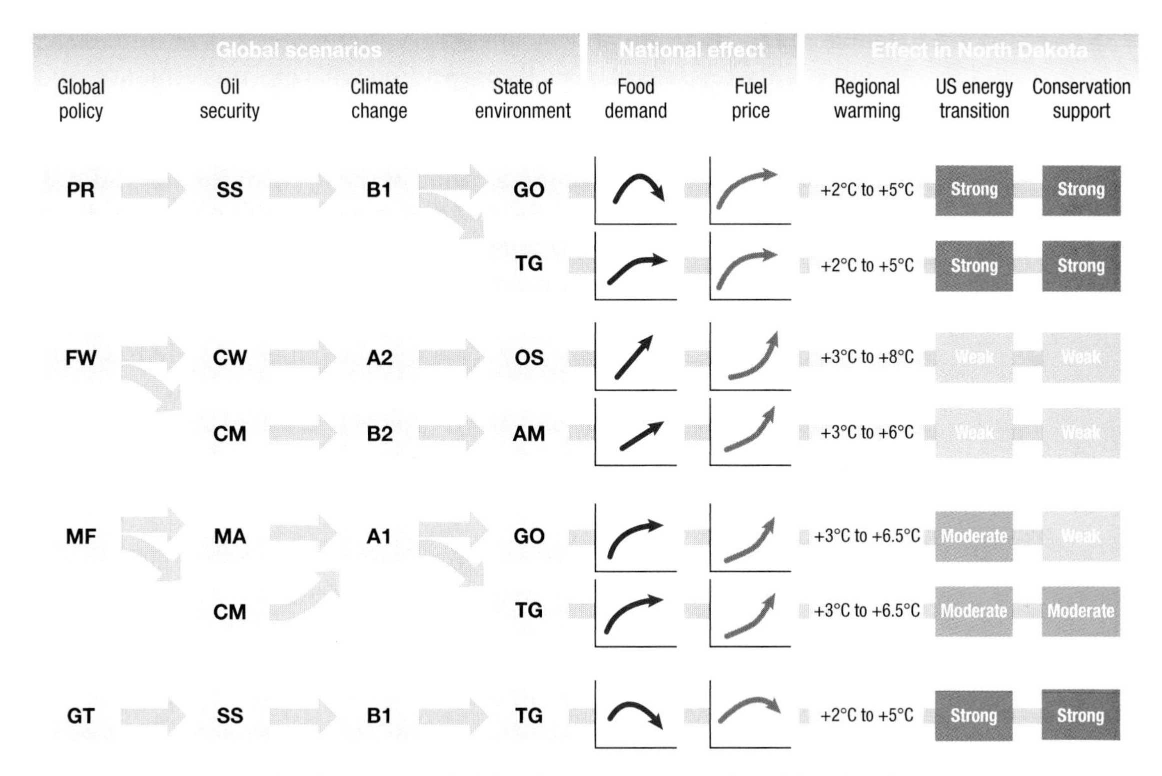


FIGURE 8.8 Pathways that link a series of global development scenarios with national environmental trends in the United States, and the likely effect of these trends in the state of North Dakota. (Adapted from Hill and Olson, 2013.)

comes of wetland conservation, whereas commodity prices were the strongest driver of futures for grasslands in North Dakota.

Scenario analysis is one tool in a broader approach to future uncertainties called *strategic foresight*. We have focused on scenario analysis here because of its usefulness in making informed choices between options. In a valuable review, Carly Cook and colleagues describe some other ways of using strategic foresight to improve environmental decision making where there is uncertainty,[21] including decisions about investment in monitoring and threat identification (see Chapter 11 on implementation for more on strategic foresight).

## Risk and Risk Assessment

Uncertainties that might negatively affect achievement of a project's objectives can be characterized as **risks**. For conservation projects, risks exist in both social and environmental space, ranging from invasive species outbreaks and catastrophic climate events (such as coral bleaching) to community reactions, policy changes, legal or liability issues, and insecure or inadequate funding streams. As with all complex projects, the delivery of conservation outcomes is influenced by our ability to assess the risks associated with our investments, and to manage and respond to these risks over time.

In the context of environmental and conservation planning, the term *risk assessment* can mean a range of things. Environmental risk assessment is a substantial field whose practice influences strategic choices for many environmentally related initiatives.[22] Environmental risk assessments are typically assessments of risks to natural or community values, for instance, to help prioritize which risks or threats to manage (see Carey et al.[23]; also see the section on threats in Chapter 5). Many such assessments have concentrated on the potential for chemicals to affect human populations and natural ecosystems. Others focus on risks arising from conservation projects, such as the potential for increased fire risk as a result of protecting native vegetation and its natural fire regimes.[24]

Many conservation agencies and organizations also undertake risk assessments for legal, compliance, and reputational issues associated with potential projects (e.g., the risk of engaging with corporate partners or undertaking controversial strategies). Although these risks are undeniably bound up with long-term conservation success, they focus not on what caused the project to fail, but on the potential consequences of this failure. Here we focus on identifying and assessing risks to the success of conservation projects as we believe this is the most relevant type of risk for conservation planning. An assessment of risks to stated project objectives should influence conservation plans in a number of ways (summarized in **TABLE 8.5**).

Despite the important and well-recognized link between risks and project outcomes, in our experience the assessment of risk and use of this information has not featured prominently in conservation planning guidance or practice. As discussed in Chapters 6 and 7, it is reasonably common for conservation plans to include estimates for the probability of success (roughly the inverse of the risk of project failure) of different projects or activities,[25] but most have not further decomposed risk based on the causes for failure. Probability is a natural unit for risk, and getting estimates of success likelihood is a substantial improvement over much conservation planning practice. However, a more thorough understanding of risk is important because we are not passive hostages in the face of risk.

The economist and historian Peter Bernstein has argued compellingly that this appreciation of risk has been one of the most influential achievements of human consciousness.[26] Risks to project success are rarely immutable; for example, the risk of a conservation project being derailed because of community opposition can be reduced through more substantial engagement with communities. However, a cost is nearly always associated with ameliorating any risk.

**TABLE 8.5** How risk assessment should influence different phases of conservation planning

| Conservation planning phase | Relevant aspects of risk assessment and management | How risk assessment should influence conservation planning phase |
|---|---|---|
| Strategy evaluation and selection<br>Determining targets and expected outcomes | • Risk identification<br>• Assessment of risk consequence and likelihood<br>• Uncertainty in assessment of risk<br>• Cost of managing or ameliorating risk | Expected outcome should be adjusted based on assessment of risk. Strategy selection should be influenced by how robust a strategy is to risk and the costs required to adequately ameliorate risks to a strategy. Process offers a dedicated opportunity to elicit diverse input about strategies and their context. |
| Monitoring design and investment | • Overall risk rank<br>• Uncertainty in assessment of risk | Identifying risks that need to be actively monitored. These might include highly ranked risks as well as low or moderately ranked risks with high uncertainty in the assessment of risk. |
| Budgeting | • Cost of monitoring, managing, or ameliorating risks | Realistic costs for risk monitoring and mitigation activities need to be considered as part of the overall cost required for a strategy or project. |
| Implementation and project management | • Overall risk rank<br>• Risk management actions | Important risks should influence strategy implementation as well as selection, often in domains such as staff capacity, leadership, or partner engagement. Risk management is an important part of project management. |
| Reporting and communication | • Risk identification<br>• Assessment of risk consequence and likelihood<br>• Risk management actions | Identifying risks demonstrates both transparency and rigor. Allows crisis communication measure to be in place in case risks transpire. |
| Identifying research priorities | • Overall risk rank<br>• Uncertainty in assessment of risk | Risks that are highly ranked but with high uncertainty may be priorities for directed research. |

## Dimensions of Risk

Risk is generally conceived as having two dimensions: likelihood and consequence. Likelihood is essentially the probability that the risk will materialize (although not always expressed as quantitatively as a probability) within some specified time frame. Consequence is a measure of a risk's impact in terms of deviation from expected project outcomes should the risk occur.[27]

Although knowledge of both dimensions is important, comparative analysis of risks is often accomplished by combining the two dimensions to give an overall rating or rank. For instance, a measure of how serious a hazard is can be given by the product of likelihood and consequence. This is the basis of publicly endorsed risk assessments in most countries.[28]

## The Risk Assessment Process

In an ideal case, the likelihood and consequence of any risk could be assessed simply based on evidence of past occurrences of similar events. For example, an insurance company has a very good understanding of how likely your house is to be burgled and how much is likely to be stolen, simply based on the evidence of past burglaries. However, the complexity of the social-ecological systems in which conservation takes place means that most projects must contend with continually shifting contexts. As such, conservation projects will rarely have extensive historical data regarding the likelihood (probability) and consequence (impact) of risks to draw upon.

In some cases, conservation agencies or organizations might look to comparable experiences in other jurisdictions to inform base rates for risks, such as violation rates of conservation easements in another state. Alternatively, we could look creatively at tactically relevant data, for example, corporate noncompliance with other regulations might be used as a base rate for the risk that a company fails to live up to its expectation under some conservation agreement. More likely is that most conservation project risk assessments will involve a subjective or semi-quantitative assessment of risks. So this is what we focus on here.

Risk assessment involves three general steps: (1) identifying risks, (2) estimating the consequence and likelihood for each risk, and (3) combining this information into an assessment of the importance of each risk and how to respond to it.

### Identifying Risks

Even the most rigorous assessment of risks will be of limited value unless the important suite of risks is first identified. There are many ways to build a list of risks to a conservation project,[29] but one approach that we have found particularly effective is known as a *pre-mortem.*[30] A pre-mortem simply involves asking the team of people involved in the planning process (or any other relevant people) to imagine that the project goes ahead as currently planned, but that it is now five years in the future and the project has gone disastrously wrong. Each person is then asked to describe to the group what they had imagined went wrong. This process serves the purposes of directly identifying risks and also triggering thought about the range of things that could feasibly go wrong. We have also found that it provides a valuable forum for staff to share concerns they might have about the chosen strategies or projects, something that frequently doesn't exist in routine planning processes.

Another effective approach to risk identification is through a systematic assessment of assumptions associated with the theory of change for each strategy or assumptions that may be implicit in a conceptual model or situation analysis. In our experience, it is becoming more common for conservation projects to state the assumptions underlying strategies and interventions. These assumptions are not always captured within a conservation plan, but it is a good practice to do so. These assumptions are usually articulated in a manner that makes them the flip side of a risk statement (e.g., we might assume for any particular project that there will be negligible long-term impacts from climate change although there is obviously some risk of that happening).

The initial list of risks identified will generally suffer from many of the linguistic uncertainties identified by Jan Carey and Mark Burgman[31] and introduced earlier in this chapter. For example, in a risk assessment that we worked on, one of the risks identified was the loss of dedicated government funding for Indigenous Protected Areas (IPAs). Stated like this, the risk was too vague; some participants might consider a partial loss of funding while others considered the complete loss of funding. We would expect total loss to have more severe consequences than 50% loss, but total loss is also less likely to occur. To address this vagueness, we separated the loss of funding for IPAs into categories of total loss and 50% loss. Obviously the actual extent of funding loss could fall at any point between none and total, but even two categories proved a satisfying solution for participants. It is important to remove linguistic uncertainties from risk descriptions as much as possible.

### Developing Indices of Consequence and Likelihood

Because we are concerned with risks to project success, it is important that we assess consequence and likelihood in the context of success. Success means different things to different people. A common understanding about what the project is trying to achieve and in what units this achievement will be measured is a necessary condition for comparing the risk judgments of multiple people.

TABLE 8.6 provides examples of consequence and likelihood indices that were developed to assess risks to a conservation project in northern Australia.[32] For each objective, a unique index of consequence was constructed; it included linguistic descriptions, a corresponding scenario in performance metric of that objective, and a score. The likelihood index similarly contained a linguistic description, a probability, and a score; but the same index remained constant across the objectives because the likelihood of a particular risk event occurring did not depend on a given objective.

There is no universal answer for the score assigned to each place in these indices, and for the case in question both linguistic descriptions and scores were based on the Risk Assessment and Management Process (RAMP) standards of the Institution of Civil Engineers and Faculty and Institute of Actuaries in Great Britain.[33]

| **Consequence index** | | |
|---|---|---|
| ***Description*** | ***Scenario**** | ***Score*** |
| Negligible/insignificant | 38%–40% | 1 |
| Marginal/minor | 25%–38% | 3 |
| Substantial/moderate | 10%–25% | 20 |
| Severe/major | 5%–10% | 100 |
| Disastrous/catastrophic | < 5% | 1000 |
| **Likelihood index** | | |
| ***Description*** | ***Probability*** | ***Score*** |
| Extremely unlikely | < 0.01% | 1 |
| Very unlikely | < 1% | 2 |
| Unlikely | 1%–20% | 4 |
| Fairly likely | 21%–49% | 8 |
| Likely | 50%–85% | 12 |
| Highly likely | Over 85% | 16 |

*Percentage of northern Australia lands that are subject to early dry season burning at least once every three years by 2020.

**TABLE 8.6** (Top) Risk consequence index, scenarios, and scores for the objective "40% of all northern Australian lands (including 10% of land outside the formal conservation estate) are effectively managed for conservation by 2020." (Linguistic descriptions and scores adapted from Lewin, 2002.) (Bottom) Risk likelihood index used in the same project. Much of the group discussion around risk consequence scenarios hinged on the assumed current baseline (e.g., how much of northern Australia is currently subject to early dry season burning, with the severity of consequence being heavily dependent on this starting point). (Adapted from Game et al., 2013.)

### Estimating Risks

Estimating risks against the indices of consequence and likelihood is an exercise in expert elicitation. As such, all the advice about expert elicitation contained in Chapter 5 is relevant here. For instance, the Delphi process described in Chapter 5 will work well for getting risk judgments.

### Prioritizing and Responding to Risks

Four pieces of information are particularly helpful in assessing risks and what, if anything, a project should do about them:

1. The overall score and rank (generally obtained by multiplying the scores for consequence and likelihood together)

2. The consequence and likelihood scores on their own (as opposed to the overall score)
3. The level of agreement between participants in the risk assessment
4. The level of uncertainty in risk estimates

In another publication, we have provided a detailed description of how each of these four elements can be easily analyzed.[34] Although these analyses will be comfortably within reach of all conservation planners, Mark Burgman and colleagues have made life even easier by providing freely available Subjective Risk Assessment software that automates the analysis of three of these four elements.

So what to do with this sort of information about risks? Conservation projects must inevitably live with risk; not all risks can be eliminated. Rather, risk assessment methods like those described here can help conservation projects prioritize which risks to try to manage. Broadly speaking, risk management involves four options: (1) actively ameliorate through strategic adjustment, (2) gather information to better understand a risk, (3) monitor a risk to respond rapidly when it occurs, or (4) do nothing. Which of these options is most appropriate depends on both the level of risk and the uncertainty around the estimates of this risk. FIGURE 8.9 illustrates how these four responses are aligned along axes of overall risk score and relative uncertainty about that score.

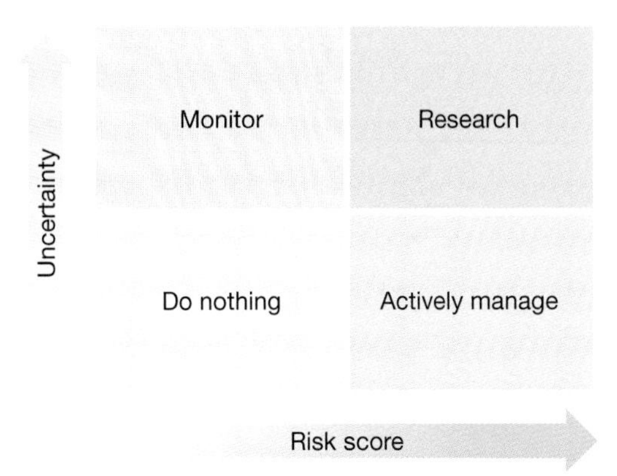


FIGURE 8.9 Four responses to conservation project risks depending on the relative importance of the risk (risk rank) and the level of uncertainty around assessment of the risk. (Data from Game et al., 2013.)

If we look at uncertainty versus overall risk score for the set of risks identified in the northern Australia exercise mentioned before (FIGURE 8.10), we can see that a project manager would probably say that "Inability to raise sustainable funding," with its high risk and relatively low uncertainty, would be a priority for immediate and active management. On the other hand, the impact of "Gamba grass," an invasive species in northern Australia, which received a low risk score but has high uncertainty, might be a good candidate for monitoring.

Explicit identification, prioritization, and where possible, management of risks are important elements of using conservation resources in an informed and accountable manner. Informed and accountable resource use is also an ideal of conservation planning. We believe that an assessment of risks to conservation success should be considered a core part of conservation planning, and one that can be accomplished rapidly using the approach illustrated here.

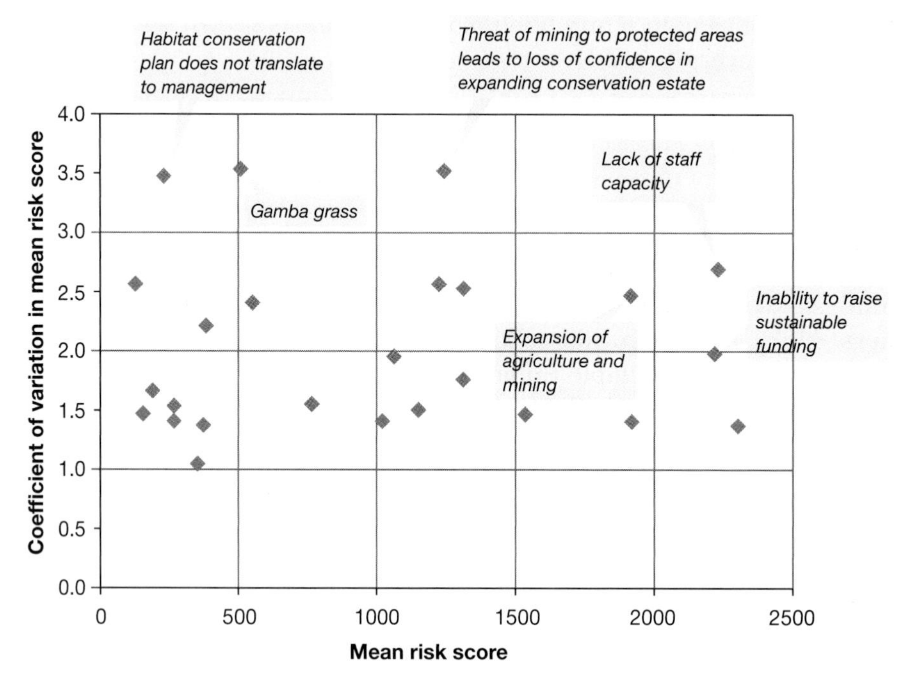


**FIGURE 8.10** Uncertainty around a risk (coefficient of variation in mean risk scores across all objectives) plotted against the mean risk scores for a series of risks to a conservation project in northern Australia. Labeled risks are those with the highest scores and the most uncertainty. This information can help prioritize the appropriate response to a risk. (Data from Game et al., 2013.)

## KEY MESSAGES

- There is no aspect of conservation planning free from uncertainty, and try as we may, it cannot be eliminated completely. It is imperative that conservation planners be able to recognize uncertainty, manage it appropriately, and understand what it means for the decisions they're informing.
- Even a highly uncertain decision can still be an informed one if uncertainty is managed well.
- Some of the ways uncertainty can be acknowledged in a conservation plan include using probabilities, using intervals, estimating the likelihood of success, and being transparent about risks.
- Uncertainty can be dealt with in three general ways in a conservation plan: (1) it can be minimized, (2) it can be compensated for, or (3) we can search for the solution that is most robust to the uncertainty that exists. A range of tools are available to help identify robust solutions.

- Scenario analysis can be a powerful tool for exploring the impact of highly uncertain futures over which there is very limited control, for example, global climate.
- Scenarios should be distinguished from alternatives or options, and they should be designed to cover as much of the uncertainty as possible without being incoherent.
- Uncertainties that could lead to negative outcomes for our objectives can be considered risks. Risks should be characterized by both their likelihood and consequence.
- A great deal of useful information about risks and the best responses to them can be gained from even a rapid and subjective approach to risk assessment. Risk assessment should be a core part of conservation planning.

## References

1. McGinnis, M., and E. Ostrom. *Social-ecological system framework: Initial changes and continuing challenges.* Ecology and Society, 2014. 19(2): 30.
2. Langford, W. T., A. Gordon, and L. Bastin, *When do conservation planning methods deliver? Quantifying the consequences of uncertainty.* Ecological Informatics, 2009. **4**(3): 123–135.
3. Regan, H. M., M. Colyvan, and M. A. Burgman, *A taxonomy and treatment of uncertainty for ecology and conservation biology.* Ecological Applications, 2002. **12**(2): 618–628.
4. Game, E. T., P. Kareiva, and H. P. Possingham, *Six common mistakes in conservation priority-setting.* Conservation Biology, 2013. **27**: 480–485.
5. Joseph, L. N., R. F. Maloney, and H. P. Possingham, *Optimal allocation of resources among threatened species: A project prioritization protocol.* Conservation Biology, 2009. **23**: 328–338.
6. Carey, J. M., and M. A. Burgman, *Linguistic uncertainty in qualitative risk analysis and how to minimize it.* Annals of the New York Academy of Sciences, 2008. **1128**(1): 13–17.
7. Allison, G. W. et al., *Ensuring persistence of marine reserves: Catastrophes require adopting an insurance factor.* Ecological Applications, 2003. **13**(1 Supplement): S8–S24.
8. Tulloch, V. J. et al., *Incorporating uncertainty associated with habitat data in marine reserve design.* Biological Conservation, 2013. **162**: 41–51.
9. Lourival, R. et al., *Planning for reserve adequacy in dynamic landscapes; maximizing future representation of vegetation communities under flood disturbance in the Pantanal wetland.* Diversity and Distributions, 2011. **17**: 297–310.
10. Polasky, S. et al., *Decision-making under great uncertainty: Environmental management in an era of global change.* Trends in Ecology & Evolution, 2011. **26**(8): 398–404.

11. Bohensky, E. et al., *Future makers or future takers? A scenario analysis of climate change and the Great Barrier Reef.* Global Environmental Change, 2011. **21**(3): 876–893.

12. Rowland, E. L., J. Cross, and H. Hartmann, *Considering Multiple Futures: Scenario Planning to Address Uncertainty in Natural Resource Conservation.* 2014, Washington, DC: U.S. Fish and Wildlife Service.

13. Mörtberg, U. et al., *Urban ecosystems and sustainable urban development—analyzing and assessing interacting systems in the Stockholm region.* Urban Ecosystems, 2012: 1–20.

14. Baker, J. P. et al., *Alternative futures for the Willamette River basin, Oregon.* Ecological Applications, 2004. **14** (2): 313–324.

15. Hamilton, M. C. et al., *Case studies of scenario analysis for adaptive management of natural resource and infrastructure systems.* Environment Systems & Decisions, 2012: 1–15.

16. McKenzie, E. et al., *Developing Scenarios to Assess Ecosystem Service Trade-Offs: Guidance and Case Studies for InVEST Users.* 2012, Washington, DC: World Wildlife Fund.

17. Bohensky, E. L., B. Reyers, and A. S. Van Jaarsveld, *Future ecosystem services in a Southern African river basin: A scenario planning approach to uncertainty.* Conservation Biology, 2006. **20**(4): 1051–1061.

18. Montibeller, G., and L. A. Franco, *Raising the bar: Strategic multi-criteria decision analysis.* Journal of the Operational Research Society, 2010. **62**(5): 855–867.

19. Carpenter, S. R. et al., Millennium Ecosystem Assessment. *Ecosystems and Human Well-Being. Volume 2: Scenarios.* 2005, Washington, DC: Island Press. 596.

20. Hill, M. J., and R. Olson, *Possible future trade-offs between agriculture, energy production, and biodiversity conservation in North Dakota.* Regional Environmental Change, 2013: 1–18.

21. Cook, C. N. et al., *Strategic foresight: How planning for the unpredictable can improve environmental decision-making.* Trends in Ecology & Evolution, 2014. **29**(9): 531–541.

22. Burgman, M. A., *Risks and Decisions for Conservation and Environmental Management.* 2005, Cambridge, UK: Cambridge University Press. 504.

23. Carey, J. M. et al., *Risk-based approaches to deal with uncertainty in a data-poor system: Stakeholder involvement in hazard identification for marine national parks and marine sanctuaries in Victoria, Australia.* Risk Analysis, 2007. **27**(1): 271–281.

24. Norman, S. P. et al., *Assessing Risks to Multiple Resources Affected by Wildfire and Forest Management Using an Integrated Probabilistic Framework.* General Technical Report PNW-GTR (802, Part 2). 2010, Portland, OR: U.S. Forest Service Pacific Northwest Research Station. 361–370.

25. For example, see Joseph et al., *Optimal allocation of resources among threatened species.* (See reference 5.)

26. Bernstein, P. L., *Against the Gods: The Remarkable Story of Risk.* 1996, New York: John Wiley & Sons. 383.
27. Cagno, E., F. Caron, and M. Mancini, *A multi-dimensional analysis of major risks in complex projects.* Risk Management, 2007. **9**(1): 1–18.
28. Burgman, *Risks and Decisions for Conservation and Environmental Management.* (See reference 22.)
29. Carey, J. M. et al., *An application of qualitative risk assessment in park management.* Australasian Journal of Environmental Management, 2005. **12**(1): 6–15.
30. Schlesinger, L. A., and C. F. Kiefer, *Just Start: Take Action, Embrace Uncertainty, Create the Future.* 2012, Boston: Harvard Business School Press; and Kahneman, D., D. Lovallo, and O. Sibony, *Before you make that big decision.* Harvard Business Review, 2011. **89**(6): 50–60.
31. Carey and Burgman, *Linguistic uncertainty in qualitative risk analysis and how to minimize it.* (See reference 6.)
32. Game, E. T. et al., *Subjective risk assessment for planning conservation projects.* Environmental Research Letters, 2013. **8**(4): 045027.
33. Lewin, C., *RAMP Risk Analysis and Management for Projects.* 2002, London: Institution of Civil Engineers and the Faculty and Institute of Actuaries in association with Thomas Telford.
34. Game et al., *Subjective risk assessment for planning conservation projects.* (See reference 32.)

# Part Two

## Special Topics in Conservation Planning

# 9

# Moving Beyond Natural: Adapting Conservation Plans to Climate Change

## Overview

With global efforts to reduce emissions progressing slowly, it is imperative that we take actions to adapt conservation projects and initiatives to a changing climate. In this chapter we outline key adaptation concepts, articulate major adaptation planning steps that link to other chapters, review the major strategic approaches to adaptation, and suggest avenues for overcoming barriers to implementing these approaches.

## Topics

- *Climate change impacts*
- *Risk and uncertainty*
- *Vulnerability assessments*
- *Ecosystem-based adaptation*
- *Social adaptation*
- *Scenario analyses*
- *Conserving the stage*
- *Land facets*
- *Connectivity*
- *Ecological function and process*
- *Refugia*
- *Climate smart conservation*
- *Adaptation barriers*

In the late 1980s, Dr. James Hansen, one of the world's foremost experts on climate change, used the analogy of a six-faced die to explain the probability of temperature extremes. If the extremes followed a normal distribution, two sides of the die would represent below-normal temperatures, two sides would represent near-average temperatures, and two would represent above-average temperatures. In recent analyses, Hansen and his colleagues[1] reported that 25 years later, the world is now twice as likely to experience above-average temperatures relative to a 1951–1980 baseline, and that in essence we need to add two additional above-average faces to the die to represent these above-average temperatures.

Moreover, Hansen and his team presented compelling evidence that recent extreme temperature events around the globe are the direct result of interactions between specific weather patterns that favor extremes and a warming climate, principally due to human-caused increases in atmospheric greenhouse gases. The Intergovernmental Panel on Climate Change (IPCC), the world's foremost authoritative body on climate change, released its Fifth Assessment Report in 2013.[2] The report concluded that

> warming of the climate system is unequivocal and since the 1950s, many of the observed changes are unprecedented over decades to millennia. The atmosphere and ocean have warmed, the amounts of snow and ice have diminished, sea level has risen, and the concentrations of greenhouse gases have increased.

As for the cause, the Fifth Assessment Report states matter-of-factly that "it is extremely likely that human influence has been the dominant cause of the observed warming since the mid-20th century." Although recent advances in climate science demonstrate that natural recurrent changes in the Earth's climate "have been a significant agent of physical, ecological, and cultural change" for millennia,[3] the Fifth Assessment Report concludes that human-caused climate change (often referred to by climate scientists as "anthropogenic forcing") over the last 50-plus years is directly influencing a variety of weather events around the world. And some impacts of this change are here to stay (because once released, greenhouse gases persist in the atmosphere for a thousand years), even if society takes dramatic and unprecedented steps to reduce the burning of fossil fuels.

Unfortunately, the world is not yet taking those dramatic actions, even though the 2014 bilateral agreement between China and the USA to establish $CO_2$ emission reduction targets for 2030 was a step in the right direction. Although the global response to reducing greenhouse emissions continues to fall short, the conservation community has responded to the existing and expected impacts of climate change on species and ecosystems (FIGURE 9.1) and given increasing attention

**FIGURE 9.1** Although the polar bear has become the poster child for species being affected by climate change (among other things, by melting of sea ice as shown in this photo), literally hundreds of impacts of climate change on species and ecosystems are being reported in the scientific literature. The IUCN Red List of Threatened Species documents the range of climate-related threats and the percentage of threatened species affected by these threats. Wendy Foden and colleagues have documented some of the world's species that are most vulnerable to climate impacts. (Data from Foden, W. B. et al., *Identifying the world's most climate change vulnerable species: A systematic trait-based assessment of all birds, amphibians, and corals.* PLoS ONE, 2013. **8**(6): e65427.)

to climate adaptation measures. Scientists such as Tom Lovejoy, recently of George Mason University, and Pat Halpin of Duke University have been writing and speaking about climate adaptation and climate change–related risks since the 1980s. From 2010 onward, an explosion of interest in adaptation has been addressed through conferences, training workshops, funding programs, peer review and gray literature, books, and high-profile publications.

One of the more comprehensive examples of adaptation has taken place in central Europe through an initiative known as *HABIT-CHANGE: adaptive management for protected areas* (www.habit-change.eu). The major objective of this initiative was "to evaluate, enhance and adapt existing management and conservation strategies in protected sites to pro-actively respond on likely influences of CC as a threat to habitat integrity and diversity." Project staff developed a complete toolbox of adaptation tools including decision support systems, adaptation recommendations, case studies, a management handbook, monitoring guidelines, climate impact analyses, habitat change maps, scenario analyses, and more. Many of the initiative's findings are summarized in a recent book, *Managing Protected Areas in Central and Eastern Europe under Climate Change*.[4]

Although there is no shortage of review papers and conceptual frameworks for developing adaptation strategies, at least two recent surveys and an assessment of U.S. National Wildlife Refuges suggest that there is limited progress in implementing adaptation-oriented conservation projects.[5] There has even been a question raised as to whether the world's major conservation organizations have the capacity or are structurally organized in a sufficient manner to respond to adaptation.[6] Consequently, there are still too few examples of on-the-ground conservation projects that have explicitly taken adaptation into account in developing strategies and actions[7] (**Box 9.1** provides an exception to this rule).

## Box 9.1 Putting Adaptation Principles into Action for Nature Reserves of the Royal Society for the Protection of Birds (RSPB)

The RSPB manages a network of 213 nature reserves in the United Kingdom.[a] Analyses were conducted to understand the long-term impacts of climate change on the individual nature reserves and at the scale of the entire network of preserves. These studies, which informed the 25-year vision for the network, revealed that the most significant impacts to preserves were loss of intertidal habitat, loss of low-lying islands due to sea-level rise, and inundation of wetlands.

Each nature reserve in the RSPB network has developed its own adaptation strategies and actions through conservation plans that are updated every five years. Many reserves are now carrying out adaptation measures. Some commonly implemented measures are re-creating intertidal habitat, re-creating freshwater wetlands away from vulnerable coastal areas, reducing the artificial drainage of peatland areas, and making efforts to increase freshwater supply for grassy wetland areas. These collective efforts suggest that the RSPB network is a showcase for turning adaptation planning into action.

[a] Ausden, M., *Climate change adaptation: Putting principles into practice.* Environmental Management, 2014. **54**(4): 685–698.

FIGURE 9.1.1 Wallasea Island RSPB Nature Reserve near Essex, England, where intertidal habitats are being re-created as an adaptation measure.

In this chapter, we review a few critical concepts to understanding climate impacts and adaptation, outline a set of steps for incorporating adaptation strategies into conservation plans, highlight the major strategic approaches to adaptation, confront the major barriers to implementation, and direct practitioners to important case studies of on-the-ground implementation. Although we consider climate adaptation planning an important enough topic to devote an entire chapter to it, we have also sprinkled notions about adaptation planning throughout the book.

## Defining Adaptation and Adaptation-Related Concepts

The Intergovernmental Panel on Climate Change (IPCC)[8] defines adaptation as "the process of adjustment to actual or expected climate and its effects. In human systems, adaptation seeks to moderate or avoid harm or exploit beneficial opportunities. In some natural systems, human intervention may facilitate adjustment to expected climate and its effects." A more straightforward, working definition of adaptation is "preparing for, coping with, or adjusting to climatic changes and associated impacts."[9]

It is worth noting in the IPCC definition that adaptation has a natural or ecological component as well as a social or human component. And consistent with our approach elsewhere in this book, here we delve into both aspects, including the linkages and feedback loops in social-ecological systems. Climate change itself is a complex physical process, and a thorough description of the processes for modeling and analyzing the impacts of climate change is beyond the scope of this chapter (see Kropp and Scholze[10] for a practitioner-oriented summary of Earth's climate systems, climate models, and climate impacts). In this section, we focus on adapting to climate change and discuss several concepts that are critical to understanding and planning adaptation strategies as well as implementing them.

### Risk and Uncertainty

In its 2010 report on Adapting to the Impacts of Climate Change,[11] the National Research Council suggested that adapting to climate change is "fundamentally a risk management strategy." Its report indicated that managing these risks comes down to using the best scientific data and analyses to estimate the impacts and consequences of climate change on human and natural systems and then choosing the appropriate strategic response.

While this is clearly sound advice, we don't expect practitioners to drop their current priorities and direct all their attention to addressing the risks of climate change. Our goal is for conservation planners to weigh the risks of climate change impacts against other factors that could significantly influence a project's ability to achieve its outcomes and then take the most appropriate actions. For example, in many situations habitat degradation or loss is likely a far greater risk to a project's success in the near term than are the impacts of climate change.

Indeed, a commentary in the journal *Nature* (2013) makes these very points—climate change impacts need to be considered as one of a suite of critical threats to biodiversity, but in many cases these impacts are not the most urgent ones that conservationists need to address.[12] On the other hand, it is important to remember that even projects focused primarily on other stressors need to incorporate changes in climate. Often our traditional methods, tools, and strategies have hidden assumptions of a stationary climate (e.g., we assume variation in climatic factors will continue to fall within some "normal" set of bounds), and our success over the long term depends on updating our practices to be robust to continued changes in climate drivers.

One of the challenges of assessing risks that are specific to climate change is the enormous amount of uncertainty associated with estimating the impacts of climate change as well as ascertaining how effective various strategies will be at addressing these impacts. In Chapter 8 we observed that there are four main classes of uncertainty in conservation plans. The first of these—inputs to the planning process, such as data—is a clear concern in assessing climate change. Examining FIGURE 9.2 helps explain these uncertainties.

We can see from this figure that our climate system itself is a complicated atmospheric process that is influenced by interactions with Earth's terrestrial, freshwater, and marine ecosystems. Scientists are not yet able to accurately incorporate all mechanisms, due to an incomplete understanding of factors that affect climate and create climatic patterns. In addition, the many different models of global climate each have somewhat different assumptions (e.g., how clouds will behave as temperatures rise) and perform differently when tested against historical climate data. Climate models also need estimates of the amount of greenhouse gases in the atmosphere—but this, in turn, relies on assumptions and models of population growth, emissions regulation policies, and economics—all with their own obvious and less obvious uncertainties.

While the weight of evidence from looking at multiple models suggests some key trends, there is a considerable amount of uncertainty associated with linking these projections to trends at specific locations or at a given time. These uncertainties increase with longer-term projections. Despite these uncertainties, it's important to remember that many practitioners and managers will not be relying solely on models to predict climate change. In many cases, substantial observational data on changes in temperature or precipitation, for example, will

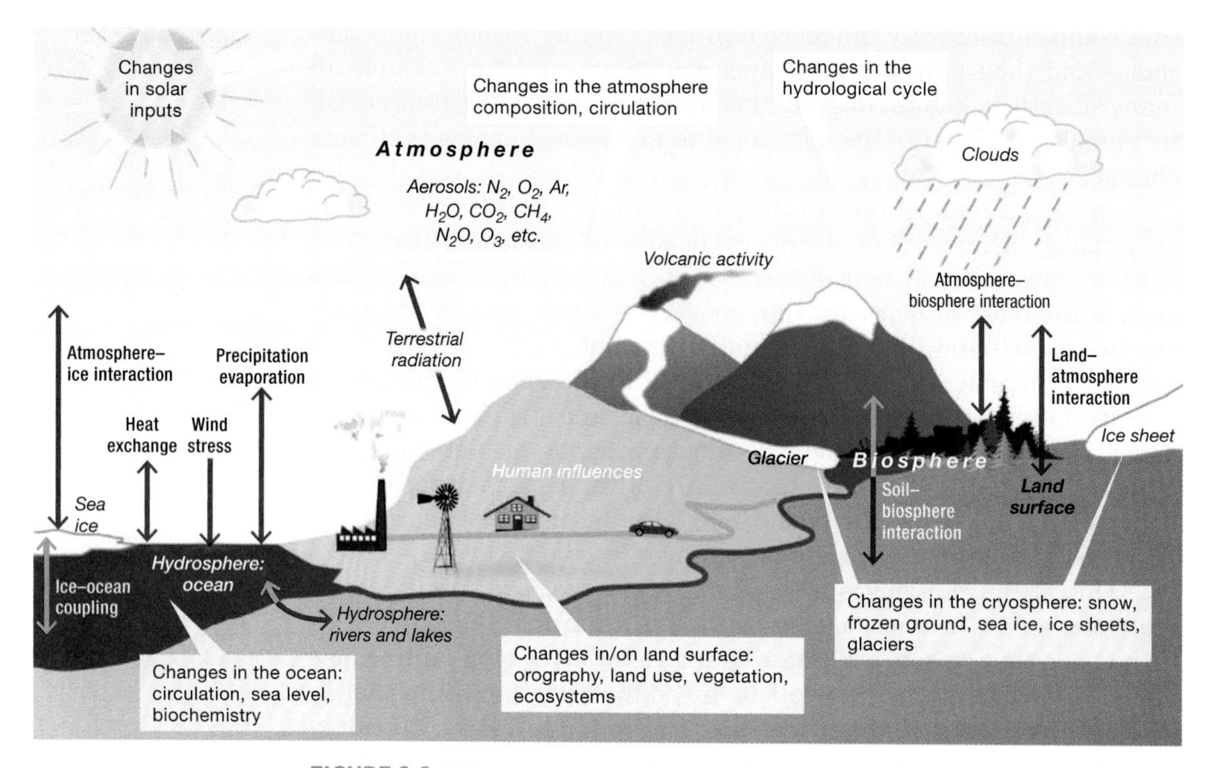


FIGURE 9.2 Major components of Earth's climate system, the physical and ecological processes of the components, and interactions between components. (Adapted from Kropp and Scholze, 2009.)

provide them with the evidence they need to move forward with conservation actions.

The interactions of the physical process of climate change (e.g., change in temperature and precipitation), in turn, have impacts on species and ecosystems, and predicting these impacts through ecological models represents a second type of uncertainty in climate impact analyses. These ecological models have their own sources of uncertainty.[13] Among them are limited or unreliable data on the natural history of species and ecosystems and their sensitivities to climate change, unknown interactions of these species and systems with non-climate stressors such as invasive species, uncertain effects of human responses to climate change,[14] and scientific disagreements on what we know about species and ecosystems and their responses to climate.

In addition to uncertainty associated with climate impact analyses, uncertainty will also be associated with the effectiveness of strategies that we use to abate and mitigate these impacts—this is the same uncertainty class we referred to in Chapter 8 when we spoke of uncertain futures for social and ecological systems. For example, in the section on strategies (later in this chapter), we explore the idea of conserving the "ecological stage." This approach assumes that most species distributions are closely tied to underlying abiotic or geophysical factors when, in fact for many species, either we don't know if this is true or we know that it is not. The important point, as with conservation strategies in general, is that many uncertainties will be associated with the effectiveness of different adaptation strategies.

As important as these climate-related uncertainties may be, we also need to acknowledge that natural resource managers and conservation practitioners have always had to deal with uncertainty in the actions they take. This was the case long before climate impacts and adaptation became part of our everyday parlance. What's different now is that climate scientists have gone to great lengths to quantify some of the uncertainties associated with climate change and projected impacts of this change. Although we don't want to diminish the importance of uncertainty and risk associated with climate change, it's important to point out that managers and conservationists now have improved tools and capabilities to assess uncertainty and risk (Chapter 8) and make informed decisions in the face of it.

## Vulnerability Assessments

The term *vulnerability assessments* can apply equally well to human and ecological communities, but in this brief section we consider only the ecological components of vulnerability assessments. The (U.S.) National Wildlife Federation (NWF), with the help of many experts, published a key report on vulnerability assessments in climate change—*Scanning the Conservation Horizon: A Guide to Climate Change Vulnerability Assessment.*[15] Vulnerability assessments have two goals: (1) identify those species and ecosystems likely to be most affected by climate change, and (2) understand why certain species and ecosystems are vulnerable so that appropriate responses can be made. As defined by the IPCC Fourth Assessment Report and NWF, vulnerability assessments have three components—sensitivity, exposure, and adaptive capacity (FIGURE 9.3).

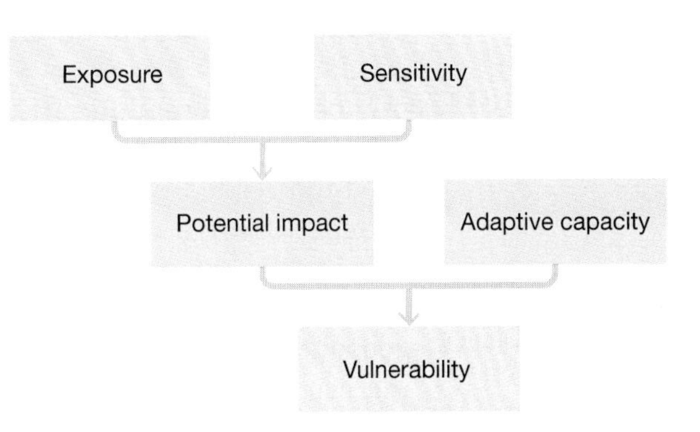


FIGURE 9.3 Key components of vulnerability assessments—exposure, sensitivity, and adaptive capacity. See text for details. (Adapted from Glick et al., 2011.)

1. **Sensitivity** refers to the degree to which a species or ecosystem is likely to be affected by a change in climate or associated factors. Innate biological characteristics such as the temperature

range within which a species can survive will influence sensitivity; some species have broad tolerances while others have narrower ones. Environmental factors will also influence sensitivity. For example, a forest that is dependent on a natural fire regime as part of its life history cycle will be sensitive to climate-induced changes in this fire regime.

2. **Exposure** is a measure of how much of a change a species or ecosystem is likely to experience. It is influenced by direct changes in climate, such as temperature and precipitation, but also by more indirect factors, such as a change in the hydrological regime or nutrient cycling. Not all populations of a species, for example, will necessarily experience the same exposure to climate change.
3. **Adaptive capacity** refers to the ability of a species or system to cope with or adjust to the effects of that change. Adaptive capacity has also been used in relationship to human communities and their abilities and capacities to help their own communities adapt as well as devote attention and resources to adaptation measures for the natural world.[16]

As we observed in Chapter 5, conservation planners have used the terms **vulnerability assessment** and *threat assessment* nearly as synonyms. In the climate adaptation literature, vulnerability assessment has taken on a slightly different meaning by the addition of this third component of adaptive capacity.

In their report *Scanning the Conservation Horizon,* Patty Glick and colleagues[17] provide examples and summary tables for factors influencing sensitivity, exposure, and adaptive capacity as well as detailed steps for conducting a vulnerability assessment. There is no single best way to conduct a vulnerability assessment, and there are a number of different quantitative and qualitative methods and tools for estimating exposure, sensitivity, and adaptive capacity. Two commonly used indices for evaluating species vulnerability are the NatureServe Climate Change Vulnerability Index (CCVI)[18] and the U.S. Forest Service System for Assessing Vulnerability of Species (SAVS).[19] Vulnerability assessments have also been conducted for ecosystems and even geopolitical units such as cities. The range of approaches for evaluating ecosystem vulnerability is diverse. As but one example, wetland scientist Joanna Ellison has assessed the vulnerability of mangrove ecosystems to climate change and sea-level rise[20] (**TABLE 9.1**).

Outputs of vulnerability assessments are as variable as the methods that produce them (**FIGURE 9.4** gives an example output). *Scanning the Horizon* includes seven case studies (all from the United States) with results of vulnerability assessments that were developed through different methods. Vulnerability assessments provide

**TABLE 9.1** Vulnerability rankings for mangrove ecosystems in Cameroon, Tanzania, and Fiji. Ellison's study provides detailed information on why various metrics were selected to evaluate the three primary components of vulnerability (exposure, sensitivity, adaptive capacity), how those metrics were measured or obtained at the field sites, and how they were subsequently assigned ranks.

| Components | Douala Estuary, Cameroon | Rufiji Delta, Tanzania | Tikina Wai, Fiji |
|---|---|---|---|
| ***Exposure*** | | | |
| Tidal range | 4 (1–1.5 m) | 1 (>3 m) | 4 (1–1.5 m) |
| Relative sea-level rise (RSLR) | 3 (site stable) | 2 (site slightly uplifting) | 4 (site slowly subsiding) |
| Sediment supply rate | 2 (fairly high) | 2 (fairly high) | 4 (fairly low) |
| Climate modeling | n/a | 2 (rainfall unchanged) | n/a |
| ***Sensitivity*** | | | |
| Mangrove condition | 2 (moderate impract) | 1 (no or slight impact) | 1 (no or slight impact) |
| Mangrove basal area ($m^2$ per hectare) | 2 (15–25) | 2 (15–25) | 1 (>25) |
| Basal area change | 1 (positive) | 1 (positive) | 1 (positive) |
| Recruitment | 2 (most species producing seedlings) | 1 (all species producing seedlings | 1 (all species producing seedlings |
| Mortality | 1 (<4%) | 1 (<4%) | 1 (<4%) |
| Litter productivity | n/d | n/d | 1 (high, including >20% fruits and flowers) |
| GIS—seaward edge retreat | 2 (some) | 2 (some) | 1 (none) |
| GIS—reduction in mangrove area | 1 (none or little) | 3 (moderate) | 1 (none or little) |
| Elevation ranges of mangrove zones | 4 (30–50 cm) | 1 (60+ cm) | 2 (50–60 cm) |
| Net accretion rates in mangroves | 3 (equal to RSLR) | 1 (>1 mm greater than RSLR) | 4 (<1 mm less than RSLR) |
| Adjacent coral reef resilience | n/a | 2 (high) | 1 (very high) |
| Adjacent sea grass resilience | n/a | n/d | 3 (moderate) |
| ***Adaptive capacity*** | | | |
| Elevations above mangroves | 3 (some migration areas available) | 1 (migration areas very available) | 3 (some migration areas available) |
| Community management capacity | 2 (fairly good) | 4 (poor) | 1 (good) |
| Stakeholder involvement | 2 (fairly good) | 3 (moderate) | 1 (good) |
| Mangrove protection legislation | 3 (moderate) | 1 (good) | 3 (moderate) |
| Total | 37 | 30 | 38 |
| Number of components | 16 | 18 | 19 |
| **Vulnerability rank** | **2.3** | **1.8** | **2.0** |

*Source:* Data from Ellison, 2015.

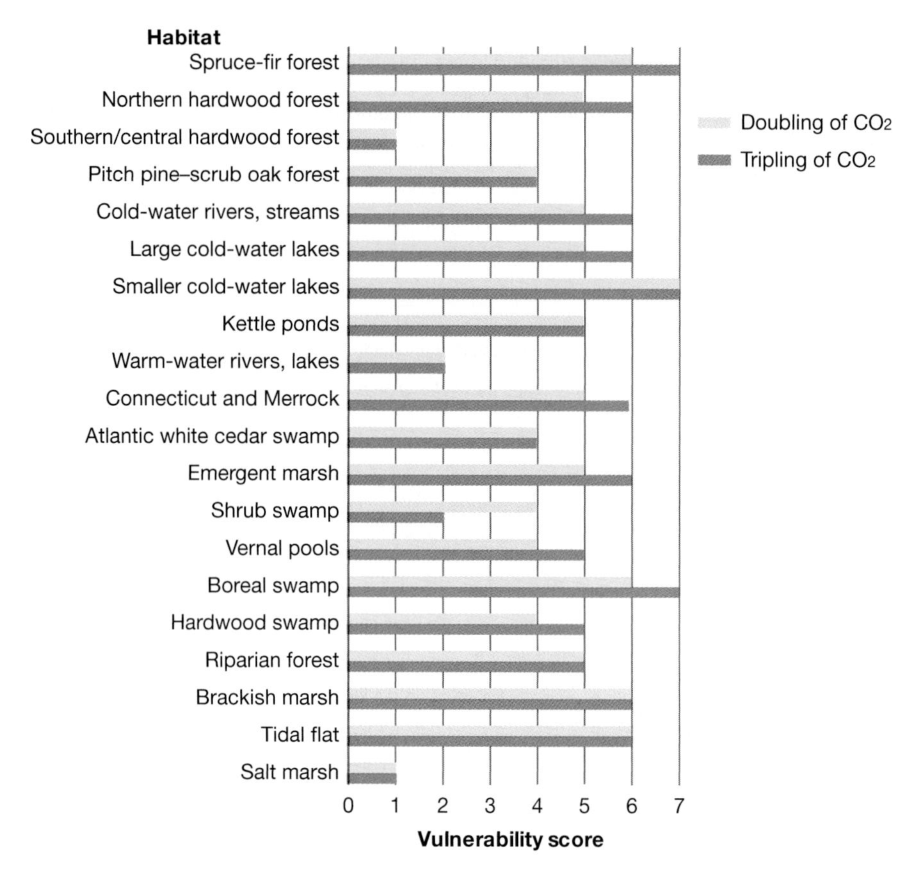


**FIGURE 9.4** The vulnerability of major habitats in Massachusetts, USA, to impacts from climate change. The green bars represent a doubling of $CO_2$ emissions; the orange bars, a tripling. Scores range from 7 (a habitat is at risk of being eliminated from the state) to 4 (habitats are unlikely to be affected) to 1 (habitats may expand in response to climate change). (Data from Glick et al., 2011.)

practitioners with an assessment of what sorts of factors put species and ecosystems at risk from climate change.[21] With that information in hand, conservation teams are better positioned to begin developing strategies that will help reduce these vulnerabilities.

## Social Adaptation and Vulnerability

Any discussion of adaptation planning would be woefully inadequate without an acknowledgment of the substantial contributions of social science to thinking and acting on climate adaptation. Indeed, the literature and knowledge base on adaptation of human communities is as great as or greater than that for natural communities. The same concepts of exposure, sensitivity, and adaptive capacity discussed earlier are equally relevant and widely used to assess the vulnerability of human communities to climate change. **Social adaptation to climate change** generally "refers to a process, action, or outcome in a system (household, community, group, sector, region, country) in order for the system to better cope with, manage or adjust to changing conditions, stress, hazard, risk or opportunity."[22]

In many parts of the world, especially the developing world, there is an increasing recognition that conservation planners need to concurrently plan for the adaptation of human and ecological communities. This notion has overlap with the concept of ecosystem-based adaptation (EBA) discussed next, with the recognition that there will not always be win-win adaptation solutions that benefit both nature and human communities. Human communities of tropical Oceania (a concentration of small island nations including Melanesia, Micronesia, Polynesia, and northern Australia) that have long been vulnerable to climatic events represent an outstanding example of social adaptation. These communities have been adapting to extreme climate events of drought, flooding, cyclones, and tsunamis for centuries through crop diversification, food storage, water rationing, and moving houses inland.[23] Much can be learned from these communities on how they have historically coped with extreme climate events and how that knowledge can be applied to today's adaptation challenges.

As we discussed in Chapter 1, conservationists have long recognized the importance of linked social-ecological systems. Social scientists working in the field of climate adaptation commonly make this link explicitly (FIGURE 9.5). Even if your plan doesn't involve humans, the actions of human communities to climate change may play an important role in influencing your conservation objectives.

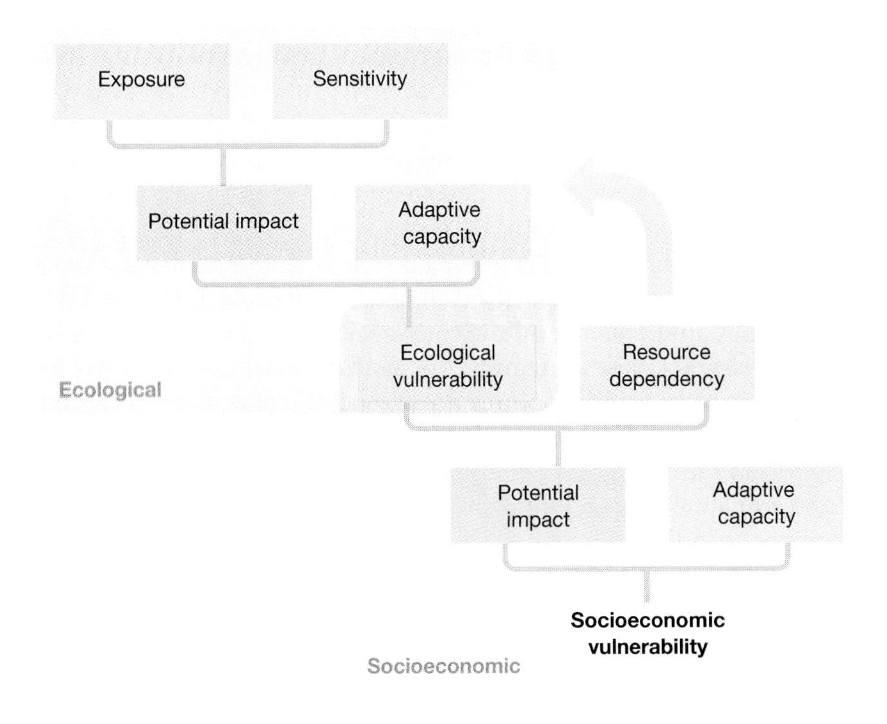


**FIGURE 9.5** Interdependency of social and ecological systems in terms of assessing vulnerability. This dependency highlights the difficulty in assessing either ecological or social-economic vulnerabilities independently. (Adapted from Marshall et al., 2010.)

In the context of human communities, exposure, sensitivity, and adaptive capacity take on similar but somewhat different meanings than those discussed for the ecological community. Social adaptation touches on a large array of disciplines including community development, risk management, disaster planning, sustainable development, food and water security, and livelihoods, just to mention a few. Several examples of incorporating climate risks and adaptation in these disciplines come from efforts to sustain tropical coastal communities that are at risk of sea-level rise, storm surges, and other impacts of climate change.

Nadine Marshall of Australia's Commonwealth Scientific and Industrial Research Organization (CSIRO) and colleagues have developed a framework with numerous case studies for social adaptation in these tropical communities.[24] They demonstrate the use of different social science tools such as SLED (Sustainable Livelihoods Enhancement and Diversification) and CRiSTAL (Community-based Risk Screening Tool—Adaptation and Livelihoods), both of which were developed by or for the International Union for the Conservation of Nature (IUCN) and other organizations partly to address the challenges of poor communities in dealing with fluctuating livelihoods and risks associated with climate change.

There is now a substantial set of experiences related to social adaptation. In their review of social adaptation, adaptive capacity, and vulnerability, Barry Smit and Johanna Wandel[25] note an important finding of these experiences: adaptation actions are rarely taken in response to climate change alone, but more commonly are taken when risks are considered in the broader frameworks of land use planning, livelihood improvements, food and water security, and other development-related initiatives. A second key finding is that local adaptation initiatives can easily be constrained by "broader social, economic, and political forces," suggesting that a development agenda may sometimes supplant well-intended adaptation actions.

Hallie Eakin and colleagues[26] divide social adaptive capacity into two forms—generic capacity that is associated with human development goals, and specific adaptive capacity that is needed to adapt to climate change. They suggest that both forms must be addressed simultaneously in order for human or social adaptation to climate change to succeed. These findings are important insights for how conservation planners and practitioners need to think about developing and implementing adaptation measures that will be effective in enhancing human communities.

## Ecosystem-Based Adaptation

The phrase **ecosystem-based adaptation** (**EBA**) (also sometimes referred to as nature-based adaptation) originated in the halls of United Nations climate negotiations and among IUCN staffers.[27]

Ecosystem-based adaptation is a form of climate adaptation, but it is not synonymous with adaptation for nature conservation purposes. The Secretariat of the Convention on Biological Diversity, as part of the negotiations related to the United Nations Framework Convention on Climate Change (UNFCCC), defines EBA as follows:

> The use of biodiversity and ecosystem services as part of an overall adaptation strategy to help people to adapt to the adverse effects of climate change. Ecosystem-based adaptation uses the range of opportunities for the sustainable management, conservation, and restoration of ecosystems to provide services that enable people to adapt to the impacts of climate change. It aims to maintain and increase the resilience and reduce the vulnerability of ecosystems and people in the face of the adverse effects of climate change.

EBA can include activities related to or complementary to disaster risk reduction, biodiversity conservation, habitat restoration, carbon sequestration, sustainable water management, and food security, among others. One of the most frequently cited examples of EBA is the conservation of mangrove ecosystems that can help abate the effects of storm surges, often helping to decrease the vulnerability of human communities living near and working in these ecosystems.

Although EBA actions are influenced primarily by the needs of human communities, in some cases they will clearly help conserve natural communities as well. A recent study along the U.S. coastline is a prime example of the potential of EBA. In that assessment,[28] researchers estimated that 67% of the U.S. coastline is protected by natural habitat—that, if lost, would double the number of poor families, elderly people, and total property value in the areas at highest risk from coastal hazards such as storm surges (FIGURE 9.6).

Similarly, a global analysis of the role and cost-effectiveness of coral reefs in risk reduction suggests that they can reduce wave energy by 97% and that at least 100 million people may have risks from storms reduced by the presence of coral reefs.[29] The IUCN has published a set of case studies of EBA projects from around the globe, such as one involving Integrated Water Resources Management in the Pangani River Basin of Tanzania (FIGURE 9.7), where conservation organizations and local communities are working to improve catchment flows that benefit human communities experiencing long-term drought as well as freshwater ecosystems.[30]

Many new tools are being developed to help implement EBA projects, and one of the most promising is Coastal Resilience 2.0 (www.coastalresilience.org). This interactive suite of decision-making tools can assist local governments and cities in developing risk reduction solutions by evaluating storm surge, sea-level rise, natural resources, and economic infrastructure at risk.

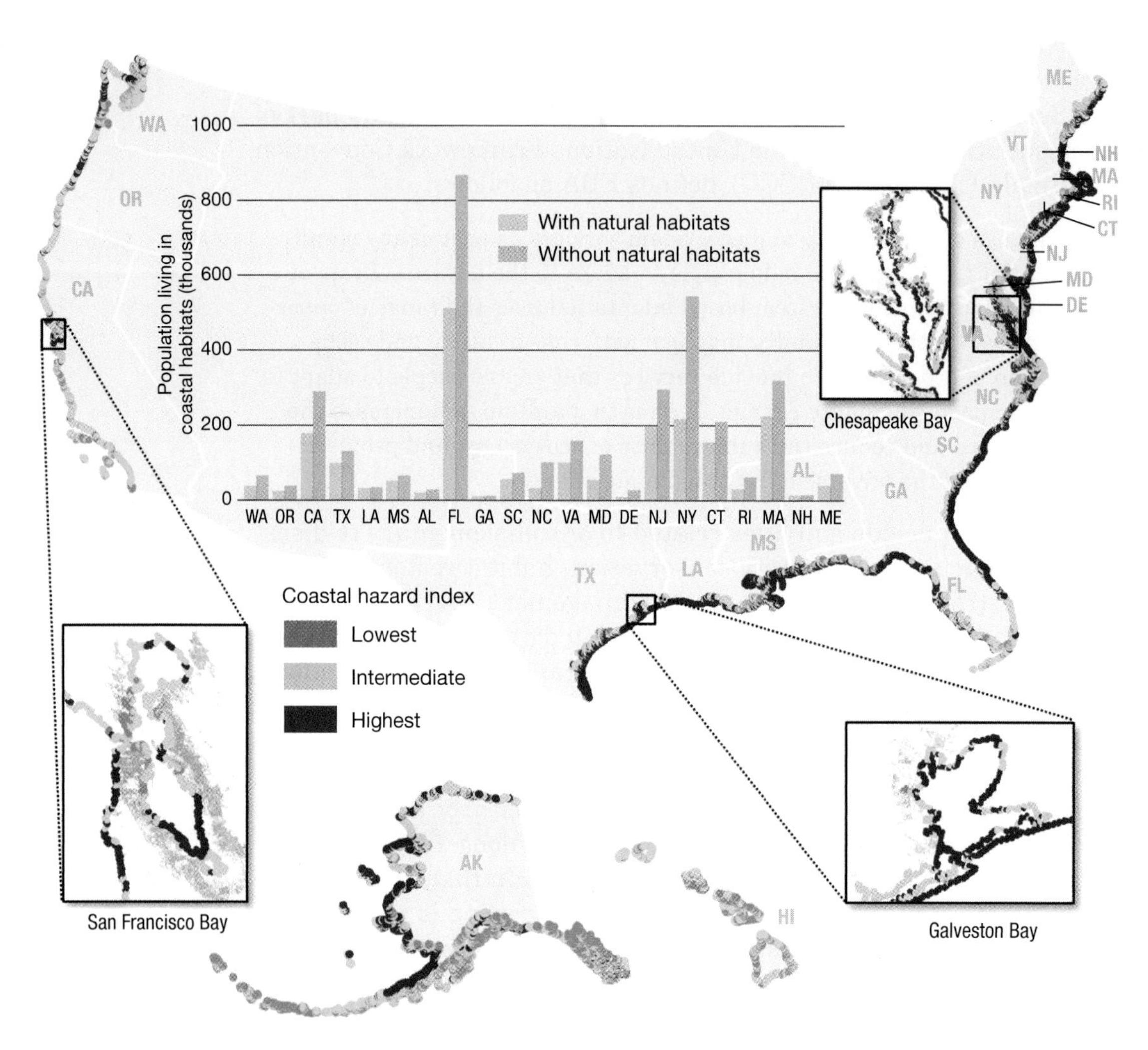


FIGURE 9.6 Exposure of the U.S. coastline and human populations to sea-level rise and increase in storm frequency in 2100. Bar graphs indicate human populations exposed to climate change impacts with or without natural habitats present. (Adapted from Arkema et al., 2013.)

## Scale Considerations

In Chapter 5, we first discussed the importance of spatial scale in conservation planning. Scale, both temporal and spatial, plays a particularly important role in analyzing climate change impacts and evaluating adaptation strategies.

First, we should state the obvious. Projections about climate impact are likely to be more accurate and more probable in the medium term (in the next 15–30 years) than in the longer term (50–100 years). In part, that's because it is difficult to predict what humans are going to do about controlling emissions; moreover, whether the rate of green-

FIGURE 9.7 Water users in the Pangani River Basin, Tanzania, where an ecosystem-based adaptation project is working to improve flows for people and freshwater ecosystems. (Adapted from Colls et al., 2009.)

house gas emissions declines, levels off, or increases will have significant bearing on the severity and frequency of climate impacts. Longer-term projections about climate change are also more difficult to make because of the uncertainties in trying to model the complex interactions between the atmosphere and the terrestrial and aquatic environments on Earth.

Of more direct relevance to conservation planners is the interplay of spatial scale, climate impacts, and adaptation strategies and actions. Adaptation planning is taking place at scales ranging from small protected areas to nearly the globe if we consider the efforts of institutions like the World Bank or its Global Environment Facility (GEF). The strategies and actions that can be planned for and implemented will obviously be influenced by the spatial scale being considered. Scale will also influence estimates of climate impacts on a particular project area.

Landscape ecologist John Wiens and climate scientist Dominique Bachelet have raised a number of issues related to the mismatch in scale between conservation planning and climate change impact analyses.[31] Projections about global climate change have largely relied on **General Circulation Models (GCMs)** of the atmosphere and oceans. These models can forecast change on the scale of hundreds of kilometers. Most conservation strategies and actions, however, take place at the scale of tens of kilometers or less. The broad-scale projections of climate change from GCMs may therefore have limited relevance to conservation strategies and actions that are taking place at much finer scales.

As a result, there is now a trend to use climate projections that have been **downscaled** through a variety of statistical and modeling techniques to produce Regional Circulation Models (RCMs) that have a finer scale of resolution. These downscaled projections of climate change are more closely aligned with the scale of actions that take place in the implementation of conservation plans. Unfortunately, because of model limitations and error propagation involved in downscaling, the uncertainty of climate projections from downscaled circulation models is much greater than GCMs, especially in regions with considerable topography. Consequently, planners need to be aware of this greater uncertainty and consider steps outlined in Chapter 8 for responding to it.

The good news is that even as we write this book, the accuracy and resolution of many downscaled climate models continue to improve, and they will be increasingly useful at the local scales (tens of $km^2$) of many conservation projects. In addition, practitioners can combine projections from downscaled models with observational data on climate from larger regions and assess what is consistent between the two to help determine the most serious climate impacts on their project.

## Resilience

Probably no single word has received more attention in the social or ecological climate adaptation literature than **resilience**.[32] Some recent papers have sought to clarify its usage,[33] while others have called for overhauling the "paradigm of resilience" in relation to climate impacts.[34] Unfortunately, it is being used to imply a variety of different meanings. In their review of the climate adaptation literature (*Moving the Conservation Goalposts*), Patty Glick and colleagues at the National Wildlife Federation wrote, "The term resilience is being used so broadly and indiscriminately—and proffered so often as an adaptation panacea—that its utility as a meaningful conservation goal is being undermined."[35] They go on to suggest that practitioners must ask "resilience of what, to what" and define the core attributes or functions of a system that make it resilient. Jodi Hilty and colleagues make a similar point in their 2012 book *Climate and Conservation*[36] that the term *resilience* often lacks any explicit meaning.

Long before its application in the climate adaptation field, resilience was used in ecology to refer to an ecosystem that could return to its initial state following disturbance.[37] At times, it has a similar meaning in adaptation—that an ecosystem impacted by climate change is resilient if it can return to its pre-impact state. As Bruce Stein and colleagues suggest in *Climate Smart Conservation*,[38] this interpretation of resilience implies an effort of trying to maintain the status quo conditions of an ecosystem. At other times, resilience in an adaptation context may imply a broader definition that includes transitions to other ecological or sociological states of condition.[39]

Although we will return to this term in the strategies section later in this chapter, the take-home lesson for practitioners is to pay careful attention to the context in which it is used so as to not perpetuate further misunderstandings. For a broader view of the term *resilience* in both social and ecological systems, practitioners and planners may want to engage with the Resilience Alliance (www.resalliance.org), a research network of scientists and practitioners who investigate the dynamic nature of social-ecological systems as a basis for advancing sustainability.

In this first section of Chapter 9, we have highlighted several important concepts that planners, scientists, and practitioners need to understand to plan appropriately for adaptation. In the next section, we put many of these concepts into practice as we outline the important steps for incorporating adaptation into conservation planning.

## Incorporating Adaptation into Conservation Planning

Several recent frameworks have been put forth for either conducting "adaptation planning" or for incorporating adaptation considerations into conservation planning. Two of these frameworks—one by Bruce Stein and colleagues[40] (FIGURE 9.8) and another by Molly Cross and colleagues[41]—have many similarities with the planning steps outlined in this book and in the *Open Standards for the Practice of Conservation,* which we have referred to previously (see Table 5.1 in *Climate Smart Conservation* for a list of step-by-step adaptation planning approaches).

Some of the planning steps are specific to adaptation, whereas others are planning steps that will be familiar to any readers of the earlier chapters in this book (e.g., establishing objectives). Although these frameworks could be used to initiate an adaptation-oriented conservation plan from inception or to modify or "climatize" an existing conservation plan so that it accommodates adaptation, the former use is emphasized in the supporting documentation for these frameworks.

In this section of the chapter, we highlight and summarize aspects of the major steps of conservation planning that are unique to considering climate change impacts and adaptation strategies. Although we have mentioned many of these aspects in earlier chapters, here we delve into them in more detail and make recommendations for

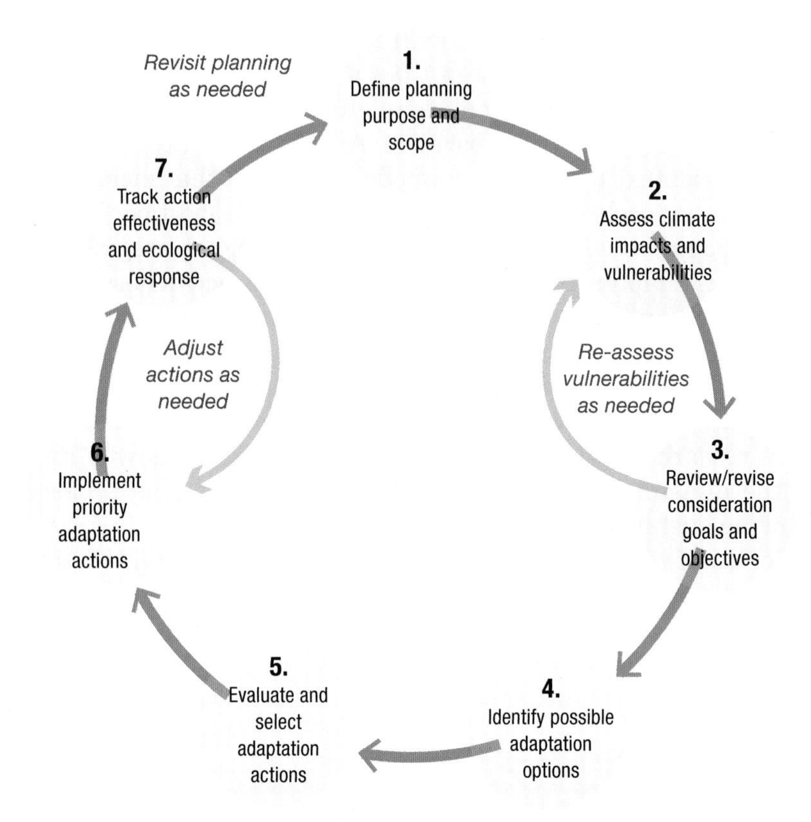


**FIGURE 9.8** An example of an adaptation planning framework from *Climate Smart Conservation*. Note the similarities to the adaptive management cycles of the *Open Standards* or Conservation Action Planning that we discussed in chapter 1 as well as to the planning steps that we have outlined in this book. (Adapted from Stein et al., 2014.)

incorporating adaptation. We also present the concepts and tools necessary to consider climate impacts and adaptation in a more coherent view of this important topic.

In most cases, these recommendations, concepts, and tools should apply across terrestrial, freshwater, and marine realms. Nevertheless, some adaptation recommendations for freshwater and marine environments deserve special attention, and we have highlighted these in **Box 9.2**. We reference the earlier book chapter in parentheses, so that readers will clearly be able to follow our guidance for adaptation planning as it relates to planning steps detailed throughout the book. See Figure 1.13 for an overview of these planning steps and as an aid in understanding how these specific adaptation steps fit into the overall planning process that we have advocated in this book. As a result, this guidance should prove useful to practitioners and planners initiating a new conservation planning effort or revising an existing one.

## Establishing or Revising Objectives (Chapter 3)

Climate impact analyses and expert opinion help us determine whether a project or program is likely to experience some impacts and give us insights into the nature of those impacts (see Box 9.4 in

## Box 9.2 Adaptation Planning in Freshwater and Marine Environments

Although most of the steps outlined earlier in this chapter and the strategic approaches discussed later apply across terrestrial, freshwater, and marine environments, some aspects of planning for adaptation in aquatic environments deserve special mention.

For example, marine biologist Elizabeth McCleod and colleagues[a] provided guidance on the design of marine protected areas in the face of climate change. Their advice included recommendations on size, shaping, risk spreading, critical areas, connectivity, and ecosystem function. The general headings are equally relevant to terrestrial environs, but the advice they provide on each of these topics is more specific to the marine environment—such as the details of larval dispersal of marine organisms. In 2012, McCleod and colleagues[b] broadened their advice to conservation planners working in the marine environment in a synthesis paper that centered on incorporating vulnerability into marine planning. Their recommendations focused on sea-surface temperature vulnerability, sea-level rise and impacts on coastal ecosystems, and vulnerability related to ocean acidification. They reviewed tools, the appropriate scale to apply them, and the advantages and limitations of these tools for addressing all three forms of vulnerability.

In 2014, Ken Anthony and a group of coral reef experts[c] developed an Adaptive Resilience-Based Management (ARBM) framework that should better enable marine planners and practitioners to manage for increased ecological resilience of reefs while reducing reef vulnerability to climate change impacts. ARBM relies on a set of management levers (e.g., improved management of sewage) to reduce key stressors. A combined strategy of reducing risk and supporting resilience is needed to address these stressors and will be critical as climate change impacts increase the vulnerability of reefs globally. FIGURE 9.2.1 illustrates the ARBM framework: management levers are marked as 1–7, and drivers, actions, and conditions of the social-ecological system are labeled 8–11.

As we observed at the start of this chapter, increasing attention is now being given to the role that coastal ecosystems can play in reducing risks to people in coastal areas (e.g., **wave attenuation** and erosion control) from extreme storm events, many of which are linked to climate change impacts. Conservation of coastal ecosystems, or green infrastructure, can in some cases be more effective—both cost-wise and in terms of reducing risks—than so-called built or gray infrastructure. Although built solutions like seawalls remain the preferred alternative in most cases for local governments and communities, conservation of coastal ecosystems is gaining some traction.

Marine biologist Mark Spalding and colleagues[d] have outlined 10 recommendations for improving the incorporation of coastal ecosystem conservation in policy, planning, and funding for hazard reduction. Here are some of their most important recommendations: (1) quantify risk reduction benefits of ecosystems at the site level; (2) develop models and plans to account for risk reductions from multiple different types of ecosystems that could occur in a project area; and (3) ensure that loans for disaster risk reduction are informed by the benefits of natural infrastructure and the costs and lost opportunities from gray infrastructure development. Earlier in this chapter, we mentioned Coastal Resilience 2.0 (www.coastalresilience.org) as an important decision support tool that can help local communities reduce the risks of hazards from sea-level rise and storm surge.

Not surprisingly, the impacts of climate change on freshwater supplies are receiving some attention from water planners and policy makers at all levels of government worldwide.

*(continued)*

## Box 9.2 Adaptation Planning in Freshwater and Marine Environments *(continued)*

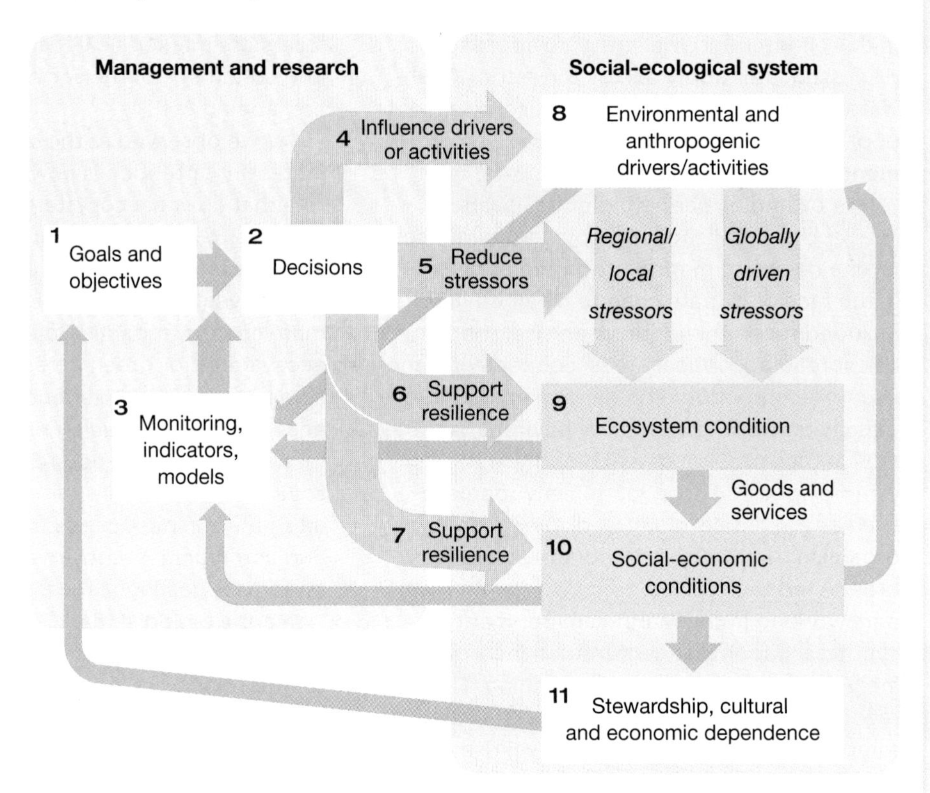


**FIGURE 9.2.1** Adaptive Resilience-Based Management (ARBM) framework combines an approach of reducing risk with supporting resilience to address climate stressors on reefs.

Drought and its associated water scarcity, intensified by climate change, are already seriously affecting agricultural and municipal water supplies—and these impacts are projected to worsen in the years ahead for many parts of the world. Far less attention is being paid to the impacts of climate change on freshwater ecosystems and the species they support and adaptation strategies that could mitigate and abate these impacts.

Allison Aldous and colleagues[e] developed adaptation strategies for three case study sites where climate impacts are anticipated: upper Klamath Basin, USA; Murray-Darling Basin, Australia; and Savannah River Basin, USA. Reconnecting floodplain wetlands and protecting groundwater recharge zones are proposed strategies for the Klamath Basin, where reduced snowpack and late season flows are affecting the system. New infrastructure to help deliver environmental flows may help the drought-plagued Murray Basin. Severe droughts are also anticipated in the biodiverse-rich Savannah Basin in the southeastern United States, where conservation and natural resource managers have been working for several years to reprogram the operation of dams and as a consequence improve downstream flows for environmental purposes.

At a broader (global) scale, the World Wildlife Fund and the World Bank[f] have taken a synthetic view of impacts, vulnerabilities, and adaptation strategies for freshwater ecosystems in natural resource management and biodiversity conservation. Their recommendations are extensive and at multiple organizational levels. At the project level, for example, their focus is on strengthening environmental flows and on the design, siting, and operation of water infrastructure projects in the future. They recommend that management objectives for freshwater adaptation strategies focus on sufficiently functioning water management institutions (enabling preconditions) with adequate monitoring and evaluation programs, policies and implementation mechanisms to protect environmental flows, and measures to reduce existing pressures on surface and groundwater flows.

---

[a] McLeod, E. et al., *Designing marine protected area networks to address the impacts of climate change.* Frontiers in Ecology and the Environment, 2008. **7**(7): 362–370.

[b] McLeod, E. et al., *Integrating climate and ocean change vulnerability into conservation planning.* Coastal Management, 2012. **40**(6): 651–672.

[c] Anthony, K. R. N. et al., *Operationalizing resilience for adaptive coral reef management under global environmental change.* Global Change Biology, 2015. **21**: 48–61.

[d] Spalding, M. D. et al., *Coastal ecosystems: A critical element of risk reduction.* Conservation Letters, 2014. **7**: 293–301.

[e] Aldous, A. et al., *Droughts, floods and freshwater ecosystems: Evaluating climate change impacts and developing adaptation strategies.* Marine and Freshwater Research, 2011. **62**(3): 223–231.

[f] Le Quesne, T. et al., *Flowing forward: Freshwater ecosystem adaptation to climate change in water resources management and biodiversity conservation*, in *Water Working Notes*, WWF, editor. 2010, Washington, DC: The World Bank.

Situation Analysis step 3). Then we need to incorporate that thinking into the overall objectives of the project. Doing so may influence the objectives of a new conservation project, especially those where there is some degree of confidence about the impacts of climate change. For example, one of us (CRG) has reviewed a number of proposals focused on the conservation of salt marshes or other coastal ecosystems where sea-level rise is a virtual certainty. In cases where development or land use change has already precluded the possibility of these coastal systems migrating inward, it may mean deciding not to focus conservation actions on these coastal areas. An alternative might be to shift the objective to conservation of partially submerged intertidal ecosystems.

Scientists at the National Wildlife Federation have used the clever phrase "moving the conservation goalpost" in referring to this process of rethinking conservation objectives in the face of climate change.[42] Conservation scientists Nicole Heller and Richard Hobbs[43] have observed that for many managers and conservation practitioners there is a tight relationship between places and the historical species composition of these places—a relationship that many conservationists associate with naturalness (Chapter 4). Given the change in species composition that is resulting from climate change, they suggest that conservation goals and objectives associated with such naturalness may be increasingly unattainable and that more "open-ended goals" or objectives such as restoring the hydrological function of a site may be more appropriate. For practitioners who want more advice on this topic, *Beyond Naturalness: Rethinking Park and Wilderness Management in an Era of Rapid Change*[44] is an excellent resource that addresses the notion of revising conservation goals and objectives from varying perspectives. Rethinking project objectives will inevitably influence how we think about alternative strategies to address climate impacts, and should include the *what* (features we are trying to conserve), *why* (outcomes we hope to achieve), *where* (geographic scope of project), and *when* (time frame of project) aspects of objectives.[45]

One example of modifying project objectives in response to climate impacts comes from California, USA, where climate scientist Rebecca Shaw and colleagues[46] evaluated the economic costs of achieving current conservation goals under future climate change scenarios for a project in the Mount Hamilton area. Using climate envelope models that predict a species distribution based on correlations with environmental variables such as temperature and rainfall and estimates of future land acquisition and management costs, they estimated that it would be necessary to expend US$1.7–1.8 billion to maintain 80% of the suitable habitat identified through climate modeling for 11 focal species through 2050. Such analyses obviously make some assumptions about the future, and any reasonable set of conservation practitioners and scientists could debate these assumptions and the various models used to predict future distributions of focal species. Regardless of the assumptions, this exercise was heuristic in pointing out that there are likely substantial costs in many existing conservation projects for trying to maintain current conservation objectives in the face of climate change. In many cases, expanding the conservation network with traditional protection tools like acquisition or easements will be prohibitively expensive.

The results from this one pilot effort point to the potential need to rethink and revise conservation objectives for many existing conservation projects, especially those with species-driven objectives, as well as develop new strategies that may be more flexible and less expensive. **Box 9.3** provides an interesting test case for establishing a peer review and learning network of existing conservation projects to evaluate the need for such future revisions to project goals and objectives.

## Selecting Features (Chapter 3)

Which biodiversity features should conservation planners focus on with an eye to climate adaptation? There are three prevailing ideas, none of which is mutually exclusive. The first was recommended by ecologist Pat Halpin[47] over 15 years ago: focus on maintaining natural disturbance processes such as fire regimes. Recently, others have

## Box 9.3 The Climate Clinic

In 2009, The Nature Conservancy brought together 20 conservation project teams from around the world in a "climate clinic" to develop adaptation strategies.[a] A step-by-step approach that paralleled many of the *Open Standards for the Practice of Conservation* was used to identify conservation targets (features), predict likely impacts on those targets, and develop strategies for abating those impacts. Project teams identified over 175 impacts and developed strategies to address over 40 of these impacts (**FIGURE 9.3.1**). Twelve of the 20 projects concluded that they needed to either change project boundaries or revise species and ecosystem-related objectives as a result of these analyses. Over half of the new strategies were focused on "resisting" the impacts of climate change, whereas the others were aimed at making species and ecosystems more resilient. The applications of specific conservation guidance before and during the clinic allowed for effective peer review and learning in addition to proposed strategic improvements to individual projects.

[a] Poiani, K. et al., *Redesigning biodiversity conservation projects for climate change: Examples from the field*. Biodiversity and Conservation, 2011. **20**(1): 185–201.

**FIGURE 9.3.1** Participants in the Climate Clinic present what they estimate will be impacts to their project area from climate change, and their initial thoughts on strategic interventions for those impacts.

reiterated this idea more broadly to expend more efforts on conserving ecological processes and functions.[48] The primary challenge in placing greater emphasis on ecosystem functions and process is that many of these remain poorly understood, and we often lack data to incorporate them into conservation plans.

The second idea concerning features has received considerable attention in recent years. It is often termed "conserving the stage" and refers to the need to conserve the diversity of underlying geophysical features (the stage) that species depend on instead of the species themselves (the actors). The argument for this approach is that for many species, the physical environment is a primary factor in determining species distribution. Because many species will move around in response to climate change, it may make more sense to focus on conserving their physical environment. Paul Beier at Northern Arizona University, Mark Anderson of The Nature Conservancy, and their colleagues have been the chief proponents of this focus on geophysical features.[49] FIGURE 9.9 provides an example of the application of a conserving-the-stage approach in the northeastern United States (we first introduced this concept in Chapter 3 in relation to abiotic features). This approach, which assumes that the distribution of biological diversity is driven by geophysical diversity, will not hold for all species in all places. But it is likely to be met for some groups, like herbaceous plants that are often tied to geophysical substrates and less likely to be met for others, such as some mammalian and avian species that are more habitat generalists.

The third major idea concerning features and adaptation is vulnerability assessment that we mentioned earlier in this chapter. Typically these sorts of assessments have been done for species (e.g., see the vulnerability analyses conducted by Silvia Carvalho and colleagues for 37 herptiles on the Iberian peninsula[50]), but they also have been applied to ecosystems (e.g., vulnerability assessment of mangrove ecosystems[51]). The National Wildlife Federation's *Scanning the Horizon* report on vulnerability assessments, mentioned previously, provides a variety of methodologies for conducting such assessments as well as numerous case studies.[52] It is worth noting that vulnerability assessments can provide insights into which species may be most vulnerable to climate change impacts; but depending on its overall objectives, a conservation project may not necessarily focus on those species.

## Situation Analysis and Data (Chapter 5)

In Chapter 5 we discussed the use and importance of conceptual models as tools for conducting a situation analysis for a conservation project. Such models can also be helpful in understanding how climate change will affect the biodiversity features, ecological processes, and social processes within a conservation project. In their Adaptation for Conservation Targets (ACT) framework for incorporating adaptation into natural resource management, Molly Cross and colleagues demonstrate that conceptual models can help identify the ecological, physical, climatic, social, and economic factors that are likely to be most important in understanding climate change impacts on conservation features and in developing adaptation strategies[53] (FIGURE 9.10). Such models help us understand not only how climate is most likely to affect a conservation project but also how other existing stressors are likely exacerbating climate impacts.

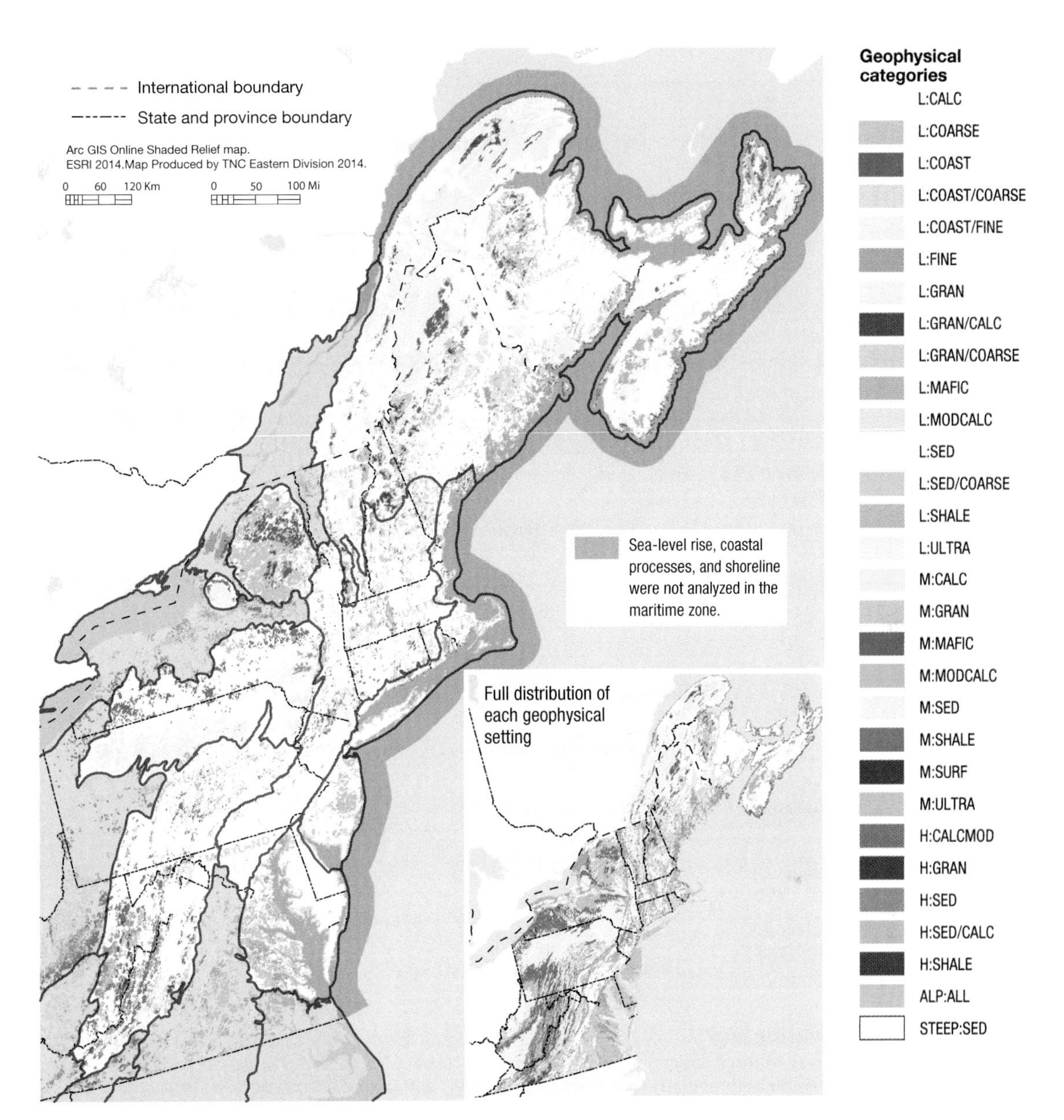


FIGURE 9.9 Geophysical settings in the northeastern United States with the highest estimated scores to relative resistance to climate change (inset shows full geographic range of geophysical settings). Full descriptions of geophysical settings and methods for estimating resilience are available in Anderson, M. et al., *Estimating climate resilience for conservation across geophysical settings.* Conservation Biology, 2014. **28**(4):959–970.

In addition to conceptual models, most project teams will gather information on possible climate impacts from some combination of expert opinion, observational data on climate, and impact analyses conducted either by the project team or by partner institutions. All these sources of information will help project teams analyze the

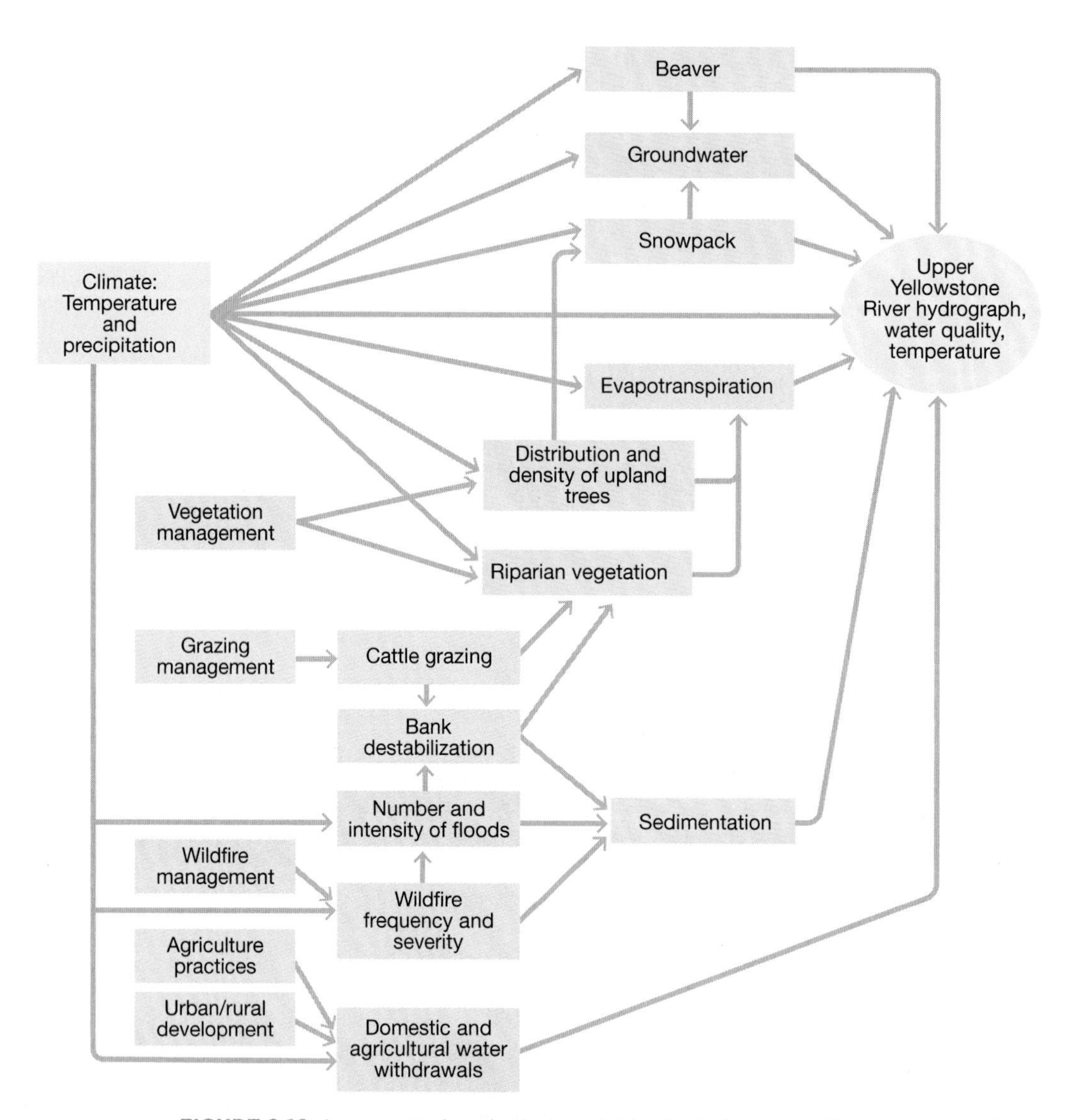


**FIGURE 9.10** A conceptual ecological model for the Yellowstone River, Montana, USA, used in the Adaptation for Conservation Targets framework. These conceptual models can be helpful in identifying points of intervention and adaptation strategies. See Chapter 5 for more information on conceptual models. (Adapted from Cross, M. et al., 2012.)

possible direct impacts to conservation features that the project hopes to conserve. But it's also important to consider how climate change may alter human land uses and behavior, aspects of climate change that are rarely considered in adaptation planning.[54] As we alluded to earlier, conducting climate impact analyses often requires some modeling expertise and some understanding of global greenhouse gas emissions (GHG) scenarios, global circulation models (GCMs), and downscaled Regional Circulation Models (RCMs), and how the results

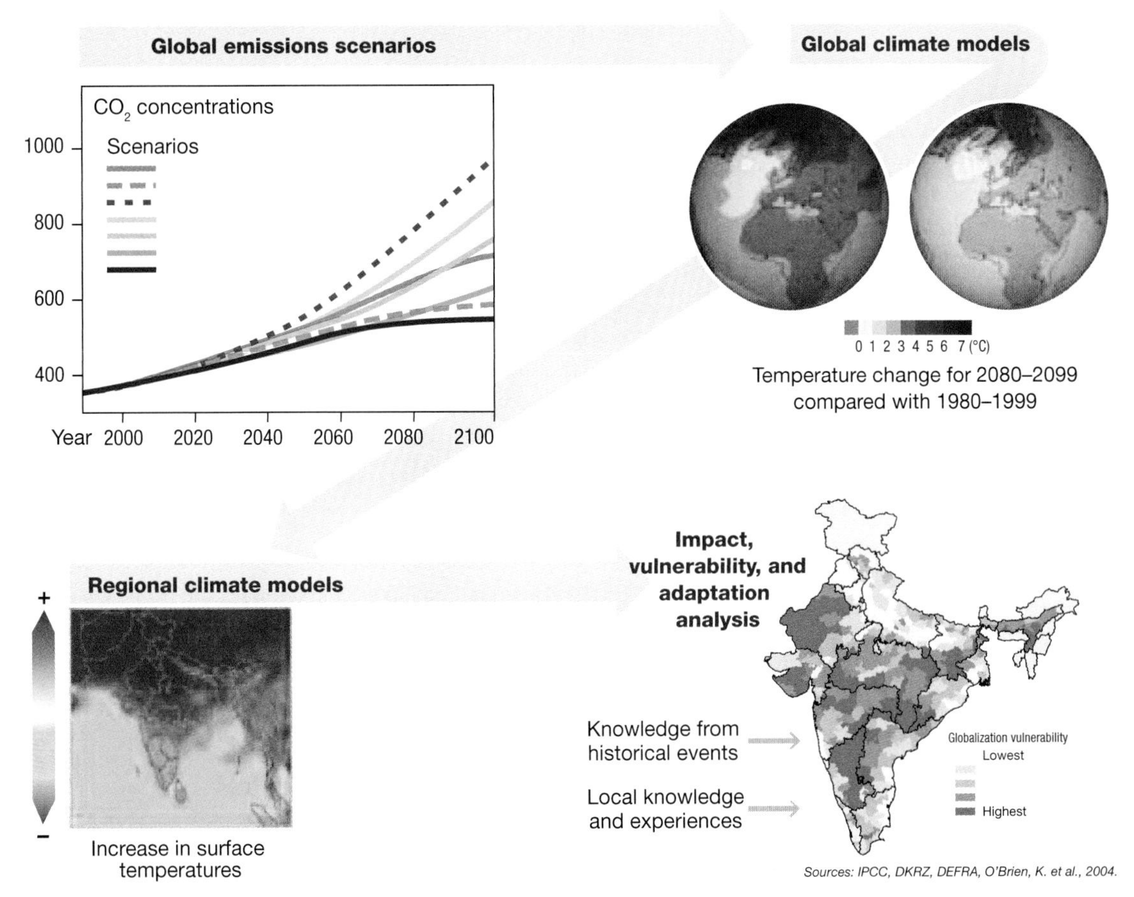


FIGURE 9.11 Steps in generating global climate models, regional climate models, and adaptation and impact assessments. (Adapted from Kropp and Scholze, 2009.)

of these modeling efforts can be used to project impacts on targeted species, ecosystems, and ecological processes and services (FIGURE 9.11).

Outputs from GCMs and RCMs that indicate the degree to which an area will get hotter or colder or wetter or drier are typically used as inputs to additional models such as dynamic vegetation models that can predict, for example, shifts in vegetation communities. Fortunately for conservation planners, there is a relatively new tool—Climate Wizard (**Box 9.4**)—that can help project teams gain an understanding of the likely climate changes that will occur in their project area.

## Developing Alternative Options (Chapter 6)

It is possible to approach adaptation from different conceptual points of view, and depending on which point of view is preferred, it will influence the strategies and actions of the project. For example, Constance Millar, a scientist with the U.S. Forest Service, and her

## Box 9.4 Climate Wizard

Climate Wizard[a] is a tool that combines web-based mapping technologies and statistical analysis platforms in a GIS environment to support practical climate change analyses. It is designed for a variety of users, ranging from a nontechnical audience of conservation practitioners and planners to advanced modelers and statisticians. It allows users to examine historical climatic change, project future climate change, and analyze extreme climatic events at sites virtually anywhere in the world.

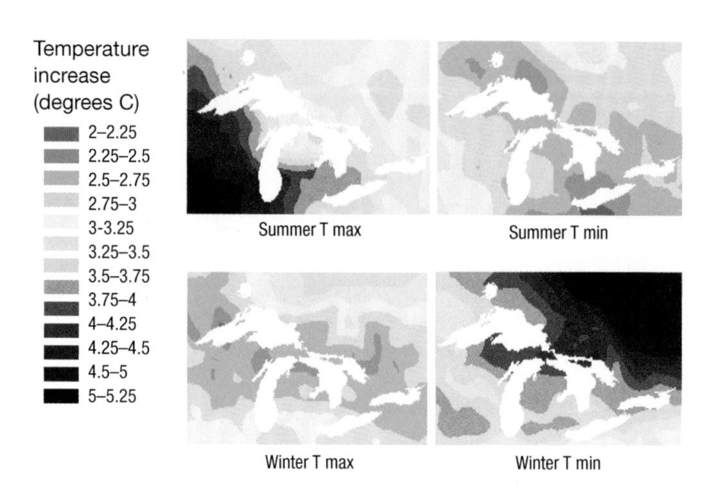


**FIGURE 9.4.1** Projected increase in temperature from climate change in the Great Lakes region by 2080 as predicted by an ensemble of three IPCC models and the A2 emissions scenario (see Figure 8.6 for explanation of emission scenarios).

Users can choose among different emissions scenarios and ensembles of General Circulation Models to run in the analyses. All that's required is a specific geographic area, a specific period for conducting the analyses, a time unit for summarizing climate data (e.g., season or annual), and a list of climate variables of interest. Output from Climate Wizard includes tables, graphs, and maps designed to answer specific climate change questions for specific areas. For example, analyses from Climate Wizard were used in the Great Lakes program of The Nature Conservancy to look at projected increases in temperature in winter and summer (see **FIGURE 9.4.1**) to evaluate climate-related risks for a conservation plan in the Great Lakes region. In this example,[b] these data suggest that in addition to increases in summer maximum temperatures, some of the strongest impacts in this region will be felt as increases in the winter minimum temperatures, an important driver of freeze-thaw cycles and frost damage.

As part of its Fifth Assessment Report in 2014, the IPCC released an updated set of future climate models that, in turn, have been incorporated into the newest version of the Climate Wizard.

The Climate Wizard project has also collaborated with the World Bank to make derived variables based on climate GCM outputs (e.g., changes in growing season length, changes in heat waves) that are relevant to climate impacts more readily available to World Bank clients and constituents (**FIGURE 9.4.2A**). Another example of a derived variable is changes in spring phenology along bird migration routes in the central and eastern United States (see **FIGURE 9.4.2B**).

---

[a] Girvetz, E. H. et al., *Applied climate-change analysis: The Climate Wizard tool.* PLoS ONE, 2009. 4(12): e8320.

[b] From the World Climate Research Programme (WCRP) Coupled Model Intercomparison Project, phase 3 (CMIP3) multi-model data set.

[c] Gibson, W. P. et al., *Development of a 103-year high-resolution climate data set for the conterminous United States.* Proceedings, American Meteorological Society, Portland, OR, May 13–16, 2002. 181–183. http://www.prism.oregonstate.edu/pub/prism/docs/appclim02103yr_hires_dataset-gibson.pdf.

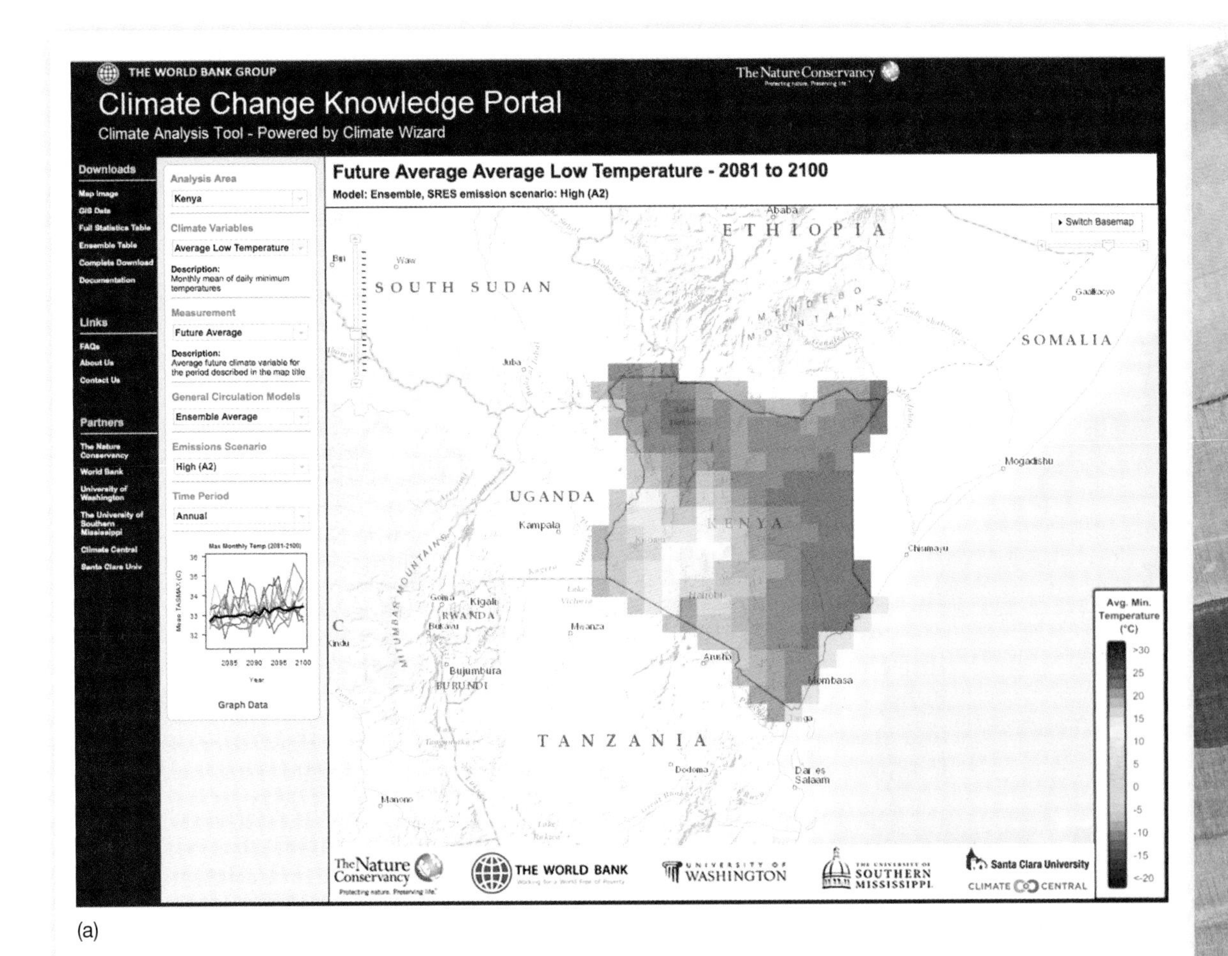


(a)

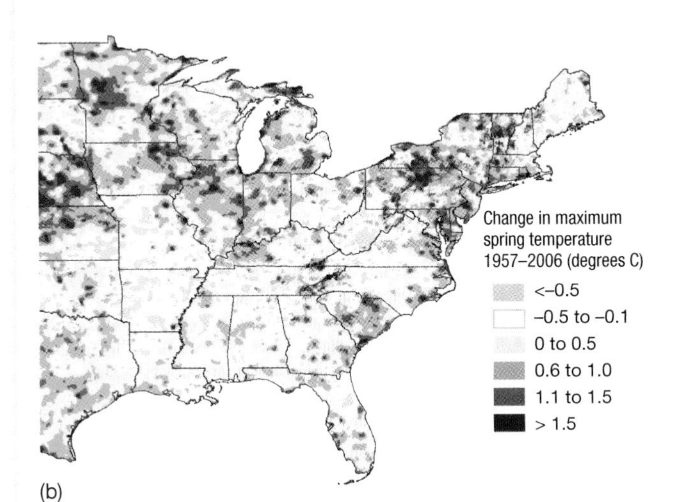


(b)

FIGURE 9.4.2 (a) Example of derived variables of climate change (e.g., average low temperature) from Climate Wizard analyses that are designed for specific clients such as the World Bank. (b) Variation in amount and direction of change in maximum spring temperature (averaged across March–May) over a 50-year period (1957–2006) as a proxy for changes in spring flowering phenology along migratory bird routes in the central and eastern United States. The temperature data are from the PRISM data set at 4-km resolution.[c]

colleagues were among the first to articulate these different conceptual approaches or options.[55] They referred to these broad categories of options to adaptation in relationship to forest management as *resistance, resilience, response,* and *realignment* options (**FIGURE 9.12**).

Resistance refers to managing ecosystems with the aim of forestalling or resisting the impacts of climate change. For example, a resistance strategy might include removing undergrowth and crowning trees to reduce fire hazards around the urban–forest interface. This is happening around many cities in the western United States that are located adjacent to forested land, as well as in the Blue Mountains of Australia where similar fire hazards exist.

Resilient management options imply activities that will result in accommodating change related to climate but tend to return an ecosystem to a prior condition after disturbance (see our discussion on resilience earlier in this chapter). An example might be improving riparian habitats along stream systems to help cool waters so that fish species are able to cope with overall warming conditions.

Response options refer to management alternatives that facilitate change in ecosystems as environmental conditions change. Managing production forests in an asynchronous manner so that different age classes of timber are harvested, as suggested by Millar and colleagues, would be a response approach that could reduce the vulnerability of these managed forests.

Finally, realignment refers to using restoration techniques in a dynamic manner to allow for the persistence of ecosystem processes and functions for highly disturbed ecosystems. For example, in Minnesota (USA) forests, The Nature Conservancy is experimenting with forest restoration techniques that involve planting a species composition that is more likely to adapt to future climate conditions.

**FIGURE 9.12** Different types of management strategies or approaches to adaptation are used in the management of natural resources. The U.S. Forest Service, for example, has broadly characterized these different approaches as resistance, resilience, response, and realignment. These are useful characterizations in general for adaptation responses in nature conservation projects and initiatives. See text for details. (Adapted from Peterson, D. et al., *Responding to Climate Change in National Forests: A Guidebook for Developing Adaptation Options.* 2011, Portland, OR: U.S. Forest Service Pacific Northwest Research Station.)

Such restoration efforts rarely involve a return to historic conditions (assuming that such conditions will be favorable under future climate change) and focus instead on adjusting systems to future conditions that are arrived at via modeling of anticipated climate impacts.

Within any given option, several alternative activities may be considered. In Chapter 7 we provided methods and tools for evaluating these alternatives. Depending on overall objectives of the conservation project, there may be a mix of adaptation activities. Some may be more resistance oriented, and others may be more focused on resilience, response, or realignment. Stein and colleagues[56] have referred to the continuum between resistance and response as managing for persistence (of biodiversity features) or change. For example, at least in the short term it may be necessary to establish objectives and targets (Chapter 4) for some features that are aimed at persistence, such as endangered species that may be regulated by law or policy. For other features, it may be more appropriate to manage for change. The previously mentioned example of a coastal marsh ecosystem where sea-level-rise models suggest that the current marsh will be inundated in the next 30 years presents a situation that should likely be managed for change.

Most adaptation projects to date have philosophically been more on the persistence end of this spectrum, but as the evidence for the rapidity and intensity of climate change impacts grows, there will be a need to focus more attention on the change or response end of this spectrum. Unless societal efforts are ultimately able to curb emissions, the impacts of climate change on species and ecosystems will continue to increase, and it will ultimately prove impractical to maintain most species and ecosystems in historical conditions, or what many of us once referred to as natural conditions.

As the title of this chapter suggests, the future of adaptation for many species and ecosystems will require strategies and actions that move beyond our preconceptions of natural. A consideration of temporal scale will be critical in helping practitioners think through the various adaptation options. While some options that are resistant or resilient may be appropriate for the short term (e.g., the next 5 to 10 years), response and realignment options will likely be more practical and realistic as long-term responses to adaptation.

## Prioritizing and Taking Actions (Chapter 7)

Once different options for responding to climate change have been explored (Chapter 6), a variety of methods can be used to identify and prioritize adaptation actions. The Adaptation for Conservation Targets (ACT) framework developed by Cross and colleagues uses a two-day workshop setting of scientists and practitioners to target specific intervention points, usually based on a conceptual model of the adaptation problem, and then identifies a small, concrete list of possible actions that could be taken in the near future.[57] TABLE 9.2 provides some examples of actions identified in workshops in which the ACT framework was used.

A slightly different method is provided by the example of the Climate Clinic, where teams from landscape-level conservation projects

**TABLE 9.2** Potential adaptation strategies and actions formulated based on applying the Adaptation for Conservation Targets (ACT) framework in an expert workshop setting

| Landscape | Conservation feature | Management goal | Actual or hypothesized climate change effects | Strategic adaptation actions |
|---|---|---|---|---|
| Jemez Mountains, New Mexico, USA | Natural stream flow regime | Maintain sufficient water in the system to support aquatic species and riparian vegetation | Reduced snowpack and greater variability in precipitation; reduced stream base flows | Restore beaver to streams<br>Build artificial structures to increase floodplain aquifer recharge<br>Apply forest thinning treatments that maximize snowpack retention and provide optimal shade to minimize sublimation and evaporation losses |
| Gunnison River Basin, Colorado, USA | Gunnison Sage-Grouse (*Centrocercus minimus*) | Increase and maintain the Gunnison population of Sage-Grouse at >3500 individuals and the Crawford population at >200 individuals | Loss of nesting habitat due to increased fire frequency, cheatgrass invasion, and sagebrush dieback; decreased habitat quality due to a decline in forbs and perennial grasses; reduced recruitment | Improve or restore nesting and wintering habitats<br>Improve or reestablish leeward mountain shrub habitats (e.g., snowberry, serviceberry) via fencing and planting<br>Maintain and expand perennial grass and forb cover by planting and fencing; abate or prevent cheatgrass encroachment by spraying |
| Four Forests Restoration Initiative area, Arizona, USA | Ponderosa pine (*Pinus ponderosa*) forest watershed function | Maintain or improve watershed function in systems dominated by ponderosa pine by maintaining and improving water quality, quantity, and timing of flow for surface and ground water; soil productivity; and recharge-to-runoff ratio | Increased temperature leads to increased potential evapotranspiration and decreased recharge; increased moisture stress for plants and lower base flows in rivers and streams that affect aquatic species | Apply forest restoration treatments (e.g., thinning, controlled burns) to reduce fire risk and drought-induced tree mortality, increase herbaceous ground cover, and enhance infiltration, soil moisture, and recharge<br>Plan for 6-year (on average) fire rotation to maintain water yield benefits |
| Bear River Basin, Utah/ Wyoming/ Idaho, USA | Bonneville cutthroat trout (*Oncorhynchus clarki utah*) | Maintain or expand the number of viable populations of Bonneville cutthroat trout in the Bear River Basin by maintaining or restoring Bonneville cutthroat trout habitat, ecology, and life history | Higher air temperatures increase evapotranspiration, decrease summer base flow, and raise summer water temperature, resulting in an expansion of uninhabitable reaches | Restore connectivity between river main stem and tributaries by rewatering streams to facilitate trout dispersal<br>Protect habitat in reaches that provide thermal refugia<br>Lower the depth of water outflow from hydropower and irrigation reservoirs to reduce downstream water temperature |

*Source:* Data from Cross et al., 2013.

developed two or three different hypotheses of change for how climate change would affect their project and then used feasibility, cost, and benefit analysis to prioritize a set of possible actions for different potential climate impacts.[58] The *Climate Smart Conservation* guide[59] suggests a variety of approaches to prioritizing actions such as focus groups, literature reviews, and eliciting expert opinion. It also recommends criteria for evaluating alternative actions, including how closely the actions will achieve the project objectives, how well the actions might achieve objectives for sectors other than nature conservation, how feasible it will be to implement the actions, and whether the actions are "climate smart" (e.g., linking actions to impacts or actions that reduce the carbon footprint).

The United Kingdom's Climate Impacts Program[60] (UKCIP) has also outlined a set of techniques that could be used to generate adaptation strategies, whether for human or ecological communities, and suggested a set of factors that should be considered in developing alternative actions. These factors include the level at which a strategy will be implemented (policy, program, or project), non-climate impacts of concern, climate impacts, decision maker's attitude to risk, and the risk and uncertainty involved in taking actions. TABLE 9.3 provides an example from the UKCIP report on various adaptation options that might be considered.

## Considering Uncertainty (Chapter 8)

As we outlined in Chapter 8, there can be many sources of uncertainty. In a project that is focused on adaptation strategies, that uncertainty might arise within climate impact analyses, from vulnerability assessments, or from the actions taken to implement a strategy, as well as the uncertainty of how conservation features or human communities will react to those actions. In addition to applying tools like scenario analysis (Chapter 8) to help address uncertainty, probably the best advice we can offer practitioners is to understand the different sources of uncertainty in a conservation project, document those sources, estimate where possible the magnitude of the uncertainty, and be prepared to revise your understanding of uncertainty as new information is generated during the lifetime of a conservation project.

Scenario planning is a process for evaluating different plausible ranges of climate impacts and developing alternative strategies to address those impacts. Once an agreed-on set of plausible scenarios for the impacts of climate change are generated, then a potential set of actions in response to these scenarios can be developed. In a recent handbook on scenario planning, *Considering Multiple Futures: Scenario Planning to Address Uncertainty in Natural Resource Conservation,*[61] Erika Rowland and her colleagues provide several case studies of scenario planning in a climate adaptation context. For example, a case study from the Crown of the Continent ecosystem along the western Canada–western U.S. border

**TABLE 9.3** Categorical types of adaptation responses for either human or natural communities and examples of each

| Adaptation type | Description/examples of application identified from UKCIP studies |
|---|---|
| Share loss | Insurance-type strategies<br>Use other new financial products that mitigate the risk<br>Diversify |
| Bear loss | Where losses cannot be avoided:<br>Certain species of montane fauna and flora (e.g., some arctic alpine flora may disappear from the UK)<br>Loss of coastal areas to sea-level rise and/or increased rates of coastal erosion |
| Prevent the effects: structural and technological (usually dependent on further investment) | Hard engineering solutions and implementation of improved design standards:<br>Increase reservoir capacity<br>Increase transfers of water<br>Implement water efficiency schemes<br>Scale up programs of coastal protection<br>Upgrade wastewater and stormwater systems<br>Build resilient housing<br>Modify transport infrastructure<br>Install or adopt crop irrigation measures<br>Create wildlife corridors |
| Prevent the effects: legislative, regulatory, and institutional | Find new ways of planning that cut across individual sectors and areas of responsibility (integration)<br>Change traditional land use planning practices, to give greater weight to new factors such as flood risk and maintaining water supply–demand balance and security of supply<br>Adopt new methods of dealing with uncertainty<br>Provide more resources for estuarine and coastal flood defense<br>Revise guidance notes for planners<br>Factor climate change into criteria for site designation for biodiversity protection<br>Amend design standards (e.g., building regulations) and enforce compliance |

*(continued)*

identified three major scenarios—climate complacency, race to refuge, and wheel spinning (TABLE 9.4). Through workshop settings, scientists and managers evaluated these scenarios and then took the first steps to develop potential adaptation strategies that included pinpointing major restoration efforts, translocating some fish stocks north out of the ecosystem, moving other fish stocks from the south into the Crown ecosystem, building some new dams (to deal with water shortfalls), and "letting go" of some prairie ecosystems in the Crown (presumably to transition into new or novel ecosystems).

## Monitoring, Evaluating, and Revising (Chapter 12)

Nowhere in the conservation field is there a greater call for monitoring both climate change impacts and the evaluation of strategies to abate those impacts than in climate adaptation projects. A similar plea for

| Adaptation type | Description/examples of application identified from UKCIP studies |
|---|---|
| Avoid or exploit changes in risk: change location or other avoidance strategy | Migration of people away from high-risk areas<br>Grow new agricultural crops<br>Change location of new housing, water-intensive industry, tourism<br>Improved forecasting systems to give advance warning of climate hazards and impacts<br>Contingency and disaster plans |
| Research | Use research to:<br>Look at long-term issues<br>Provide better knowledge of relationship between past and present variations in climate and the performance of environmental, social, and economic systems (e.g., fluvial and coastal hydrology, drought tolerance and distribution of flora and fauna, economic impacts on key industrial sectors and regional economies)—i.e., reduce uncertainty about the consequences of climate for receptors and decision makers<br>Improve short-term climate forecasting and hazard characterization<br>Produce higher-resolution spatial and temporal data on future climate variability from model-based climate scenarios<br>Provide more information on the frequency and magnitude of extreme events under climate change<br>Find better regional indicators of climate change<br>Develop more risk-based integrated climate change impact assessments |
| Education, behavioral | Lengthen planning time frames (need to consider not just the next two to five years, but 2020s, 2050s, and beyond)<br>Reduce uneven stakeholder awareness on climate change<br>Increase public awareness to take individual action to deal with climate change<br>(e.g., on health, home protection, flood awareness) and accept change to public policies (e.g., on coastal protection, landscape protection, biodiversity conservation) |

*Source:* Data from Willows and Connell, 2003.

monitoring and evaluation comes from those who are focused on adaptation of human communities to climate change.[62] The reasons for this, as we noted earlier in this chapter, are quite simple—the uncertainty surrounding the assessment of climate impacts, the unknown efficacy of many adaptation strategies, and the pace at which the impacts of climate change are being realized necessitate that we monitor and evaluate the effectiveness of actions intended to abate and mitigate these impacts.

As University of Washington climate scientists Josh Lawler and colleagues have pointed out,[63] all natural resource management actions have some uncertainty surrounding them; but those directed at specific climatic change impacts will often have more uncertainty (FIGURE 9.13), largely due to the challenges of predicting these future climatic changes.

It's reasonable to consider here what makes monitoring and evaluation of climate adaptation different from monitoring and evaluation

**TABLE 9.4 Scenarios developed as part of adaptation planning for the Crown of the Continent ecosystem in the U.S.–Canadian Rocky Mountains. This type of scenario planning helped managers and practitioners make decisions on which adaptation actions to take.**

| |
|---|
| Climate Complacency—Is Anyone Out There? This scenario features local-scale climate volatility and ecosystem diversification, and increasing growth pressures due to climate change consequences occurring elsewhere. Lack of national leadership and inflexible policies, combined with public attention being focused on challenges elsewhere, severely restrict external assistance for the Crown of the Continent. The region must rely on its own creativity, flexibility, initiative, and resources. |
| Colorado Creeps North—Wheel Spinning. This scenario features steady regional trends toward dryness and increasing growth pressures due to severe climate change consequences occurring elsewhere. While societal concern is focused elsewhere, national leadership and policies support a wide variety of options for adaptation. |
| Race to Refuge—Big Problems, Big Solutions. This scenario features rapid climate change leading to transformative ecosystem changes in all parts of the Crown of the Continent region. This scenario used A1B (see Figure 8.6) climate projections from 2050 to represent conditions in 2020, and 2100 entries for 2050, with concomitant strong impacts on southwestern drought and sea-level rise producing extreme pressures on food production and human migration. However, society is focused on the region as the "last best place," and national leadership and policies support any innovations the region desires. |

*Source:* Adapted from Rowland et al., 2014.

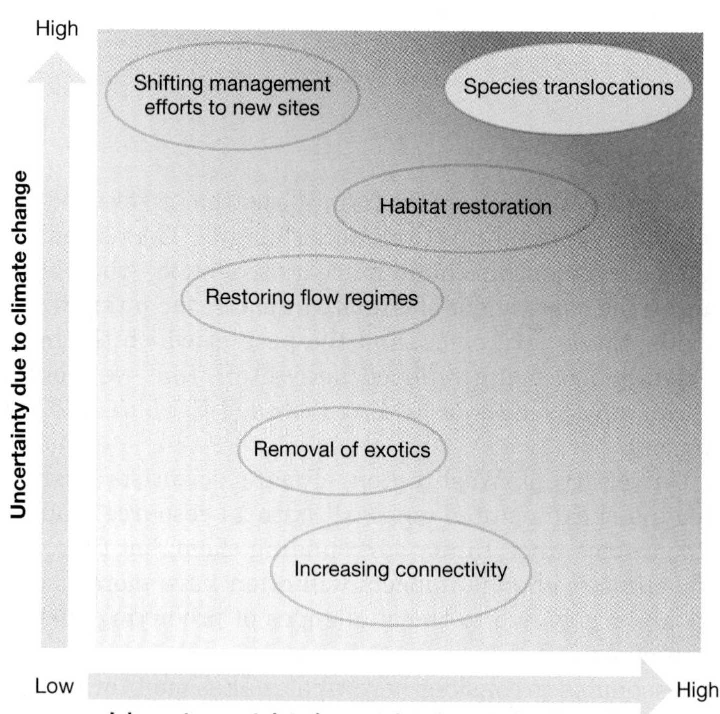


**FIGURE 9.13** Different types of management or adaptation strategies may have different levels of uncertainty associated with them, whether they are used in relationship to climate change or as conservation strategies independent of climate change. The strategies here are intended as a generalization. Those near the top of the y-axis are likely more influenced by uncertainty associated with future climate change. (Adapted from Lawler et al., 2010.)

efforts for conservation projects in general. As Stein and colleagues suggest in *Climate Smart Conservation,* climate impacts and adaptation efforts may change what we monitor, where we monitor, when we monitor, and possibly who is involved. For example, changes in the environment may necessitate changes in what we monitor. Changes in species distributions or ecological processes may necessitate changes in where we monitor. Climate impacts on processes like phenology may influence when we monitor. Finally, as climate impacts are already being felt across large landscapes and seascapes, we may need to create new partnerships or organizations to best address monitoring over such large areas.

Of course, all the information in the world from monitoring and evaluation won't help improve a conservation project unless it is actually used to revise the strategies and actions being undertaken. As we discuss further in Chapter 12 on monitoring for results, this is an underappreciated step in adaptively managing a conservation project. If a particular strategy is not working or no longer appropriate, often a certain degree of reluctance or inertia gets in the way of revising our thinking. It's essential that project teams take the time to regularly check in and use the data and information from monitoring and evaluation efforts to make midcourse corrections that will improve the chances of effectively adapting to climate change.

In the next section of this chapter, we expand on the diversity of adaptation strategies and actions that the global conservation community is taking to mitigate and abate the impacts of climate change.

## Strategic Adaptation Approaches

A considerable amount of literature on adaptation has been published during the last two decades, and since approximately 2010 we have seen an explosion of papers addressing strategic approaches to adaptation.[64] Some approaches are robust to a broad range of anticipated climate change impacts (e.g., improving ecosystem function), while others are contingent on specific impacts (e.g., assisted migration). Here, we summarize the major approaches, offer some examples of each, and point the reader to more detailed information on tools for implementing these approaches as well as their pros and cons.

Due to the variety of situations worldwide in which adaptation approaches are likely to be needed and taken, it's safe to say that all of these approaches will find use somewhere. Which strategic approach will be most useful will depend on the type of climate impacts, the vulnerabilities of various conservation features, the scale of the project, its objectives, and the financial and staff resources that are available as well as the levels and types of uncertainty involved.

As we consider these broad strategies, most readers will recognize that some have been used in nature conservation for decades. Improving connectivity and expanding protected areas are the two most obvious approaches that nature conservationists have long deployed. What's different about the use of these approaches in adaptation is what the *Climate Smart Conservation* authors have referred to as **intentionality**—the need to be purposeful, deliberate, and transparent as we consider how these strategic approaches may help us achieve our conservation objectives and outcomes in the face of specific climate impacts. The linking of these strategies and actions to these specific impacts in a transparent, "show your work"[65] way ultimately will distinguish the use of these approaches for achieving adaptation.

## Improving Connectivity

No other approach to adaptation has received more attention in the biodiversity conservation community than that of increasing connectivity (usually among protected areas or other conservation areas).[66] In part that is because establishing corridors and linkages was already a desirable conservation strategy, and one that received considerable scientific and practitioner scrutiny before climate change and adaptation became the important topics that they are today. Increasing connectivity has long been viewed as an important conservation approach to help overcome the negative effects of habitat fragmentation on many species.

Increasing connectivity is also seen as an effective strategy for many species whose distributions are changing in response to a changing climate. At the same time, we need to think about connectivity somewhat differently from a climate change perspective. For example, Molly Cross and colleagues[67] suggest three key ways that addressing connectivity within a climate change context differs from traditional connectivity practices:

1. Planners need to think about what habitats for a species may need to be connected in the future and that those habitats may differ from what is needed today.
2. While wildlife connectivity issues to date have primarily been focused on wide-ranging species, in the future it will be necessary to consider a broader array of species that may need to move in response to climate.
3. How we think about the temporal and spatial scale needs for connectivity today may be quite different from how we need to think about it in the future under various climate change scenarios. For example, in some cases a corridor that was established primarily for wildlife movement in response to habitat fragmentation may no longer be a functional corridor in a changing climate.

For the practitioner who wants a thorough background on corridors and connectivity, Jodi Hilty and colleagues' book *Corridor Ecology* is a definitive reference.[68] More recently, Tristan Nuñez and colleagues[69] in the Pacific Northwest of the United States have developed a simplified connectivity planning approach specific to climate change that is based on spatial gradients of temperature. In addition, the Wildlife Conservation Society has published an excellent guidebook to connectivity that explores the different methods and approaches for analyzing connectivity and establishing linkages.[70] Finally, along with several colleagues, we published a paper that outlines the general uses of connectivity for adaptation in a regional planning context, discusses some of the assumptions behind taking a connectivity approach, and evaluates some of the trade-offs in their use.[71]

Whether improving larval dispersal in a marine environment, enhancing fish migration in riverine systems, or establishing local migration corridors for large ungulates, connectivity is a useful adaptation approach for marine, freshwater, and terrestrial realms. **Box 9.5** illustrates an on-the-ground example of improving connectivity for climate adaptation.

### Box 9.5 Improving Connectivity in Vernal Pools, New Jersey, USA

Along the coast of New Jersey, climate change is altering the hydrographs of vernal pools, and through sea-level rise, it is inundating many pools as well. These pools are critical wetland habitat for many species.

In response, the Conserve Wildlife Foundation of New Jersey is creating new vernal pools that will increase the connectivity among this complex of vernal pool communities and help in the colonization of upland vernal sites by species of amphibians, reptiles, and invertebrates that have limited dispersal capability. Sites for new vernal pools are being established above the anticipated area where sea-level rise will affect the pool system. Such a strategy should allow for the long-term adaptive management of a vernal pool complex in the face of climate change.

**FIGURE 9.5.1** Vernal pools, New Jersey, USA.

## Identifying and Protecting Refugia

It is possible to identify and conserve those areas least likely to be affected by climatic change, often due to geological, topographical, or hydrological factors. Mountain environments, for instance, are often examples of such places due to the variability in microclimate that can occur in mountains over short distances. Alternatively, conservation planners and practitioners may also identify those areas where plants and animals may move in the future because they will be the most suitable places in the landscape.

In both of these situations, such places are known as climatic **refugia**. In the latter case, one of us (ETG) and his colleagues identified potential climate refugia in Papua New Guinea on the basis of expected changes in seven climate variables[72] (FIGURE 9.14). There are a number of different ways for identifying refugia[73] depending on the scale of climate data used and whether or not analyses are focused on climatic stability, habitat stability, or both. As with any approach to

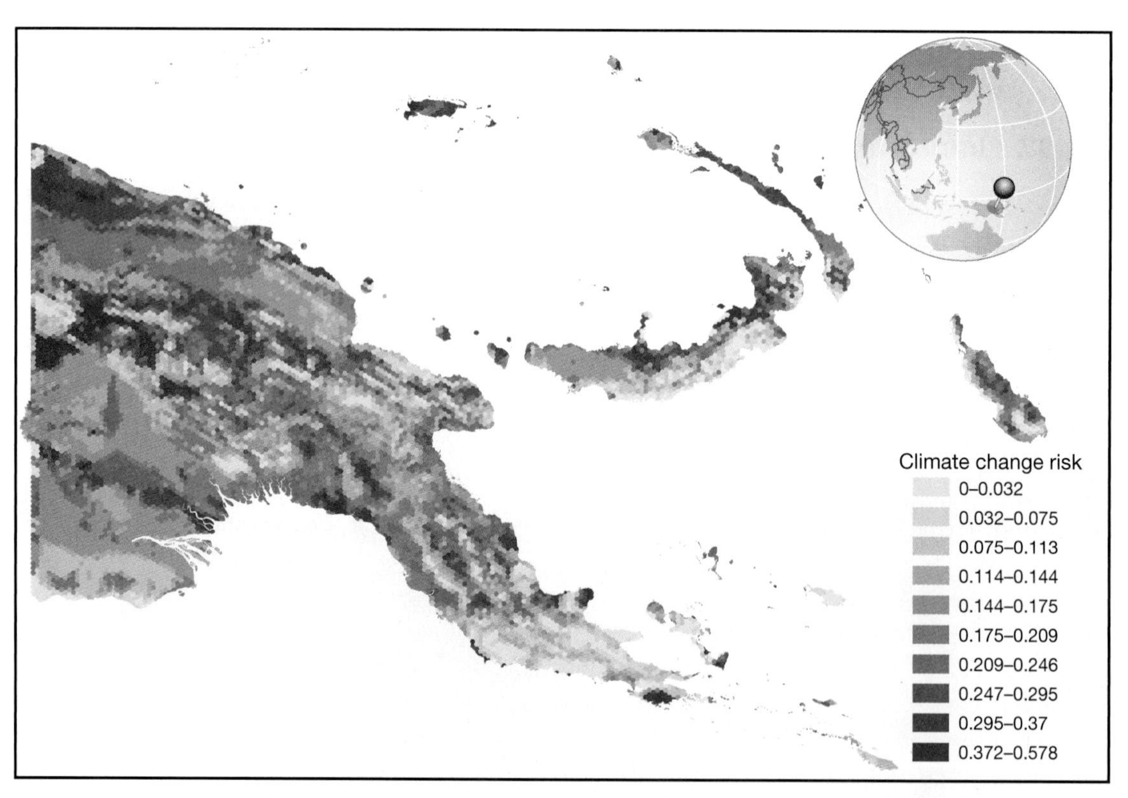


FIGURE 9.14 Identification of potential climate refugia (areas of relatively lower climate change risk based on the modeling of seven climate variables in Papua New Guinea). Climate change risk values represent the difference between climate variables defined on the basis of PRISM data from 1960 to 1991 and projected changes in these variables in the year 2100, based on climate emission scenario A2.

adaptation, there are pros, cons, and trade-offs, several of which we and our colleagues identified in a 2012 paper on adaptation planning.[74]

## Expanding Protected and Conservation Area Networks

As species shift their distributions in response to climate change, one logical approach to adaptation is to expand individual protected and conservation areas. New areas also would be added to enlarge the overall network, although such actions need to be conducted as a strategic response to climate impacts and not in a business-as-usual approach.

Several authors in recent years have called for expanding the world's protected area network as a climate adaptation response.[75] The International Union for the Conservation of Nature (IUCN), through the World Commission on Protected Areas, has taken a leadership role in this regard with the publication of *Natural Solutions—Protected Areas: Helping People Cope with Climate Change*[76] and another IUCN guidance document to be released in 2016—*Responding to Climate Change: Guidance for Protected Area Managers and Planners.*[77]

As the authors of *Natural Solutions* have noted, protected areas can play a significant role in adaptation through two different mechanisms: (1) "maintain ecosystem integrity, buffer local climate, reduce risks[78] and impacts from extreme climatic events," and (2) "maintain essential ecosystem services that help people cope with changes in water supplies, fisheries, incidence of disease and agricultural productivity caused by climate change." They outlined six key management and policy changes that are needed for protected areas to play these roles, including more and larger protected areas, connecting protected areas at landscape and seascape levels, improved management, and management strategies aimed specifically at adaptation.

Some recent research provides support for this protected area approach. Alison Johnston and colleagues[79] in the United Kingdom examined internationally important bird populations at a network of protected areas in northwestern Europe. Their spatial climate–abundance modeling suggests that although species distributions are expected to change on many of these protected areas, most protected areas will continue to have a sufficient number (albeit often different) of important species to retain their legal status. More importantly, because of the underlying habitat complementarity represented by this set of protected areas, it is expected to remain an important conservation network even with considerable climate change impacts and species turnover resulting from those impacts.

In 2012, an International Conference on Managing Protected Areas under Climate Change (IMPACT) was held in Europe. The conference covered six major thematic areas,[80] and the most important recommendations were (1) adaptation actions need to be "tailored to specific protected area and local conditions"; (2) management of protected areas in the face of climate change must take into account the wider perspective of land use (or water use) that happens in areas adjacent to and surrounding protected areas; and (3) protected areas can't be managed in isolation but instead need to be managed over larger areas that contain multiple protected areas. Among other objectives, the detailed recommendation stemming from this conference should help guide the Natura 2000 network of European protected areas[81] in developing a more coordinated and comprehensive approach to climate impacts and adaptation.

## Conserving the Stage—Conserving Land Facets

Earlier in this chapter, we discussed how various strategic approaches to adaptation can influence which biodiversity features planners and practitioners will focus on. One such approach that we discussed has been referred to as "conserving the stage" or conserving land facets. In short, this strategic approach focuses on conserving the underlying geophysical environment in which species exist. That geophysical environment can be characterized by **land facets**—recurring landscape units defined on the basis of geology, soil, and topography.

Recognizing that not all species will be conserved by this approach under climate change scenarios, Paul Beier and colleagues also recommend complementing land facets with planning approaches that focus on land cover and selected species as conservation features.[82] The notion of using geophysical factors as a conservation feature in planning has been used in conservation planning for decades and is really just another version of the coarse-filter approach to conservation (see Chapter 3). The primary assumption of this approach is that geophysical factors can serve as adequate surrogates for biodiversity under climate change scenarios. Although some data support this approach for the northeastern United States, much work remains to be done to determine its adequacy in other parts of the world.

A special section of *Conservation Biology* in 2015 (vol. 29, no. 3) contains numerous papers on this topic with a wide range of views on when and under what conditions it makes sense to use this approach. The strategy of conserving the stage should be particularly important in situations where there is high uncertainty about climate impacts and species responses. It is essentially a bet-hedging strategy to conserve all different types of physical environments. This approach can also be helpful in informing the connectivity and protected area adaptation strategies that we just discussed.

## Assisted Migration—Managed Relocation

Helping species adapt to climate change by physically moving them outside their historic distribution is referred to as assisted migration/colonization, or managed relocation. Much has been written on this topic, including reviews that tackle its scientific, regulatory, and ethical underpinnings.[83] This approach is contentious because projects to relocate or translocate species historically have a bad track record.

Moreover, the introduction of species outside their range has sometimes resulted in ecosystem-level disasters. Some of the most notable examples are introductions of invasive fish species, which then have considerable negative impacts on native fish species assemblages or other species groups such as amphibians. On the other hand, increasingly sophisticated and thoughtful approaches are being used, for example, to assist migration in forest ecosystems. These new methods, which are resulting from genetic analyses and frameworks, may prove essential in restoration and reforestation of species and ecosystems that have been negatively impacted by climate change.[84]

**Box 9.6** summarizes some of the key questions to consider when developing a managed relocation adaptation strategy. In addition to these key questions, several Australian scientists have developed the first quantitative framework for evaluating whether to translocate a species, which species should be prioritized for assisted migration, and when and where to move them.[85] Practitioners should find this framework helpful when confronted with often difficult decisions about assisted migration, and it should complement an existing set of IUCN guidelines on species reintroductions.

## Improving Ecosystem Function and Process

We now know that many species are already shifting their distributions as a result of climate change—there are scores of published examples of these changes around the globe. Because much of biodiversity conservation has traditionally been focused on conserving these patterns of biodiversity (the species inhabiting a particular place), many conservation practitioners and scientists are suggesting that under climate change scenarios, we should place a greater emphasis on maintaining the processes and functions that help maintain this biodiversity and pay less attention to the actual patterns or distributions.

In many respects, conservation planners have been paying increased attention to the function and process of ecosystems for quite some time now (see Chapter 4). Climate impacts only add to the urgency and importance of paying more attention to ecosystem process and function. Many of the actions to improve ecosystem function and

## Box 9.6 Key Questions for Considering Assisted Colonization or Managed Relocations as an Adaptation Strategy

**Ethical questions**

What are the goals of conservation, and why do we value those goals?

Which conservation goals take ethical precedence over others and why?

What is the ethical responsibility of humans to protect biodiversity (genotypic, population, species, ecosystem)?

Is there an ethical responsibility to refrain from activities that may cause irreversible impacts, even if restraint increases the risk of negative outcomes?

How does society make decisions in consideration of divergent ethical perspectives?

**Legal and policy questions**

Do existing laws and policies enable appropriate managed relocation actions?

Do existing laws and policies inhibit inappropriate managed relocation actions?

Do the existing implementation policies of environmental laws provide the guidance for resource managers to fulfill their obligations for climate change adaptation?

What is the process for managers, stakeholders, and scientists to work collaboratively to make managed relocation decisions?

Who pays for managed relocation, including the studies needed to support an action, monitoring, and the outcomes of the management action?

**Ecological questions**

To what extent do local adaptation, altered biotic interactions, no-analog climate space, and the persistence of suitable microhabitats within largely unsuitable landscapes mitigate the extinction risk (and managed relocation need) of species listed as vulnerable?

What evidence suggests that species are absent from climatically suitable locations because of dispersal limitations that could be addressed by managed relocation?

What are the limits of less dramatic alternatives to managed relocation, such as increasing habitat connectivity?

How well can we predict when management must address interacting suites of species rather than single species?

How well can we predict when relocated species will negatively affect host system species or ecosystem functioning (e.g., nutrient flux through food webs, or movement of individuals)?

How well can we predict the likelihood of a species' successful long-term establishment in light of a changing climate?

**Integrated questions**

What are the priority taxa, ecosystem functions, and human benefits for which we would consider invoking managed relocation?

What evidence of threat (extinction risk, loss of function, loss of benefit to people) triggers the decision process?

What is adequate evidence that alternatives to managed relocation are unavailable and that the probability that managed relocation will succeed is adequate?

What constitutes an acceptable risk of harm and what are adequate assurances for the protection of recipient ecosystems?

Who is empowered to conduct managed relocation, and what is their responsibility in the event that the consequences are not those predicted?

*Source*: Data from Schwartz et al., 2012.

process are likely to be related to hydrological regimes and the challenges of balancing water use across human and ecological communities. Stream and lake ecosystems are already showing the impacts of climate change and will continue to do so through, for example, reduced snowpack, early snowmelt, or increases in evapotranspiration and water temperature. Reconnecting floodplains to reduce anticipated flooding from climate change and restoring shellfish as a tactic for shoreline stabilization and coastal defense against rising sea levels are but two of many examples of how ecosystem structure and function can be improved to help abate the effects of climate change. **Box 9.7** provides a species-level example of ecosystem process restoration.

## Reducing Non-Climate Stressors

Climate change rarely acts alone to impact species and ecosystems. As anyone who has worked in natural resource management or conservation knows all too well, often a whole suite of factors are negatively impacting the species, habitats, and ecosystem processes we are trying to conserve. Many of these factors act in concert with climate change impacts to make matters worse, and they are usually referred to as non-climate stressors. Reducing these stressors is an often cited strategy in the adaptation literature (e.g., Hansen and Hoffman[86]).

### Box 9.7 Beavers as Adaptation Agents

In the Colorado Plateau of the southwestern United States, the Grand Canyon Trust is reintroducing beaver to drought and flood-ridden streams that have been affected by climate change. Beavers are helping to restore riparian habitat and improving water storage and flow capabilities on several streams on the Colorado Plateau, as you can see in **FIGURE 9.7.1**. The beaver reintroduction is a demonstration of improving ecological process and function, and it also exemplifies a resilience strategy.

**FIGURE 9.7.1A** Before beaver reintroduction.

**FIGURE 9.7.1B** After beaver reintroduction.

Invasive species, pollution, habitat fragmentation, or loss of an ecological process such as a natural flow regime are some of the more commonly cited stressors whose combined or cumulative effects on many species and ecosystems can be exacerbated by climate change. Therefore, reducing those non-climate stressors that are most closely tied to the projected impacts of climate on a conservation project may reduce the overall impacts of climate change and/or increase the adaptive capacity of a system. Practitioners should undertake these actions with some forethought of how specific non-climate stressors are interacting with the potential impacts of climate change and not assume a priori in a business-as-usual manner that simply reducing these stressors will be an appropriate climate adaptation strategy.

We have provided an overview of some of the most important strategic approaches to climate adaptation. While not intended to be comprehensive, this overview should encompass most of the strategic interventions that practitioners, scientists, and planners will consider. Moreover, our overview should point the reader to important references where he or she can gain more detailed information and data about how these strategic approaches can be best applied. Fortunately, most of these approaches are robust to specific climate change impacts; indeed, most of these approaches are being deployed in many conservation projects around the world because they are tried-and-true strategies for nature conservation more broadly but are especially pertinent to climate adaptation.

## Getting Over the Adaptation Hurdles

Many of the hurdles to planning for and implementing adaptation strategies are similar to challenges in general for conservation planning, a topic that we will cover in more detail in Chapter 11. At the same time, we need to recognize that the uncertainties surrounding climate change and its impacts as well as the need to get serious about managing for change pose significant challenges for the conservation and natural resource community.

In a 2010 article, Susanne Moser and Julia Ekstrom[87] developed a framework and tool for identifying and overcoming the barriers to adaptation. Their framework divides the adaptation process into three components—the actors, the system of concern, and the greater context in which these operate. These are the sources from which barriers arise. The authors go on to identify common barriers in each of the major parts of the adaptation process (e.g., **TABLE 9.5** addresses barriers in the management phase).

Finally, Moser and Ekstrom provide a simple matrix to help identify intervention points for overcoming barriers. Collectively, the

**TABLE 9.5** Common barriers in the management phase of adaptation projects

| Phase and process stages: Managing | Barriers |
|---|---|
| Implement option(s) | Threshold of intent<br>Authorization<br>Sufficient resources (fiscal, technical, etc.)<br>Accountability<br>Clarity/specificity of option<br>Legality and procedural feasibility<br>Sufficient momentum to overcome institutional stickiness, path dependency, and behavioral obstacles |
| Monitor outcomes and environment | Existence of a monitoring plan<br>Agreement, if needed, and clarity on monitoring targets and goals<br>Availability and acceptability of established methods and variables<br>Availability of technology<br>Availability and sustainability of economic resources<br>Availability and sustainability of human capital<br>Ability to store, organize, analyze, and retrieve data |
| Evaluate effectiveness of option | Threshold of need and feasibility of evaluation<br>Availability of needed expertise, data, and evaluation methodology<br>Willingness to learn<br>Willingness to revisit previous decisions<br>Legal limitations on reopening prior decisions<br>Social or political feasibility of revisiting previous decisions |

*Source:* Data from Moser and Ekstrom, 2010.

nature of the barrier, its source, and the possible point of intervention provide an overall direction for overcoming adaptation barriers—which, as Moser and Ekstrom suggest, may constitute a major portion of the initial parts of any adaptation conservation project.

Although there has been limited research on overcoming adaptation barriers, more recent work suggests that the presence of a preexisting policy or set of actions that are now also contributing to adaptation is an important enabling condition[88] that could be helpful in overcoming barriers. Another important enabling condition is the mainstreaming of adaptation actions and strategies into other sectoral policies such as water management or urban planning.

One of the most significant challenges in implementing adaptation strategies will be confronting the human responses to climate change. As coral reef scientist Tim McClanahan and colleagues point out, this challenge may arise in part due to the lack of human social capacity to develop adaptation measures.[89] In many cases, the responses of human communities are likely to make the conservation of biodiversity and ecosystem services even more difficult.

Conservation International scientist Will Turner and colleagues look at historical responses of human communities to climate changes and also project future indirect impacts of humans on conservation-oriented adaptation efforts.[90] For example, widespread droughts or floods could result in loss of agricultural productivity in some areas; in turn, this could lead to development of nearby natural habitats to make up for that loss in productivity. The challenge to conservationists will be to proactively anticipate and react to these changes, especially when some of them could involve large-scale land use changes related to an expanded agricultural frontier.

In some cases, the human responses that conservationists will need to react to will be human adaptations (e.g., building gray infrastructure along the coast in response to sea-level rise). At other times, conservationists will need to evaluate and respond to complex interactions between responses to emission mitigation and adaptation, which may well involve trade-off analyses between emission reduction efforts, biodiversity conservation, and development objectives. In some cases, carefully planned REDD+ projects (Reduced Emissions from Deforestation and Forest Degradation), which we first mentioned in Chapter 1, may provide an opportunity to achieve emissions reductions and biodiversity conservation that includes adaptation strategies (e.g., protected area improvements) while avoiding conflicts with development sectors.[91]

## KEY MESSAGES

- Frameworks for adaptation planning and literature reviews of adaptation strategies have dramatically increased, but on-the-ground examples of implemented actions for adaptation are just beginning to test these frameworks and strategies in the context of climate change.
- Most conservation projects will likely plan for adaptation not as a distinct effort, but one that is integrated with a broader conservation planning approach. Nevertheless, several concepts outlined in this chapter (e.g., resilience) will be important for practitioners to understand as well as some specific tasks in the conservation planning cycle that are particular to climate adaptation (e.g., climate impact analyses).
- Although this chapter is largely concerned with adaptation from the perspective of nature conservation, there is a vast literature and growing set of experiences of social or human adaptation to

climate change. Conservation planners need to be familiar with both; and a challenge, if not expectation, of the future will be to link the efforts in adaptation for human and ecological communities. There are already several global examples of these types of projects, which are referred to as ecosystem-based adaptation.

- Any new conservation initiative should consider the possible impacts of climate change through its planning efforts; it is equally important for existing conservation projects to reconsider whether existing conservation objectives are still appropriate and to modify plans and actions accordingly.
- Although there are many similarities between adaptation planning and the major steps of any conservation planning effort, confronting the uncertainty surrounding climate change impacts as well as the effectiveness of adaptation interventions is a particularly important attribute of adaptation planning. Monitoring of these aspects of the effort will be especially important in adaptation projects because of this uncertainty.
- There are many different options and approaches for undertaking adaptation measures in a conservation project (e.g., resistance, resilience, response, realignment). Within any one project, it is possible to deploy more than one of these approaches. For instance, some strategies might emphasize a resistance approach, while others are focused on response or realignment. Which approaches and strategies are selected will depend on the project's objectives and desired outcomes. Fortunately, whatever overall options and approaches are chosen for adaptation, there are increasingly a variety of specific strategies and actions that can be implemented and case study examples of their implementation.

## References


1. Hansen, J., M. Sato, and R. Ruedy, *Perception of climate change.* Proceedings of the National Academy of Sciences, 2012. 109(37): E2415–E2423.
2. IPCC, Summary for policy makers, in *Climate Change 2013: The Physical Science Basis.* Contribution of Working Group I to the Fifth Assessment Report of the Intergovernmental Panel on Climate Change, T. F. Stocker et al., editors. 2013, Cambridge, UK: Cambridge University Press.
3. Millar, C., and L. Brubaker, *Climate Change and Paleoecology: New Contexts for Restoration Ecology,* in *Restoration Science*, M. Palmer, D. Falk, and J. Zedler, editors. 2006, Washington, DC: Island Press. 315–340.
4. Rannow, S., and R. Marco, editors. *Managing Protected Areas in Central and Eastern Europe under Climate Change.* Advances in Global Change Research Series. Vol. 58, 2014. New York: Springer.

5. Archie, K. M. et al., *Climate change and western public lands: A survey of U.S. federal land managers on the status of adaptation efforts.* Ecology and Society, 2012. **17**(4): 20; Biagini, B. et al., *A typology of adaptation actions: A global look at climate adaptation actions financed through the Global Environment Facility.* Global Environmental Change, 2014. **25**: 97–108; and Fischman, R. L. et al., *Planning for adaptation to climate change: Lessons from the U.S. National Wildlife Refuge System.* BioScience, 2014. **64**(11): 993–1005.
6. Armsworth, P. R. et al., *Are conservation organizations configured for effective adaptation to global change?* Frontiers in Ecology and the Environment, 2015. **13**: 163–169.
7. Andrade Perez, A. et al., *Building Resilience to Climate Change: Ecosystem-based Adaptation and Lessons from the Field in Ecosystem Management.* Series No. 92010. Gland, Switzerland: IUCN. 164.
8. Field, C. B. et al., *Climate Change 2014: Impacts, Adaptation, and Vulnerability. Part A: Global and Sectoral Aspects.* Contribution of Working Group II to the Fifth Assessment Report of the Intergovernmental Panel on Climate Change. 2014, Cambridge, UK, and New York: Cambridge University Press. 1132.
9. Stein, B. A. et al., *Climate Smart Conservation: Putting Adaptation Principles into Practice.* 2014, Washington, DC: National Wildlife Federation. 262.
10. Kropp, J., and M. Scholze, *Climate Change Information for Effective Adaptation: A Practitioner's Manual.* 2009, Eschborn, Germany: Deutsche Gesellschaft fur Technische Zusammenarbeit (GTZ).
11. National Research Council, *Adapting to the Impacts of Climate Change.* 2010, Washington, DC: National Academies Press. 292.
12. Tingley, M. W., L. D. Estes, and D. S. Wilcove, *Ecosystems: Climate change must not blow conservation off course.* Nature, 2013. **500**(7462): 271–272.
13. Glick, P., B. A. Stein, and N. A. Edelson, *Scanning the Conservation Horizon: A Guide to Climate Change Vulnerability Assessment.* 2011, Washington, DC: National Wildlife Federation. 176.
14. Turner, W. R. et al., *Climate change: Helping nature survive the human response.* Conservation Letters, 2010. **3**(5): 304–312.
15. Glick et al., *Scanning the conservation horizon.* (See reference 13.)
16. See, for example, McClanahan, T. R. et al., *Conservation action in a changing climate.* Conservation Letters, 2008. **1**(2): 53–59.
17. Stein et al., *Climate Smart Conservation.* (See reference 9.)
18. Young, B. et al., *A natural history approach to rapid assessment of plant and animal vulnerability to climate change,* in *Wildlife Conservation in a Changing Climate,* J. Brodie, E. Post, and D. Doak, editors. 2012, Chicago: University of Chicago Press. 129–152.
19. Bagne, K. E., and D. M. Finch. *Developing a tool to assess wildlife species vulnerability to climate change,* in *New Mexico Forestry and Climate Change Workshop.* 2008, Albuquerque, NM: U.S. Forest Service.

20. Ellison, J., *Vulnerability assessment of mangroves to climate change and sea-level rise impacts.* Wetlands Ecology and Management, 2015. **23**: 115–137.
21. Pacifici, M. et al., *Assessing species vulnerability to climate change.* Nature Climate Change, 2015. **5**(3): 215–224.
22. Smit, B., and J. Wandel, *Adaptation, adaptive capacity and vulnerability.* Global Environmental Change, 2006. **16**(3): 282–292.
23. Grantham, H. S. et al., *Ecosystem-Based Adaptation in Marine Ecosystems of Tropical Oceania in Response to Climate Change.* Pacific Conservation Biology, 2011. **17**: 1–18.
24. Marshall, N. A. et al., *A Framework for Social Adaptation to Climate Change; Sustaining Tropical Coastal Communities and Industries.* 2009, Gland, Switzerland: IUCN.
25. Smit and Wandel, *Adaptation, adaptive capacity and vulnerability.* (See reference 22.)
26. Eakin, H. C., M. C. Lemos, and D. R. Nelson, *Differentiating capacities as a means to sustainable climate change adaptation.* Global Environmental Change, 2014. **27**(0): 1–8.
27. Colls, A., N. Ash, and N. Ikkala, *Ecosystem-Based Adaptation: A Natural Response to Climate Change.* 2009, Gland, Switzerland. IUCN: 16.
28. Arkema, K. K. et al., *Coastal habitats shield people and property from sea-level rise and storms.* Nature Climate Change, 2013. **3**(10): 913–918.
29. Ferrario, F. et al., *The effectiveness of coral reefs for coastal hazard risk reduction and adaptation.* Nature Communications, 2014. **5**.
30. Colls et al., *Ecosystem-Based Adaptation.* (See reference 27.)
31. Wiens, J. A., and D. Bachelet, *Matching the multiple scales of conservation with the multiple scales of climate change.* Conservation Biology, 2010. **24**(1): 51–62.
32. Myers-Smith, I. H., S. A. Trefry, and V. Swarbrick, *Resilience: Easy to use but hard to define.* Ideas in Ecology and Evolution, 2012. **5**: 44–53.
33. Standish, R. J. et al., *Resilience in ecology: Abstraction, distraction, or where the action is?* Biological Conservation, 2014. **177**: 43–51.
34. Linkov, I. et al., *Changing the resilience paradigm.* Nature Climate Change, 2014. **4**(6): 407–409.
35. Glick, P., H. Chmura, and B. A. Stein, *Moving the Conservation Goalposts: A Review of the Climate Change Adaptation Literature.* 2011, Washington, DC: National Wildlife Federation. 1–25.
36. Hilty, J. A., C. C. Chester, and M. S. Cross, editors, *Climate and Conservation: Landscape and Seascape Science, Planning, and Action.* 2012, Washington, DC: Island Press.
37. Holling, C. S., *Resilience and stability of ecological systems.* Annual Review of Ecology and Systematics, 1973. **4**(1): 1–23.
38. Stein et al., *Climate Smart Conservation.* (See reference 9.)
39. Folke, C., *Resilience: The emergence of a perspective for social–ecological systems analyses.* Global Environmental Change, 2006. **16**(3): 253–267.

40. Stein et al., *Climate Smart Conservation.* (See reference 9.)

41. Cross, M. S. et al., *The Adaptation for Conservation Targets (ACT) framework: A tool for incorporating climate change into natural resource management.* Environmental Management, 2012. **50**(3): 341–351.

42. Glick et al., *Moving the Conservation Goalposts.* (See reference 35.)

43. Heller, N. E., and R. J. Hobbs, *Development of a natural practice to adapt conservation goals to global change.* Conservation Biology, 2014. **28**(3): 696–704.

44. Cole, D. N., and L. Yung, editors, *Beyond Naturalness: Rethinking Park and Wilderness Stewardship in an Era of Rapid Change.* 2010, Washington, DC: Island Press. 304.

45. Stein et al., *Climate Smart Conservation.* (See reference 9.)

46. Shaw, M. R. et al., *Economic costs of achieving current conservation goals in the future as climate changes.* Conservation Biology, 2012. **26**(3): 385–396.

47. Halpin, P. N., *Global climate change and natural area protection: Management responses and research direction.* Ecological Applications, 1997. **7**(3): 828–843.

48. Groves, C. et al., *Incorporating climate change into systematic conservation planning.* Biodiversity and Conservation, 2012. **21**(7): 1651–1671.

49. Anderson, M. G., and C. E. Ferree, *Conserving the stage: Climate change and the geophysical underpinnings of species diversity.* PLoS ONE, 2010. **5**(7): e11554; and Brost, B. M., and P. Beier, *Use of land facets to design linkages for climate change.* Ecological Applications, 2011. **22**(1): 87–103.

50. Carvalho, S. B. et al., *From climate change predictions to actions—conserving vulnerable animal groups in hotspots at a regional scale.* Global Change Biology, 2010. **16**(12): 3257–3270.

51. Ellison, J. et al., *Climate Change Vulnerability Assessment and Adaptation Planning for Mangrove Systems.* 2012, Washington, DC: World Wildlife Fund.

52. Glick et al., *Scanning the Conservation Horizon.* (See reference 13.)

53. Hilty et al., editors, *Climate and Conservation.* (See reference 36.)

54. Chapman, S, et al., *Publishing trends on climate change vulnerability in the conservation literature reveal a predominant focus on direct impacts and long time-scales.* Diversity and Distributions, 2014. **20**(10): 1221–1228.

55. Millar, C. I., N. L. Stephenson, and S. L. Stephens, *Climate change and forests of the future: Managing in the face of uncertainty.* Ecological Applications, 2007. **17**(8): 2145–2151.

56. Stein et al., *Climate Smart Conservation.* (See reference 9.)

57. Cross, M. S. et al., *Accelerating adaptation of natural resource management to address climate change.* Conservation Biology, 2013. **27**(1): 4–13.

58. Poiani, K. A. et al., *Redesigning biodiversity conservation projects for climate change: Examples from the field.* Biodiversity Conservation, 2011. **20**: 185–201.

59. Stein et al., *Climate Smart Conservation.* (See reference 9.)

60. Willows, R., and R. Connell, editors, *Climate Adaptation: Risk, Uncertainty, and Decision-Making.* UKCIP Technical Report. 2003, Oxford, UK: UK Climate Impacts Programme.

61. Rowland, E., M. Cross, and H. Hartmann, *Considering Multiple Futures: Scenario Planning to Address Uncertainty in Natural Resource Conservation.* 2014, Shepardstown, WV: U.S. Fish and Wildlife Service. 172.

62. Spearman, M., and H. McGray, *Making Adaptation Count: Concepts and Options for Monitoring and Evaluation of Climate Change Adaptation.* 2011, Eschborn, Germany: GIZ and World Resources Institute.

63. Lawler, J. J. et al., *Resource management in a changing and uncertain climate.* Frontiers in Ecology and the Environment, 2010. **8**(1): 35–43.

64. Lawler, J. J., *Climate change adaptation strategies for resource management and conservation planning.* Annals of the New York Academy of Sciences, 2009. **1162**(1): 79–98; Heller, N. E., and E. S. Zavaleta, *Biodiversity management in the face of climate change: A review of 22 years of recommendations.* Biological Conservation, 2009. **142**(1): 14–32; and Mawdsley, J. R., R. O'Malley, and D. S. Ojima, *A review of climate-change adaptation strategies for wildlife management and biodiversity conservation.* Conservation Biology, 2009. **23**(5): 1080–1089.

65. Stein et al., *Climate Smart Conservation.* (See reference 9.)

66. Heller and Zavaleta, *Biodiversity management in the face of climate change: A review of 22 years of recommendations.* (See reference 64.)

67. Cross, M. S. et al., *From connect-the-dots to dynamic networks: Maintaining and enhancing connectivity to address climate change impacts on wildlife,* in *Wildfire Conservation in a Changing Climate,* J. Brodie, E. Post, and D. Doak, editors. 2012, Chicago: University of Chicago Press. 307–329.

68. Hilty, J. A., W. Z. Lidicker Jr., and A. M. Merenlender, *Corridor Ecology: The Science and Practice of Linking Landscapes for Biodiversity Conservation.* 2006, Washington, DC: Island Press. 344.

69. Nuñez, T. A. et al., *Connectivity planning to address climate change.* Conservation Biology, 2013. **27**(2): 407–416.

70. Aune, K. et al., *Assessment and Planning for Ecological Connectivity: A Practical Guide.* 2011, Bozeman, MT: Wildlife Conservation Society. 78.

71. Groves et al., *Incorporating climate change into systematic conservation planning.* (See reference 48.)

72. Game, E. T. et al., *Informed opportunism for conservation planning in the Solomon Islands.* Conservation Letters, 2011. **4**(1): 38–46.

73. Ashcroft, M. B., *Identifying refugia from climate change.* Journal of Biogeography, 2010. **37**: 1407–1413.

74. Groves et al., *Incorporating climate change into systematic conservation planning.* (See reference 48.)

75. Hannah, L. et al., *Protected area needs in a changing climate.* Frontiers in Ecology and the Environment, 2007. **5**(3): 131–138.

76. Dudley, N. et al., *Natural Solutions—Protected Areas: Helping People Cope with Climate Change.* World Commission on Protected Areas Best Practice Series. 2009, Gland, Switzerland: IUCN.
77. Gross, J. et al., *Responding to climate change: Guidance for Protected Area Managers.* World Commission on Protected Areas Best Practice Series, C. R. Groves, editor. 2016, Gland, Switzerland: IUCN.
78. Murti, R., and C. E. Buyck, *Safe Havens: Protected Areas for Disaster Risk Reduction and Climate Change Adaptation.* 2014, Gland, Switzerland: IUCN. 168.
79. Johnston, A. et al., *Observed and predicted effects of climate change on species abundance in protected areas.* Nature Climate Change, 2013. **3**(12): 1055–1061.
80. Rannow, S. et al., *Managing protected areas under climate change: Challenges and priorities.* Environmental Management, 2014. **54**(4): 732–743.
81. Kati, V. et al., *The challenge of implementing the European network of protected areas Natura 2000.* Conservation Biology, 2014. **29**(1): 260–270.
82. Brost and Beier, *Use of land facets to design linkages for climate change.* (See reference 49.)
83. Schwartz, M. W. et al., *Managed relocation: Integrating the scientific, regulatory, and ethical challenges.* BioScience, 2012. **62**(8): 732–743; Lawler, J. J., and J. D. Olden, *Reframing the debate over assisted colonization.* Frontiers in Ecology and the Environment, 2011. **9**(10): 569–574; and Neff, M. W., and B. M. H. Larson, *Scientists, managers, and assisted colonization: Four contrasting perspectives entangle science and policy.* Biological Conservation, 2014. **172**: 1–7.
84. Gray, L. K., and A. Hamann, *Strategies for reforestation under uncertain future climates: Guidelines for Alberta, Canada.* PLoS ONE, 2011. **6**(8): e22977.
85. Rout, T. M. et al., *How to decide whether to move species threatened by climate change.* PLoS ONE, 2013. **8**(10): e75814.
86. Hansen, L., and J. Hoffman, *Climate Savvy: Adapting Conservation and Resource Management to a Changing World.* 2010, Washington, DC: Island Press. 256.
87. Moser, S. C., and J. A. Ekstrom, *A framework to diagnose barriers to climate change adaptation.* Proceedings of the National Academy of Sciences, 2010. **107**(51): 22026–22031.
88. Eisenack, K. et al., *Explaining and overcoming barriers to climate change adaptation.* Nature Climate Change, 2014. **4**(10): 867–872.
89. McClanahan et al., *Conservation action in a changing climate.* (See reference 16.)
90. Turner et al., *Climate change: Helping nature survive the human response.* (See reference 14.)
91. Venter, O. et al., *Using systematic conservation planning to minimize REDD+ conflict with agriculture and logging in the tropics.* Conservation Letters, 2013. **6**(2): 116–124.

# 10

# Planning for Ecosystem Services: Building a Bridge to Human Well-Being

## Overview

Interest in the conservation of ecosystem services (ES) is steadily increasing, as evidenced by research papers, conferences and trainings, multilateral platforms and assessments, and tools and methods for estimating these services. Despite this interest, incorporating ES in conservation planning has proven challenging. Lack of data and easily applied methods for measuring and mapping services have been major limitations, and much of the work has been theoretical. We review the basic methods for quantifying, valuating (attaching financial value to a service), and mapping ES. We outline the basic steps for planning for the conservation of ES, and suggest how such planning can be better integrated with biodiversity conservation. Finally, we discuss some of the most important criticisms of ES and highlight some of the most important lessons learned to date in ES conservation efforts.

## Topics

- *Millennium Ecosystem Assessment (MEA)*
- *The Economics of Ecosystems and Biodiversity (TEEB)*
- *Intergovernmental Science-Policy Platform on Biodiversity and Ecosystem Services (IPBES)*
- *Ecosystem service (ES) mapping*
- *ES valuation*
- *Human well-being*
- *ES integration with biodiversity*
- *Planning steps*
- *Critiques of ES concepts*
- *Lessons learned*

The early years of conservation planning were largely focused on the patterns of biodiversity and on trying to ensure that species of concern and the spectrum of communities and ecosystems were adequately represented in conservation plans. Over time, conservation planners and biologists grew more concerned about the function and process of ecosystems and began to incorporate these components into their plans. Meanwhile, urban and landscape planners and architects, as well as engineers and agricultural experts, have long been focused on ecosystem function as they constructed plans on the most appropriate places for development in urban and nonurban areas alike.

For example, in the 1960s, Australian engineers and scientists undertook a large-scale rangeland rehabilitation program in the Ord River catchment. Their goal was to help reduce erosion from severe overgrazing that could add significant sediment loads to a reservoir (Lake Argyle) and dam (Ord River Dam) that were being planned as part of an agricultural and irrigation development project in the Kimberley region of northwest Australia[1] (FIGURE 10.1). Eventually that grassland restoration project became one of the most successful large-scale restoration projects in Australia; the dam and reservoir were completed in the early 1970s, and many of the rehabilitated lands were added to Australia's protected area estate in the region of Purnululu National Park.

FIGURE 10.1 Ord River restoration project in northwestern Australia. In anticipation of a planned hydroelectric project and reservoir, the Australian government initiated a large-scale rangeland restoration project that would ultimately help reduce sediment loading to the planned dam and reservoir. By removing livestock and undertaking revegetation efforts, the project succeeded, and it demonstrated an early concern and focus on ecological processes (desertification, sedimentation) and ecosystem services (irrigation development from dam and reservoir).

Today, planners across the range of conservation planning disciplines, including government (**Box 10.1**) and nongovernmental organizations (NGOs), are increasingly paying attention to the benefits that the species, structure, and functions of ecosystems provide to people—ecosystem services (ES)—and how those services can best

## Box 10.1 Ecosystem Services and Planning in the U.S. Forest Service and Federal Government

In its 2012 Planning Rule, which revises the regulations for development of national forest land management plans in the United States, the role of ecosystem services in U.S. Forest Service planning received considerable attention. The planning rule emphasizes, for example, that the purpose of plans is to result in national forests that are managed so that they

> *have the capacity to provide people and communities with ecosystem services and multiple uses that provide a range of social, economic, and ecological benefits for the present and into the future. These benefits include clean air and water; habitat for fish, wildlife, and plant communities; and opportunities for recreational, spiritual, educational, and cultural benefits.*[a]

The planning rule specifically mandates the agency to include ecosystem services in its assessment phase of forest planning ("benefits people obtain from the NFS planning area [ecosystem services]"). The 2012 Planning Rule essentially set forth broad goals and policy direction for which national forests are intended to be managed.

At the same time, the agency has developed a more detailed set of planning directives that are intended to help implement the new rule. Within those directives, there is preliminary guidance on how to incorporate ecosystem services into the assessment phase of forest planning. That advice includes identifying and evaluating

- Key ecosystem services provided by the plan area that may be influenced by the land management plan. This evaluation should include the condition and trend of these key ecosystem services and the ability of the plan area to provide these ecosystem services in the future.
- The geographic scale at which the plan area contributes to ecosystem services (e.g., watersheds, counties, regional markets, or ecoregions).
- Drivers likely to affect future demand for and availability of key ecosystem services.
- The stability or resiliency of the ecosystems or key characteristics of ecosystems that currently maintain the plan area's key ecosystem services.
- The influence of non-NFS lands or other conditions beyond the authority of the Forest Service that influence the plan area's ability to provide ecosystem services.

The actual plan itself for an individual national forest unit must "describe the desired conditions for the key ecosystem services. Desired conditions may describe different mixes of ecosystem services from different management or geographic areas." The objectives of any single national forest plan

*(continued)*

**Box 10.1 Ecosystem Services and Planning in the U.S. Forest Service and Federal Government** *(continued)*

need to "consider the linkage between the key ecosystem services and how plan objectives contribute to the intended achievement of the level, quality, or delivery to the public of the key ecosystem services."

Given that the 2012 Planning Rule is the first time that ecosystem services have been explicitly included in forest planning, the requirements outlined here are a tall order for any institution. The U.S. Forest Service is in the process of developing additional guidance and tools to assist forest-planning teams in incorporating ecosystem services into their planning processes. A recent report by the Forest Service's Pacific Northwest research station on trade-offs among ecosystem services is a good example of such planning guidance.[b]

---

[a] Forest Service, USDA [accessed August 3, 2015]. *National Forest System Land Management Planning.* http://www.fs.usda.gov/Internet/FSE_DOCUMENTS/stelprdb5362536.pdf.

[b] Kline, J. D., and M. J. Mazzotta, *Evaluating tradeoffs among ecosytem services in the management of public lands*, in *General Technical Reports*. 2012, Olympia, WA: Pacific Northwest Research Station.

be incorporated into both conservation and development planning (FIGURE 10.2). Indeed, in late 2014 a special issue of the journal *Landscape Ecology* was devoted to this very issue—integrating ecosystem services into landscape-level planning.[2]

An appreciation of these services to human well-being continues to grow. Topics range from the importance of ES of reefs (e.g., wave attenuation) and other coastal habitats in reducing risks from sea-level rise and storm surge to property and vulnerable people,[3] to the consequences of degraded ecosystem services for human health and the spread of disease,[4] to the impacts of broad-scale land use change on services such as carbon storage and food production.[5] As these and other examples in this chapter suggest, the discipline of ecosystem services and their conservation is a rapidly evolving one.

Although ecologists and economists have been analyzing, discussing, and debating the concept of ecosystem services for decades, at least two major publications, both published in 1997, helped propel the concept into the mainstream of nature conservation. First, Stanford biologist Gretchen Daily edited the influential *Nature's Services: Societal Dependence on Natural Ecosystems,*[6] which made explicit ties between biodiversity conservation, ES, and human well-being. At the same time, Robert Costanza and colleagues[7] published a seminal paper that estimated the global value of 17 ES in 16 biomes at approximately $US33 trillion (nearly twice the global gross national product at that time), and made the argument that these services were critical to Earth's "life-support systems." Then, in 2000 the United Nations called for a global assessment of ecosystem services, which was

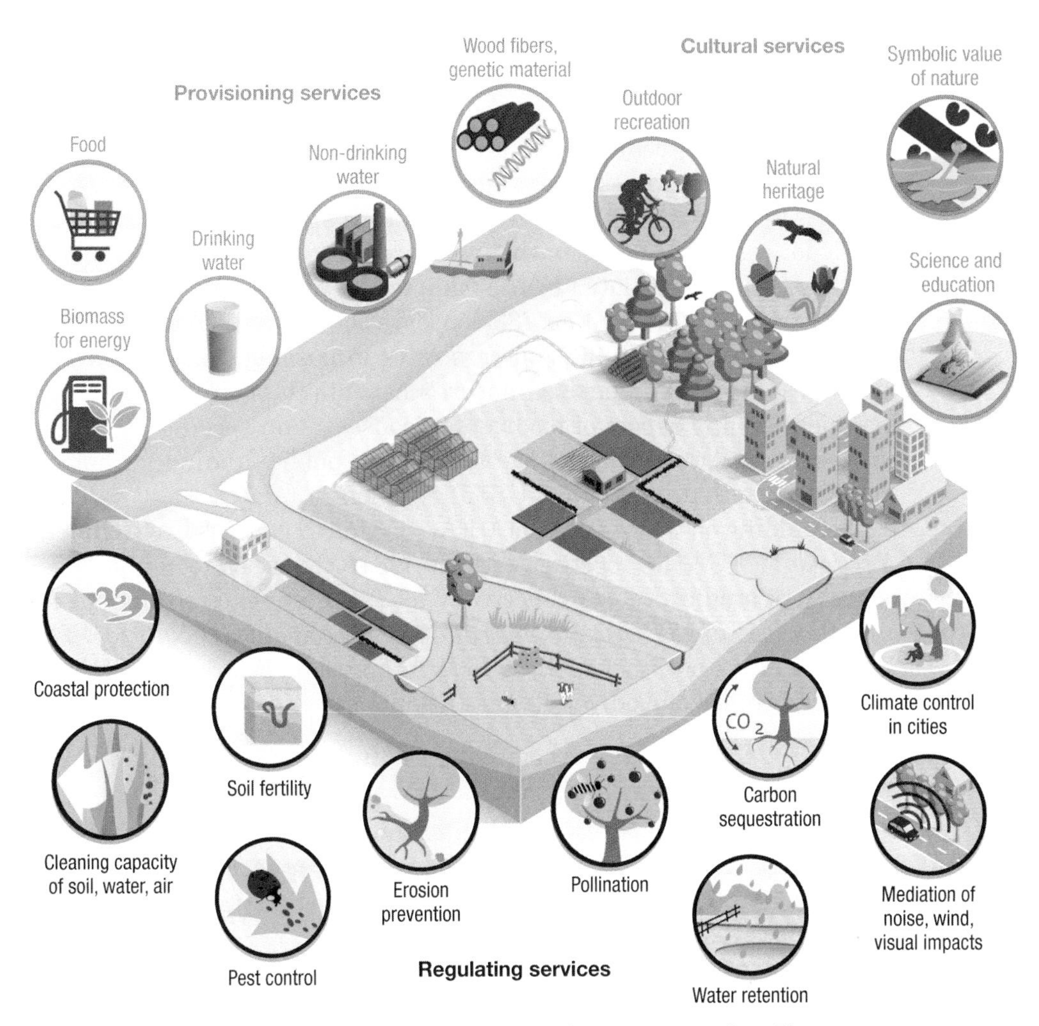


FIGURE 10.2 Conceptual diagram of the major types of ecosystem services. Those involved in nature conservation efforts are paying increasing attention to ecosystem services. These services are the result of a natural ecological process in nature, such as the flow of fresh water in a river, but they necessarily must flow from nature to a human beneficiary to be defined as an ecosystem service. More detailed information on the different types of services is described in the text. (Data from PBL Netherlands Environmental Assessment Agency.)

subsequently conducted under the auspices of the United Nations Environment Programme. In 2005, this highly touted effort was published as the *Millennium Ecosystem Assessment* (MEA).[8]

Among the many key findings of MEA was a striking one: 60% of 24 ecosystem services examined globally were being degraded and/or used unsustainably (e.g., freshwater and air purification), often as a consequence of meeting demands for fuel, fiber, water, and timber. That rate of degradation was thought to be increasing and could in some cases be potentially irreversible (e.g., dead zones in coastal oceans resulting from inland water quality issues). Moreover, the negative impacts of these degradations were in many cases being borne by the world's poorest people.

Since publication of the *Millennium Ecosystem Assessment*, several global initiatives, conferences, and books have served to advance the concept of ES. In addition, there is evidence that a focus on ecosystem services in conservation projects and initiatives can broaden the constituency for conservation, even though there is considerable debate about this topic (discussed later in this chapter).[9] For example, The Nature Conservancy has worked on 183 conservation-related ballot measures in 23 U.S. states (with 90% success rates in passing these initiatives) that have largely emphasized nature's benefits to people and often do not mention any aspect of biodiversity conservation specifically. These successful ballot measures have resulted in public funding for nature conservation that has totaled US$51.8 billion.

Building on the work of the MEA, the United Nations Environment Programme (UNEP) together with the G8+5 countries* commissioned a series of studies known collectively as *The Economics of Ecosystems and Biodiversity (TEEB): Mainstreaming the Economics of Nature.*[10] The TEEB studies outlined a three-tiered approach to incorporating the economic values of biodiversity and ES into societal decision making:

1. *Recognizing value*—recognizing the full range of ES that affect different stakeholder groups across society
2. *Demonstrating value*—analyzing the costs and benefits of different ecosystem services and how different decisions about use of biodiversity and ES will affect these costs and benefits for different stakeholder groups
3. *Capturing value*—using a variety of economic tools to better incorporate the value of ES into decision making

This three-tiered approach was then applied to a variety of case studies from around the globe to demonstrate its value to decision making for ecosystems (forests), cities, and the business community (mining).

Following in the footsteps of the TEEB report, UNEP in association with a second UN program (International Human Dimensions Programme) collaborated to produce the *Inclusive Wealth Report 2012: Measuring Progress Towards Sustainability,*[11] another major effort to demonstrate the value of incorporating ES into societal decisions about economic development. The most important objectives of the Inclusive Wealth project were (1) to develop an index (the Inclusive Wealth Index) with particular emphasis on natural capital that would help government measure progress toward sustainability and the

---

*Canada, France, Germany, Italy, Japan, United States, United Kingdom, and Russia plus Brazil, China, India, Mexico, South Africa

**green economy**, and (2) to provide analysis that would help countries improve environmental, social, and economic planning.

As further evidence of the increasing prominence of ES in the global conservation and development community, a new body known as the **Intergovernmental Science–Policy Platform on Biodiversity and Ecosystem Services (IPBES)**[12] was established under the auspices of the United Nations Environment Programme to "strengthen the science–policy interface for biodiversity and ecoystem services for the conservation and sustainable use of biodiversity, long-term human well-being and sustainable development" (FIGURE 10.3).

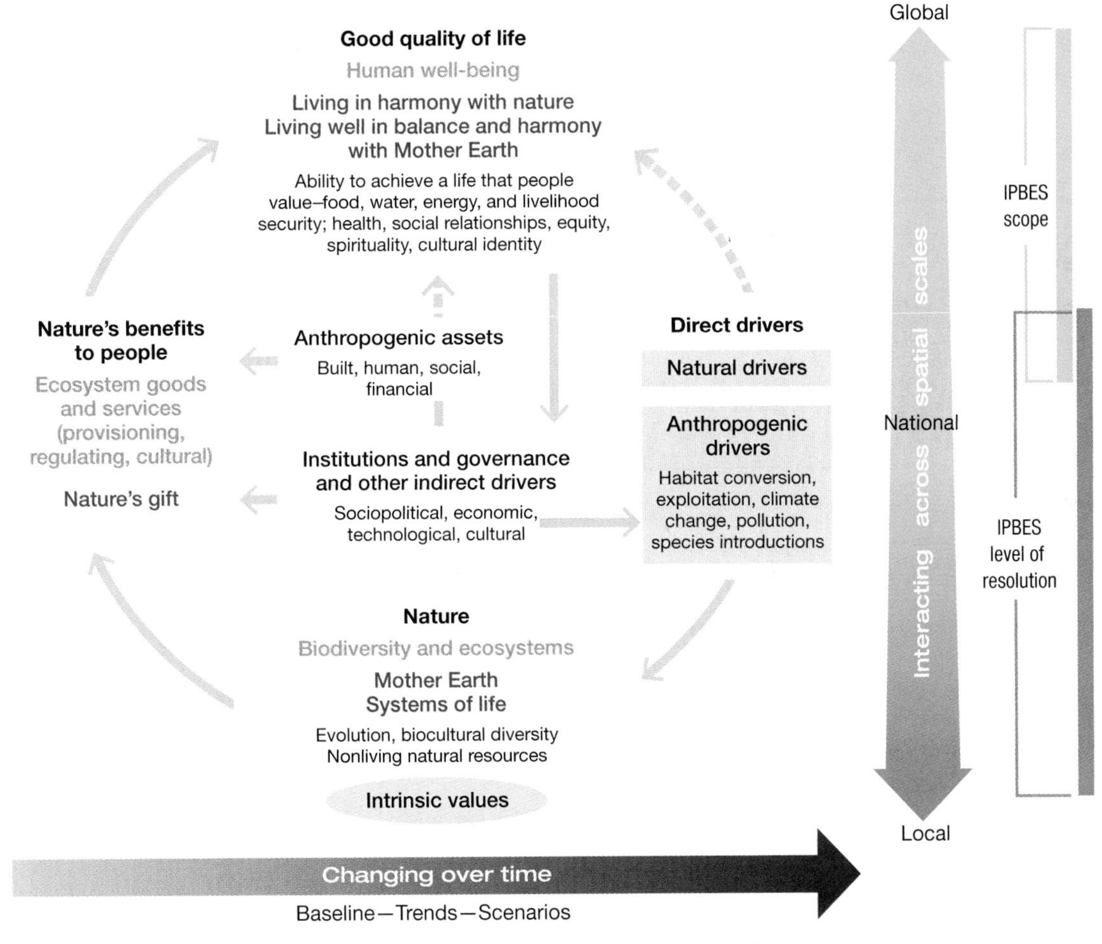


FIGURE 10.3 Conceptual framework of the Intergovernmental Science–Policy Platform on Biodiversity and Ecosystem Services (IPBES). The platform's conceptual framework includes six interlinked elements constituting a social-ecological system that operates at various scales in time and space: nature, nature's benefits to people (ecosystem services), anthropogenic assets, institutions and governance systems and other indirect drivers of change, direct drivers of change, and quality of life. (Adapted from United Nations Environment Programme, *Report of the second session of the Plenary of the Intergovernmental Science–Policy Platform on Biodiversity and Ecosystem Services.* Ankara, Turkey, 2014 [accessed May 23, 2015].

In many respects, the IPBES is a follow-up effort to the *Millennium Ecosystem Assessment.* As of 2014, the organization has 199 member states as signatories and has just established its first program of work aimed at achieving the principal objectives. Although its effectiveness as a body remains untested,[13] the existence of IPBES will continue to draw governments' attention to the interlinkages of ES and biodiversity conservation and the needs to conserve both. In the same way that the IPCC has supported the policy mechanism of the **United Nations Framework Convention on Climate Change** (**UNFCCC**), the IPBES is meant to be a policy-relevant supportive platform for the Convention on Biological Diversity and other multilateral environmental agreements.

Three important aspects of ES are essential to understanding how we can best incorporate them into conservation planning—classifying ES, mapping them, and valuating them. In the sections ahead, we explore those essential aspects, review a set of steps for conservation planning for ES, discuss some of the most important critiques of ES related to nature conservation, and conclude with some lessons learned.

As we probe more deeply into planning approaches that incorporate ES, now is an opportune time for an important footnote to everything we recommend in this chapter: planning for ES is still in its early years in many respects, there is no one right or highly recommended approach to go about doing this work, and there is a great deal of ongoing research as to the most appropriate methods, tools, and overall approaches.

## Classifying Ecosystem Services

To incorporate ES into conservation planning, we need an appreciation of the different types of services. Here we use the most widely cited typology, which was developed as part of the MEA and defined four broad categories of ES (FIGURE 10.4):

1. **Provisioning**—products obtained from ecosystems such as food, fiber, fresh water, and fuel. More direct examples include fisheries, crops, and timber.
2. **Regulating**—benefits that come from the regulation of ecosystem processes such as water purification, pollination, erosion control, pest control, and climate regulation.
3. **Cultural**—intrinsic and nonmaterial benefits that people receive from ecosystems, such as aesthetic values, spiritual values, recreation, and cultural heritage.
4. **Supporting**—services that more indirectly benefit people but are necessary for all other ecosystem services. In many

**FIGURE 10.4** Four types of ecosystem services. (a) Pollination is an example of a regulating service, (b) food production is an example of a provisioning service, (c) recreation is an example of a cultural service, and (d) nutrient cycling is an example of a supporting service. See text for details on these ecosystem service types.

respects, these services are synonymous with ecosystem processes—photosynthesis, nutrient cycling, and soil formation are prime examples.

More detailed explanations of these services can be found in **TABLE 10.1**, which also provides examples of indicators of these services. These indicators are critical for planning because these variables themselves, or proxies for them, can be used to map and model ecosystem services and ultimately incorporate them in conservation plans.

Provisioning and regulating services have received the most attention, in part because we either depend on them for our survival (provisioning) or because they have been the focus of considerable ecological research for decades (regulating). Although ecosystem service scientists argue that cultural services have received the least attention in the discipline,[14] ironically some of these services—such as areas important for recreation—have been incorporated in conservation plans and land use decisions. In any event, there are signs that this oversight is changing, including a 2014 issue of the journal

**TABLE 10.1 Examples of different types of ecosystem services and indicators of these services**

| | Services comments and examples | Ecological process and/or component providing the service (or influencing its availability) = functions | State indicator (how much of the service is present) |
|---|---|---|---|
| ***Provisioning*** | | | |
| 1 | *Food* | Presence of edible plants and animals | Total or average stock in kg/ha |
| 2 | *Water* | Presence of water reservoirs | Total amount of water ($m^3$/ha) |
| 3 | *Fiber and fuel and other raw materials* | Presence of species or abiotic components with potential use for timber, fuel, or raw material | Total biomass (kg/ha) |
| 4 | *Genetic materials:* genes for resistance to plant pathogens | Presence of species with (potentially) useful genetic material | Total "gene bank" value (e.g., number of species and sub-species) |
| 5 | *Biochemical products and medicinal resources* | Presence of species or abiotic components with potentially useful chemicals and/or medicinal use | Total amount of useful substances that can be extracted (kg/ha) |
| 6 | *Ornamental species and/or resources* | Presence of species or abiotic resources with ornamental use | Total biomass (kg/ha) |
| ***Regulating*** | | | |
| 7 | *Air quality regulation:* (e.g., capturing dust particles) | Capacity of ecosystems to extract aerosols and chemicals from the atmosphere | Leaf area index NOx-fixation, etc. |
| 8 | *Climate regulation* | Influence of ecosystems on local and global climate through land cover and biologically mediated processes | Greenhouse gas balance (esp. C-sequestration); land cover characteristics, etc. |
| 9 | *Natural hazard mitigation* | Role of forests in dampening extreme events (e.g., protection against flood damage) | Water-storage (buffer) capacity in $m^3$ |
| 10 | *Water regulation* | Role of forests in water infiltration and gradual release of water | Water retention capacity in soils, etc. or at the surface |
| 11 | *Waste treatment* | Role of biota and abiotic processes in removal or breakdown of organic matter, xenic nutrients, and compounds | Denitrification (kg N/ha/y); Immobilization in plants and soil |
| 12 | *Erosion protection* | Role of vegetation and biota in soil retention | Vegetation cover root matrix |
| 13 | *Soil formation and regeneration* | Role of natural processes in soil formation and regeneration | For example, bioturbation |
| 14 | *Pollination* | Abundance and effectiveness of pollinators | Number and impact of pollinating species |
| 15 | *Biological regulation* | Control of pest populations through trophic relations | Number and impact of pest control species |

*(continued)*

*Ecology and Society*[15] that featured a variety of articles on cultural ecosystem services. All of these services can be strongly linked to many elements of human well-being (see Figure 10.3). They also represent important linkages to the social-ecological system that we described in Chapter 1, although clearly the "ecological" portion of the SES system has received most of the attention in how ES are defined, mapped, and measured.

Much work remains to be done to assess ES on the basis of the interaction between social and ecological systems and in how changes in human well-being feed back into the supply of ES[16] (e.g., a response to drought by ranchers with increases in irrigation efficiency will affect

| | Services comments and examples | Ecological process and/or component providing the service (or influencing its availability) = functions | State indicator (how much of the service is present) |
|---|---|---|---|
| ***Habitat or supporting*** | | | |
| 16 | *Nursery habitat* | Importance of ecosystems to provide breeding, feeding, or resting habitat for transient species | Number of transient species and individuals (esp. with commercial value) |
| 17 | *Gene pool protection* | Maintenance of a given ecological balance and evolutionary processes | Natural biodiversity (esp. endemic species); habitat integrity (irt min. critical size) |
| ***Cultural and amenity*** | | | |
| 18 | *Aesthetic:* appreciation of natural scenery (other than through deliberate recreational activities) | Aesthetic quality of the landscape, based on, e.g., structural diversity, "greenness," tranquility | Number/area of landscape features with stated appreciation |
| 19 | *Recreational:* opportunities for tourism and recreational activities | Landscape features, attractive wildlife | Number/area of landscape and wildlife features with stated recreational value |
| 20 | *Inspiration for culture, art, and design* | Landscape features or species with inspirational value to human arts, etc. | Number/area of landscape features or species with inspirational value |
| 21 | *Cultural heritage and identity:* sense of place and belonging | Culturally important landscape features or species | Number/area of culturally important landscape features or species |
| 22 | *Spiritual and religious inspiration* | Landscape features or species with spiritual and religious value | Presence of landscape features or species with spiritual value |

*Source:* Data from DeGroot, R. et al., *Challenges in integrating the concept of ecosystem services and values in landscape planning, management and decision making.* Ecological Complexity, 2010. **7**(3): 260–272.

the local and regional supply of freshwater ES). At the same time, we recognize that many other factors contribute to human well-being—religious beliefs, improvements in technology (e.g., smartphones), changes in governance, and improvement in global trade relations, just to name a few. Nevertheless, the importance of conserving ecological processes and the ecosystem services derived from them is gaining additional support as the impacts of climate change such as sea-level rise and extreme storm events are both increasingly obvious and are impacting human communities throughout the world. As a result, focusing on maintaining and restoring ecosystem processes and services will be a critical response to climate change in the future (Chapter 9).

## Mapping Ecosystem Services

The ability to map ES is essential; if we can't assign some sort of spatial priority to a service, then it will be difficult to figure out how to plan for its conservation. Jan Schägner and colleagues[17] from the European Union have extensively reviewed the methods for mapping ecosystem service values in monetary terms. As they point out, mapping ES values is useful in several arenas, including

- *Green accounting.* Taking environmental costs into account in the operations of a business or government; the *Inclusive Wealth Report* previously mentioned is an example of green accounting.
- *Land use planning.* Examining different land uses across regions and trade-offs among them, including those that support various ecosystem services.
- *Resource allocation.* Mapping ES locations provides information on where and where not to invest in conservation or development or trade-offs between the two.
- *Payments for ecosystem services (PES).* With the increasing popularity of this incentive-based conservation strategy, mapping values can provide helpful baseline information for analyses to determine where and in what manner to undertake PES schemes.

Mapping ES values generally involves two processes: determining what the service is, and determining what monetary value should be attached to that supply (although as we will explain, it's possible and common to map ES without attaching a monetary valuation). By far the most common approach to mapping ES is to use a single biophysical variable, most often from land cover/land use data, to map the supply of an ES across some planning or analysis unit.[18]

The classic study to use this method was Robert Costanza and colleagues' 1997 *Nature* paper,[19] which used a global land cover/land use map to identify the world's major ecosystem types. The researchers then assigned mean monetary values per unit area of each of 17 different ecosystem services to each of these major ecosystem types. This combination of land cover/land use proxies with unit monetary values of ES is referred to as the **benefit transfer method**. The name is derived from the fact that the method uses an empirical estimate of the values produced from one habitat or ecosystem type and then transfers that benefit to similar habitats or ecosystems elsewhere. Costanza and colleagues' 1997 global estimates of ecosystem services were based on this approach, and they have been more recently updated.[20]

A breakthrough in mapping ecosystem services occurred with the development of **ecological production functions**, which are the foundation of a widely used ES tool—**InVEST (Integrated Valuation of Ecosystem Services and Tradeoffs)**—and are carefully articulated with many examples in the 2011 book *Natural Capital: Theory and Practice of Mapping Ecosystem Services.*[21] These production functions "define how the spatial extent, structure, and functioning of ecosystems determine the production of ecosystem services."[22]

The typical production function has an output that can be mapped in biophysical terms (e.g., crop yield per unit area or sediment retention per unit area). These biophysical variables can be combined with appropriate market prices or other means of monetary valuation to estimate the economic value of an ES when such data are available. The major advantage of using ecological production functions is that they overcome the assumption of constant value of an ES per unit area that is associated with the benefit transfer methods. **Box 10.2** provides an example of using InVEST and marine ecological production functions to assist the Belize government in management of coastal development and marine areas.

## Box 10.2 Using InVEST to Help Develop an Integrated Coastal Zone Management Plan in Belize

InVEST—Integrated Valuation of Ecosystem Services and Tradeoffs—is a decision support tool developed by the Natural Capital Project, a collaboration of TNC, WWF, Stanford University, and the University of Minnesota. InVEST combines biophysical models of ecosystem services (ecological production functions) with economic information about ecosystem services to assist in decision making about conservation and natural resource management issues at appropriate scales. InVEST helps practitioners and planners envision different plausible scenarios of future development and conservation, it can reveal relationships among different services, and it can be used with fairly simple models that require relatively small amounts of data as well as more complicated models that involve more parameters and necessitate more time and expertise to apply.

The Natural Capital Project (NatCap) and WWF have worked closely with the Coastal Zone Management Authority and Institute (CZMAI) of Belize to develop an Integrated Coastal Zone Management Plan to guide future development and conservation of the coastal ecosystems of Belize. The team first worked to create models and maps of the current distribution of six important ecosystem services—fisheries biomass, aquaculture biomass, coastal protection, carbon storage/sequestration, wave energy, and recreation—and to estimate the economic value of these services[a] (FIGURE 10.2.1).

Then, working with policy makers, stakeholders, and scientists throughout Belize in a second major component of this project, three future scenarios were developed: conservation, informed management, and development. These scenarios essentially represented different zoning schemes for conservation and development. The conservation scenario emphasized protection of existing ecosystems and was intended to represent the views of the environmental community. The development scenarios placed a premium on coastal development and extractive industries without zoning guidance. The third scenario, informed management, blended development and conservation interests into a "science-based zoning scheme" that balanced economic returns from coastal development with minimizing environmental impacts.

In a third component of this work, NatCap and CZMAI scientists used four of the five ecosystem service models originally developed for the project (for habitat risk, coastal protection, spiny lobster fishery, and tourism/recreation) to assess how these services would fare under the different scenarios. FIGURE 10.2.2 is an example

(continued)

## Box 10.2 Using InVEST to Help Develop an Integrated Coastal Zone Management Plan in Belize *(continued)*

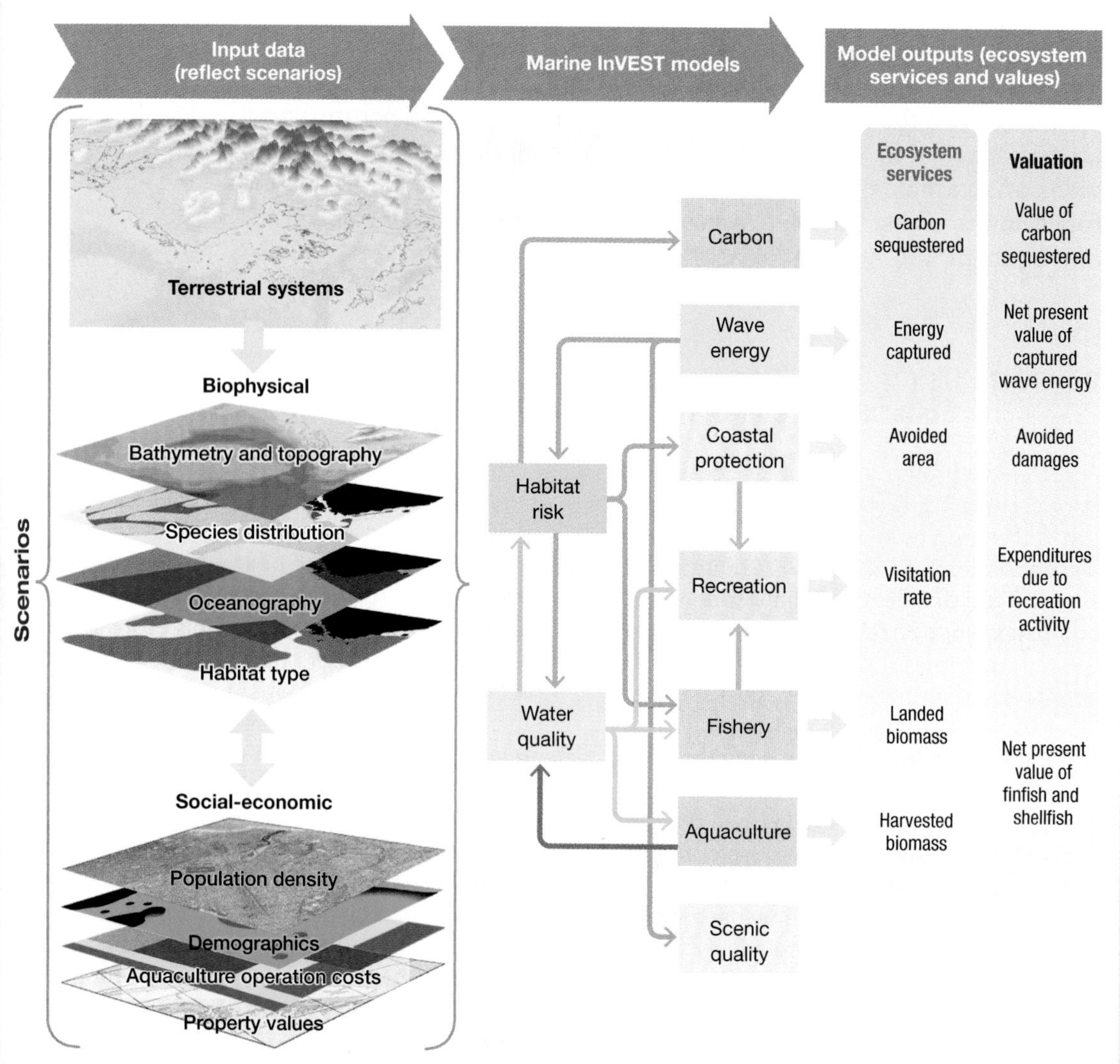


**FIGURE 10.2.1** An example of the data, ecosystem services, their models, and model outputs used to inform development and conservation of Belize's coastal ecosystems. (Adapted from Guerry et al., 2012.)

of how the spiny lobster fishery catch and revenue would be affected in the different planning regions under the three different scenarios.[b]

Through additional public feedback and InVEST modeling, the informed management scenario was selected as the preferred scenario. It was then revised and improved to provide greater economic return and greater return on ecosystem services, largely by moving development activities to less environmentally sensitive areas. This scenario was released for public review, and after additional modification, it has served as the basis for the Final Integrated Coastal Management Zone Plan for

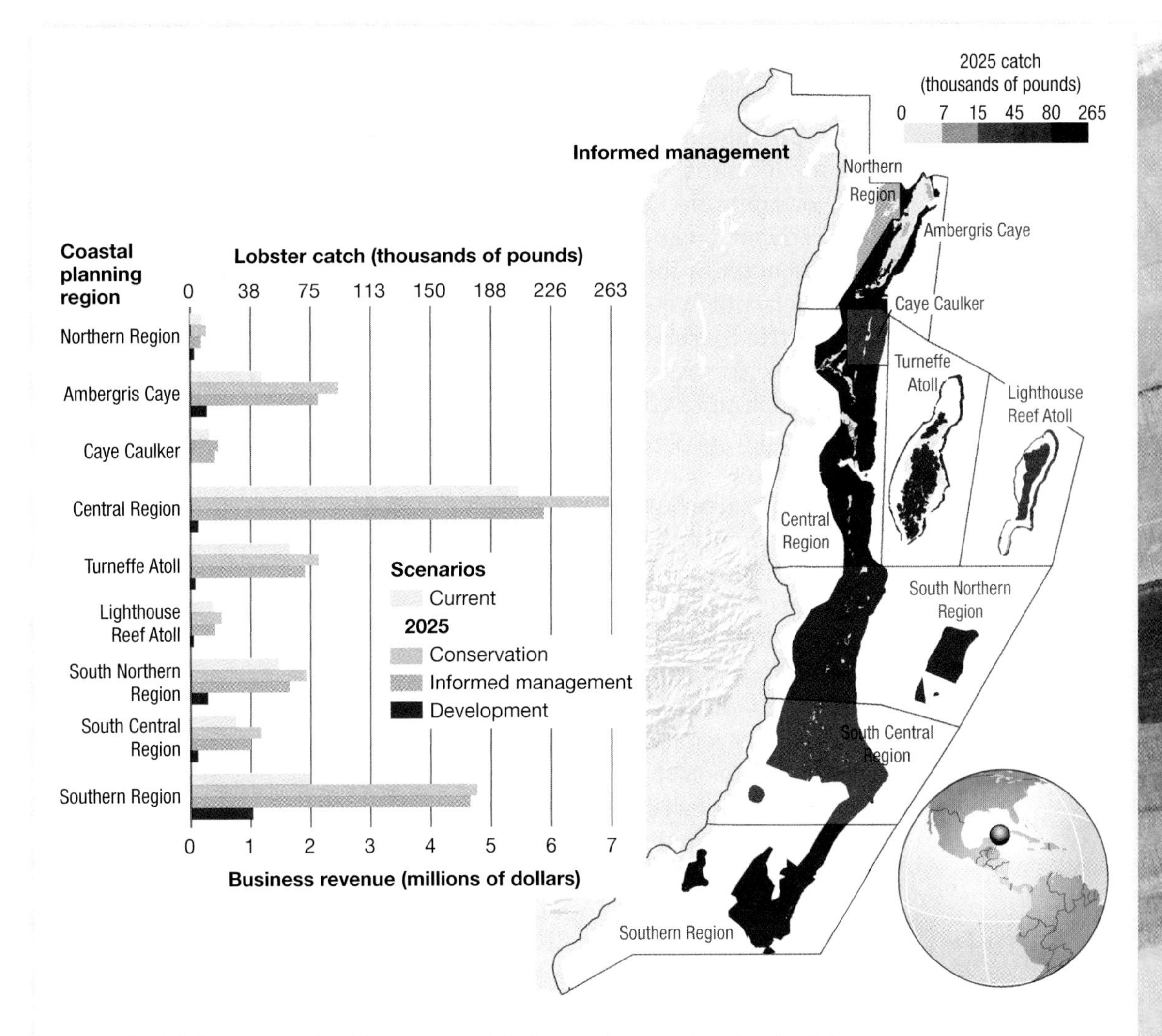


FIGURE 10.2.2 An example of how one modeled ecosystem service—lobster fishery—would far under different scenarios of development in various coastal regions of Belize. (Adapted from Clark et al., 2013.)

Belize. As of summer 2014, the plan is pending approval by the Belize government.

Further details on the use of InVEST in this marine conservation planning process can be obtained through several NatCap publications that are maintained on the group's website .[c]

---

[a] Guerry et al. *Modeling benefits from nature: using ecosystem services to inform coastal and marine spatial planning.* International Journal of Biodiversity Science, Ecosystem Services & Management, DOI:10.1080/21513732.2011.647835.

[b] Clarke, C., et al., *Belize Integrated Coastal Zone Management Plan.* 2013, Belize City: Coastal Zone Management Authority and Institute. 423.

[c] See, for example, Arkema, K. K., editor, *Coastal Development in Belize: Comprehensive Nationwide Coastal and Marine Spatial Planning.* 2014, Stanford, CA: Natural Capital Project, Stanford University.

Despite the considerable advances in mapping ecosystem services, questions remain about the accuracy and precision of mapped values.[23] One recent review suggests that fewer than a third of all ES studies provide an estimate of the error associated with precision or accuracy of maps, and when those estimates are provided they are often considerable. One veteran ES scientist relayed to us that the most frequent question he gets in working with government or non-government planning teams concerns the accuracy of production function models and maps of these functions. In addition to these concerns, there is little consensus on which mapping method is most appropriate to a specific purpose or context. The inconsistency in mapping methods across studies and assessments can be an obstacle to using ES data within a conservation plan and to comparisons and syntheses of methods across plans.[24]

A second review of methods to map ecosystem services builds upon these findings.[25] Most published projects that map ES consider only a few services, often at scales greater than 1000 km$^2$. Importantly, most of the mapping is concerned with where the service is generated and pays little attention to how the service flows to beneficiaries, interactions with other services, or impacts on various stakeholder or beneficiary groups.

It is a critical and often overlooked point that ecosystem services by definition need to benefit people. Thus mapping and conserving these services needs to include not only the supply side of ES (where they come from) but the delivery function as well. Mapping and conservation efforts for ES pay considerable attention to the supply of a service (freshwater for municipal uses), but often short shrift to how that service will get delivered or flow to the beneficiaries (drinking water for people),[26] although recent advances and tools for estimating supplies and flows are beginning to overcome these hurdles (**Box 10.3**). In practice, the type of ES assessed, the available data, the characteristics of the planning area, funding and staff expertise, and the decision-making context of the planning effort will be important factors in determining which mapping method to apply.[27]

## Valuation of Ecosystem Services

The monetary value of an ecosystem service is generally determined by two factors: (1) the supply of that service, which is usually related to an ecosystem process (e.g., the flow regime of a river); and (2) the demand for that service.[28] The former is usually the realm of ecologists and physical scientists such as hydrologists, while the latter is the expertise of economists. Estimating the supply of a service involves quantifying in it some manner, a topic we addressed in the previous section on

## Box 10.3 Mapping ES Supplies, Flows, and Beneficiaries—An Example from the Puget Sound, USA

Using the Artificial Intelligence for Ecosystem Services (ARIES) decision support system for ecosystem services, U.S. Geological Survey scientist Kenneth Bagstad and colleagues[a] estimated theoretical and actual services provided for five ES in the Puget Sound area, USA: (1) carbon sequestration and storage, (2) scenic viewsheds, (3) proximity to open space for homeowners, (4) flood regulation in developed land, and (5) sediment regulation for reservoirs. ARIES combines models of ES supply and demand with network flow models that help quantify ES. It can assess differences in theoretical ES supply and actual use because the provision or flow of each ES to beneficiaries is accounted for individually and parceled out to non-overlapping beneficiary groups. Because some loss of the service will occur as it flows to users, the actual use or demand and the theoretical supply are always different.

The four maps in **FIGURE 10.3.1** demonstrate the model difference between the ES supply of flood regulation and the actual delivery of that supply to beneficiaries. The difference between the theoretical supply of floodplain mitigation (map b) and the theoretical delivery of flood sinks and floodplain development to help regulate floods (map d) is considerable.

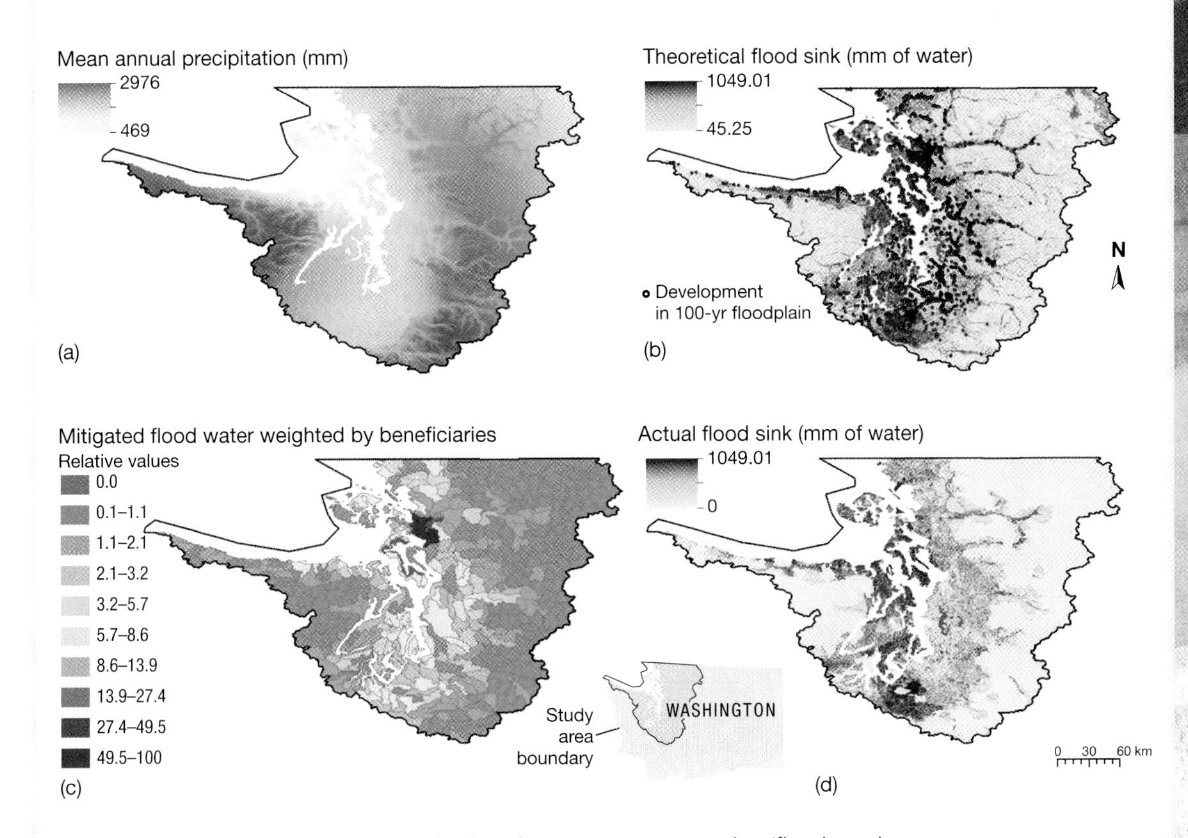


**FIGURE 10.3.1** Theoretical and actual values for an ecosystem service (flood regulation) in the Puget Sound Basin, WA (USA): (a) sources of the service, (b) sinks for the service, (c) actual values for the service, and (d) actual values for the service. (Data from Bagstad et al., 2014.)

*(continued)*

**Box 10.3** Mapping ES Supplies, Flows, and Beneficiaries—An Example from the Puget Sound, USA *(continued)*

Over 50% of the sub-watersheds in the Puget Sound lack any floodplain development. Sub-watersheds that can provide only a limited amount of flood mitigation (flood sinks that can absorb, intercept, and detain floodwater) and have few beneficiaries receive a low actual flood mitigation score compared to the theoretical scores. Sub-watersheds with large sink values and large numbers of beneficiaries receive relatively greater mitigation scores.

Using the ARIES methods and tools, Bagstad and colleagues demonstrate just how easy it is to overestimate ES without a careful consideration of beneficiaries and where those beneficiaries are located. Such was the case with four of the five services they examined: actual provision ranged from 16% to 66% of the theoretical.

This approach to assessing and mapping ES has several advantages over the traditional mapping of ES supplies. First, it can identify regions critical to maintaining the supply and flow of a service to specific beneficiaries. Second, it can inform Payment for Ecosystem Services (PES) schemes by identifying areas where ES flows are degraded as well as areas where particular beneficiary groups will have access to a service. Finally, conservation and restoration actions can be prioritized and focused on those areas where ecosystem service flows will best be maintained or enhanced.

[a] Bagstad, K. J. et al., *From theoretical to actual ecosystem services: Mapping beneficiaries and spatial flows in ecosystem service assessments.* Ecology and Society, 2014. **19**(2).

mapping ES. In this section, we focus on the demand for the service, which involves estimating the monetary value per unit of service.

The concept of **willingness to pay** underlies most valuations of ES. It is an amount that people are willing to pay or sacrifice to derive some level of satisfaction from a service. In some cases, this willingness to pay can be measured directly by market prices (such as your willingness to pay a particular price for a favorite fresh seafood item); in other cases, economists must use more indirect methods to determine this willingness to pay.

The values of different types of ecosystem services are determined with different methods (TABLE 10.2). Economists Lawrence Goulder and Donald Kennedy[29] have divided the methods into two main categories—those dealing with direct use values and indirect use values. Drinking water provided by streams to a city is an example of a direct use value. Within direct use values, there are both consumptive use values (e.g., bottled spring water) and nonconsumptive values (e.g., whitewater rafting). For direct consumptive use values, the value of an ecosystem service can often be determined by market values such as the price we pay for a produce item. Services that have direct value tend to fall in the provisioning and cultural use categories, whereas indirect use values typically are applied to supporting and regulating services.

TABLE 10.2 Valuation methods for ecosystem services

| Services | Types of values offered | Valuation method |
|---|---|---|
| ***Provisioning services*** | | |
| Sustenance of plants and animals | Direct use values | |
| | —Consumptive | Direct valuation based on market prices |
| | —Nonconsumptive | Indirect valuations (revealed expend-iture methods, contigent valuation method) |
| | Indirect use values | (No valuations necessary if plants/animals with direct values are counted) |
| ***Regulating services*** | | |
| Water filtration, flood control, pest control, pollination, climate stabilization | Direct and indirect use values | Estimation of service's contribution to profit (holding all else constant) |
| ***Other services*** | | |
| Generation of spiritual, aesthetic, and cultural satisfaction | Existence value | Indirect valuations (contingent valuation method) |
| | Direct, nonconsumptive use value | Indirect valuations (revealed expend-iture methods, contigent valuation method) |
| Recreational services | Nonconsumptive direct use value (e.g., from bird watching) | Indirect valuations (revealed expend-iture methods, contingent valuation method) |
| Generation of option value* | Option value | Empirical assessments of individual risk aversion |

* Option value represents a component of the overall value offered by a potential future ecosystem service, supplementing other values attributed to this potential service. See discussion in text.

*Source:* Data from L. H. Goulder and D. Kennedy (2011).

For nonconsumptive services such as bird watching, it is generally not possible to use market prices to estimate the values of these services. In these cases, economists must infer the value of an ES from methods such as **revealed expenditures** or survey methods. Goulder and Kennedy suggested that visiting a park to observe wildlife is a good example of the revealed expenditure approach. When one of us takes visitors to nearby Yellowstone National Park in the western United States to observe wolves, we first pay an entrance fee to get into the park. During the day, we are also accruing other costs such as the fuel it takes to drive to the park and back, the food we purchase while we are recreating, the new spotting scope we bought to see wolves from long distances, and perhaps the cost of time taken (income lost) to visit the park. Collectively, these expenses (or the portion of them that can be attributed to benefits obtained from wildlife viewing) represent the amount that park visitors might be willing to pay to observe wildlife in the park.

Survey methods can also be used to estimate these nonconsumptive values. **Contingent valuation methods** are a common form of such survey methods in which individuals are asked what price they are willing to pay to preserve a particular service such as an environmental feature or to prevent the loss or be compensated for the loss of a particular service. In the previous example, interviewees might be asked how much they would be willing to pay to go on a wildlife viewing trip in Yellowstone National Park if that was the only way to observe wildlife there. Like all of the methods described in this section, there are pros and cons of using this method. For example, some economists are critical of contingent valuation methods because there is often a poor correlation between values revealed by surveys and the true preferences of individuals. Nevertheless, in many situations survey methods are the only possible way of estimating values of certain services. That was the case for the infamous *Exxon Valdez* oil spill in Alaska in which conservationists used contingent valuation methods to quantitatively estimate damages from the spill.

There may be cases where it is necessary and helpful to estimate non-use values, the values that individuals derive from ecosystems just by recognizing their existence. Such values often have spiritual, moral, or cultural meanings. Economists refer to these types of values as existence values, and they can be estimated through survey methods such as those described earlier.

Many different economic methods can be used to assign values to ES. Our purpose here is simply to introduce you to a few of the better-known methods. In fact, scores of papers and books have been written about these methods. Although it is beyond the scope of this book to tackle this topic in any detail, a quick Internet search will reveal more references to these concepts and methods; Goulder and Kennedy[30] provide a good overview as well.

But before we leave this section, one important concept—**marginal and total value** is worth mentioning. In most cost-benefit analyses for environmental projects, economists are not concerned with the total value of an ecosystem service, but rather its marginal value. In her foundational book on ES, Stanford biologist Gretchen Daily[31] described the difference between marginal and total value with an example from a household's water consumption and utility bill (**FIGURE 10.5**).

Consumers are likely willing to pay a substantial sum for the first few cubic feet of water they receive each month, but for each subsequent cubic foot, the amount consumers are willing to pay declines. A graph of the amount of money that consumers are willing to pay over a one-month period pinpoints the marginal value of water—the amount people are willing to pay for each successive increment of an ES. Most market prices therefore represent the marginal value of a commodity and not its total value. Although total value of ES continues to be used in many ES analyses, the downside is that these figures tend to be static, one-time estimates. As ES scientists Taylor Ricketts and Eric Lonsdorf recently demonstrated with pollinators, coffee plantings, and remnant tropical forest patches in Costa Rica,[32] marginal values are more dynamic and in many situations are more useful in determining conservation and restoration priorities because they better reflect the types of planning alternatives that decision makers are considering.

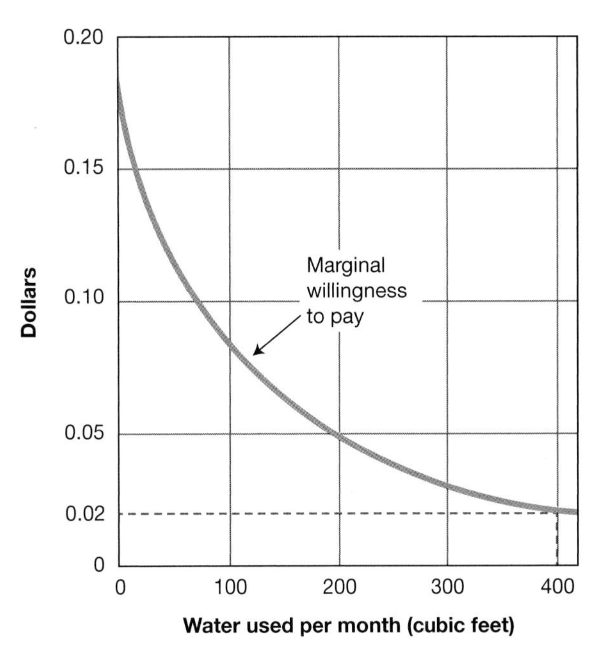


FIGURE 10.5 The marginal value of an ecosystem service (household water). Marginal value is what one more unit or the next unit of a good is worth to a person. For a household, the first few cubic feet of water on a monthly basis are worth more than subsequent uses because water is perceived to be in plentiful supply in most places. (Data from Daily, 1997.)

A bewildering array of tools are available to quantify and valuate ES. Fortunately, Kenneth Bagstad of the U.S. Geological Survey and his colleagues have assessed and tested 17 of these tools against eight evaluation criteria such as time required to use it, application to different scales, level of development, and how the tool deals with uncertainty.[33] TABLE 10.3 summarizes the researchers' evaluation of these tools and gives planners a good starting point for considering which tools their teams might find most useful in particular circumstances.

In addition, ES scientist Rolf de Groot and colleagues[34] from the Netherlands have developed a searchable Ecosystem Service Value Database consisting of over 1000 value estimates of 22 different services in 10 different biomes. The database contains information on the valuation method, year of valuation, the value type, original currency, discount rate, and other details on the calculation method. They also discuss a number of important caveats to using these valuations. Perhaps their best advice is that the values of ES in monetary units are best used not as decision-making tools in themselves, but as additional information to complement ES assessments and provide decision makers with data that can be used in trade-off analyses. An example of such a trade-off might be the cost of building a new water treatment facility versus the cost of conserving an intact watershed that provides some portion of that water treatment function as part of a natural ecological process.

**TABLE 10.3** An evaluation of different ecosystem service tools including information on uncertainty, time requirements for using the tool, scale that it can be applied, whether it is monetized, and how it can be integrated with other tools

| Tool | Quantifiable, approach to uncertainty | Time requirements | Capacity for independent application | Level of development and documentation | Scalability | Generalizability | Nonmonetary and cultural perspectives | Affordability, insights, integration with existing environmental assessment |
|---|---|---|---|---|---|---|---|---|
| ESR | Qualitative | Low,depending on stakeholder involvement in the survey process | Yes | Fully developed and documented | Multiple scales | High | No valuation component | Most useful as a low-cost screening tool |
| InVEST | Quantitative, uncertainty through varying inputs | Moderate to high, depending on data availability to support modeling | Yes | "Tier 1" models fully developed and documented; "Tier 2" documented but not yet released | Watershed or landscape scale | High, though limited by availability of underlying data | Biophysical values,can be monetized | Spatially explicit ecosystem service trade-off maps; currently relatively time consuming to parameterize |
| ARIES | Quantitative, uncertainty through Bayesian networks and Monte Carlo simulation | High to develop new case studies, low for preexisting case studies | Yes, through web explorer or standalone software tool | Fully documented; case studies complete but global models and web tool under development | Watershed or landscape scale | Low until global models are completed | Biophysical values,can be monetized | Spatially explicit ecosystem service trade-off, flow, and uncertainty maps; currently time consuming for new applications |
| LUCI | Quantitative, currently does not report uncertainty | Moderate; tool is designed for simplicity and transparency, ideally with stakeholder engagement | Yes, though website is under development and more detailed user guidance is presumably forthcoming | Initial documentation and case study complete; follow-up case studies in development | Site to watershed or landscape scale | Relatively high; a stakeholder engagement process is intended to aid in "localizing" the data and models | Currently illustrates trade-offs between services but does not include valuation | Spatially explicit ecosystem service trade-off maps; designed to be relatively intuitive to use and interpret |
| MIMES | Quantitative, uncertainty through varying inputs (automated) | High to develop and apply new case studies | Yes, assuming user has access to SIMILE modeling software | Some models complete but not documented | Multiple scales | Low until global or national models are completed | Monetary valuation via input–output analysis | Dynamic modeling and valuation using input–output analysis; currently time consuming to develop and run |

| Tool | Quantifiable, approach to uncertainty | Time requirements | Capacity for independent application | Level of development and documentation | Scalability | Generalizability | Nonmonetary and cultural perspectives | Affordability, insights, integration with existing environmental assessment |
|---|---|---|---|---|---|---|---|---|
| EcoServ | Quantitative, uncertainty through varying inputs | High to develop new case studies, low for existing case studies | Yes, pending release of web explorer | Under development, not yet documented | Site to landscape scale | Low until global or national models are completed | Biophysical values,can be monetized | In development, will offer spatially explicit maps of ecosystem service trade-offs |
| Co$ting Nature | Quantitative | Low | Yes | Partially documented | Landscape scale | High | Outputs indexed, bundled ecosystem service values | Rapid analysis of indexed, bundled services based on global data, along with conservation priority maps |
| SolVES | Quantitative, no explicit handling of uncertainty | High if primary surveys are required, low if function transfer approach is used | Yes, assuming user has access to ArcGIS | Fully developed and documented | Watershed or landscape scale | Low until value transfer can be shown to successfully estimate values at new sites | Nonmonetary preferences (rankings) of relative values for stakeholders | Provides maps of social values for ecosystem services; time consuming for new studies but lower cost for value transfer |
| Envision | Quantitative | High to develop new case studies | Yes | Developed and documented for Pacific Northwest case study sites | Landscape scale | Place-specific | Allows nonmonetary trade-off comparison, also supports monetary valuation | Cost-effective in regions where developed; time consuming for new applications |
| EPM | Quantitative | High to develop new case studies, low for existing case studies | Yes, through web browser | Developed and documented fo three case study sites | Watershed or landscape scale | Place-specific | Ecological, economic, and quality-of-life attributes could support nonmonetary valuation | Cost-effective in regions where developed; time consuming for new applications |
| InFOREST | Quantitative | Low, accessed through online interface | Yes, through web browser | Developed and documented only for Virginia | Site to landscape scale | Currently place-specific | Designed as a credit calculator, no economic valuation | Cost-effective in regions where developed; time consuming for new applications |

(continued)

**TABLE 10.3** An evaluation of different ecosystem service tools including information on uncertainty, time requirements for using the tool, scale that it can be applied, whether it is monetized, and how it can be integrated with other tools (*continued*)

| Tool | Quantifiable, approach to uncertainty | Time requirements | Capacity for independent application | Level of development and documentation | Scalability | Generalizability | Nonmonetary and cultural perspectives | Affordability, insights, integration with existing environmental assessment |
|---|---|---|---|---|---|---|---|---|
| EcoAIM | Quantitative | Relatively low for basic mapping, greater for nonmonetary valuation | No | Public documentation unavailable | Watershed or landscape scale | High | Incorporates stakeholder preferences via modified risk analysis approach | Spatially explicit ecosystem service trade-off maps; relatively time consuming to run |
| ESValue | Quantitative, uncertainty through Monte Carlo simulation | Relatively high to support consultant stakeholder valuation process | No | Public documentation unavailable | Watershed or landscape scale | High | Nonmonetary preferences via ranked analysis of trade-offs by stakeholders | Stakeholder-based relative ecosystem service value assessment; relatively time consuming |
| EcoMetrix | Quantitative | Relatively low to support field visits and data analysis | No | Public documentation unavailable | Site scale | High, where ecological production functions are available | Designed as a credit calculator, no economic valuation | One method for site-scale ecosystem services assessment |
| NAIS | Quantitative, reports range of values | Variable depending on stakeholder involvement in developing the study | No | Developed but public documentation unavailable | Watershed or landscape scale | High, within limits of point transfer | Dollar values only | Point transfer for "ballpark numbers," building awareness of values |
| Ecosystem Valuation Toolkit | Quantitative, reports range of values | Assumed to be relatively low | Yes | Under development | Watershed or landscape scale | High, within limits of point transfer | Dollar values only | Point transfer for "ballpark numbers," building awareness of values |
| Benefit Transfer and Use Estimating Model Toolkit | Quantitative, uncertainty through varying inputs | Low | Yes | Fully developed and documented | Site to landscape scale | High | Dollar values only | Low-cost approach to monetary valuation |

*Source:* Data from Bagstad et al. (2013).

## Planning Steps for Ecosystem Services

Over a decade ago, one of us wrote how conservation planning focused on biodiversity was just beginning to consider how to best incorporate ecosystem functions and processes.[35] Because these functions and processes are the foundational components of ES, it is welcome news that conservation planning has made tremendous progress over the last decade in better incorporating them into conservation plans.[36] It is critical, however, to understand that progress in conservation planning related to ecosystem process and function is not the same as progress in planning for ES.

Ecosystem process and function essentially represent the supply side of an ES; to incorporate ES into planning, we must also include the demand side—delivering the service to a human beneficiary. In 2007, South African biologist Benis Egoh and colleagues[37] reviewed the degree to which ES had been included in conservation assessments. Of 100 randomly selected, published assessments, only 7 had included ES in them; another 13 referred to ES without specifically including them. In 2005, marine conservation scientist Heather Leslie conducted an in-depth analysis of 27 marine conservation planning projects from around the globe and found that ecosystem services were rarely included in them.[38] Although ES were beginning to garner more attention in conservation planning, Egoh and colleagues suggested that the lack of data and tools for mapping ES was a major constraint to progress. The challenges of identifying beneficiaries for various services and including their use of services in mapping and planning was also viewed as a major challenge.

In 2010 Egoh, along with a cast of well-known Australian and South African conservation planners, published one of the first major case studies to examine the synergies and trade-offs in conserving biodiversity and ES in the Little Karoo region of South Africa.[39] They evaluated three ES: carbon storage, water recharge, and fodder provision along with spatial data on biodiversity across several planning scenarios (FIGURE 10.6). Trade-off analyses suggested that reducing biodiversity targets by a few percent could gain 20%–40% improvements for the ES targets. Although able to integrate ES and biodiversity in planning, Egoh and colleagues observed several practical challenges to implementation, including among others capturing the scale of benefit flows and the absence of markets for most ES.

At the same time that South African conservation planners like Egoh and Belinda Reyers were working to incorporate ES into conservation assessments, Kai Chan (now at the University of British Columbia) and colleagues[40] were doing the same in North America. Using the decision support system Marxan (Chapter 7), they

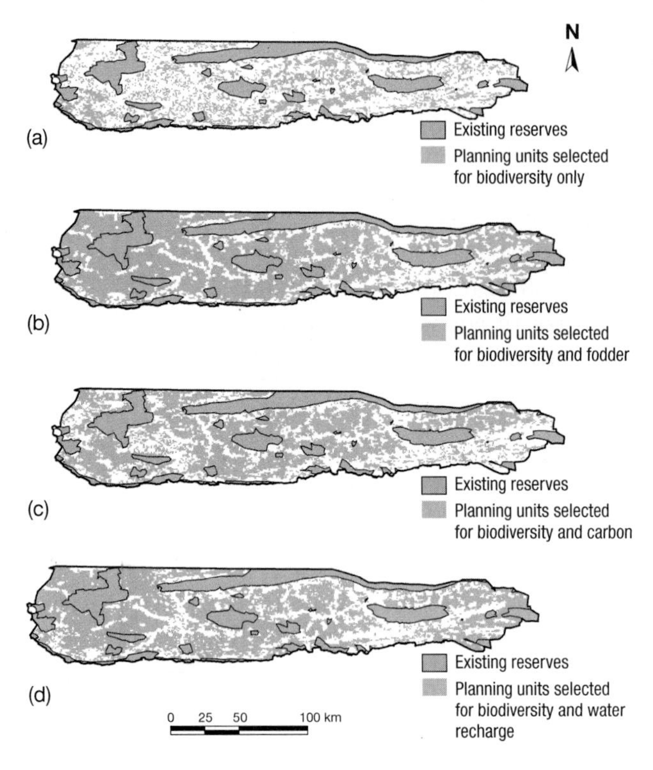


**FIGURE 10.6** Trade-off analyses among the targeting of biodiversity and ES in the Little Karoo region, South Africa. Conservation priorities from four different spatial scenarios of areas important for conservation: (a) only biodiversity features, (b) biodiversity and grass fodder (ecosystem service), (c) biodiversity and carbon, and (d) biodiversity and water recharge. (Data from Egoh et al., 2010.)

explored the trade-offs and opportunities between biodiversity conservation and six ES (carbon storage, crop pollination, flood control, forage production, outdoor recreation, water provision) in the Central Coast ecoregion of California, USA. Like Egoh and colleagues' work in South Africa, Chan's team found that targeting biodiversity alone provided considerable co-benefits for all ES while targeting the ES alone resulted in substantial trade-offs for biodiversity, primarily related to the two agriculturally associated services (FIGURE 10.7).

A number of other investigations have been made of the trade-offs between biodiversity and ecosystem services in both marine and terrestrial sea and land use planning situations. Some of these investigations tend to converge on an important conclusion: it is possible to promote plans for land and sea use and management actions that will simultaneously advance both biodiversity and ecosystem service objectives[41] (in other words, "win-win" situations).

A more recent analysis of ecosystem services trade-offs and synergies challenges this conclusion to some degree. Caroline Howe and several UK colleagues[42] conducted a meta-analysis of over 1300 reports that analyzed possible win-win outcomes for conserving ES that benefit the environment and that benefit human well-being. Their results suggested that trade-offs occurred nearly three times more often than synergies and were more likely to happen when (1) one of the stakeholders had a private interest in an available natural resource, (2) provisioning ES were involved, and (3) at least one stakeholder was acting at the local level. Importantly, their analyses also revealed that synergies or win-win situations were more likely to happen when participants had some understanding of the factors responsible for trade-offs between environmental and human well-being outcomes.

In summary, the evidence is mixed and inconsistent regarding when and how there will be trade-offs between biodiversity or

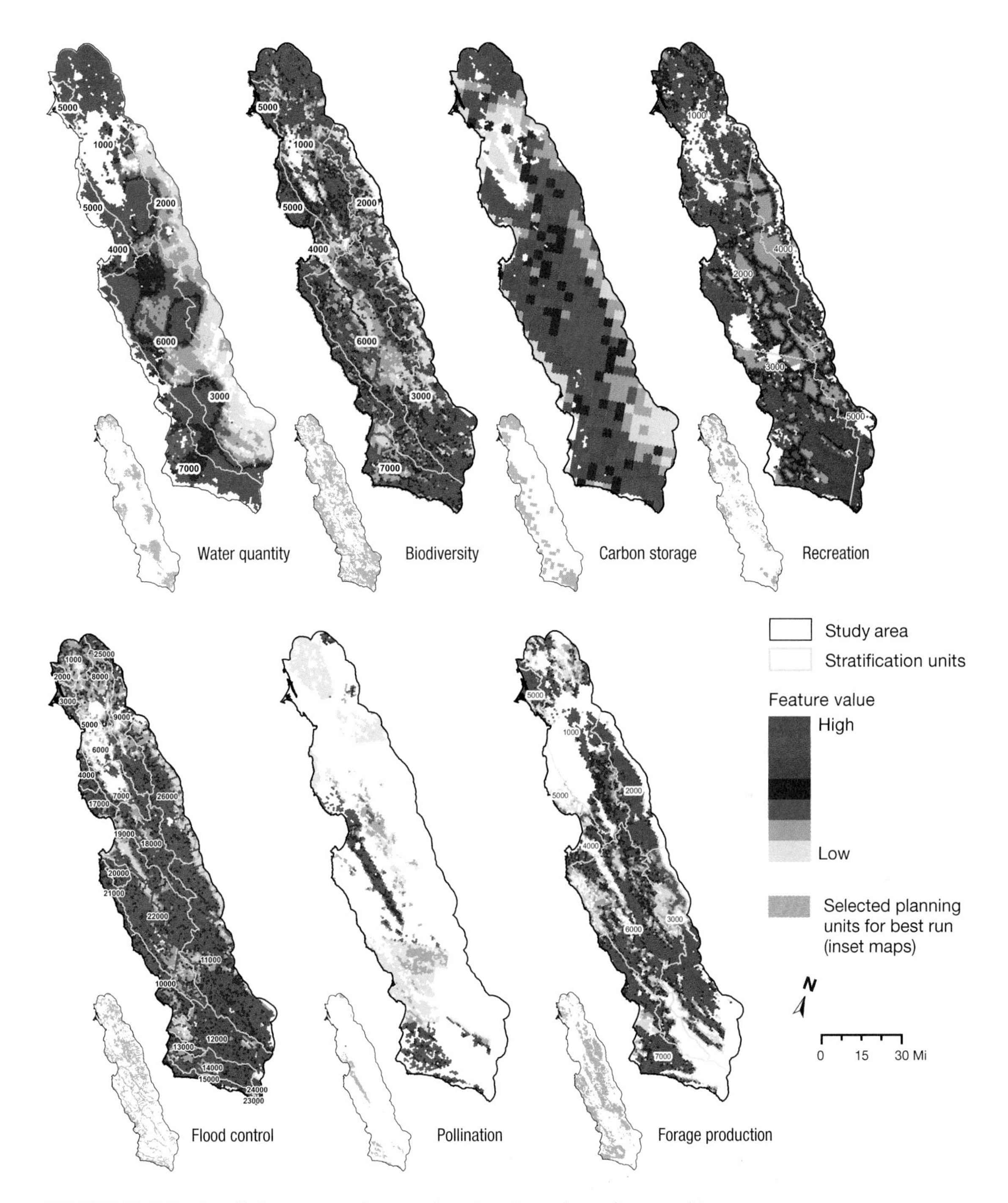


FIGURE 10.7 Trade-offs in conservation area locations based on a focus on biodiversity features and six ecosystem services in the Central Valley, California, USA. (Data from Chan et al., 2006.)

environmentally oriented outcomes and those focused more on human well-being. Nevertheless, the analyses that have been done provide planners with some sound advice about which circumstances are more likely to result in trade-offs than synergies, and this advice can be heeded as we engage in the planning steps outlined below.

In a 2013 book on conservation planning, Chan and colleagues Natalie Ban and Robin Naidoo[43] outlined a set of steps for incorporating ES into conservation plans. We have modified and adapted these steps here, and suggest that conservation planners, scientists, and practitioners alike will find the steps to be useful guidance on incorporating ES into conservation plans. To make the steps easier to understand within the context of this book, each heading identifies the planning step as we have articulated it in the book (e.g., "Establishing Objectives..."), and the first text line identifies the ES planning step envisioned by Chan and colleagues (e.g., "Identify relevant ES . . .").

### Step 1: Establishing Objectives and Conservation Features (Chapter 3)

*Identify relevant ES, benefits, and constituents-beneficiaries.* Determining which ES to include in a plan is best accomplished at the same time as identifying the fundamental objectives of the plan and its conservation features. Many possible ES could be incorporated into a plan, but this is the time to select those that match up well with the overall objectives of the conservation project, are of interest to key stakeholders (which can be determined in stakeholder analyses, outlined in Chapter 2), and for which data or models can be brought to bear to analyze these services (Chapter 5).

### Step 2: Situation Analysis and Data (Chapter 5)

*Characterize relationships among ES, benefits, values, and stakeholders.* As the sections on valuation and mapping of ES suggest, we can quantify and map ES in many different ways. This needs to be done in a manner that will be most meaningful to a project's stakeholders but also in a relevant manner to strategies and actions that focus on conserving specific ES benefits. Chan and colleagues[44] have suggested a framework for accomplishing this second step (FIGURE 10.8).

This characterization also needs to differentiate among the many different ES that may be provided in a planning area and include the temporal component of when an ES is most important to beneficiaries. For most conservation plans, the totality of ecosystem services is not what matters; it is the differences in those services that will be delivered depending on different planning scenarios, alternative strategies and actions that might be taken, and different land use

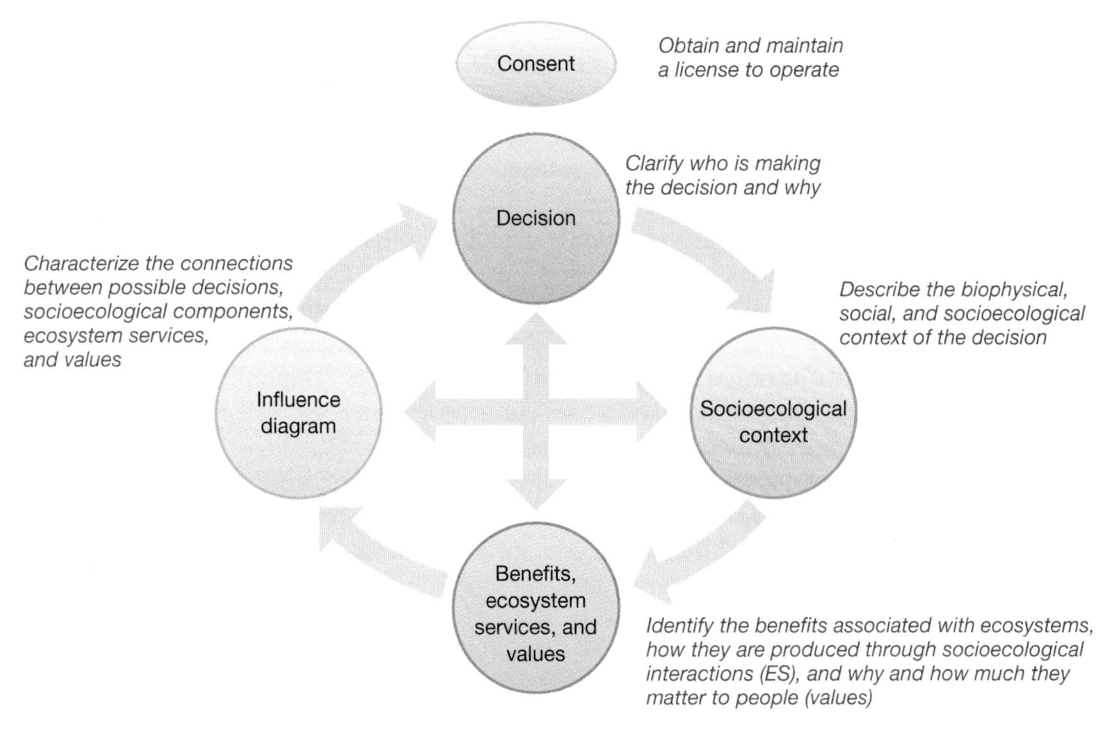


FIGURE 10.8 A framework for understanding ecosystem services. The value placed on ecosystem services will depend in part on the social-ecological context in which the service is provided and the beneficiaries of that service. Policies and decision makers will also have influence on how ecosystem services are defined and valued, and on who benefits from them. (Data from Chan et al., 2012.)

alternatives that could be selected. As a result, it's important for planners to analyze and understand these differences in benefits. Developing a conceptual model of the project as part of the situation analysis and assessing what data are available for various planning aspects is the appropriate time to characterize ES relationships.

## Step 3: Framing and Solving Conservation Planning Problems (Chapters 6–7)

*Map and valuate ecosystem services.* It's difficult to conserve an ES or an element of biodiversity if it can't be mapped and located. Earlier in this chapter, we outlined some of the different methods for mapping ES. One of the most innovative approaches is to develop ecological production functions or benefit functions. Such functions help model the benefit of ES, either individually or in **bundles**, so that planners can analyze how these benefits will vary spatially and by alternative scenarios. This is equivalent to the prediction of consequence that we introduced in Chapter 6.

Planners will need to ensure that the spatial scales of ES align with planning units. While most conservation planning analyses are conducted at one spatial scale, it's important for planners not to assume that a single spatial scale will be relevant to all ES. Many ES, such as those related to water provision, will occur across many individual planning units, but other services may occur only within the relatively small areas of individual planning units (e.g., pollination).

This step is also the time to decide whether to estimate the monetary values of various services. Earlier we reviewed some of the different approaches for valuating ES. Depending on the type of ES (e.g., cultural or provisioning), planners will need to identify the most appropriate valuation technique. On the other hand, many ES are not easily assigned a monetary value, and as previous examples have demonstrated, it is often not necessary to assign a monetary value to sufficiently capture that service's value in a conservation plan. Although some constituents and stakeholders will care about whether a value is assigned, others will be more interested in the trade-offs in ES across different land use scenarios. In these cases, it will be sufficient to express an ES as a biophysical variable (**FIGURE 10.9**).

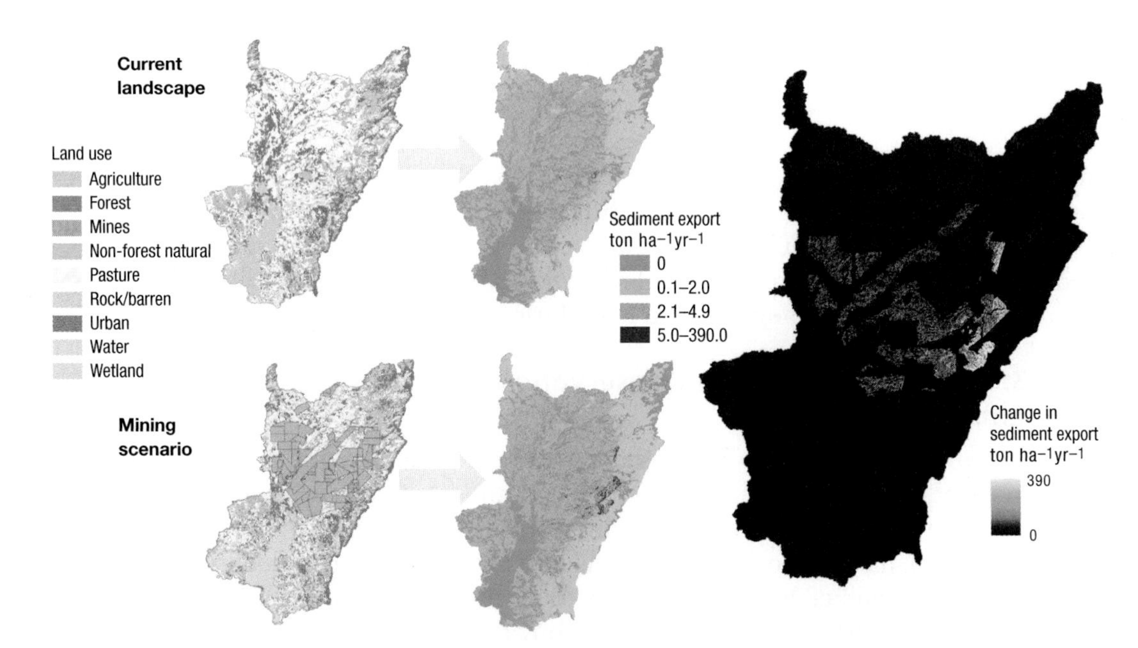


**FIGURE 10.9** Modeling an ecosystem service with the InVEST erosion model for Cesar Department (state) in northern Colombia. The mining scenario is for full build-out of all proposed mine permits. Ecosystem service impact is estimated as lost sediment retention capacity compared with the current landscape.

### Step 4: Solving Conservation Planning Problems (Chapter 7)

*Integrate ES into planning analysis.* Once ES have been mapped, they will usually need to be incorporated into some sort of decision support system (see Chapter 7) for further analyses. In some cases it may desirable to treat ES as features (Chapter 3) and set targets for them. That is likely the case for most cultural services, such as recreation. In other situations, it may be most appropriate to consider ES as part of the **costs** or **cost surface** (Chapter 5) in the decision support tool and GIS being used for the analyses.

Chan and colleagues evaluate the pros and cons of treating ES as costs or as planning features.[45] This step is also when trade-off analyses will be conducted,[46] often comparing outcomes of scenarios that might emphasize conserving ecosystem services with scenarios that emphasize biodiversity or some other land use (FIGURE 10.10).

While they are comparing spatial scenarios of the locations of various biodiversity and ecosystem features, planners are likely considering alternative strategies for conserving these features. As various theories of change for strategies are articulated, it's an opportune time to consider their consequences for various stakeholders,[47] including those strategies that might emphasize ES. For example, the development of hydropower in the western Amazon basin (an ecosystem service in western Amazonia) could negatively impact migratory fish in that ecosystem, which in turn, could impact the diet and income

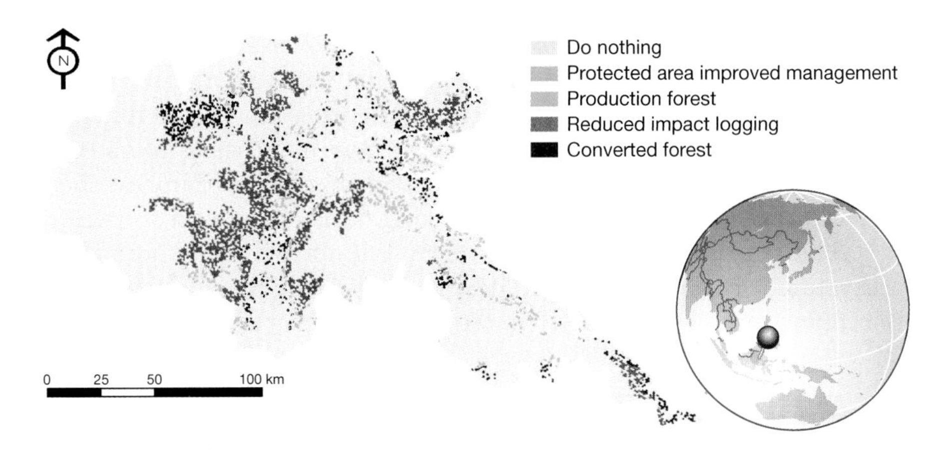


FIGURE 10.10 Meeting multiple objectives of biodiversity and ecosystem services in a REDD (Reduced Emissions from Deforestation and Forest Degradation) project in Berau, Indonesia. (Data from Venter, O., et al., *Using systematic conservation planning to minimize REDD+ conflict with agriculture and logging in the tropics.* Conservation Letters, 2013. **6**: 116–124.)

FIGURE 10.11 Migratory fish being harvested in western Amazonia. A SNAP (Science for Nature and People, http://www.snap.is) working group is investigating the potential impacts of hydropower and other infrastructure on freshwater resources (including migratory fish) and local people in the western Amazon. Both hydropower and migratory fish can be considered an ecosystem service, although they clearly are provisioned to people in different ways and have different beneficiaries.

from fishing of local people in western Amazonia (FIGURE 10.11). On the other hand, some stakeholders stand to gain electricity and possibly jobs in rural locations of Amazonia and would clearly derive benefits from hydropower development. Examples like these help emphasize the importance of trade-off analyses.

### Step 5: Assessing Risk and Uncertainty (Chapter 8)

*There are many assumptions, risks, and uncertainties in conservation planning.* This is especially the case with ecosystem services because of the need in most cases to model the outputs of various services. Being explicit about the assumptions behind these models is critical because many decision makers will gloss over such assumptions unless they are conveyed in language they will understand. As we suggested earlier, efforts can also be made to estimate the precision and accuracy of mapping ES. Finally, it may be possible to do some **sensitivity analyses** by varying values in ES models or in how those models are used in conservation planning decision support software.

There are many sources of information for learning more about ecosystem services in general and incorporating them into conservation plans more specifically. We recommend that readers explore the following initiatives and their websites for more information.

- The Natural Capital Project is a collaboration of Stanford University, The Nature Conservancy, the University of Minnesota, and the World Wildlife Fund (www.naturalcapitalproject.org).
- The Ecosystem Services Partnership (www.es-partnership.org) is a global network of practitioners from government agencies, NGOs, academia, and the business community. It is an excellent

resource for up-to-date information on all aspects of ecosystem services. The partnership includes thematic working groups on a range of topics including ES planning and management, indicators, assessment, mapping, valuation, modeling, and more. Case studies, available on the website, cover a wide variety of topics and list global experts to contact as well. In addition, the partnership is connected to a new journal—*Ecosystem Services*—that publishes a large number of papers relevant to including ES in conservation planning (several of those papers are cited in this chapter).

- The National Ecosystem Service Partnership in the United States launched in early 2015 an extensive online *Federal Resource Management and Ecosystem Services Guidebook* (https://nespguidebook.com/), which should provide a wealth of information to conservation planners on how to better incorporate ES into conservation planning.

## The Other Side of the Story: Critiques and Integration

Despite the rapid rise in interest in ES and the proliferation of scientific papers and methods on them, skepticism remains about the effectiveness of focusing on ES for nature conservation. Much of the skepticism relates specifically to their valuation and the effectiveness of **Payment for Ecosystem Services** (**PES**) schemes or strategies. For example, in an editoiral in the journal *Nature* that generated considerable buzz in the scientific community, ecologist Douglas McCauley[48] argued that market-based solutions to conservation such as those promoted by ecosystem service advocates have four serious limitations. Among his most significant concerns was that fluctuation in market values makes PES schemes risky and that monetizing nature in this manner may imply that nature is only worth conserving when it is profitable to do so. Others have raised similar and additional concerns about market-based solutions and the promise of the PES approach,[49] including the need to improve on the science behind PES projects.[50]

Recent research in psychology and human behavioral science also supports the notion that payments may actually undermine the conservation motivations of some groups. In a report from WWF-UK,[51] author Tom Crompton suggests that tackling environmental and humanitarian issues through appeals to individual self-interests only reinforces the perceived importance of these interests, simultaneously diminishing and undermining the value basis of concern about bigger-than-self issues. A recent empirical study published in *Nature Climate Change*[52] also supports this notion that appealing to economic self-interests may not be the best way to promote behavioral changes that are good for conservation. Finally, in a scintillating book entitled *What Money Can't Buy,* Harvard professor Michael Sandel reaches similar conclusions about many incentive programs.[53] Paying students, for example, to make better grades has mixed results at best—unless those payments also influence the attitudes of individuals toward achievement, which turns out to be more important than the financial reward itself.

Some criticisms of the ES approach extend beyond the economic valuation and commodification of nature. Kent Redford and William Adams,[54] among others, point out that some ecosystem processes and services may be bad for people but good for nature or biodiversity conservation. Floods, for example, are often bad for human communities (despite some recent recognition that reestablishing floodplains may be a good long-term climate adaptation strategy in many places) but are essential for the life-history strategies of many species. Carbon may sometimes best be sequestered in timber plantations, but such plantations are rarely considered beneficial for biodiversity more broadly. Along similar lines, Redford and Adams observed that many ecosystem services can be sustained by non-native species (e.g., forests in the state of Hawaii, USA, may be dominated by non-natives but still perform a watershed conservation function). In some cases this situation could provide a perverse incentive to replace native species with non-natives.

A group of Dutch scientists recently synthesized the critiques of ES, the counterarguments to these critiques, and a possible path forward in resolving these issues.[55] Along with the issues already noted, their overview highlights several additional critiques, including the vagueness and unwarranted optimism of the concept and ethical considerations that the ES concept excludes intrinsic values of nature (see **TABLE 10.4** for a summary of these critiques and possible resolutions).

Taken as a whole, most critiques of ES as a conservation approach or broad strategy concede that the approach should be part of the conservationist's toolbox. They raise legitimate issues about focusing too singularly on ES as a conservation tool, the trade-offs among ES approaches and other aspects of nature conservation, and problems related to economic valuations of nature and market-based approaches. New methods and tools for incorporating ES in conservation plans are being developed as we write this book, and we anticipate that ES approaches to conservation will continue to gain significance in the conservation and development arenas, especially as corporations and some segments of society attach greater importance to sustainability.

Several conservation scientists and practitioners, including the "critiques" synthesis paper noted earlier, have called for the closer integration of ES conservation efforts with projects focused on biodiversity conservation. At a philosophical level, biodiversity conservation

**TABLE 10.4** A synthesis of critiques of ecosystem services, the arguments for and against these critiques, and suggested ways to resolve these criticisms

| Critique | Arguments | Counterarguments | Way forward |
|---|---|---|---|
| Environmental ethics | The ES concept excludes intrinsic value of nature.<br>Nature conservation should be based on intrinsic instead of anthropocentric values. | The ES concept bundles valid anthropocentric arguments.<br>The cultural ES domain includes values with elements of intrinsic values, for instance, existence value. | Anthropocentric framing could be used for broad argumentation in support of conservation and sustainable use of ecosystems.<br>Stronger acknowledgment of existence aspects within the cultural services domain could bring different worldviews together |
| Human–nature relationship | The focus on ES could promote an exploitative human–nature relationship.<br>This might contradict holistic perspectives of indigenous people. | The ES concept could reconnect society to nature.<br>Nonmaterial values can be covered in the cultural ES domain, to include people's values and needs. | The ES concept offers a "platform" for bringing people and their different views and interests together.<br>Attention is needed to move beyond the Western origin of the ES concept. |
| Conflicts with the concept of biodiversity | The ES concept might replace biodiversity protection as a conservation goal.<br>There is inconclusive evidence of a "win-win" scenario between biodiversity and ES.<br>ES might not safeguard biodiversity, but instead divert attention and resources. | There are conceptual overlaps between ES and biodiversity.<br>There is a growing body of evidence that biodiversity underpins the ecosystem functions that give shape to ES.<br>Current initiatives based on ES lead to a broad perspective on land management and conservation. | Indirect inclusion of biodiversity in several ES categories can pave the way for potential "win–win" scenarios.<br>Further research and monitoring are needed to clarify the relationships between biodiversity and ES. |
| ES valuation | The ES concept comprises economic framing.<br>ES assessments often involve economic valuation. | Monetary valuation provides additional information in decision-making processes.<br>ES assessments do not necessarily involve valuation and valuation does not necessarily involve monetization. | Develop both biophysical and sociocultural value indicators of ES to explain human–nature relationships. |
| Commodification and PES | The ES approach is based on the assumption that payment for ES will ensure their provision. | Assessing ES in monetary terms does not necessarily equate to using market instruments. | Focus on ES approaches that include nonmarket instruments. |
| Vagueness | ES has become a "catch-all" phrase because of its many vague definitions. | Imprecision of the ES concept can spur creativity and refinement of definitions.<br>Use of the ES concept can facilitate multiple societal actors to interact without consensus on the precise meaning and can foster transdisciplinary research. | ES offer common ground for debate and methodological progress in different scientific fields.<br>Use of the ES concept can build bridges between science and practice, enabling for integrated, transdisciplinary approaches to solve "wicked problems." |
| Optimistic assumptions and normative aims | The ES concept is too optimistic.<br>Ecosystem outputs may not always be beneficial to humans. | Positive terminology shows the optimistic intentions and research interests.<br>ES is one of the many normative concepts used within environmental science.<br>Total value freedom is impossible for science embedded in sociocultural contexts. | Scientists should be explicit and transparent about whether research aims and provided information are normative.<br>ES scientists are challenged to find ways to systematically consider implicit assumptions and perceptions of stakeholders and practitioners on ES and connected values. |

*Source:* Data from M. Schröter et al. (2014).

efforts are often associated with the recognition that nature has intrinsic value, which is an end in itself for conservation (a biocentric view). In contrast, ES are associated with instrumental values—those that serve the ends of human needs (an anthropogenic view). This dichotomy has in part led to debates about the relative merits of conserving ES as part of nature conservation projects.

South African conservation planner Belinda Reyers and colleagues[56] and Jérôme Cimon-Morin and colleagues[57] have suggested that these debates have been useful to illuminate critical differences in concepts, motivations, goals, and values, but they also have the real potential to unnecessarily divide the conservation and natural resource community. Conservation scientist Georgina Mace and colleagues[58] have suggested that there is confusion among many conservation practitioners and scientists about the relationship between the fields of biodiversity conservation and ES, and that this confusion is leading to incoherent public policy in some cases. They concluded that to effectively manage ecosystems, an interdisciplinary approach will be required that includes conservation biologists, ecosystem scientists, resource economists, and others—a point we made about conservation planning teams in Chapter 2.

We concur with Reyers and others that the most useful path forward for conservation is to have a better understanding of the differences and trade-offs between efforts directed toward biodiversity and ES, but appreciate that there is also considerable common ground between the two concepts. We would emphasize that a critical part of the understanding is to appreciate not only the overlap between biodiversity and ES priorities, but the most likely reasons for overlap or areas of difference. In an extensive review of the synergies between biodiversity and ES in conservation planning, Cimon-Morin and colleagues concluded that the type and number of ES mapped, the data available to map them, the methods applied, and the metrics used to measure biodiversity will all influence the degree of overlap between areas important to both biodiversity and ES. As a result, it will be prudent for most projects to make specific efforts to incorporate ES in planning and not rely on theoretical overlaps with biodiversity to achieve ES objectives.

In summary, ES provides an additional rationale for nature conservation beyond the intrinsic values of biodiversity conservation. It may broaden the constituency for nature conservation in some situations regardless of whether those services are actually valued in monetary terms and despite some legitimate concerns about the role of ES in nature conservation. At the end of the planning day, the emphasis that a conservation project team places on ES and their conservation will depend on what the team and its stakeholders really value (in other words, its fundamental objectives; see Chapter 3). Given the close tie between ecological processes and ES, and building on the recommendations of Reyers, Cimon-Morin, and others, we encourage conservation planners, scientists, and practitioners to incorporate ES into conservation plans wherever possible, analyze and understand how conserving these services will contribute to the goals and objectives of a project, and ultimately help build a bridge between biodiversity conservation efforts and ES that provides direct benefits to human well-being.

## Lessons Learned

Mary Ruckelshaus and her colleagues[59] at NatCap have harvested some of the most important lessons they have learned in nearly a decade of work in this arena at over 20 demonstration sites around the globe. Their number one lesson is familiar to many planners and conservationists—the process is at least as important as the tools used in the plan and the products derived from the planning process. They urge practitioners to use an approach that incorporates both biodiversity and ES into an assessment, does so in iterative, flexible formats in terms of the type of information that is used, and builds trust and mutual understanding among stakeholders, scientists, planners, and policy makers. Taking such an approach requires that a certain level of scientific, planning, and policy expertise be engaged in the assessment and planning process, and it can often be a relatively long and time-consuming engagement.

A second important lesson from Ruckelshaus and her colleagues is also one that will resonate with many who have been involved in conservation planning—keep the planning tools and the planning process as simple as possible. The bottom line here is that most decision makers want simple tools with which they can visualize how different policy decisions will lead to different outcomes for biodiversity and ES. The tools and methods should be well documented and their limitations and assumptions understood and transparent—including the various forms of uncertainty (Chapter 8) associated with ES models, a topic in the ES field that has not received sufficient attention. And when local experts who are involved in the assessment and planning process can be trained to use the tools, of course the buy-in by these local experts and decision makers is likely to be even greater.

Earlier in this chapter, we explored the economic valuation of ES. It is a mistake to assume that consideration of ES in planning requires monetizing nature conservation or various choices regarding land and water use and management. Estimating the monetary value of ES can be an important activity in any planning or assessment process and is sometimes an eye-opener for many policy makers when actual financial values are attached to a service. But as we have just seen, it is neither necessary nor sometimes desirable to estimate the monetary worth of ES. As Ruckelshaus and her colleagues cleverly observed, "It is not always about the money."

What decision makers are often most interested in is how different policies or decisions about land and sea use will affect the delivery of ES, and the beneficiaries or recipients of these services. For example, in a city where one of us resides, the federal agency responsible for managing the watershed that supplies our municipal drinking water is considering a set of forest management activities (including harvest and associated road building) to reduce the risk of wildfire and subsequent sedimentation, in part because of its possible impacts to municipal water supplies. City managers and the federal natural resource agency, on the other hand, need to balance that consideration with the fact that the same area is heavily used for multiple public recreation activities. Many of those recreationists are opposed to the road-building and timber-harvesting activities that would be involved and are questioning the return on investment (ROI) for sediment reduction.

These are exactly the sorts of issues that call for a good assessment of ecosystem services. It would provide decision makers with a range of scenarios as well as ROI estimates for different scenarios (e.g., an expenditure of $x$ funds to conduct timber harvest and road building would result in $y$ reduction in fire hazard and $z$ impact on public recreation, which might be measured in user-days). In this case, there is no real need to attach a monetary value per se to the public water supply that comes from the forest. The more important pieces of information are actually about estimates of reduced risk of wildfire and reduced recreational use from the proposed activity in comparison to other scenarios that might also reduce risks to municipal water supplies, including a business-as-usual one (no timber harvest or road building).

As a final but important note, we would be remiss if we did not observe that bringing ecosystem services into conservation plans will make the planning process more complicated compared to a plan that focuses only on, for example, biodiversity features. Additional decisions will need to be made, such as selecting particular services to focus on in the planning effort and determining the beneficiaries of those services. In most cases, modeling of ES will be necessary to quantify the service, map where it occurs, and examine alternative scenarios for the provisioning of these services. Some thought may also need to be given to whether data are available to assign a monetary value to a service, or if this information is even desirable given the constituents and decisions to be made within the planning context. Finally, some challenging decisions may need to be made in terms of trade-offs in the conservation of ES with other objectives. As we have observed in this chapter, whether ES are included in the planning process will largely depend on the overall objectives and values that are driving the conservation project.

## KEY MESSAGES

- Interest in the conservation of ES continues to grow since the publication of the *Millennium Ecosystem Assessment* (MEA), as indicated by several follow-up global initiatives, a new journal (*Ecosystem Services*), new methods and tools for assessing ES, and a dramatic increase in publications on the topic.
- Attaching monetary value to ES can be helpful from a conservation perspective, but it is not critical to including ES in conservation and other development-related planning processes. Assigning other nonmonetary values to ES can be useful.
- Incorporating ES in conservation plans can help identify trade-offs and synergies among conservation and human well-being aspects of conservation projects.
- The ability to quantify and map ES is critical to including them in conservation initiatives, and the development of ecological production functions was a breakthrough in mapping ES and developing scenarios for how changes in land use affect ES.
- There is no single best methodology for incorporating ES in conservation plans, but there is an evolving set of useful steps that planners can follow.
- ES are distinguished from ecosystem functions and processes in that they must include a human beneficiary and a mechanism for the service to flow to the beneficiary.
- Practical and philosophical concerns about conserving ES remain among key constituents in the conservation community.
- Unlike biodiversity conservation planning methods, much research and development is still needed to develop better data sets, methods, and tools for mainstreaming the conservation of ES and integrating it with biodiversity priorities.

## References


1. Novelly, P. E., and I. W. Watson, *Successful grassland regeneration in a severely degraded catchment: A whole of government approach in North West Australia.* Environmental Science and Engineering, 2007: 469–486.
2. Albert, C. et al., *Integrating ecosystem services in landscape planning: Requirements, approaches, and impacts.* Landscape Ecology, 2014. **29**(8): 1277–1285.
3. Arkema, K. K. et al., *Coastal habitats shield people and property from sea-level rise and storms.* Nature Climate Change, 2013. **3**(10): 913–918.
4. Myers, S.S. et al., *Human health impacts of ecosystem alteration.* Proceedings of the National Academy of Sciences, 2013. **110**(47): 18753–18760.
5. Lawler, J. J. et al., *Projected land-use change impacts on ecosystem services in the United States.* Proceedings of the National Academy of Sciences, 2014. **111**(20): 7492–7497.
6. Daily, G., ed. *Nature's Services: Societal Dependence on Natural Ecosystems.* 1997, Washington, DC: Island Press. 391.
7. Costanza, R. et al., *The value of the world's ecosystem services and natural capital.* Nature, 1997. **387**(6630): 253–260.
8. United Nations Environment Programme, *Millennium Ecosystem Assessment, Ecosystems and Human Well-Being: Synthesis.* 2005, Washington, DC: Island Press.
9. Goldman, R. L. et al., *Field evidence that ecosystem service projects support biodiversity and diversify options.* Proceedings of the National Academy of Sciences, 2008. **105**(27): 9445–9448; Marvier, M., and H. Wong, *Resurrecting the conservation movement.* Journal of Environmental Studies and Sciences, 2012. **2**(4): 291–295.
10. TEEB, *The Economics of Ecosystems and Biodiversity: Mainstreaming the Economics of Nature: A Synthesis of the Approach, Conclusions and Recommendations of TEEB.* 2010, United Nations Environment Programme (UNEP). 39.
11. UNU-IHDP and UNEP, *Inclusive Wealth Report 2012: Measuring Progress towards Sustainability.* 2012, Cambridge, UK: Cambridge University Press.
12. United Nations Environment Programme, *Report of the Second Session of the Plenary of the Intergovernmental Science–Policy Platform on Biodiversity and Ecosystem Services,* Antalya, Turkey, December 9–14, 2013.
13. Vohland, K. et al., *How to ensure a credible and efficient IPBES?* Environmental Science & Policy, 2011. **14**(8): 1188–1194.
14. Chan, K. M. A. et al., *Where are cultural and social in ecosystem services? A framework for constructive engagement.* BioScience, 2012. **62**(8): 744–756.
15. Plieninger, T. et al., *Sustaining ecosystem services in cultural landscapes.* Ecology and Society, 2014. **19**(2).
16. Reyers, B. et al., *Getting the measure of ecosystem services: A social–ecological approach.* Frontiers in Ecology and the Environment, 2013. **11**(5): 268–273.

17. Schägner, J. P. et al., *Mapping ecosystem services' values: Current practice and future prospects.* Ecosystem Services, 2013. **4**: 33–46.

18. Ibid.

19. Costanza et al., *The value of the world's ecosystem services and natural capital.* (See reference 7.)

20. Costanza, R. et al., *Changes in the global value of ecosystem services.* Global Environmental Change, 2014. **26**: 152–158.

21. Kareiva, P. et al., *Natural Capital: Theory and Practice of Mapping Ecosystem Services.* 2011, Oxford, UK: Oxford University Press. 365.

22. Tallis, H., and S. Polasky, *Assessing multiple ecosystem services: An integrated tool for the real world,* in *Natural Capital: Theory and Practice of Mapping Ecosystem Services,* P. Kareiva et al., editors. 2011, Oxford, UK: Oxford University Press. 34–50.

23. Schägner et al., *Mapping ecosystem services' values.* (See reference 17.)

24. Maes, J. et al., *Mapping ecosystem services for policy support and decision making in the European Union.* Ecosystem Services, 2012. **1**( (1): 31–39.

25. Pagella, T., and F. Sinclair, *Development and use of a typology of mapping tools to assess their fitness for supporting management of ecosystem service provision.* Landscape Ecology, 2014. **29**(3): 383–399.

26. Bagstad, K. J. et al., *From theoretical to actual ecosystem services: Mapping beneficiaries and spatial flows in ecosystem service assessments.* Ecology and Society, 2014. **19**(2); Mitchell, M. G. E. et al., *Reframing landscape fragmentation's effects on ecosystem services.* Trends in Ecology & Evolution. **30**(4): 190–198.

27. Schägner et al., *Mapping ecosystem services' values.* (See reference 17.)

28. Ibid.

29. Goulder, L. H., and D. Kennedy, *Interpreting and estimating the value of ecosystem services,* in *Natural Capital: Theory and Practice of Mapping Ecosystem Services,* P. Kareiva et al., editors. 2011, Oxford, UK: Oxford University Press. 15–33.

30. Ibid.

31. Daily, ed. *Nature's Services: Societal Dependence on Natural Ecosystems.* (See reference 6.)

32. Ricketts, T. H., and E. Lonsdorf, *Mapping the margin: Comparing marginal values of tropical forest remnants for pollination services.* Ecological Applications, 2013. **23**(5): 1113–1123.

33. Bagstad, K. J. et al., *A comparative assessment of decision-support tools for ecosystem services quantification and valuation.* Ecosystem Services, 2013. **5**: 27–39.

34. de Groot, R. et al., *Global estimates of the value of ecosystems and their services in monetary units.* Ecosystem Services, 2012. **1**( (1): 50–61.

35. Groves, C., *Drafting a Conservation Blueprint: A Practitioner's Guide to Planning for Biodiversity.* 2003, Washington, DC: Island Press. 457.

36. Pressey, R. L., R. M. Cowling, and M. Rouget, *Formulating conservation targets for biodiversity pattern and process in the Cape Floristic Region, South Africa.* Biological Conservation, 2003. **112**(1–2): 99–127.

37. Egoh, B. et al., *Integrating ecosystem services into conservation assessments: A review.* Ecological Economics, 2007. **63**(4): 714–721.

38. Leslie, H. M., *A synthesis of marine conservation planning approaches.* Conservation Biology, 2005. **19**(6): 1701–1713.

39. Egoh, B. N., et al., *Safeguarding biodiversity and ecosystem services in the Little Karoo, South Africa.* Conservation Biology, 2010. **24**(4): 1021–1030.

40. Chan, K. M. A. et al., *Conservation planning for ecosystem services.* PLoS Biology, 2006. **4**(11): e379.

41. Lester, S. E. et al., *Evaluating tradeoffs among ecosystem services to inform marine spatial planning.* Marine Policy, 2013. **38**: 80–89; Polasky, S. et al., *Where to put things? Spatial land management to sustain biodiversity and economic returns.* Biological Conservation, 2008. **141**(6): 1505–1524.

42. Howe, C. et al., *Creating win-wins from trade-offs? Ecosystem services for human well-being: A meta-analysis of ecosystem service trade-offs and synergies in the real world.* Global Environmental Change, 2014. **28**: 263–275.

43. Chan, K. M. A., N. C. Ban, and R. Naidoo, *Integrating conservation planning with human communities, ecosystem services, and economics,* in *Conservation Planning: Shaping the Future,* F. L. Craighead and C. L. Convis Jr., editors. 2013, Redlands, CA: Esri Press. 21–50.

44. Chan et al., *Where are cultural and social in ecosystem services? A framework for constructive engagement.* (See reference 14.)

45. Chan, K. M. A., L. Hoshizaki, and B. Klinkenberg, *Ecosystem services in conservation planning: Targeted benefits vs. co-benefits or costs?* PLoS ONE, 2011. **6**(9): e24378; Chan et al., *Integrating conservation planning with human communities, ecosystem services, and economics.* (See reference 43.)

46. Martinez-Harms, M. J. et al., *Making decisions for managing ecosystem services.* Biological Conservation, 2015. **184**: 229–238.

47. Ibid.

48. McCauley, D. J., *Selling out on nature.* Nature, 2006. **443**(7107): 27–28.

49. Muradian, R. et al., *Payments for ecosystem services and the fatal attraction of win-win solutions.* Conservation Letters, 2013. **6**(4): 274–279.

50. Naeem, S. et al., *Get the science right when paying for nature's services.* Science, 2015. **347**(6227): 1206–1207.

51. Crompton, T., *Common Cause: The Case for Working with Our Cultural Values,* 2010, Surrey, UK: WWF-UK. 16.

52. Bolderdijk, J. W. et al., *Comparing the effectiveness of monetary versus moral motives in environmental campaigning.* Nature Climate Change, 2013. **3**(4): 413–416.

53. Sandel, M. J., *What Money Can't Buy: The Moral Limits of Markets.* 2012, New York: Farrar, Straus, and Giroux. 256.

54. Redford, K. H., and W. M. Adams, *Payment for ecosystem services and the challenge of saving nature.* Conservation Biology, 2009. **23**(4): 785–787.

55. Schröter, M. et al., *Ecosystem services as a contested concept: A synthesis of critique and counter-arguments.* Conservation Letters, 2014: **7**(6):514–523.

56. Reyers, B. et al., *Finding common ground for biodiversity and ecosystem services.* BioScience, 2012. **62**(5): 503–507.

57. Cimon-Morin, J., M. Darveau, and M. Poulin, *Fostering synergies between ecosystem services and biodiversity in conservation planning: A review.* Biological Conservation, 2013. **166**: 144–154.

58. Mace, G. M., K. Norris, and A. H. Fitter, *Biodiversity and ecosystem services: A multilayered relationship.* Trends in Ecology & Evolution, 2012. **27**(1): 19–26.

59. Ruckelshaus, M. et al., *Notes from the field: Lessons learned from using ecosystem service approaches to inform real-world decisions.* Ecological Economics, 2015. **116**: 11–21.

# Part Three

## Implementation and Monitoring of Conservation Plans

# 11

# From Planning to Action and Communication: The Art of Implementation

## Overview

In this chapter, we address what it takes to successfully implement a conservation plan. We review some of the major challenges to implementation and draw on lessons that have been learned about why plans succeed or fail. We emphasize the importance of sound project management, including budgeting and work planning. We describe three approaches to implementation that may lead to greater conservation impact. Finally, we offer a few key nuggets of wisdom from the ever-growing body of knowledge on how scientists (and planners) can communicate better about the practice of conservation.

## Topics

- *Knowing-doing gap*
- *Checklist of plan contents*
- *Project management*
- *Budgeting and work planning*
- *Implementation challenges and lessons learned*
- *Tools for implementation*
- *Leveraging and "scaling-up" strategies*
- *Integrating conservation and land use planning*
- *Conservation opportunities*
- *Working at multiple scales*
- *Strategic communication*

## Implementation—Influencing Conservation Decisions

Most people involved in nature conservation or natural resource management are doers—they are taking actions, implementing plans in one fashion or another. Many of our colleagues in The Nature Conservancy, for example, spend their days working on land acquisitions or easements, helping landowners improve the ecological stewardship of their land, providing technical support to government and nongovernmental partners, or working with a government agency or multilateral organization to improve a policy or legislate a new one. Whether patrolling for poachers in an African park, working with grazing permittees to improve range management, or overseeing compliance with forest laws in Brazil, many of our colleagues in natural resource agencies are similarly taking conservation action—often in response to a natural resource management plan.

Although much has been made in conservation planning of what conservation biologist Andrew Knight and colleagues from South Africa have referred to as the **knowing-doing gap**,[1] in fact a lot of "doing" is taking place in conservation and natural resource management. This is not to deny that many plans, assessments, and planning methods described in the peer-reviewed literature fail to be implemented. Certainly such a gap exists. But at least equally important are planning efforts in which implementation occurs but falls short in one way or another. In either situation—no implementation or inadequate implementation—to help overcome these challenges, clear guidance is needed on how conservation plans can actually influence conservation decisions. That is the focus of this chapter.

There are many ways to improve a plan's chances of being successfully implemented. By this we mean acting on the strategies and actions outlined in a plan, and in so doing influencing key decisions that relate to the project or initiative and that positively affect the specified objectives. Perhaps the single most important piece of advice we can offer is to think about implementation from the moment a planning effort gets off the ground, and continue thinking about it throughout the planning process. For this reason, we have mentioned aspects of implementation throughout the book.

For example, it's essential to understand who the most important clients and stakeholders are for any particular planning effort (Chapter 2), ensure they are appropriately engaged in the planning process, and create products from the planning process that will enhance the enabling conditions for conservation for these stakeholders. Recent analysis of stakeholders in a conservation initiative in the Fitz-Stirling region of Western Australia underscores the importance of understanding stakeholders by demonstrating the influence of social networks among stakeholders in reaching conservation outcomes, especially across different geopolitical scales and units.[2]

In this chapter we explore an array of recommendations, tools, and techniques that can help a plan to achieve its desired influence. We start with some of the "nuts and bolts" that help make plans successful, first by looking at a hypothetical table of contents of a completed conservation plan to ensure that all the parts are in place. We then examine two critical but underappreciated components of conservation planning—annual work planning and budgeting, where the details of available funds, fund-raising, budget preparation, and who does what all come into play. We also touch on the role that good project management[3] plays in implementation.

To understand what makes a plan successful, we need to appreciate what has caused many planning efforts to fall short. We'll review the most important hurdles to implementation, but importantly spend more time understanding the enabling conditions, guiding principles, and lessons learned for successful implementation techniques from governmental, nongovernmental, and private sector organizations. We include some outstanding examples of planning efforts that have had real conservation impacts.

We conclude the chapter with some thoughts on strategic communication and its essential role in successful implementation of a plan and ultimately in the conservation impact of a plan. Done right, good communication can lead to all sorts of great results. Done poorly, it can set a conservation project back and even lead to failure.

## A Table of Contents

From your own experience, you likely have some good ideas about the components of a quality conservation plan. And though there's some truth in the cliché that "the process is at least as important as the plan," any team will be challenged to successfully implement a conservation project without a clear idea of the anticipated outcomes, a strong set of strategies to achieve those outcomes, the budget to do it with, and an understanding of success (or failure) when they see it.

**Box 11.1** outlines a table of contents for a hypothetical conservation project. Although the names aren't critical, the major components are all ones that we have identified in the planning process that forms the foundation for this book and that should be part of a written plan in one form or another. Some teams may choose to include all of these components in one document, while others may separate some components into additional plans (e.g., a monitoring plan or an implementation plan).

Although some material may need to be confidential and available only to the planning team or organization in charge (e.g., fund-raising

### Box 11.1 Proposed Table of Contents for a Conservation Plan

- *Executive summary.* Perhaps the most important section of the plan because many readers won't get beyond it; also a useful section for fund-raising and outreach.
- *Planning context.* Purpose of the plan, decisions to be made, decision makers, audience, constraints, or sideboards from previous planning efforts or law and policy.
- *Planning team and process.* Members, skill sets, organizations involved, team charter, management process and roles.
- *Situation analysis.* Economic, social, ecological, political trends and opportunities within the social-ecological system; usually includes a conceptual model and assessment of threats to conservation features; may also include some analysis of enabling conditions for conservation and likely barriers to implementation.
- *Project scope.* Strategic, geographic, and temporal "boundaries" of the project.
- *Fundamental objectives and desired outcomes.* The ultimate outcomes in a conservation project that we hope to achieve: the ends, not the means; those things we care most about—healthier forests, more native species, improved livelihoods.
- *Conservation features.* The elements of biodiversity, ecosystem processes, and social systems (human well-being) that are the focus of the planning efforts and, where appropriate, the quantitative targets (or goals) that have been set for these features.
- *Range of strategies.* The different strategies or major interventions that are under consideration for use in a conservation project or program and a rationale for how decisions will be made to focus on certain strategies and not others.
- *Strategy selection and theory of change.* The strategies selected for implementation by a project or program and a rationale for how and why those strategies will be implemented.
- *Data and knowledge.* Summary of the types of data, knowledge (expert, local, traditional) and associated meta-data that were used in the plan.
- *Risks.* Summary of those factors considered most likely to influence the successful implementation of strategies.
- *Monitoring program.* A plan for what actions will be taken during the project to measure progress and evaluate the effectiveness of strategies and actions.
- *Work planning.* Detailed timeline of actions and tasks required to implement the plan, who is responsible, and proposed deadlines.
- *Budgeting and fund-raising.* Detailed assessment of the staff and financial resources needed to implement the strategies and actions and a realistic fund-raising plan to ensure that these resources are in place.
- *Communication.* Summary of the different types of internal and external communications that will take place related to the project (websites, press releases, blogs, field trips).
- *Operational or implementation plan.* Details on how the plan will be implemented.

information), we advocate keeping this sort of information to a minimum and making the vast majority of planning information transparent and openly accessible to the public and all stakeholders. Detailed information is best placed in appendices, and these are sometimes located in a separate document to keep the plan from becoming unwieldy. Finally, the level of detail of any planning effort should be tailored to the needs of the audience and decisions at hand, but the core elements of the plan should be as succinct as possible. Advice we have consistently received is that relatively shorter, to-the-point plans are likely to be easier to communicate to a wide variety of audiences, more likely to be read, and as a result, more likely to be implemented.

In the following sections, we delve a little deeper into two components of this table of contents that are critical to successfully implementing a plan—work planning, and budgeting and fund-raising—as well as the importance of project management in ensuring that all these components of the planning process are addressed coherently and efficiently.

## Figuring Out What to Do When—Work Planning

Many agencies and nongovernmental organizations (NGOs) that undertake the development of conservation plans also use some sort of annual work plan. These plans detail the tasks, activities, and responsibilities for implementing the various strategies and actions that a conservation or natural resource management project will undertake—essentially the "what, who, when, and how much."[4] Some practitioners may find this level of planning to just be too much (as the cartoon in FIGURE 11.1 suggests), but for complex projects with many moving pieces, it is essential. No successful business would take on a similarly complex project without a detailed work plan.

FIGURE 11.1 Conservation (strategic) plans, operational plans for implementing them, and detailed work plans for assigning tasks and deadlines can easily overwhelm a conservation project or program. These different forms of planning should be scaled to the size and complexity of a conservation project.

Sometimes a specific project or program will have a stand-alone work plan. At other times, these work plans of individual conservation projects will need to be integrated as part of a larger overarching, annual work plan that includes numerous projects and has many individuals working on multiple projects. Work plans distill what can be complex projects into manageable chunks that help in setting priorities, estimating costs, establishing a schedule, and managing a project to keep it on track.

Depending on project complexity, planners can use a number of software packages that are designed for work planning and project management. Miradi, which we mentioned in earlier chapters, is a desktop strategic planning software package for conservation projects, and it also has a built-in work planning feature. For less complicated projects, it is not necessary to use software packages—especially expensive and sophisticated project management software, which can involve steep learning curves. Adequate work plans can be developed in fairly simple spreadsheets, and for many projects, they may be all that you need. **TABLES 11.1** through **11.3** provide some examples of work plans.

In Chapter 3, we explored how to establish fundamental objectives for a conservation project—the things we really care about, value, and want to accomplish over the long term. We also noted that a plan usually needs to identify nearer-term objectives that are focused on the *means* of achieving longer-term objectives (ends). These nearer-term objectives are often referred to as **intermediate objectives**, and they form the basis of annual work plans. For example, in Table 11.2,

**TABLE 11.1** A sample work plan for a conservation project produced by the adaptive management software Miradi, a cooperative venture of the Conservation Measures Partnership and Sitka Technology Group (http://www.miradi.org).

| | | | | Work Units | | | Projected Expenses | | | Budget Totals | | |
|---|---|---|---|---|---|---|---|---|---|---|---|---|
| Item | Progress | Who | When | 2012 | 2013 | Total | 2012 | 2013 | Total | 2012 | 2013 | Total |
| Marine Site | | JH, GdR, LE, EM, MIM, AT, LE2 | | 820 | 564 | 1384 | 99,500 | 66,500 | 171,2... | 270,9... | 148,1... | 424,3... |
| 1. Campaign to stop shark fin soup | | | | | | | | | | | | |
| 1. Campaign to stop shark fin soup | Scheduled | AT, EM, GdR, JH, MIM | 2012-01-01 - 2013-12-31 | 360 | 152 | 512 | 5,000 | 3,000 | 13,000 | 50,050 | 19,750 | 74,800 |
| Research local impact of shark fishing | Scheduled | JH, MIM | Q1 FY12 - Q2 FY12 | 101 | | 101 | 5,000 | | 5,000 | 15,100 | | 15,100 |
| Plan campaign with CAI | Scheduled | AT, EM, GdR, JH, MIM | Q1 FY12 - Q2 FY12 | 62 | | 62 | | | 5,000 | 14,200 | | 19,200 |
| Develop materials for restaurants and consumers | Scheduled | AT, MIM | Q2 FY12 - Q3 FY12 | 50 | | 50 | | | | 4,500 | | 4,500 |
| Implement campaign (radio, TV, print) | Scheduled | AT, EM, MIM | 2012-07-01 - 2013-12-31 | 95 | 100 | 195 | | | | 8,500 | 9,000 | 17,500 |
| Evaluate reach and uptake of message | Scheduled | AT, JH | 2012-10-01 - 2013-12-31 | 25 | 25 | 50 | | | | 2,250 | 2,250 | 4,500 |
| Adapt campaign as needed | Scheduled | AT, EM, GdR, JH, MIM | 2012-10-01 - 2013-12-31 | 27 | 27 | 54 | | | | 5,500 | 5,500 | 11,000 |
| SHARK1. # incidents per year of shark fishing boats illegally fish | Scheduled | JH | 2012-01-01 - 2013-12-31 | 5 | 5 | 10 | | | | 500 | 500 | 1,000 |
| SHARK1a. % of restaurants that are active participants | Scheduled | JH | 2012-01-01 - 2013-12-31 | 10 | 10 | 20 | | | | 1,000 | 1,000 | 2,000 |
| SHARK1b. % of consumers surveyed in urban markets that can | Scheduled | JH | 2012-01-01 - 2013-12-31 | 10 | 10 | 20 | | | | 1,000 | 1,000 | 2,000 |
| 2. Promotion of sustainable fishing techniques | | | | | | | | | | | | |
| 3. Promote spill mitigation techniques | | | | | | | | | | | | |
| 3. Promote spill mitigation techniques | Scheduled | EM, GdR, JH, MIM | 2012-01-01 - 2013-12-31 | 50 | 12 | 62 | | | | 10,400 | 3,900 | 14,300 |
| Compile funding info on spill mitigation | Scheduled | EM, GdR | 2012-01-01 - 2013-09-30 | 7 | 12 | 19 | | | | 2,100 | 3,900 | 6,000 |
| Identify potential vessels | Scheduled | EM, JH | FY12 | 10 | | 10 | | | | 1,250 | | 1,250 |
| Research and identify best practices for spill mitigation | Scheduled | EM, GdR, JH | FY12 | 12 | | 12 | | | | 2,150 | | 2,150 |
| Initial individual contacts with vessel owners | Scheduled | GdR, JH, MIM | Q2 FY12 - Q3 FY12 | 21 | | 21 | | | | 4,900 | | 4,900 |
| Hold series of workshops with interested vessel owners | Scheduled | EM, GdR, JH, MIM | FY12 | 0 | | 0 | | | | 0 | | 0 |
| PUFF. Number of breeding pairs of ruby crested puffins | Scheduled | | | | | | | | 200 | | | 200 |
| 4. Strengthen law enforcement | | | | | | | | | | | | |
| Strengthen law enforcement | Scheduled | GdR, LE2, LE | 2012-01-01 - 2013-12-31 | 10 | 0 | 10 | 35,000 | 25,000 | 60,000 | 90,000 | 25,000 | 115,0... |
| 5. Other costs | | | | | | | | | | | | |
| Operational costs (overhead) | Not Specified | | | | | | 36,000 | 25,000 | 61,000 | 36,000 | 25,000 | 61,000 |
| Travel and other major expenses | Not Specified | | | | | | 13,500 | 13,500 | 27,000 | 13,500 | 13,500 | 27,000 |

**TABLE 11.2** Part of a hypothetical annual work plan for a conservation project in the Bering Sea. The plan consists of a set of tasks and activities that fall under one of the major objectives of the project—reducing mortality in seabirds in the Bering Sea—and includes who will do what, the deadline, the time period, and the cost of the action.

# BERING SEA WORK PLAN FOR OBJECTIVE 3a*

| Activity/ Task # | Work breakdown for Commercial Fisheries **Objective 3a:** Reduce the number of albatross and other seabirds caught in longlines & nets by 90% by 2010 in US waters and by 50% by 2015 in Russian waters. | Complete | Who | Deadline | Cost | Nov | Dec | Ja |
|---|---|---|---|---|---|---|---|---|
| **3a.1** | **Expand tori line use in Russian longline fleet** | | | | | | | |
| 3a.1.1 | Expand education program with fishermen | ✓ | K. Alvarez (WWF) | 27-Nov-02 | 1,000 | | | |
| 3a.1.2 | Secure funding for tori lines and other equipment; purchase and ship | ✓ | J. Miller (TNC) | 10-Jan-03 | 2,000 | | | |
| 3a.1.3 | Tori lines distribution between main Russian longline fishing companies | ✓ | L. Ko (EPA) | 10-Jan-03 | 2,000 | | | |
| 3a.1.4 | Other mitigating equipment promotion (integrated weight line, etc.) | | K. Alvarez (WWF) | 1-Feb-03 | 3,000 | | | |
| **3a.2** | **Improve understanding of interactions between US fisheries and incidental seabird take** | | | | | | | |
| 3a.2.1 | Better quantify seabird/gear interaction rate in trawl fisheries (Lead = USFWS) | ✓ | K. Alvarez (WWF) | 10-Jan-03 | 1,000 | | | |
| 3a.2.2 | Coordinate development of database of spatial and temporal distribution of all fishing effort in the Bering Sea ecoregion | ✓ | J. Miller (TNC) | 15-Jan-03 | 2,000 | | | |
| 3a.2.3 | Quantify drop-off rate for seabirds caught on longlines (those that go under and don't come up) (Lead = USFWS) | | L. Ko (EPA) | 9-Mar-03 | 2,000 | | | |
| **3a.3** | **Obtain ban on high seas driftnet fisheries in Russia** | | | | | | | |
| 3a.3.1 | Conduct an analysis of current high-seas driftnet practices | ✓ | K. Alvarez (WWF) | 11-Feb-03 | 1,000 | | | |
| 3a.3.2 | Lobby Russian government | ✓ | J. Miller (TNC) | 19-Feb-03 | 2,000 | | | |
| 3a.3.3 | Lobby Japanese government with assistance of the WWF Japan | | L. Ko (EPA) | 30-Apr-03 | 2,000 | | | |
| **3a.4** | **Establish observer program in Russia** | | | | | | | |
| 3a.4.1 | Develop and implement program | ✓ | J. Miller (TNC) | 18-Apr-03 | 1,000 | | | |
| 3a.4.2 | Propose changes for Russian legislation, regulations & system of observers activities | | L. Ko (EPA) | 30-Apr-03 | 2,000 | | | |

| KEY | |
|---|---|
| | Major Deadlines |
| | Tasks |
| | Phases |
| | Holidays |

*This is a hypothetical work plan; only the actions and tasks are actuals.

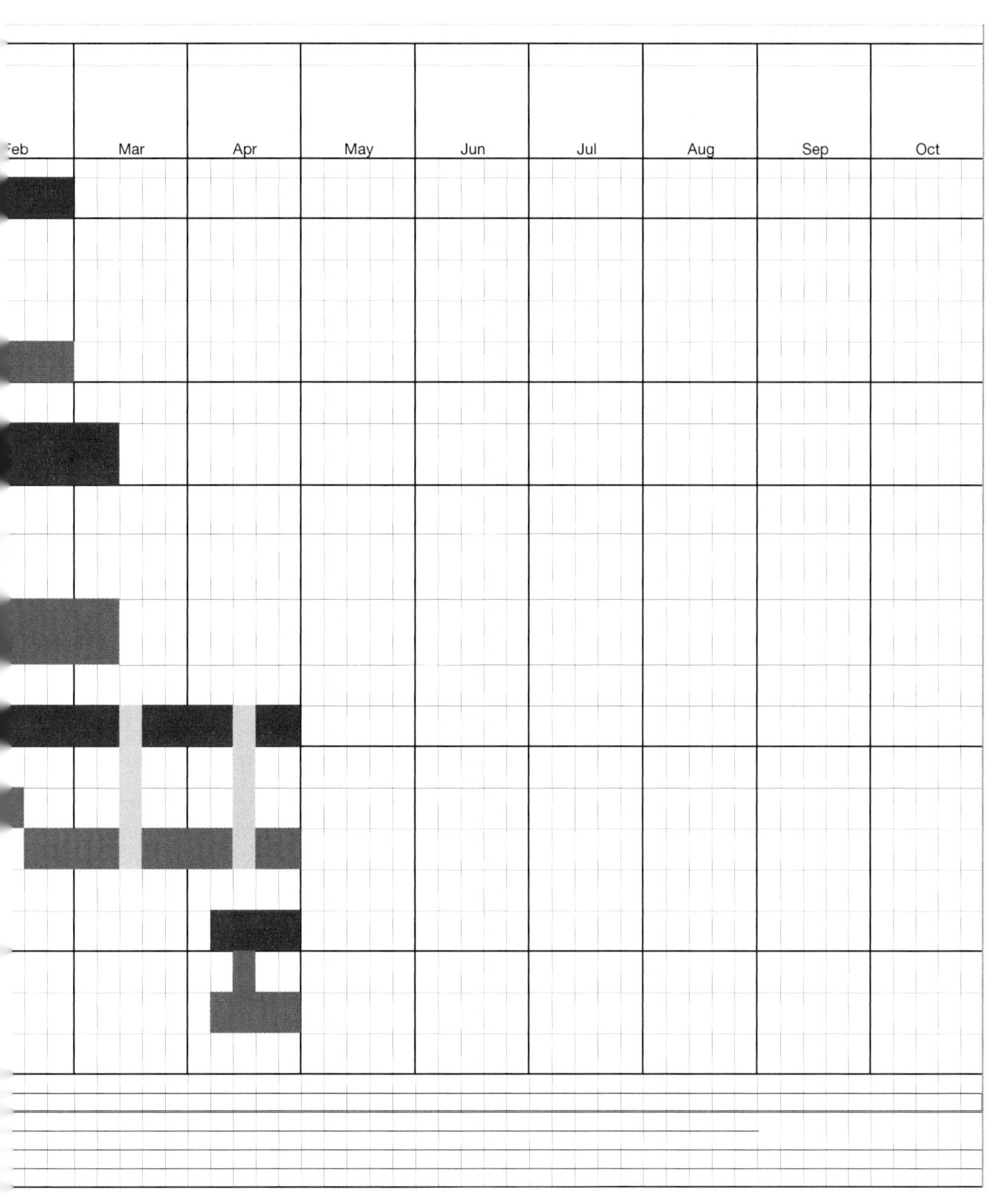


*Source:* Data from Mikov et al. (2007).

**TABLE 11.3** Example of a Gantt chart used to help identify who is responsible for undertaking particular tasks, the timeline for doing so, and how responsibilities overlap for teammates involved in various activities of a conservation project.

**C. Spreadsheet Gantt Chart (preferred format)**

This format can be done in a spreadsheet or text document. If you use a spreadsheet, you can manipulate the information more flexibly. For example, you can:

- Add a column of unit cost figures and quickly estimate the cost of a task.
- Sort all tasks assigned to an individual and calculate the allocation of that person's time.

| Work breakdown: activities and tasks | Who | Dates | | | | Deliverables |
|---|---|---|---|---|---|---|
| | | Jun | Jul | Aug | Sep | |
| **Objective A** | | | | | | Completion of nest boxes |
| **Activity A** | | | | | | Half of nest boxes installed |
| - Task A1 | Jose | | | | | Remaining nest boxes in |
| - Task A2 | Mary | | | | | |
| - Task A3 | Mary | | | | | Monitoring plan complete |
| **Activity B** | | | | | | Monitoring cameras set up |
| - Task B1 | Andreas | | | | | |
| - Task B2 | Elena | | | | | Ongoing photos of nests |
| **Monitoring indicator M** | | | | | | Analysis of nesting success |
| - Task M1 | Jose | | | | | |
| - Task M2 | Elena | | | | | Complete long-term M plan |
| **Monitoring indicator N** | | | | | | |
| - Task N1 | Mary | | | | | Two funding proposals sent |
| **Operational work X** | | | | | | Completion of nest boxes |
| - Task X1 | Elena | | | | | Half of nest boxes installed |

*Source:* Data from Mikov et al. (2007).

objective 3a (concerned with reducing the number of seabirds caught in longlines and nets) could be considered an intermediate objective. In this case, the fundamental objective is related to viable populations of seabirds in the Bering Sea in which the threats from longlines and net fishing had been reduced to an acceptable level. Beneath the intermediate objective in Table 11.2 are a series of the activities and tasks necessary to achieve that objective. It's important not to get too caught up in semantic differences that you will see across various conservation plans. Some plans will refer to intermediate objectives as *intermediate results* or even *intermediate outcomes* or just *outcomes*. The important point is that some nearer-term objectives or results are established as part of a project's overall theory of change (see Chapter 6) and that they serve as the basis for annual work plans.

## Ensuring That Resources Are Sufficient: Budgeting and Fund-Raising

While preparing budgets and fund-raising may not be the motivating components of conservation, everyone understands the importance of this part of the planning and project cycle—without adequate funds, conservation and management is mostly just talk. For NGOs that often rely on project-to-project fund-raising, this component of planning is not only mandatory but frequently gets a project team into trouble. There is a fine line to walk in being ambitious about the funds it really takes to implement a project and practical about the funds that can realistically be raised. When that gap between ambition and practicality gets too wide, many projects may find themselves in financially unsustainable waters (FIGURE 11.2). The reality is that for many planning situations, budgeting is essentially work planning—what gets budgeted for is what gets done.

"I can complete the project under budget and ahead of schedule, but you'll need to allocate additional time and money for that."

**FIGURE 11.2** It's easy for conservation projects to get overly ambitious in the strategies and actions they want to take and not have sufficient funding to get the job done.

It is probably obvious, but good to review, why you need to devote serious time to budget preparation. Although the most important reason is to ensure you have adequate funds to get the work done, budgets are also helpful if not essential for preparing fund-raising proposals, helping teams reach clarity over what resources are needed, providing transparency to donors, and identifying gaps between expenses and income.[5]

For most conservation projects, two types of budgets need to be prepared. The first is a more generalized budget for a multiyear project based on the conservation plan for that project. Human resources (staffing) are often the most significant component of any budget, as are any large capital expenses (e.g., vehicles), so it's good to keep these factors in mind. One useful way to develop this multiyear budget is to parcel it into the different strategies that the project will undertake and estimate costs for each of those strategies.

Another approach to the multiyear budget is to project it based on the detailed costs of a single-year budget, which is the second type of budget that most projects prepare. Work plans often estimate the costs of each activity, and these costs can in turn be used to develop a detailed budget. TABLE 11.4 shows a sample budget template that is based on costs derived from a work plan and broken into typical budget categories of staff, consultants, travel and meetings, communication, office costs, and field costs (along with some useful tips for completing the budget). Chapter 5 provides guidance on estimating many of the costs in a conservation plan that can in turn be used to prepare annual and multiyear budgets.

**TABLE 11.4** A sample multiyear budget for a conservation project broken into typical categories of staff time, consultant time, travel, communication, office administration, and capital expenses.

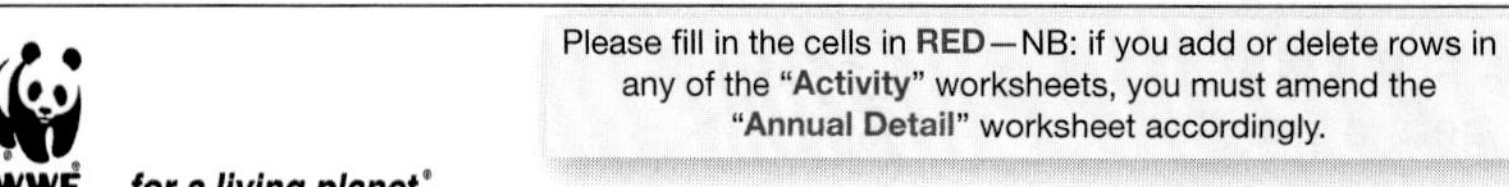


WWF Network Standards

**Note:** The project number and title will be automatically reported in all other worksheets.

**Note:** The date, author's name and office's name will be automatically reported in all other worksheets.

Project Nr: [xxxxxx]
Project Title: [xxxxxx]
Activity Full Title: [xxxxxx]

Budget issued on: [date]
Budget prepared by: [author's name]
Office name: [name of WWF office]

**DETAILED ACTIVITY SCHEDULE FORMAT: Note that this is illustrative both in format and in the G/L codes included - Activity 1**

In [Currency]

**Note:** Budget currency will be automatically reported in all other worksheets.

**Note:** Don't forget to include the costs of the "Project Administrator" (or "Project Finance Analyst").

**Note:** Budget financial years will be automatically reported in all other worksheets.

| WWF G/L CODE REF. | CATEGORY | RATE | UNIT | ACTIVITY 1 | | | | | | | | |
|---|---|---|---|---|---|---|---|---|---|---|---|---|
| | | | | YEAR 1 | | YEAR 2 | | YEA | YEAR 5 | | PROJECT TOTAL | |
| | | | | # Units | Cost | # Units | Cost | # Unit | # Units | Cost | # Units | Cost |
| 50 | STAFF COSTS | | | | | | | | | | | |
| | Salaries & Benefits, Outposted staff | | | | | | | | | | | |
| | 1 | | | | 0.00 | | | | 0% | 0.00 | 0% | 0.00 |
| | 2 | | | | 0.00 | | | | 0% | 0.00 | 0% | 0.00 |
| | 3 | | | | 0.00 | | | | 0% | 0.00 | 0% | 0.00 |
| | Salaries & Benefits, Local Hires | | | | | | | | | | | |
| | 1 | 0.00 | Year | 0% | 0.00 | 0% | 0.00 | 0% | 0% | 0.00 | 0% | 0.00 |
| | 2 | 0.00 | Year | 0% | 0.00 | 0% | 0.00 | 0% | 0% | 0.00 | 0% | 0.00 |
| | 3 | 0.00 | Year | 0% | 0.00 | 0% | 0.00 | 0% | 0% | 0.00 | 0% | 0.00 |
| 50 | *TOTAL - STAFF COSTS* | | | | 0.00 | | 0.00 | | | 0.00 | | 0.00 |
| 51 | THIRD PARTY FEES | | | | | | | | | | | |
| | Consultatnt Fees | | | | | | | | | | | |
| | 1 | 0.00 | Day | 0.0 | 0.00 | 0.0 | 0.00 | 0.0 | 0.0 | 0.00 | 0 | 0.00 |
| | 2 | 0.00 | Day | 0.0 | 0.00 | 0.0 | 0.00 | 0.0 | 0.0 | 0.00 | 0 | 0.00 |
| | 3 | 0.00 | Day | 0.0 | 0.00 | 0.0 | 0.00 | 0.0 | 0.0 | 0.00 | 0 | 0.00 |
| | Consultant expenses | | | | | | | | | | | |
| | 1 | 0.00 | Day | 0.0 | 0.00 | 0.0 | 0.00 | 0.0 | 0.0 | 0.00 | 0 | 0.00 |
| | 2 | 0.00 | Day | 0.0 | 0.00 | 0.0 | 0.00 | 0.0 | 0.0 | 0.00 | 0 | 0.00 |
| | 3 | 0.00 | Day | 0.0 | 0.00 | 0.0 | 0.00 | 0.0 | 0.0 | 0.00 | 0 | 0.00 |
| 51 | *TOTAL - THIRD PARTY FEES* | | | | 0.00 | | 0.00 | | | 0.00 | | 0.00 |
| 52 | OTHER GRANTS & AGREEMENTS | | | | | | | | | | | |
| | Grants (to non-WWF entities) | | | | | | | | | | | |
| | 1 | n/a | n/a | | 0.00 | | 0.00 | | | 0.00 | | 0.00 |
| | 2 | n/a | n/a | | 0.00 | | 0.00 | | | 0.00 | | 0.00 |
| | Agreements (to non-WWF entities) | | | | | | | | | | | |
| | 1 | n/a | n/a | | 0.00 | | 0.00 | | | 0.00 | | 0.00 |
| | 2 | n/a | n/a | | 0.00 | | 0.00 | | | 0.00 | | 0.00 |
| 52 | *TOTAL - GRANTS & AGREEMENTS* | | | | 0.00 | | 0.00 | | | 0.00 | | 0.00 |
| 53 | TRAVEL, MEETING & TRAINING COSTS | | | | | | | | | | | |
| | **Staff Travel:** | | | | | | | | | | | |
| | Staff - Air Fare | | | | | | | | | | | |
| | 1 | 0.00 | R/trip | 0 | 0.00 | 0 | 0.00 | 0 | 0 | 0.00 | 0 | 0.00 |
| | 2 | 0.00 | R/trip | 0 | 0.00 | 0 | 0.00 | 0 | 0 | 0.00 | 0 | 0.00 |
| | 3 | 0.00 | R/trip | 0 | 0.00 | 0 | 0.00 | 0 | 0 | 0.00 | 0 | 0.00 |
| | Staff - Local Travel & Subsistence | | | | | | | | | | | |
| | 1 | 0.00 | Day | 0 | 0.00 | 0 | 0.00 | 0 | 0 | 0.00 | 0 | 0.00 |
| | 2 | 0.00 | Day | 0 | 0.00 | 0 | 0.00 | 0 | 0 | 0.00 | 0 | 0.00 |
| | 3 | 0.00 | Day | 0 | 0.00 | 0 | 0.00 | 0 | 0 | 0.00 | 0 | 0.00 |
| | **Workshops & Training:** | | | | | | | | | | | |
| | Workshops/Training: Registration Fees | | | | | | | | | | | |
| | 1 | 0.00 | p/W | 0 | 0.00 | 0 | 0.00 | 0 | 0 | 0.00 | 0 | 0.00 |
| | 2 | 0.00 | p/W | 0 | 0.00 | 0 | 0.00 | 0 | 0 | 0.00 | 0 | 0.00 |
| | Workshops/Training: Meals & Facility | | | | | | | | | | | |
| | 1 | 0.00 | Day | 0 | 0.00 | 0 | 0.00 | 0 | 0 | 0.00 | 0 | 0.00 |
| | 2 | 0.00 | Day | 0 | 0.00 | 0 | 0.00 | 0 | 0 | 0.00 | 0 | 0.00 |
| | Participant Travel | | | | | | | | | | | |
| | 1 | 0.00 | R/Trip | 0 | 0.00 | 0 | 0.00 | 0 | 0 | 0.00 | 0 | 0.00 |
| | 2 | 0.00 | R/Trip | 0 | 0.00 | 0 | 0.00 | 0 | 0 | 0.00 | 0 | 0.00 |
| | Participant Stipends/Expenses | | | | | | | | | | | |
| | 1 | 0.00 | Day | 0 | 0.00 | 0 | 0.00 | 0 | 0 | 0.00 | 0 | 0.00 |
| | 2 | 0.00 | Day | 0 | 0.00 | 0 | 0.00 | 0 | 0 | 0.00 | 0 | 0.00 |
| | Hospitality | 0.00 | Day | 0 | 0.00 | 0 | 0.00 | 0 | 0 | 0.00 | 0 | 0.00 |
| | Speaker's Fees & Expenses | 0.00 | Day | 0 | 0.00 | 0 | 0.00 | 0 | 0 | 0.00 | 0 | 0.00 |
| 53 | *TOTAL - TRAVEL, MEETING & TRAINING COSTS* | | | | 0.00 | | 0.00 | | | 0.00 | | 0.00 |

*for a living planet*®

Project Nr: [xxxxxx]
Project Title: [xxxxxx]
Activity Full Title: [xxxxxx]

Budget issued on: [date]
Budget prepared by: [author's name]
Office name: [name of WWF office]

**DETAILED ACTIVITY SCHEDULE FORMAT: Note that this is illustrative both in format and in the G/L codes included - Activity 1**

In [Currency]

| WWF G/L CODE REF. | CATEGORY | RATE | UNIT | ACTIVITY 1 YEAR 1 # Units | YEAR 1 Cost | YEAR 2 # Units | YEAR 2 Cost | YEA # Units | YEAR 5 # Units | YEAR 5 Cost | PROJECT TOTAL # Units | PROJECT TOTAL Cost |
|---|---|---|---|---|---|---|---|---|---|---|---|---|
| 54 | COMMUNICATION & FUNDRAISING COSTS | | | | | | | | | | | |
| | Research Materials and publications | | n/a | | 0.00 | | 0.00 | | | 0.00 | | 0.00 |
| | Other | | n/a | | 0.00 | | 0.00 | | | 0.00 | | 0.00 |
| 54 | *TOTAL - COMMUNICATION & FUNDRAISING COSTS* | | | | 0.00 | | 0.00 | | | 0.00 | | 0.00 |
| 55 | MISCELLANEOUS COSTS | | | | | | | | | | | |
| | Miscellaneous | | n/a | | 0.00 | | 0.00 | | | 0.00 | | 0.00 |
| 55 | *TOTAL - MISCELLANEOUS COSTS* | | | | 0.00 | | 0.00 | | | 0.00 | | 0.00 |
| 56 | OFFICE RUNNING COSTS | | | | | | | | | | | |
| | Rental/Lease of office premises | 0.00 | Month | 0 | 0.00 | 0 | 0.00 | 0 | 0 | 0.00 | 0 | 0.00 |
| | Rental/Lease of office equipment | 0.00 | Month | 0 | 0.00 | 0 | 0.00 | 0 | 0 | 0.00 | 0 | 0.00 |
| | Office maintenance, Repairs & Cleaning | n/a | n/a | | 0.00 | | 0.00 | | | 0.00 | | 0.00 |
| | Office utilities & Insurance | n/a | n/a | | 0.00 | | 0.00 | | | 0.00 | | 0.00 |
| | Telecommunications (phone, fax, email, videoconf) | n/a | n/a | | 0.00 | | 0.00 | | | 0.00 | | 0.00 |
| | Postage & Freight | | | | 0.00 | | 0.00 | | | 0.00 | | 0.00 |
| | Stationary & Supplies | | | | 0.00 | | 0.00 | | | 0.00 | | 0.00 |
| | Other office running costs | | | | 0.00 | | 0.00 | | | 0.00 | | 0.00 |
| 56 | *TOTAL - OFFICE RUNNING COSTS* | | | | 0.00 | | 0.00 | | | 0.00 | | 0.00 |
| 57 | FIELD RUNNING COSTS | | | | | | | | | | | |
| | Field equipment | 0.00 | Ea. | 0 | 0.00 | 0 | 0.00 | 0 | 0 | 0.00 | 0 | 0.00 |
| | Vehicles | | | | 0.00 | 0 | 0.00 | 0 | 0 | 0.00 | 0 | 0.00 |
| | Field infrastructure | | | | 0.00 | 0 | 0.00 | 0 | 0 | 0.00 | 0 | 0.00 |
| | Land | | | | 0.00 | 0 | 0.00 | 0 | 0 | 0.00 | 0 | 0.00 |
| | Other Field Costs | | | | 0.00 | 0 | 0.00 | 0 | 0 | 0.00 | 0 | 0.00 |
| 57 | *TOTAL - FIELD RUNNING COSTS* | | | | 0.00 | | 0.00 | | | 0.00 | | 0.00 |
| 58 | CAPITAL ASSET COSTS | | | | | | | | | | | |
| | Land & Buildings | | n/a | | 0.00 | | 0.00 | | | 0.00 | | 0.00 |
| | Office Equipment | 0.00 | Ea. | 0 | 0.00 | 0 | 0.00 | 0 | 0 | 0.00 | 0 | 0.00 |
| | Field Equipment | 0.00 | Ea. | 0 | 0.00 | 0 | 0.00 | 0 | 0 | 0.00 | 0 | 0.00 |
| | Vehicles | 0.00 | | | | | | 0 | 0 | 0.00 | 0 | 0.00 |
| | Other Capital Outlays | | | | | | | | | 0.00 | | 0.00 |
| | Depreciation of Capitalized Assets | | | | | | | | | 0.00 | | 0.00 |
| | Gain/Loss on Fixed Asset Disposal | | | | | | | | | 0.00 | | 0.00 |
| 58 | *TOTAL - CAPITAL ASSET COSTS* | | | | | | | | | 0.00 | | 0.00 |
| 59 | MANAGEMENT COSTS | | | | | | | | | | | |
| | Management Costs | 12.50% | n/a | | 0.00 | | 0.00 | | | 0.00 | | 0.00 |
| 59 | *TOTAL - MANAGEMENT COSTS* | | | | | | | | | 0.00 | | 0.00 |
| 60 | FUNDING TO WWF NETWORK | | | | | | | | | | | |
| | Contributions to WWF Network | | n/a | | | | | | | 0.00 | | 0.00 |
| 60 | *TOTAL - FUNDING TO WWF NETWORK* | | | | | | | | | 0.00 | | 0.00 |
| | TOTAL DIRECT COSTS (Without 59) | | | | 0.00 | | 0.00 | | | 0.00 | | 0.00 |
| | **TOTAL PROJECT ACTIVITY COSTS (Incl. 59)** | | | | **0.00** | | **0.00** | | | **0.00** | | **0.00** |

Lease/rental
Repairs and maintenance
Electricity and fuel
Insurance and taxes

Insurance and Taxes

Field supplies
Field clothing
Field rations/allowances

**Note:** Network standard management fee rate as per WWF network cost recovery standard—can be lower as primary donor's restrictions prevail. In this case, the relevant portion of the management costs must be charged directly to the project in the appropriate budget lines.

**Budget line mostly used by donor NOs or home offices.**
**Important note:** Amounts in this line must include the recipient office's management costs (i.e., 8.5%). The 4% management costs for the transfer agent must be indicated in the budget line 59—"Management Costs."

Activity 1 | Activity 2 | Activity 3 | Activity 4 | Activity 5 | Activity 6 | Activity 7 | Activity 11 | Activity 12

*Source:* Data from Mikov et al. (2007).

Many conservation projects are supported by cause-specific fund-raising or discretionary government spending. Like project management, fund-raising, whether from the perspective of a natural resource agency or a nongovernmental conservation organization, is a huge topic. Numerous books have been written on it, courses and workshops are offered on it, and a vast array of resources available on the Internet alone are related to this topic. Although it's beyond the scope of this book to delve into fund-raising in any detailed way, a few important points are worth making.

1. Most conservation organizations and, increasingly, natural resource agencies have philanthropy or development staff members who are dedicated to fund-raising. It's essential to involve these fund-raising staff members in the planning process so that they can clearly understand the goal to be achieved and can advise on what resources may be raised for it. As discussed in Chapter 6, a project's theory of change can be a powerful tool for fund-raisers, and we have found development staff to be one of the best and most critical checks of program logic.
2. If a project is not entirely new, then it is worth looking into what funding sources have been used to support what sorts of strategies and actions in the past, which of those have been well funded, and which have fallen short. Project directors, managers, and fund-raisers will want to know how much the project will rely on discretionary funding (in a government agency, these are usually dedicated public funds used to support a wide range of activities), foundation support, government grants, or private funding. They will also want to know which of these funds are "in hand" versus "committed," "requested," or a planned request.[6] **FIGURE 11.3** provides information on the uses and sources of funds for a hypothetical conservation project.
3. An important point about fund-raising is that it may be helpful and necessary to prepare a fund-raising prospectus that is largely excerpted from the plan. These are usually brief; they succinctly describe what the conservation project will be doing, and most importantly, they state its short- and long-term conservation outcomes.

All the best strategic thinking that a team can bring to bear on a conservation problem is wasted if sufficient funds are not raised to support the project. Because of the uncertainty in budgets and fund-raising, it's good practice for teams to do a little what-if thinking if part of the anticipated funds for a project are not raised, or—and this occasionally happens—more funds are raised than were originally planned. In either case, it is useful for a project team to think through contingencies: Where would budget cuts be made? How would

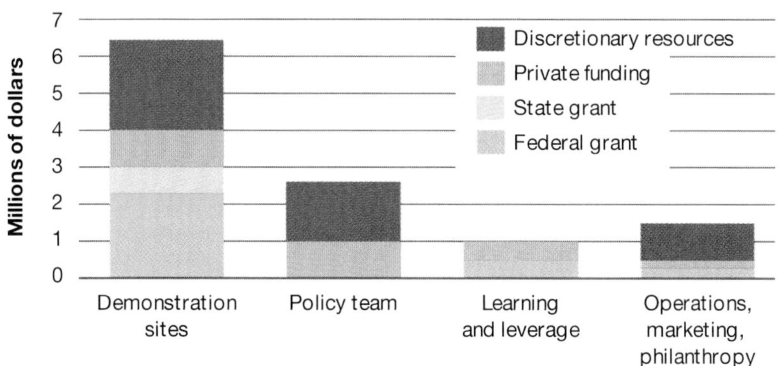


**FIGURE 11.3** A hypothetical conservation project and the sources of funding being used to accomplish various major components of the project, including its administration (operations, marketing, and philanthropy).

such cuts affect products that need to be delivered? How might they affect the team's ability to achieve the project outcomes? What would a team do with additional funds and capacity?

It's worth noting here that some agencies and organizations develop separate operational plans. These plans typically contain information on funding needs, human resources, risk factors, and even in some cases an exit strategy for the project. In short, where we have suggested that separate sections of a conservation plan will contain information on risks, budgets, and work planning, an operational plan would place this information in a separate document. For those who might find it useful to develop a separate operational plan, the WWF Programme Standards provide more detailed information.[7]

## Staying on Track and Getting the Plan Done: Project Management

Ensuring that the planning process moves as smoothly and effectively through the various components we outlined in the table of contents (see Box 11.1) is the realm of good project management. It is an important skill in its own right. Although team charters, team building, Gantt charts, work plans, leadership considerations, and occasional training courses are increasingly tools of the conservation trade, those of us who studied some aspect of natural resources in school may not realize that entire postgraduate degrees in project management are available. Project management has a dazzling array of resources: institutes are devoted to it, online courses taught on it, thick reference books written about it, certifications available for it, complicated software systems for implementing it, and cartoon strips that poke fun at how bad companies and organizations can be at doing it. It's beyond the scope of this book to offer much advice on project management, but in **Box 11.2** we have tried to capture some points about project management that we have found most useful.

## Box 11.2 Project Management Basics

There is a sea of advice on project management. Here we have distilled a small number of key points that we have found most useful.

- Pay particular attention to building a strong team with good leadership and being clear about the context for planning (Chapter 2). Doing so will help ensure that project management gets off to the right start.
- A detailed work plan that uses something like a Gantt chart (see Table 11.3) to manage all the activities and deadlines will help keep a project on time and within budget.
- Become familiar with the well-known project management triangle: cost, scope, and schedule are the three major constraints to project management. Changes in one of these constraints almost always mean changes in the others. For example, increasing the project scope likely means that it will cost more and take longer to complete. An alternative but similar view of the project management triangle is known as "good, fast, and cheap" (**FIGURE 11.2.1**). The familiar maxims are that you can achieve only two of these constraints at any one time—if you want something fast and cheap, for example, it probably won't have good quality.

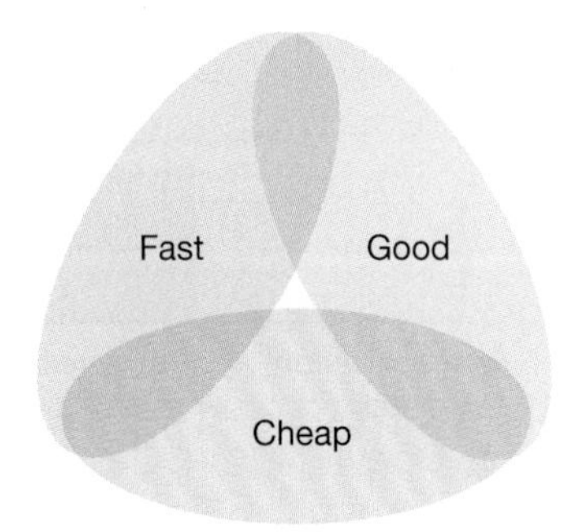


**FIGURE 11.2.1** A diagrammatic look at the project management triangle representing the three major constraints to conducting any project.

Depending on how complicated the project is and on the team's level of experience in project management, it may be worth perusing one of the "bibles" of project management. James Lewis's *The Project Manager's Desk Reference* (2006) would be a good starting point.[a]

[a] Lewis, J., *The Project Manager's Desk Reference.* 3rd ed. 2006, New York: McGraw-Hill.

## The Challenges of Implementation

As far back as the mid-1990s, Emily Talen, now an urban planner at Arizona State University, was encouraging the professional planning community to systematically evaluate why planning succeeds or fails.[8] More recently, Madeleine Bottrill and Bob Pressey have done the same for the community of conservation planners.[9] Although they provided a framework for evaluating the effectiveness of planning and reviewed the evidence to date, the sum of Bottrill and Pressey's review is that a limited amount of information is available on the effectiveness of conservation planning and on its implementation successes. As a discipline, conservation planning has undertaken inadequate self-reflection about whether plans are successfully implemented, and there has been a distinct lack of systematic evaluation of planning.

Despite this shortcoming, anecdotally we understand quite a bit about why plans are or are not successfully implemented. For example, the U.S. Fish and Wildlife Service (USFWS) prepares comprehensive conservation plans (CCPs) for its National Wildlife Refuge units, and in 2012 an internal committee evaluated implementation and other aspects of this planning process.[10] Two of their most important findings were that (1) conservation planners need to develop a strong linkage between broader landscape conservation plans that cover multiple land ownerships and the CCPs; and (2) they also need to develop a standard tracking system for understanding and assessing the degree to which CCPs have actually been implemented.

Later in this chapter, we address this topic of moving from broader landscape or regional plans to finer-scale plans such as those of CCPs. This challenge is clearly not restricted to the USFWS. In a parallel study of CCPs conducted by Indiana University, the authors noted that most of the refuge plans they analyzed provided extensive descriptions of the biological and physical resources and the threats to them but fell far shorter on the details of prescriptive actions that should be taken to help conserve these resources.[11] We have noted a similar trend in conservation plans in general. Not surprisingly, trained biologists and ecologists will go to great lengths to describe what they know best—the ecology of species and habitats and what threatens them—and pay less attention to the more challenging strategies and actions needed to conserve them. It is one of the reasons we have labored in this book to focus so much attention on framing the conservation planning problem and the development of strategies and actions to help solve the problem.

It is important to distinguish between considering why a plan is successfully implemented or not, and why the conservation project or program that is the focus of the plan may fail or succeed. A well-developed plan may be successfully implemented, yet the project itself may be somewhat less than successful. A well-executed plan might fail because the strategies and actions didn't have the anticipated impacts. For example, we are aware of a freshwater conservation project in which upstream dam operations were altered to more closely resemble a natural flow regime downstream that would benefit a migratory fish. Unfortunately, the initial efforts in this project to improve stream conditions for fish migration had the opposite effect—the fish retreated farther downstream (subsequent research has demonstrated that an incoming weather front was likely responsible for the downstream fish migration). This example notwithstanding, there are often close ties—and lessons to be learned—between successes or failure at plan implementation and success or failure in the project itself.

Gunnar Keppel and colleagues reviewed why conservation efforts in Pacific tropical countries have a poor track record.[12] They concluded that limited engagement of private landowners (who are important for conservation in the region), a poor understanding of social and cultural dynamics of local communities, and limited collaboration between conservation organizations, government agencies, and academicians were some of the most important reasons for failed conservation efforts. These shortcomings could potentially be traced to problems in conservation planning related to inadequate engagement of stakeholders, a limited social assessment, and weak partnership development.

Along similar lines, a group of U.S. Geological Survey (USGS) scientists[13] examined 11 broad conservation partnerships and programs spanning 29 countries. They identified several important challenges to these broad-scale conservation programs that are relevant to implementation:

- Staff turnover and communication across partner organizations, as well as continued funding of the partnership
- Reaching agreement across diverse partners with varying missions on the fundamental objectives of the program
- Mismatch across partners between policies agreed to on paper and actual implementation on the ground
- Making implementation decisions across a complex group of stakeholders and political jurisdictions

As suggested by the USGS example, there is considerable evidence and a growing body of literature that limited collaboration with important stakeholders will undermine the success of planning efforts. An editorial in the journal *Conservation Biology*[14] compellingly makes this point, arguing that collaborative engagement with people who have a stake in the outcome of conservation is an essential ingredient of success (**FIGURE 11.4**). Success involves not only appropriate engagement across stakeholders who often strongly disagree but also a serious and transparent commitment to implementation once agreements are reached. And unless those involved in leading the planning process understand the social and cultural context they are working in, successful engagement is unlikely. European landscape planners have reached similar conclusions about the importance of collaborating with stakeholders (see, e.g., Opdam et al.[15]), especially for achieving conservation at local scales.

Our own evaluation of planning implementation in The Nature Conservancy, as well as informal discussions with planners in governmental natural resource agencies, suggest a few additional challenges:

- Many plans are too long and too detailed to be understood, read, or supported by stakeholders.
- Too few plans give serious thought to operationalizing them—what financial and staff resources are really needed to bring a

**FIGURE 11.4** A stakeholder workshop. These participants are both learning about a conservation project and providing input into its design and implementation. There is growing evidence that collaborating with a range of stakeholders in a conservation project can be a key to its success and that failing to do so can undermine the ability to successfully implement a project.

plan to action? In some cases, the strategies and actions may be legitimate based on the best available science but impractical to implement if they are too costly.[16]

- Many plans are never completed and are missing key components such as a prioritizing of strategies and actions, a fund-raising component, or a monitoring and evaluation section. It's easy to understand why these partially completed plans may fall short on implementation.
- Some plans lack engagement from agency or conservation organization program leaders. Especially when these plans suggest a new strategic direction or emphasis or they attempt to answer questions in which managers have shown little interest, this lack of engagement may result in limited support for implementation.
- Many conservation plans fail to fully articulate the conservation problem(s) they are trying to address—that is, they fail to establish a set of measurable objectives and develop a set of actions that can be prioritized to reach those objectives.[17] Some planning efforts compound this problem by having long lists of strategies and actions with no sense of priorities.

Finally, it's worth mentioning that getting conservation work done is challenged by the interplay of human and ecological communities, and all the uncertainty they bring. As one colleague noted to us, great plans and even significant funding cannot necessarily overcome the chaos that ensues from frequent turnover in partners. It's these less obvious factors that make achieving some measures of success more difficult in some circumstances and challenge the best of practitioners to work with and not be overwhelmed by the level of complexity. Still, many practitioners and projects overcome these challenges, and in the next section we focus on some of their success factors.

## Lessons Learned in Successfully Implementing and Managing Conservation Projects

Daniel Mazmanian and Paul Sabatier,[18] two public policy scientists from California, analyzed the successful implementation of the California Coastal Commission's land use planning and then evaluated the most important criteria from that effort and several other case studies. They concluded that six criteria determined implementation success:

1. Clear objectives
2. Links between objectives and actions (theory of change, in our language)
3. An agency with adequate resources and authority
4. Skilled implementation managers
5. Stakeholder support
6. A supportive socioeconomic and policy environment

More recently, Chris Joseph and colleagues[19] analyzed those factors leading to implementation success of several broad-scale land use plans in British Columbia. Through systematic interviews of implementers, they concluded that 18 criteria contributed to implementation success. Although their study was limited in sample size, they reached some of the same conclusions that Mazmanian and Sabatier had. Among their most important success factors were a clearly understood problem, strong leadership, adequate resources and authority, clear role of stakeholders, and a sound monitoring program.

Building on the key success factors elaborated in these case studies as well as our experiences in developing conservation plans, we can make some recommendations for increasing the likelihood of successfully implementing a plan. Although it is not exhaustive, here is our "top 10" list:

1. *Engage senior management and project directors in developing a conservation plan from its inception.* Plans need to address problems or questions that managers and directors perceive as being important to their agency or organization. Those that do are more likely to garner implementation support from those same leaders.

2. *Have an experienced, strong leader in place to direct the planning effort* (see Chapter 2 on building an effective planning team). Strong leadership may be even more important in implementing a plan where patience, focused attention, and the relentless pursuit of goals often make the difference between failure and success.
3. *Don't disconnect implementation of the plan from development of the plan.* Make implementation a core part of the planning process from beginning to end, and help make sure the team "keeps its eyes on the prize" of working to achieve the project's fundamental objectives.
4. *Evaluate a range of alternative interventions and actions, but also ensure that you have assembled the right expertise to evaluate the feasibility of these interventions.* It's particularly important to estimate the costs of these alternatives (Chapter 7); too many plans fall short because they give inadequate attention to the cost of implementation.
5. *Engage stakeholders.* Most successful models of implementation point to the importance of stakeholder support.[20] A vast body of literature over the last 20 years has addressed the topic of collaborative (or participatory) planning that goes well beyond planning for conservation or natural resource management (**Box 11.3** gives an example that includes participatory planning).
6. *After evaluating alternative strategies* (Chapter 6), *try to pinpoint a small number of high-priority ones the project can undertake that will likely make a real difference in progress toward a project's fundamental objectives.* Steady progress toward a project's ultimate objectives by aiming for intermediate objectives will help ensure that the project continues to be successfully implemented. Conservation rarely happens overnight.
7. *Pay attention to the planning context* (Chapter 2). Why did your team undertake this planning effort, who wants the plan, what decisions will be made from it, and who will make those decisions? As simple as these ideas are, too few plans give these points sufficient attention.
8. *Tailor communication to key audiences.* Any particular conservation plan has many different audiences—the team, senior staff, donors, partners, stakeholders, and others. If a team wants these different audiences to pay attention to a conservation plan, it needs to design communications with these audiences in mind. We provide more advice later in this chapter on communicating about a conservation plan.

## Box 11.3 A Conservation Planning Success Story—Marine Protected Area Network in North-Central California

In 2009, the California Fish and Game Commission established a new marine protected area (MPA) network in north-central California (**FIGURE 11.3.1**). The network included 22 new MPAs covering 20% of the planning region in relatively strictly protected areas.[a] This network was the result of a highly successful planning effort that involved public, private, and nongovernmental organization (NGO) staff members and engaged a wide range of stakeholders.

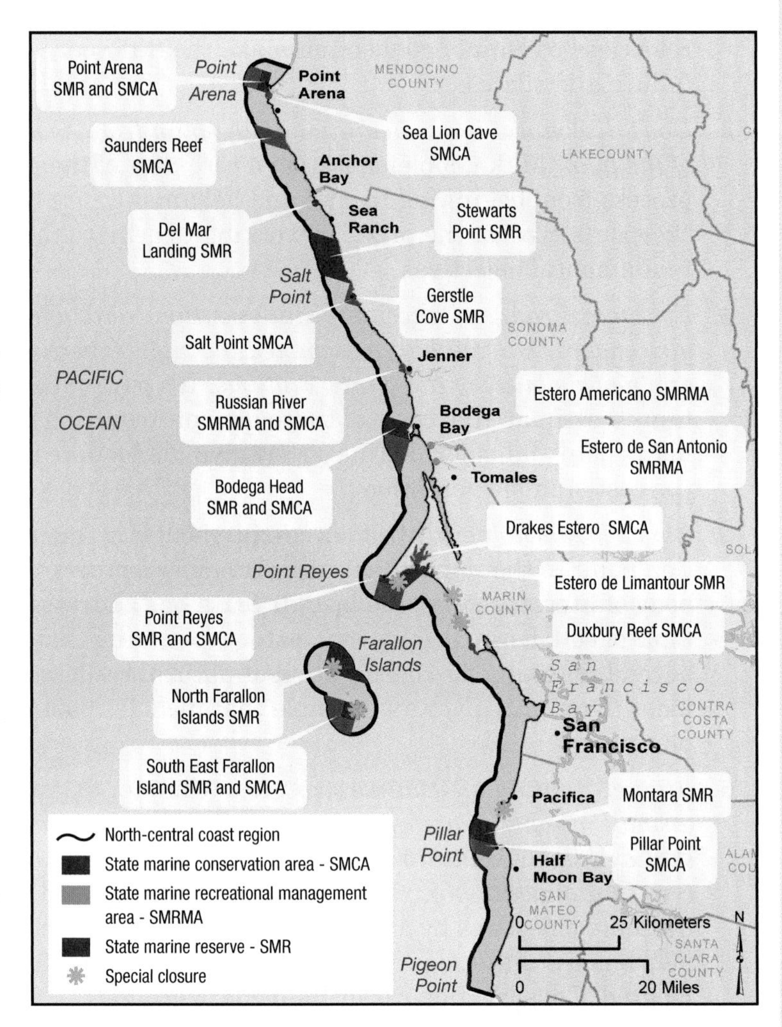


Fortunately, marine biologist Mary Gleason and her colleagues documented in detail what made this planning effort successful, and there is much to be learned from it. From their detailed analysis, it is clear that several important factors contributed to the successful establishment of California's first major network of MPAs:

- *Clearly defined roles and responsibilities in MPA planning and implementation.* Underlying legislation (the Marine Life Protection Act) and a public-private political body with clout (Marine Life Protection Act Initiative) were responsible for making this happen.
- *Facilitated stakeholder engagement and transparent public process.* Broad, deep, consistent, and professionally facilitated stakeholder engagement with clearly documented outcomes helped build trust across stakeholders.
- *Clearly defined goals and objectives.* Lessons learned from previous MPA planning processes helped establish mutually acceptable goals and objectives early in the process.
- *Strong science underpinnings and decision support system.* Scientific underpinnings of MPA design are strong, and the full depth and breadth of this knowledge was brought to bear by an experienced and credible set of scientists in this planning process.

- *The design of alternative networks through consensus-based approach.* Stakeholders themselves, with input from a scientific advisory group, developed an iterative set of alternative MPA network designs that included social-economic considerations.
- *Transparent decision making.* A significant investment in transparency throughout the process helped keep stakeholders engaged and built support for the final outcome.
- *Underlying legislation that supported the planning process.* A strong piece of legislation provided the need for the MPA planning process and the establishment of the network; it was a critical factor to its success and to having the resources to conduct the process.
- *Well-resourced and well-funded planning effort.* Without staff time and considerable financial resources to invest in the planning process, the final outcome—which had wide public support—might not have been reached.

---

[a] Gleason, M. et al., *Science-based and stakeholder-driven marine protected area network planning: A successful case study from north-central California.* Ocean & Coastal Management, 2010. **53**(2): 52–68.

9. *Monitor for results.* We have devoted an entire chapter to this topic (Chapter 12), but it's enough to say here that a plan that monitors its progress toward objectives and evaluates the effectiveness of its strategies to achieve these objectives is almost certainly going to be more successful than one that doesn't.
10. *Design products and processes from conservation planning to engage decision makers.* In the conservation community, there is a rule that all conservation is local. Although this "rule" clearly has exceptions, it certainly applies to a large proportion of conservation decisions and actions taking place at the levels of local and community government.

Few projects have given recommendation 10 as much thought as the STEP project—Subtropical Thicket Ecosystem Planning[21] in South Africa. For example, the team developed a *STEP Mapbook* (FIGURE 11.5) and a *STEP Handbook* providing municipal-level decision makers with fine-scale maps of proposed conservation areas, along with detailed management recommendations for these areas.[22]

More recently, this same project team used social marketing techniques to interview politicians in municipalities of the Subtropical Thicket ecosystem (decision makers) and improve the "products" from the planning exercise that are delivered to policy makers.[23] **Box 11.4** provides an additional example of designing conservation planning products to be effective with specific decision makers—in this case, managers of Danish forests.

Our suggestions, as simple as they may be in some cases, will help to increase the likelihood of successfully implementing any

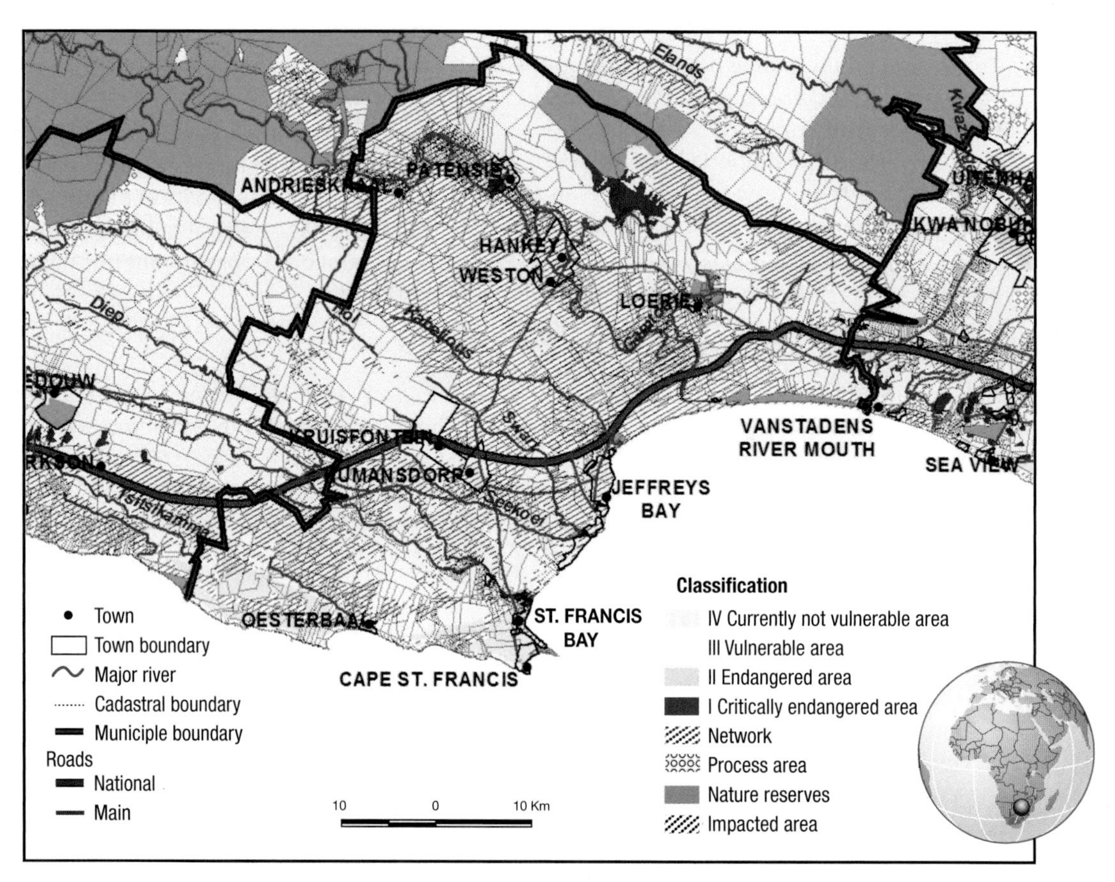


FIGURE 11.5 Detailed map of a potential conservation area in the Subtropical Thicket Ecosystem Plan (STEP). Stakeholders and decision makers in the STEP process were provided with a *STEP Mapbook* of detailed maps, like this one, that provided critical information necessary to implement conservation actions at these local scales. (Adapted from Knight et al., 2011.)

conservation planning effort. At the same time, implementation can proceed and succeed in many different ways. Ultimately, we want implementation to have as much conservation impact as possible.

## Three Implementation Approaches for Greater Conservation Impact

In this section, we broadly highlight three different approaches to achieving successful implementation beyond the boundaries of any single conservation project or initiative: (1) leveraging strategies and actions for systemic impacts; (2) hitting the "conservation sweet spot," where the results of one conservation planning effort intersect with other similar efforts; and (3) taking advantage of various opportunities.

## Box 11.4 Management of Danish Forests

In 2002 the Danish government decided to explicitly manage Danish state forests for multiple objectives, including conservation and recreation in addition to timber production. To that end, they implemented an approach called "Close-to-Nature Forest Management."[a] The approach was designed to provide better habitat for native plants and animals, and allow a range of recreational opportunities while also improving the sustainability of the forestry economy, especially in the face of storms and the impacts of climate change.

The key attributes of Close-to-Nature Forest Management are forest stands that contain mixed age classes of native, site-appropriate species compositions, and avoiding clear-cutting. This approach represented a major shift in the management of state forests, which previously were managed through clear-cutting stands that were of uniform age and species. Consequently, implementation of Close-to-Nature Forest Management plans proved difficult among forest managers. To address this issue, the government took the long-term objectives for forest stand structure and dynamics and created illustrations to reflect these. This resulted in illustrations of 19 forest development types (FDTs), such as that in FIGURE 11.4.1 for beech forests.

Each FDT was then matched to a set of site characteristics based on traits such as geology and soil types, nutrient and water supply, and drainage. The combination of these illustrations and matched site characteristics allowed forest managers to better understand what the outcomes they were trying to achieve looked like, and therefore what was needed to implement the plan effectively. Because Close-to-Nature Forest Management laid out a long-term vision for Danish forests, additional illustrations were developed that represent the process of conversion that forests were likely to undergo as they transitioned from traditional forestry to the new model. For example, Figure 11.4.1 illustrates one of 19 Forest Development Types being managed by the Danish forestry agency with a close-to-nature method that in this case focuses on maintaining a native beech-dominated forest for both timber production and biodiversity conservation purposes.

[a] Larsen, J. B., *Close-to-Nature Forest Management: The Danish approach to sustainable forestry,* in *Sustainable Forest Management: Current Research,* J. M. G. Garcia and J. J. Diez Cas, editors. 2012, InTech (open access book). 478.

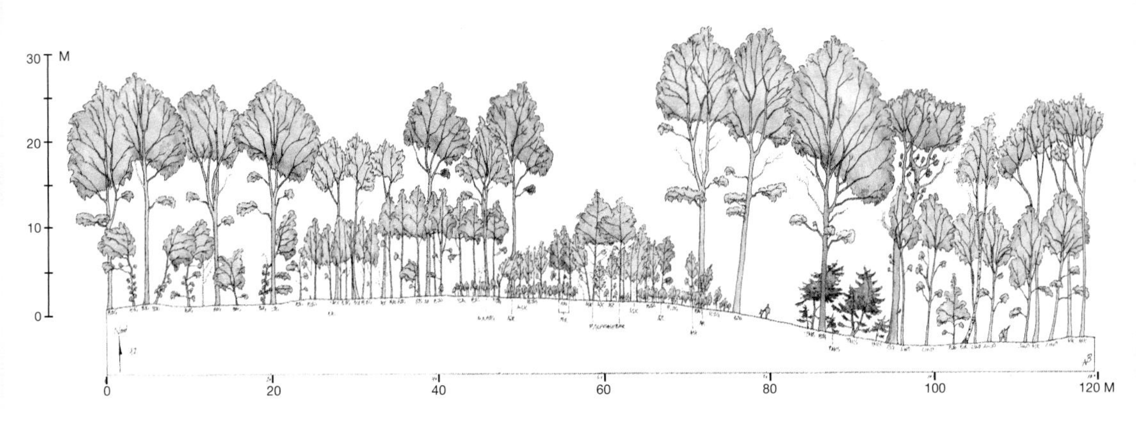


FIGURE 11.4.1 Forest development type 12: Beech with ash and sycamore.

## Systemic Impact for Conservation—Leveraging and Scaling Up

Much successful work of nature conservation is local and specific—one local issue at a time. Ideally, as planners we want to achieve systemic change with our conservation plans and strategies. When systemic change has happened in conservation, it's usually the result of implementing a plan, and specifically a strategic option or action at one scale (that might be a place, or it could be a policy) that has positive impacts over a much larger scale. It's easy to talk about this sort of leverage or upscaling of conservation, but it turns out to be very hard to do.

Some colleagues from The Nature Conservancy have given this issue considerable thought, drawing on models and frameworks from the social sciences for spreading an innovation or idea.[24] They suggest that there are three basic scaling approaches in conservation:

1. *Product orientation.* A successful program, tool, method, or idea gets passed on to a targeted group of users.
2. *Partnership orientation.* Different parties work together to tackle a conservation problem, which in turn results in the solution being disseminated across multiple organizations.
3. *Influence orientation.* The knowledge or solution to a problem exists, but certain impediments must be removed or conditions need to change before the knowledge or solution can have impact.

Product orientation can occur through replication—the copying of a model, idea, or program that has been effective in one place and could be effective in others. For example, conservation work with the U.S. Army Corps of Engineers to improve water flows below a dam for endangered mussels and fishes on the Green River in Kentucky, USA, led to a nationwide program to improve flows at a pilot suite of Corps-operated dams[25] (**FIGURE 11.6**). Product orientation can also occur through dissemination of an idea via presentations, articles, and other media. The rapid proliferation of "water fund" projects in Latin America and beyond,[26] a type of PES (Payment for Ecosystem Services), is an example of scaling through dissemination.

Partnership orientation might occur via two different avenues—networking or strategic affiliations. Through networking, different parties come together for the purpose of achieving some common conservation good. The Conservation Measures Partnership and the Conservation Coaches Network (CCNet), working together to improve the *Open Standards,* referenced many times in this book, is a good example of a partnership designed to spread best practices for the planning, implementation, and management of conservation projects and programs.

FIGURE 11.6 Green River Lake Dam, Kentucky. The Nature Conservancy worked with the U.S. Army Corps of Engineers to change the operations of this dam so that flows below the dam were more likely to conserve endangered freshwater mussels and fishes. The work at this individual dam eventually grew into a much larger partnership with the U.S. Army Corps to improve the ecological sustainability of many of its dam operations across the United States. This is an example of scaling up a conservation project and strategies that are effective at one site and implementing them at multiple places.

The European Union's Natura 2000 initiative[27] is both a network and a strategic affiliation with governments in Europe (FIGURE 11.7). It is a network of protected areas aimed at conserving some of Europe's most threatened species and habitats. Strategic affiliation is a partnership formed around a specific set of deliverables such as the Natura network of protected areas.

Another example of strategic affiliation is the formal partnerships developed by the Wildlife Conservation Society with several African countries such as Gabon. The society assists these countries in establishing, designing, and managing their national parks. In many parts of the world, partnerships developed around connectivity or corridor initiatives are great examples of strategic affiliations. See *Linking Australia's Landscapes* (2013) for specific case studies.[28]

Finally, influence orientation refers to changes in policy, markets, or campaigns for change that all can bring about scaling in conservation. One of the best examples of this sort of scaling is the World Wildlife Fund (WWF) Market Transformation Initiative.[29] Through this initiative, WWF works with major companies around the world "to change the way key global commodities are produced, processed, consumed, and financed" and to reduce their impacts on the

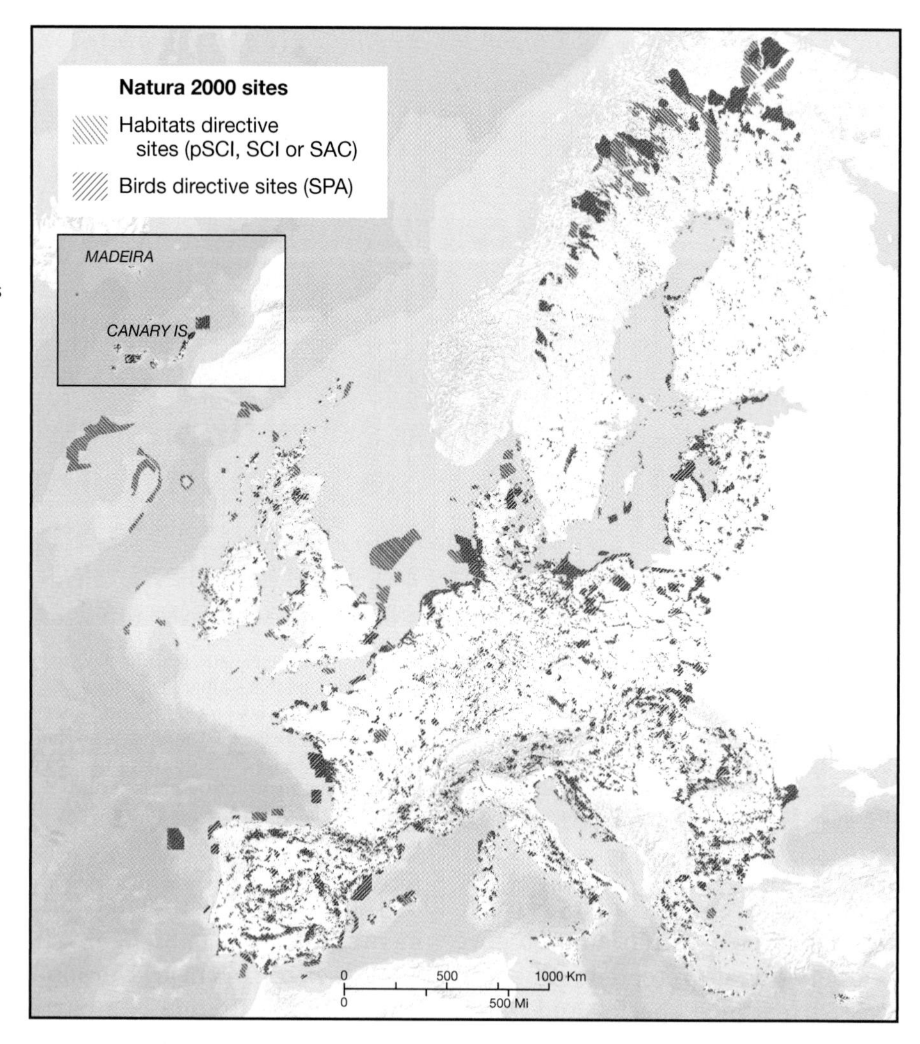


**FIGURE 11.7** Natura 2000 protected areas. This network of protected areas, established by policy of the European Union (EU), focuses on ensuring the long-term survival of Europe's most threatened species and habitats. The Natura 2000 network currently consists of over 22,000 sites covering over 1 million km$^2$. (Data from European Commission Environment.)

environment. Their partners include corporations from the fishing, forestry, aquaculture, agriculture, and bioenergy industries.

## Hitting Nature's Sweet Spot—Integrating Conservation Planning with Land Use and Landscape Planning

Tennis players are well aware of the "sweet spot" in their rackets when they hit a powerful and effective stroke. In conservation planning, the sweet spot involves connecting efforts that identify important places and strategies to conserve nature with the policy and legal framework of land use planning that would provide a strong enabling and implementation environment.

Conservation planning from the perspective of conservation biologists involves identifying important places to conserve and the strategies for achieving outcomes (the "where" and "how" of conservation planning mentioned in Chapter 1). Professional planners who often work for some form of government are generally involved in land use or landscape planning, which focuses primarily on where and what types of development should take place. Their planning efforts are to some degree the mirror image of efforts by conservation planners.

Unfortunately there has been limited professional overlap, either in collaborations or educational curriculum, between the traditional conservation (biology) planning community and the professional land planning community—although there are now at least a few examples of attempts at this kind of collaboration.[30] Conservation planner and landscape ecologist Frank Davis has used word clouds to demonstrate the overlap and lack of connectivity between these disciplines (FIGURE 11.8). He also notes how connecting these two communities of

FIGURE 11.8 Word clouds about planning created using Wordle.net. Figure 11.8(a) is taken from a chapter in a book focused on conservation planning. Figure 11.8(b) is from a more traditional land use planning handbook. (Adapted from Davis, 2013.)

planners will help make the type of planning we outline in this book more relevant to the broader world and presumably more influential in making conservation decisions.

In the USA, the Environmental Law Institute published a compelling set of essays on how conservation science should play a more significant role in land use planning processes.[31] One of the best examples of this connection was planning conducted by Carl Steinitz of Harvard University for the Camp Pendleton region, a U.S. Department of Defense installation in the midst of highly developed San Diego County, California.[32] In another example, the Wildlife Conservation Society has worked closely for a long time with local governments in Fiji in the planning, design, and management of a network of marine protected areas,[33] especially to make the network more resilient to the impacts of climate change.

Scientists have long observed that conservation areas and efforts are disproportionately located in places that are largely unsuitable for development.[34] Conversely, they have noted that many of the world's most endangered habitats and species occur in those areas that have high human population numbers and high levels of development. In these areas where development and conservation collide, future collaborations between more traditional conservation planning efforts and land use planning processes will be critical for successful conservation. Arguably, as the globe becomes more urbanized and cities grow in political influence, mainstreaming the practice and practitioners of conservation planning as outlined in this book into public planning processes may be one of the most effective overall approaches for promoting and achieving nature conservation.

Better integration of traditional conservation planning with land use planning methods could also help to elucidate the trade-offs between conservation and development, and between conservation and human well-being.[35] Despite much debate and discussion about win-win solutions for conservation and development, it appears that trade-offs between conservation and development are far more common than win-win solutions. As international experts on conservation and development, Tom McShane and colleagues have acknowledged that "hard choices" have to be made and that the conservation and development communities "need to engage in a social process that allows for compromise and explicit acknowledgment of risks and costs, while at the same time gaining ever more clarity and purpose regarding those things that should not be traded off."[36] Throughout this book, we maintain that the evaluation of alternative strategic options, the analysis and acknowledgment of trade-offs, and a participatory, transparent planning process advocated by McShane and colleagues are some of the most important ingredients to successfully implementing conservation plans and initiatives.

## Being Opportunistic

For one of us, there was a time in our career when conservation planning was seen as a principal way to help move an organization away from a somewhat unfocused (and at times referred to pejoratively as "opportunistic") approach to achieving its biodiversity mission to a more systematic and presumably efficient and effective way of doing business. Decades later, the irony of including some thoughts on the importance of being opportunistic in conservation planning may be hard to miss. But the skill of knowing when to seize opportunities to achieve conservation and even to realign conservation plans around those opportunities should not be underestimated.[37]

Who would have known that Michael Fay's "megatransect" expedition (FIGURE 11.9) across the Congo Basin of Africa and the publicity it received through *National Geographic* magazine and television coverage would lead the president of Gabon to designate more than a dozen new national parks in his country? When The Nature Conservancy and the Sonoran Institute developed the Sonoran Desert Ecoregional Plan, they couldn't have anticipated that then U.S. president Clinton's administration might use information in that plan to create several new national monuments as part of the National Landscape Conservation System in the United States. And we often have heard stories about once-in-a-generation opportunities that arise when a landowner

FIGURE 11.9 Dr. Michael Fay of the Wildlife Conservation Society spent 455 days hiking over 2000 kilometers (1200 miles) across the Congo Basin (a megatransect) to document the status of wildlife and ecological conditions there. Results of those surveys helped convince the president of Gabon to create 13 new national parks in his country.

decides to sell his or her land to a conservation-minded party or bring it under protection through other means.

Our point here is that unexpected opportunities can sometimes cause us to change the priorities in our conservation plans. At times, these opportunities will lead to big conservation gains that may never be realized through other means, and they are also good reminders that the whole purpose of plans is to have more informed decisions that help advance conservation.

Such informed decision making from the context of conservation planning was first called "informed opportunism" by conservation planner Reed Noss and colleagues.[38] Examining the trade-offs in conservation sites within the Greater Yellowstone region with a vulnerability-irreplaceability framework, they correctly observed that in the real world, the most vulnerable and irreplaceable sites aren't always the ones that we can act on. Moreover, they noted that conservation practitioners need to be prepared to take advantage of opportunities to conserve places that might be, for example, irreplaceable in terms of biodiversity but not particularly vulnerable today.

Nearly a decade later, one of us (EG) developed a similar framework of informed opportunism while working with stakeholders on a conservation plan for the Solomon Islands.[39] The changing opportunities that may be present for many planning processes are a good reminder that plans shouldn't be seen as static documents and that revisiting our theory of change is a necessary part of implementation to ensure we are being as effective as possible.

Several articles in a special section of the December 2014 edition of the journal *Conservation Biology* more formally recognize the importance of conceptualizing and assessing conservation opportunity as part of the planning process in conservation and natural resource projects.[40] For example, Katie Moon and colleagues have recognized three different types of conservation opportunities:

1. *Potential.* Parties actively remove barriers to problem solving within the social-ecological system of a conservation project.
2. *Traction.* Parties take advantage of a change to the situation or context of a conservation project such as a natural disaster to move help advance conservation actions.
3. *Existing.* Opportunities may exist, but they have not yet been identified by a conservation team.

Hurricane Katrina in the southeastern United States is a prime example of a traction opportunity in which billions of dollars are now being made available for ecological restoration, mitigation, and other types of conservation actions.

Conservation practitioners may better be able to evaluate and recognize conservation opportunities by using a structured process of **strategic foresight** and several tools associated with this process such as **scanning**, scenario planning (previously discussed in Chapters 8 and 9), and **backcasting**.[41] Scanning refers to collecting a diverse amount of information on a conservation topic or topics from a variety of sources in relatively short amounts of time; many readers may be familiar with annual articles that are published on a "horizon scanning" of new conservation issues.[42] Backcasting involves starting with a conservation goal or objective and working backward to break conservation problems and barriers down into smaller pieces that are more readily addressed. Carly Cook and colleagues[43] discuss these tools in more detail and apply them to a real problem of climate change and adaptation in the state of New York, USA. These tools and others identified in the special section of *Conservation Biology* will increasingly find a place in the conservation toolbox, and practitioners will likely find it useful to become more familiar with them and begin incorporating them in the conservation planning process.

We have discussed some different approaches that may help practitioners and planners improve the implementation success of conservation plans. The scale at which the strategies and actions behind these different approaches are considered is an important issue in plan development and implementation. In the next section, we consider some important issues concerning the spatial scales at which plans are developed and implemented.

## Moving from Regional Assessments and Plans to Local Actions

Many conservation plans have been developed for a relatively large spatial region—such as an ecoregion, state, district, province, or other geopolitical unit that usually spans tens of thousands of square kilometers. These plans and assessments identify important places on the ground (or in the water) for achieving the plan's goals and objectives, but they often speak little about the strategies and actions necessary to conserve these places. The ecoregional visions and assessments of the World Wildlife Fund and The Nature Conservancy, the U.S.-based State Wildlife Action Plans, and the ecoregional assessments of Australian state governments are examples of such plans. In many cases, organizations then prepare more detailed strategic plans—usually at the smaller spatial scale of a conservation project—that outline what actions will be taken to conserve the features of the important places that were identified at the regional scale.

In Chapter 6 we discussed some of the problems that are created by answering the "where" and "how" questions through separate planning processes.[44] We observed that by separating these critical questions, we have painted ourselves into a corner thanks to an ill-defined conservation problem. By contrast, a well-defined conservation problem has some objectives (e.g., brown bear [*Ursus arctos*] persistence in isolated mountain ranges of Europe); a well-defined set of options (e.g., reducing human-caused mortality of brown bears); and a prediction of the consequence each option will have for the objectives. Regional assessments or plans that identify only the places of importance don't give us the critical information we might need to conserve those places. As a result, these plans are often unable to adequately set priorities among hundreds of possible sites for conservation action.

At the same time, the reality is that it's difficult to know what actions are needed to conserve a place without more detailed planning at individual prospective conservation areas. At the smaller spatial scale of many individual conservation areas, often more detailed data are available for conservation planning than at broad regional scales where data either are not available or are unaffordable or impractical to use at such broad scales. For example, cities or local governments may have more detailed data on property boundaries, land costs, human demography, and road densities or even fine-scale vegetation maps and locations of rare and endangered species than are generally feasible or available to use in regional planning exercises.[45] It may also be more feasible to assemble experts on more local scales to gather additional information on threats to biodiversity features. All of these additional data can help inform the development of strategies for more localized actions.

The take-home message is that regional-scale plans can be implemented at a variety of scales depending on the strategies being pursued. However, when some of these strategies involve more local actions at the scale of municipalities or local government, it may be necessary to adapt and refine the regional plan because the priorities of places will change through time as actions are taken at some places and not at others, or as additional data and information become available. We appreciate that additional data gathering and planning will be needed in many cases to develop conservation strategies at more localized scales than those for which a conservation plan was originally developed.

Regardless of the scale involved, no plan is likely to be successfully implemented without effective communication about the plan and the conservation project or program it hopes to advance. In the final section of this chapter on implementation, we discuss the important topic of communication and suggest how to most effectively talk and write about a conservation plan and project in a way that will enhance opportunities for implementation.

## Strategic Communication for Implementation

In the United States, beliefs about climate change can be predicted along ideological lines, roughly depending on whether someone is politically liberal or conservative. Dan Kahan, a law professor from Yale University, and his colleagues[46] initially hypothesized that as people gained more knowledge of climate change, the differences along ideological lines would markedly decrease. In fact, the opposite occurred: the more people knew, the more likely they were to diverge in beliefs based on ideology. A climate denier who gains more knowledge of climate change is likely to become a more committed climate denier rather than suddenly getting concerned about climate change. That result might seem counterintuitive, but in reality it has a firm basis in psychology and sociology. Intelligent, scientifically literate people are highly adept at finding the evidence that fits their worldview.

What does this have to do with conservation planning? Often scientists believe that if they can just supply people with enough facts, they will be persuaded. Evidence suggests otherwise. You need a full communications toolkit that takes into account peoples' values and beliefs. Otherwise, you will be convincing only those who are already convinced.

Scientists and conservation planners aren't always the most effective verbal communicators, especially to nonscientific audiences. Who among us hasn't seen the nearly mind-numbing PowerPoint image loaded with bulleted phrases or a similarly indecipherable graph? From the perspective of a conservation plan, anything short of effective communication can be problematic for several reasons. First, if the constituents for a conservation plan don't clearly understand it, they will interpret it in their own ways—and often not the way the planners and project directors or managers intended.

Second, most conservation plans have several audiences. And unless we fine-tune our messages for these different audiences, we may not be as effective as we otherwise could be. For instance, most of the constituents for a conservation project and its plan don't need to see the whole plan. Indeed, an executive summary or "reader's version" will work well for most audiences, and in some cases you may even need to customize a version of a summary for select audiences.

Finally, as we have observed throughout this book, conservation plans that are going to be effective in achieving conservation will be based in part on scientific information and analyses that support certain strategies and actions. And therein may lie both a communication challenge and an opportunity (how best to talk about the scientific information and analyses). Fortunately, a considerable amount of attention is now being focused on effective science communication. Much of the advice and expertise on this subject should be helpful to conservation planning teams that are in the midst of communicating about and implementing a conservation plan.

The following is a summary of what we believe are some of the most important pieces of advice for communicating about your conservation plan and project. We hope that by incorporating some of these points into your plan, you'll find the intended audiences will want to help you implement it.

1. *Understand who your audiences are, analyze them if necessary, and craft targeted communications for each of them.* The fishermen who are going to be restricted from fishing in a certain part of the new marine protected area network that your team just helped create aren't going to want to hear the same messages as the governor of the state or province whose new laws and policies called for the establishment of this MPA network. At the same time, it's important to appreciate that different audiences (Chapter 2) will come to the (implementation) table with different sets of knowledge and information. Crafting effective communication requires understanding what your audiences do and do not know, and then filling in important gaps.
2. *Highlight the information in your plan that is relevant to decision makers who could influence its implementation.* But also understand that different types of decision makers may need or want different types of information, and your team will need to do some research on what information will be most useful to which decision makers. If your project is working with a First Nation in Canada to create a new protected area, the government of the province or territory may want to know about, for example, which threatened or endangered species occur in the area or some mining opportunities that may be precluded for a large, influential multinational corporation. On the other hand, maintaining certain subsistence fishing or hunting rights may be the most important priority that comes out of a plan for the First Nations people involved. Neither the First Nation nor the province government may need to know about the detailed life history of particular boreal forest bird species.
3. *Conservation plans are as much about values as they are about science—it is important in communication to distinguish between the two and acknowledge value differences.*[47] Wolves were successfully reintroduced to Yellowstone National Park, USA, in 1995 and have been expanding for years to other public and private lands outside Yellowstone. Many of our friends and colleagues in the conservation and natural resource community appreciate and value seeing wolves and observing their

populations recover, but not all stakeholders in the Greater Yellowstone ecosystem hold those same values—especially some hunters and livestock ranchers. Even though strong laws may exist that mandate the recovery of wolves in the region, a failure to understand and acknowledge the different values of various constituents regarding wolf conservation has been a major impediment to wolf recovery in the region.

4. *Use communication to build trust in the science that underpins your conservation plan.*[48] To help build this trust, report scientific facts or results in your conservation plan in an open and transparent way. At the same time, it's important for the plan and its proponents not to confuse scientific facts (e.g., a population of endangered huemul deer [(*Hippocamelus bisulcus*] in southern Chile and Argentina needs glacial scrublands and beech forests as its primary breeding and feeding habitat) with values (the importance of conserving this habitat for the deer), because this confusion may easily disenfranchise stakeholders.
5. *Take advantage of the whole communication toolbox, not just a couple of tools you might be comfortable with.* Depending on your age and predispositions, you may conclude that effective communication about a conservation plan or program involves giving a great presentation or conducting a great interview with a reporter. Of course you are right—these are effective communication media. But what about blogs, or communicating about your work through social media? There is even advice for how to develop the most effective visualization techniques to help advance a concept, strategy, or some other aspect of a conservation plan (see, e.g., McInerny et al.[49]). The point, of course, is that the communication toolbox holds many tools, and planners need to get familiar with as many as possible and put them to use in appropriate situations.
6. *Tell a good story.* As marine ecologist Heather Leslie and her colleagues observed, good conservation stories can help decision makers understand the consequences of different options and decisions.[50] They related the story of how the father of an early occupant of Cabo Pulmo in Baja, Mexico, switched from fishing to diving tourism after watching the decline of the marine ecosystem. The authors went on to tell how the reef ecosystems recovered, and how that recovery spurred the establishment of a national park. Scientists and conservation practitioners are often in the best position to tell these stories. Today there are professional organizations that can help young scientists learn how to communicate effectively (e.g., COMPASS, in the marine

conservation arena). Along with organizations that can provide communication training, Randy Olson's book *Don't Be Such a Scientist* (2009) has some good advice and anecdotes on what it takes for a scientist to tell good stories. He writes about how not to be too cerebral (coming across too brainy) or too literal (too focused on the science facts).[51] Additional advice on telling good stories can be found in Andy Goodman's books.[52]

For more detailed guidance on communication, it's hard to beat Susan Jacobson's *Communication Skills for Conservation Professionals*.[53] For a more analytical look at how scientists can improve the effectiveness of their communication, *The Proceedings of the National Academy of Sciences* published a series of articles in 2013 and 2014 that are well worth your time.[54] Finally, from the perspective of communicating about the scientific analyses and information within a conservation plan, the advice for scientists in Nancy Baron's *Escape from the Ivory Tower* (2010) ought to be required reading.[55]

Throughout this chapter, we have implicitly or explicitly noted the importance of communication in successfully implementing a conservation plan. Its importance and our advice might best be summarized in a quote from a colleague we interviewed about improving conservation planning methods: "It's not just about changing planning methods, but it's about how to get the right information, to the right people, at the right time, in a format that is right for them." Effective communication in conservation is an acquired skill, and it should never be taken for granted.

## KEY MESSAGES

- Positioning a conservation plan for successful implementation needs to be a priority for a project team from the start of the project—especially at the early stages, when key audiences and stakeholders are identified.
- Although work plans, budgets, and fund-raising may sound like mundane project management tasks, failure to take these tasks seriously can undermine the plan's ability to be successfully implemented. Getting conservation done requires resources, and too many plans are under-resourced when it comes to implementing key strategies and actions.
- Long "laundry lists" of strategies, poor engagement with key stakeholders, limited involvement of project directors, and weak leadership are some of the leading causes of poor plan implementation.

- We provide a "top 10" list of recommendations for improving implementation that includes paying more attention to the planning context (what decisions will a plan make, who will make these decisions, what constraints will affect these decisions—Chapter 2), selecting a few high-priority strategies and actions to implement that could have the greatest impact, and designing planning products and communications that target selected audiences.
- We describe three approaches to implementation that could lead to greater conservation impact: (1) leveraging and scaling conservation actions, (2) integrating conservation planning with land use planning, and (3) taking advantage of conservation opportunities.
- No plan is likely to be successfully implemented without effective communication about that plan and its conservation project or programs. From telling a good story to understanding the difference between science and values, we offer a range of advice on how to communicate in a manner that will increase the chances for successful implementation.

## References


1. Knight, A. T. et al., *Knowing but not doing: Selecting priority conservation areas and the research–implementation gap.* Conservation Biology, 2008. **22**(3): 610–617; Hulme, P. E., *Bridging the knowing–doing gap: Know-who, know-what, know-why, know-how and know-when.* Journal of Applied Ecology, 2014. **51**(5): 1131–1136.
2. Guerrero, A. M., R. R. J. McAllister, and K. A. Wilson, *Achieving cross-scale collaboration for large scale conservation initiatives.* Conservation Letters, 2014. **8**(2): 107–115.
3. Lewis, J., *The Project Manager's Desk Reference.* 3rd ed. 2006, New York: McGraw-Hill.
4. Mikov, M. et al., *Step 3.1 Work plans and budgets,* in *Resources for Implementing the WWF Project and Programme Standards.* 2007, Gland, Switzerland: World Wildlife Fund.
5. Ibid.
6. The Nature Conservancy, *Conservation Business Planning Guidance, 2013.* Arlington, VA: The Nature Conservancy. 153. [accessed September 13, 2015]. http://www.conservationgateway.org/ConservationPlanning/BusinessPlanning/Documents/CBP_Guidance.pdf.
7. Beale, W., M. Maquet, and J. Tua, *Step 2.3 Design operational plan,* in *Resources for Implementing the WWF Project and Programme Standards.* 2007, Gland, Switzerland: World Wildlife Fund.

8. Talen, E., *Success, failure, and conformance: An alternative approach to planning evaluation.* Environment and Planning B: Planning and Design, 1997. **24**(4): 573–587.
9. Bottrill, M. C., and R. L. Pressey, *The effectiveness and evaluation of conservation planning.* Conservation Letters, 2012. **5**(6): 407–420.
10. Planning Team, *Draft Final Report.* 2012, Washington, DC: U.S. Fish and Wildlife Service.
11. Meretsky, V. J., and R. L. Fischman, *Learning from conservation planning for the U.S. National Wildlife Refuges.* Conservation Biology, 2014. **28**(5): 1415–1427.
12. Keppel, G. et al., *Conservation in tropical Pacific Island countries: Why most current approaches are failing.* Conservation Letters, 2012. **5**(4): 256–265.
13. Beever, E. A. et al., *Successes and challenges from formation to implementation of eleven broad-extent conservation programs.* Conservation Biology, 2014. **28**(2): 302–314.
14. Edwards, F. N., and M. L. Gibeau, *Engaging people in meaningful problem solving.* Conservation Biology, 2013. **27**(2): 239–241.
15. Opdam, P. et al., *Science for action at the local landscape scale.* Landscape Ecology, 2013. **28**(8): 1439–1445.
16. Cook, C. N. et al., *Achieving conservation science that bridges the knowledge–action boundary.* Conservation Biology, 2013. **27**(4): 669–678.
17. Game, E. T., P. Kareiva, and H. P. Possingham, *Six common mistakes in conservation priority setting.* Conservation Biology, 2013. **27**(3): 480–485.
18. Mazmanian, D. A., and P. A. Sabatier, *Implementation and Public Policy.* 1989, Lanham, MD: University Press of America.
19. Joseph, C., T. I. Gunton, and J. C. Day, *Implementation of resource management plans: Identifying keys to success.* Journal of Environmental Management, 2008. **88**(4): 594–606.
20. Knight, A. T., R. M. Cowling, and B. M. Campbell, *An operational model for implementing conservation action.* Conservation Biology, 2006. **20**(2): 408–419.
21. Knight, A. T. et al., *Walking in STEP: Lessons for linking spatial prioritizations to implementation strategies.* Biological Conservation, 2011. **144**(1): 202–211.
22. Pierce, S. M. et al., *Systematic conservation planning products for land-use planning: Interpretation for implementation.* Biological Conservation, 2005. **125**(4): 441–458.
23. Wilhelm-Rechmann, A., R. M. Cowling, and M. Difford, *Using social marketing concepts to promote the integration of systematic conservation plans in land-use planning in South Africa.* Oryx, 2014. **48**(1): 71–79.
24. The Nature Conservancy, *Designing for Scale.* 2012, Arlington, VA: The Nature Conservancy.

25. Konrad, C. P. et al. *Evaluating dam re-operations for freshwater conservation in the Sustainable Rivers Project.* River Research and Applications, 2012. **28**(6): 777–792.

26. Goldman-Benner, R. L. et al., *Water funds and payments for ecosystem services: Practice learns from theory and theory can learn from practice.* Oryx, 2012. **46**(1): 55–63.

27. Maiorano, L. et al., *Contribution of the Natura 2000 network to biodiversity conservation in Italy.* Conservation Biology, 2007. **21**(6): 1433–1444.

28. Fitzsimons, J., I. Pulsford, and G. Westcott, eds. *Linking Australia's Landscapes: Lessons and Opportunities from Large-Scale Conservation Networks.* 2013, CSIRO: Collingwood, Victoria, Australia. 320.

29. WWF Market Transformation Initiative, *Better Production for a Living Planet.* 2012, Gland, Switzerland: World Wildlife Fund.

30. Groves, C., *The Conservation Biologist's Toolbox for Planners.* Landscape Journal, 2008. **27**(1): 81–96; Gordon, A. et al., *Integrating conservation planning and land use planning in urban landscapes.* Landscape and Urban Planning, 2009. **91**(4): 183–194; Davis, F., *Summary: Building a broader base for conservation planning,* in *Conservation Planning: Shaping the Future,* F. L. Craighead and C. L. Convis Jr., editors. 2013, Redlands, CA: Esri Press. 397–404; and Theobald, D. M. et al., *Incorporating biological information in local land-use decision making: Designing a system for conservation planning.* Landscape Ecology, 2000. **15**: 35–45.

31. Environmental Law Institute, ed. *Lasting Landscapes: Reflections on the Role of Conservation Science in Land Use Planning,* 2007. Washington, DC: Environmental Law Institute. 89.

32. Steinitz, C., *An alternative future for the region of Camp Pendleton, California.* 1997, Cambridge, MA: Harvard University Graduate School of Design.

33. Weeks, R., and S. D. Jupiter, *Adaptive comanagement of a marine protected area network in Fiji.* Conservation Biology, 2013. **27**(6): 1234–1244.

34. Scott, J. M. et al., *Nature reserves: Do they capture the full range of America's biological dviersity?* Ecological Applications, 2001. **11**(4): 999–1007.

35. McShane, T. O. et al., *Hard choices: Making trade-offs between biodiversity conservation and human well-being.* Biological Conservation, 2011. **144**(3): 966–972; McShane, T. O., and M. P. Wells, eds. *Getting Biodiversity Projects to Work.* Biology and Resource Management Series. 2004, New York: Columbia University. 442.

36. McShane et al., *Hard choices.* (See reference 35.)

37. Radeloff, V. C. et al., *Hot moments for biodiversity conservation.* Conservation Letters, 2013. **6**(1): 58–65.

38. Noss, R. F. et al., *A multicriteria assessment of the irreplaceability and vulnerability of sites in the Greater Yellowstone ecosystem.* Conservation Biology, 2002. **16**(4): 895–908.

39. Game, E. T. et al., *Informed opportunism for conservation planning in the Solomon Islands.* Conservation Letters, 2011. **4**(1): 38–46.

40. Moon, K. et al., *A multidisciplinary conceptualization of conservation opportunity.* Conservation Biology, 2014. **28**(6): 1484–1496.

41. Cook, C. N. et al., *Using strategic foresight to assess conservation opportunity.* Conservation Biology, 2014. **28**(6): 1474–1483.

42. Sutherland, W. J. et al., *Horizon scan of global conservation issues for 2011.* Trends in Ecology & Evolution, 2011. **26**(1): 10–16.

43. Cook et al., *Using strategic foresight to assess conservation opportunity.* (See reference 41.)

44. Game et al., *Six common mistakes in conservation priority setting.* (See reference 17.)

45. Hilty, J. A., W. Z. Lidicker Jr., and A. M. Merenlender, *Corridor Ecology: The Science and Practice of Linking Landscapes for Biodiversity Conservation.* 2006, Washington, DC: Island Press. 344.

46. Kahan, D. M. et al., *The polarizing impact of science literacy and numeracy on perceived climate change risks.* Nature Climate Change, 2012. **2**(10): 732–735.

47. Dietz, T., *Bringing values and deliberation to science communication.* Proceedings of the National Academy of Sciences, 2013. **110**(Supplement 3): 14081–14087.

48. Ibid.

49. McInerny, G. J., et al., *Information visualisation for science and policy: Engaging users and avoiding bias.* Trends in Ecology & Evolution. **29**(3): 148–157.

50. Leslie, H. M. et al., *How good science and stories can go hand-in-hand.* Conservation Biology, 2013. **27**(5): 1126–1129.

51. Olson, R., *Don't Be Such a Scientist: Talking Substance in an Age of Style.* 2009, Washington, DC: Island Press. 206.

52. Goodman, A. *The Goodman Center Resources.* 2015 [accessed January 18, 2015]. http://www.thegoodmancenter.com/resources/.

53. Jacobson, S. K., *Communication Skills for Conservation Professionals.* 2009, Washington, DC: Island Press. 351.

54. Fischhoff, B., *The science of science communication.* Proceedings of the National Academy of Sciences, 2013. **110**(Supplement 3): 14033–14039; Fischhoff, B., and D. A. Scheufele, *The science of science communication II.* Proceedings of the National Academy of Sciences, 2014. **111**(Supplement 4): 13583–13584.

55. Baron, N., *Escape from the Ivory Tower: A Guide to Making Your Science Matter.* 2010, Washington, DC: Island Press. 246.

# 12

# Monitoring and Evaluation for Conservation Impact

## Overview

Monitoring and evaluation (M&E) play a critical role to assess progress in implementing the strategies and actions of a conservation plan, and to evaluate whether a project's fundamental objectives are being achieved. We clarify the sometimes confusing terminology of M&E and adaptive management, and focus on two types of monitoring—performance measurement and impact evaluation. We provide conservation practitioners and scientists with a simple set of steps to develop an effective M&E program along with tools, guidance, and examples. We encourage a practical approach that includes considering whether investments should be made in monitoring, and for what reason. We emphasize the importance of learning and improving projects based on the analysis of monitoring data, and conclude with recommendations for overcoming challenges to effective monitoring programs.

## Topics

- *Monitoring questions*
- *Types of monitoring and evaluation programs*
- *Adaptive management*
- *Steps in monitoring and evaluation programs*
- *Indicator selection*
- *Investment choices in monitoring*
- *Social and ecological metrics*
- *Learning*
- *Overcoming barriers to effective monitoring*

In March 2006, the U.S. Army Corps of Engineers experimented with new environmental flows below a dam on the Savannah River, Georgia, U.S.A. The project was designed in part to assist shortnose sturgeon (*Acipenser brevirostrum*) and other fish species with upstream migration to spawning areas.[1] Radio telemetry monitoring of sturgeon revealed that instead of improving upstream migration, the flows caused sturgeon to move downstream, possibly due to the pulse release of cold water. Initially, scientists recommended that future high-flow pulses should coincide with natural storm events that help raise water temperature. Subsequent research has revealed that an ambient (cold) weather front likely was responsible for the downstream movement of sturgeon, that river volume has no impact on sturgeon movements, and that a water temperature of 10°C is what triggers upstream migration of sturgeon.

This example from the Savannah River of using data to understand the impact of a conservation action is just what we have in mind when we use the phrase "monitoring and evaluation for conservation impact." A new management action was taken to help improve fish migration on a regulated river. Careful monitoring and evaluation of the experiment revealed unintended results, and corrective actions continue to be taken as ongoing research and monitoring helps evaluate the actions needed to achieve objectives. Subsequent land protection efforts to minimize floodplain development have allowed high seasonal environmental flows to continue.[2] This example represents one approach to monitoring and evaluation. In practice, there are a variety of approaches to monitoring and evaluation depending on the particular project circumstances and questions to be answered; we will delve into those in more detail in this chapter.

The Bill and Melinda Gates Foundation, one of the largest philanthropic foundations in the world, has articulated three primary reasons for investing in a monitoring and evaluation (M&E) program, all of them entirely relevant to nature conservation efforts:

1. *Track progress.* Hold a project accountable for what it is doing by measuring inputs, activities, and outputs.

2. *Inform strategies and actions.* Test assumptions and track achievements by measuring outputs, outcomes, and impacts, and understand why a project is succeeding or failing.

3. *Contribute to the field.* Share the results of measuring outcomes and impacts so that there is broader understanding (in the conservation community in our case) of what works and does not work (another way of saying that we should contribute to the body of knowledge on conservation evidence).

Tracking progress and measuring outcomes and impacts to inform strategies and understand whether a project is succeeding or failing are critical to making the informed decisions that are at the heart of the conservation planning process we have articulated throughout this book. In addition, monitoring data are often the basis of status and trend information that contributes significantly to situation analysis (Chapter 5), to the selection of conservation features (Chapter 3), and to the evidence base that a particular strategy or set of actions is likely to succeed (Chapter 6).

Even though an M&E program is integral to conservation planning, conservation organizations and government natural resource agencies have been notoriously poor at gathering, incorporating, and implementing monitoring. In a survey of member organizations of the Conservation Measures Partnership, Matt Muir noted that while 95% of conservation projects conducted by these member organizations intend to carry out adaptive management,[3] only 5% actually followed through with any monitoring or evaluation. Along similar lines, Carly Cook and colleagues, in a survey of over 100 protected areas in Australia, found that few managers used any kind of evidence base to evaluate their actions and that 60% relied on experience alone.[4] Indeed, the scientific literature abounds with articles that report on failures to use data and information to guide conservation actions. In this chapter, we argue that any conservation project, program, or initiative that wants to evaluate or measure its progress in near- and longer-term objectives and outcomes must plan to do so from its inception and consider this task during each component of the planning process.

Unfortunately, many conservation projects never get to the point of monitoring and evaluating progress toward outcomes. There are many reasons for this. Oftentimes, M&E is not budgeted for at all; or it is budgeted insufficiently, and so becomes a low priority. In some cases, the staff lacks the expertise to appropriately design and carry out the monitoring, analyze the results, and report back. At other times, project directors and managers may place a low priority on M&E because they are concerned that its transparency may demonstrate that they are not making adequate progress or perhaps significantly falling short of project goals and objectives. For all of these reasons and more, conservation projects and programs typically fall short of what is desired for M&E. Throughout this chapter, we provide advice to scientists and practitioners on how to buck this trend.

## Approaches to Monitoring and Evaluation

Monitoring and evaluation programs can be designed to answer different types of questions, and all of them can be valuable to the conservation planning process that we have advanced in this book.[5]

Conservation and social scientist Mike Mascia and colleagues have described and compared five major approaches to conservation M&E.[6] Their terminology relies heavily on the literature of program evaluation as practiced by the American Evaluation Association and includes the familiar lexicon of inputs, activities, outputs, outcomes, and impacts (**Box 12.1**).

In the remainder of this chapter, we have adopted the Mascia et al. terminology as our standard to help provide advice on M&E to practitioners in clear, consistent language. **TABLE 12.1** contains detailed information on the five approaches; the following are brief definitions:

1. *Ambient monitoring*—the process of systematically observing the state of social and/or environmental conditions over time, independent of a conservation intervention. Ambient monitoring is often referred to in other literature as "status assessment," "status monitoring," or "surveillance monitoring" (**FIGURE 12.1**).

**TABLE 12.1** Different types of monitoring and evaluation (M&E) programs and the types of questions, data, and audiences for each

| | Ambient monitoring | Management assessment | Performance measurement | Impact evaluation | Systematic review |
|---|---|---|---|---|---|
| ***Focal question*** | What is the state of ambient social and/or environmental conditions, and how are these conditions changing over time and space? | What are the management inputs, activities, and outputs associated with a conservation intervention, and how are these changing over time? | To what extent is a conservation intervention making progress toward its intended objectives for activities, outputs, and outcomes? | What intended and unintended impacts are causally induced by a conservation intervention? | What is the state of the evidence for the impact of an intervention, and what does this evidence say about intervention impacts? |
| ***Timing*** | Varies; often preintervention | During implementation | During and after implementation | Post-implementation, with pre-implementation baseline | Post-implementation |
| ***Scale*** | Any; often state/ province (social), landscape, ecoregion (ecological), or country (both) | One or more interventions, usually protected areas | Single project or program | Multiple projects or one or more programs, with corresponding nonintervention comparison group | Multiple projects, programs, or policies |
| ***Implementer*** | Professional researchers, citizen volunteers | Project and program managers, government agencies | Project managers | Professional researchers and evaluators | Professional researcher |

## Box 12.1 Commonly Used Terms in Monitoring and Evaluation (M&E)[a]

- *Inputs*—the financial, human, and material resources used in a conservation intervention.
- *Outputs*—the products, goods, and services that result from an intervention.
- *Activities*—actions taken or work performed related to an intervention.
- *Outcomes*—desired end that outputs are intended to achieve.
- *Impacts*—the intended and unintended consequences that are directly or indirectly caused by an intervention.

[a] Mascia, M. B. et al., *Commonalities and complementarities among approaches to conservation monitoring and evaluation.* Biological Conservation, 2014. **169**: 258–267.

It often has the intention of spotting changes in trends of a biodiversity feature that in turn may influence conservation planners and scientists to give more attention to that feature. In general, ambient monitoring addresses the question, "How is the social-ecological system doing in a planning region?" Data

| | Ambient monitoring | Management assessment | Performance measurement | Impact evaluation | Systematic review |
|---|---|---|---|---|---|
| ***Decisions supported*** | Spatial and temporal priority-setting, selection of strategies and objectives | Setting priorities among potential capacity-building investments at one or more projects | Program reporting and accountability assessments; Adapt activities and strategies to enhance performance | Adaptive management of existing and future interventions, scaling up or down future investments in said intervention | Selecting an intervention; scaling up or scaling down investments in said intervention |
| ***Practitioner audience*** | Decision makers at local to global levels | Project and program managers, donors, senior decision makers | Project and program managers, donors, senior decision makers | Project and program managers, senior decision makers, donors | Project and program managers, senior decision makers, donors |
| ***Data collection methods*** | Primary data collection; remote sensing, transects (ecological); household surveys, focus groups (social) | Expert judgment, secondary sources | Expert judgment, secondary sources, occasional primary data | Primary data collection or manipulation of secondary source data; remote sensing, transects (ecological); household surveys, focus groups, interviews (social) | Data extraction from secondary sources |
| ***Data analysis*** | Moderate to complex; may require data processing and statistical analyses | Simple; requires scoring self-administered questionnaires | Simple to moderate; may require statistical manipulation of secondary source data | Complex; requires data management and sophisticated statistical analyses | Moderate to complex; requires sophisticated data extraction and statistical analyses |

*Source:* Adapted from Mascia et al. (2014).

**FIGURE 12.1** An example of ambient monitoring. Fish sampling conducted by the Wisconsin Department of Natural Resources (WDNR) Long Term Resource Monitoring Program (LTRMP) field office in La Crosse, Wisconsin.

from such monitoring are often used as controls to evaluate the effectiveness of a management intervention.

2. *Management assessment*—the process of measuring the management inputs, activities, and outputs associated with a conservation intervention to identify management strengths and weaknesses. These assessments do not try to measure specific performance goals but instead assume that strategies and actions with sufficient staff and resource capacity are more likely to deliver desired conservation outcomes than those without. As Mascia and colleagues point out, it remains unclear whether "well-managed interventions" lead to more successful outcomes—and, at least in some cases, they do not.[7] In conservation circles, these assessments tend to focus on whether particular management processes have been followed. Increasingly, however, schools of management and management theory from the business sector are influencing how conservation initiatives are conducted. This influence is often manifested by examining the effectiveness of various management processes.
3. *Performance measurement*—the process of measuring progress toward specific project, program, or policy objectives, including desired levels of activities, outputs, and outcomes. Most perfor-

mance measurements do not compare the outcomes of an intervention to what might have happened in the absence of that intervention. As a result, researchers cannot confidently infer whether observed outcomes are really due to particular strategies and actions. This approach to monitoring is also referred to in some literature as "performance monitoring" or "performance evaluation." It is often the type of M&E associated with "performance-based," "results-based," or adaptive management cycles that are commonly used in planning and management of conservation and natural resource initiatives.

4. *Impact evaluation*—the systematic process of assessing the causal effects of a project, program, or policy by comparing what actually happened due to an intervention to what would have happened without it. Experimental or quasi-experimental methods are required for rigorous impact evaluations. Thus they often require substantial resources, and many conservation projects and programs are not conducive to this sort of experimental design. Daniela Miteva and colleagues recently reported on such evaluations for strategies related to protected areas, decentralization of forest management, and payments for ecosystem services.[8]
5. *Systematic review*—a structured process to evaluate the evidence for impacts of a particular conservation intervention that involves collating, appraising, and synthesizing information and data relevant to the effectiveness of one or more strategies and actions.[9] Such reviews have been used in the health field for quite some time. Bangor University professors Andrew Pullin and Theresa Knight first discussed systematic reviews as applicable for conservation, leading to a movement called "evidence-based conservation."[10] See websites for the Collaboration for Environmental Evidence (http://www.environmentalevidence.org) and Conservation Evidence and its associated journal (www.conservationevidence.com).

Although all five forms of M&E have application to the development and implementation of conservation plans, our primary focus in this chapter is on performance measurement and impact evaluation. These two approaches most directly allow us to determine whether the strategies and actions we are proposing and implementing as part of the planning process will have the desired impact. In many situations, these forms of M&E would be part of an adaptive management program, although as we have already noted, that happens all too rarely. In the next section we focus on the application of adaptive management, because of its overarching importance to both monitoring and managing conservation projects.

## Adaptive Management

Few topics in conservation are more complicated than **adaptive management** (**AM**). Scores of papers have been published on the subject, and thousands of conservation practitioners claim to be doing it. But understanding just what "it" is can be confusing and inconsistent. The application of AM in natural resource management can largely be traced to the early work of four individuals: C. S. Holling, Carl Walters, Ray Hilborn, and Lance Gunderson. In its simplest form, adaptive management can be defined as "learning by doing," meaning it combines the need to take conservation actions while trying to learn which actions will be most effective. Indeed, this feedback between learning and decision making about which strategies and actions to use is a defining feature of AM.

Although there are many definitions of adaptive management, a 2004 report by the U.S. National Research Council articulated the process in relatively straightforward terms:

> Flexible decision-making that can be adjusted in the face of uncertainties as outcomes from management actions and other events become well understood. Careful monitoring of these outcomes both advances scientific understanding and helps adjust policies or operations as part of an iterative learning process.[11]

From this definition, we can appreciate that monitoring is a critical component of AM. Adaptive management can be realized when information from monitoring allows a project team to learn about the effectiveness of their strategies and actions, and make improvements (**FIGURE 12.2**).

Most of the literature on AM distinguishes between **active** and **passive adaptive management**. As with AM itself, these two forms

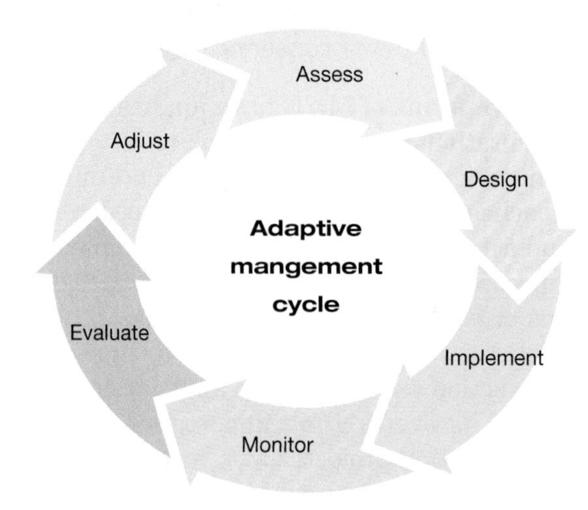


**FIGURE 12.2** A typical adaptive management cycle diagram. Many similarities between this cycle diagram and those of the *Open Standards* and Conservation Action Planning have been featured in earlier chapters. (Data from Greig, L.A. et al., *Insight into enabling adaptive management.* Ecology and Society, 2013. **18**[3]: 24.)

have been defined in various ways. Again, the National Research Council report provides a useful distinction:

> Within "passive" adaptive management, a single, preferred course of action, based on existing information and understanding, is selected. Outcomes of management actions are monitored, and subsequent decisions are adjusted based on the outcomes. This approach contributes to learning and to more effective management, but it is limited in its ability to enhance scientific and management capabilities for conditions that go beyond the course of action selected. By contrast, an "active" adaptive management approach reviews information before management actions are taken. A range of competing, alternative system models of ecosystem and related responses (e.g., population changes in a target species), rather than a single model, is then developed. Management options are then chosen based upon evaluations of these alternative models. All modes of adaptive management require outcomes of management actions to be monitored. Learning is achieved by observing system responses to management actions.

It is critical to carefully consider when it is useful to take an AM approach and when it may not be. In many conservation projects, there is often not much uncertainty about what action to take. For example, if an important natural area is on the verge of being developed, then the protection of that area by a government or nongovernmental organization may be one of the few options that remain. In other situations, there may be limited time or options for organizational learning. Byron Williams and Eleanor Brown, as detailed in an extensive U.S. Department of Interior report on AM,[12] have outlined five conditions for when an AM approach may be most useful:

1. Management actions need to be taken but their consequences are hard to predict.
2. Clear and measurable objectives are available to guide management decisions.
3. An opportunity exists to apply learning to management by considering and comparing a range of management alternatives (FIGURE 12.3).
4. Analysis and evaluating of monitoring data can be used to reduce uncertainty in the effectiveness of management actions (FIGURE 12.4).
5. Stakeholders can be actively involved throughout the adaptive management process.

The fact that several of these conditions are rarely met for many conservation projects is a good clue about why AM is not more commonly

FIGURE 12.3 Northern spotted owl (*Strix occidentalis caurina*), the central figure in the Northwest Forest Plan (USA), a set of policies and guidelines to manage old-growth and other forests in the Pacific Northwest of the United States. This plan was one of the first large-scale adaptive management efforts in the USA, designed in part to reduce uncertainty of a range of timber management actions on this threatened owl species. More recently, monitoring has revealed that competition from barred owls (*Strix varia*) has become a major factor threatening future persistence of northern spotted owls in some locations.

applied in conservation work. For example, although we know of unpublished examples of AM in The Nature Conservancy (TNC), in reality only a few of the hundreds of conservation projects that TNC engages in consider a range of conservation or management options and quantitatively compare their effectiveness (in other words, use an active AM approach). Two projects—using ecosystem management to protect an array of endangered species on Department of Defense lands in Florida, USA,[13] and a comparison of alternative agricultural best management practices in the midwestern USA[14]—stand out as exceptional applications of active AM in TNC.

In their extensive review of AM, Martin Westgate and colleagues identified 1336 published articles related to AM.[15] Only 61 of these articles actually demonstrated some implementation of AM, only 27 of those had quantitative results, and only 13 referred to taking an experimental management (active AM) approach. The bottom line is that only a few published examples demonstrate AM in practice in conservation. Westgate and colleagues highlighted a few of these from around the globe, including management of the Colorado River (Grand Canyon National Park, USA), restoration of woodland bird

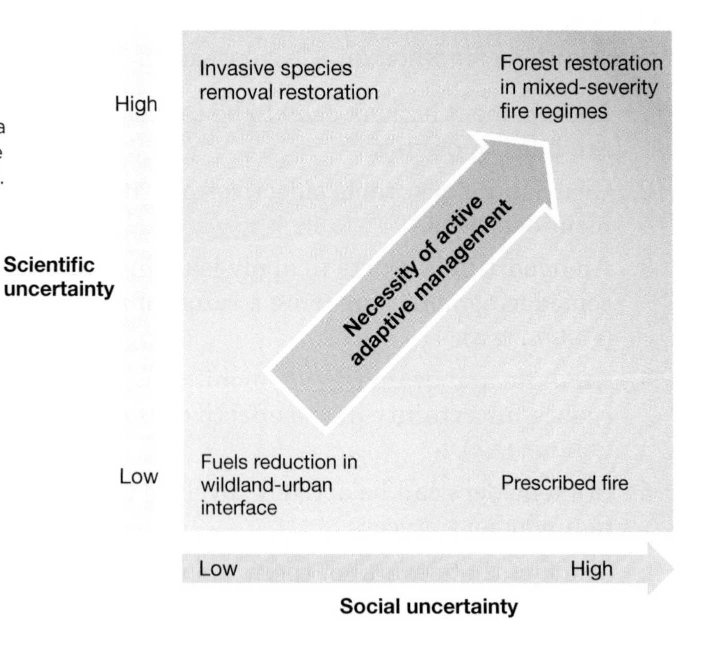


FIGURE 12.4 The relationship between uncertainty and the need to take an AM approach. Higher uncertainty, either among human communities or ecological communities with regard to proposed actions, suggests a greater need for an AM approach. (Adapted from Larson, A. et al. *Making monitoring count: Project design for active adaptive management.* Journal of Forestry, 2013. **11**(5): 348-356.

habitats in Australia, monitoring of agri-environment schemes in the United Kingdom, and management of sika deer in Japan.

One of the most vexing issues about AM is that practitioners have applied the term to a wide variety of approaches, ranging from "any management in which actions or approaches change through time (by whatever means) to a more traditional interpretation that demands integration of models, active experimentation, monitoring, decision analysis, and a mix of alternative management actions."[16] In effect, referring to AM as being applicable to such a wide variety of program approaches and situations has watered down the original concept and increased confusion for those who may want to apply it.

So why do so many AM efforts fall short of success? Although many papers address this question, Craig Allen and Lance Gunderson have taken the most comprehensive look at this situation.[17] They identified nine pathologies that lead to the failure of AM, six of which we have highlighted here as being most critical in our experience:

1. *Experiments are difficult.* Because ecosystems are often slow to respond to interventions, the scale of many environmental interventions makes repeated treatments impractical, managers who want control are resistant to experiments, and such experiments can be expensive (**FIGURE 12.5**).

**FIGURE 12.5** Establishing marine protected areas (MPAs) and limiting or closing fishing in some areas is one management action for which ecosystems are known to respond relatively quickly to a change in management.

2. *Prescriptions are not followed.* For adaptive management to be an effective process, directors and managers need to be constantly attentive to "surprises" and new information, and incorporate this information into project design instead of dogmatically following an original prescription of approaches.
3. *Learning is not used.* At times, monitoring results will demonstrate that progress toward objectives is falling short. For a whole host of reasons, a project team may be slow to respond to such results—sometimes due to the divide between scientists and managers and at other times due to the challenges of admitting that a long-term investment in strategies is simply not working and it is easier to keep doing the same thing. This is known in the business world as sunk cost bias, or the Concorde fallacy, for the continued investment in a type of jet airliner that was perpetually plagued with problems and poor return on investment (see Chapter 6 for details).
4. *Avoiding hard truths.* Making natural resource decisions for which there are considerable social, political, economic, or ecological risks is difficult; as a result, some conservation projects are risk averse. As Allen and Gunderson point out, management of large rivers provides good case studies where important steps may be taken to conserve riparian habitats, for example, even though the most significant problem (sometimes referred to as the "killer threat") is the lack of a natural flow regime below a dam that is needed to maintain these habitats.
5. *Process lacks leadership.* Throughout our careers we have seen the difference that a strong, effective leader can make in the success of a conservation project. We touched on this topic in Chapter 2, and it applies equally well to these circumstances—complex and thorny conservation problems require effective leaders who can manage different stakeholders and views, engage a team in creatively developing a range of possible solutions to complicated problems, and manage successful interventions.
6. *Focus on planning, not action.* It is easy to take actions with little thought and planning, and it is also easy to get wrapped up in endless planning that leads to paralysis and fatigue. It happens in large government bureaucracies, but it is not uncommon in conservation organizations when planning gets disconnected from project management and leadership.

In addition to these "pathologies," three additional points about AM are worth highlighting.[18] First, most AM projects are of relatively short duration, suggesting it may be difficult to sustain the required monitoring efforts for longer times in terms of funding and staff resources. Second, AM projects tend to take place at single and smaller spatial scales; few occur over large areas with multiple spatial scales and geopolitical units, probably due to the difficulty of designing adequate management alternatives in the latter situation. Third, AM depends on being able to effectively gauge performance toward objectives, something that is challenging when working in complex systems with multiple objectives. One of us (EG) has analyzed this challenge in detail.[19]

In summary, there are obvious reasons to use an AM process in conservation planning; but for a range of other reasons, it is challenging to effectively implement. For most agencies and conservation organizations, an AM approach will represent a subset of the M&E of strategies and actions that should be taking place. It will be most suitable for circumstances in which a range of management actions are taking place, there is a high degree of uncertainty about the effectiveness of those actions, there is a desire to learn more broadly about those actions, and there are sufficient resources to invest in monitoring.

## Developing a Monitoring and Evaluation (M&E) Program

Throughout this book, we have highlighted steps that are important foundations to having an effective M&E program. In this section we synthesize those steps into a more comprehensive view of an M&E program for conservation impact.

### Steps in Monitoring and Evaluation

#### A. Defining the Key Audiences

There are many different audiences who have interest in how a project is proceeding. It is important to understand these different audiences from the beginning of a project (see Chapter 2), and determine what their questions or information needs are. TABLE 12.2 highlights some of the different audiences.

The project team, local stakeholders, and the supporters of or donors to the project are likely the most important audiences. A project team must be concerned not only with long-term progress, but the accomplishment of shorter-term objectives. Important donors, on the other hand, will be more focused on the big picture. They will be more concerned, for example, about whether an elephant population is being maintained while the project team may be focused on the near term in reducing poaching or managing other elephant-human conflicts. Higher-level decision makers such as senior managers in an agency or conservation organization are a particularly important audience, and they may be a challenging one when monitoring results don't demonstrate adequate progress toward objectives. It's also important to

**TABLE 12.2** Different audiences for monitoring and evaluation (M&E) programs

| Audience | Typical information needs/interests |
|---|---|
| Project team | How is the project progressing; Are results chains assumptions valid; What is working, what is not, and why; Is your team achieving its objectives in the time frame expected; How to improve the project |
| Project partners | How is the project progressing; Are results chains assumptions valid; What is working, what is not, and why; Is your team achieving its objectives in the time frame expected; How to improve the project |
| Senior managers | Are we undertaking actions we said we would; Is the project staying on course; Is it staying within budget; Do any risks to the institution occur because of the project |
| Donors | How is the project progressing; Are projects achieving objectives in the time frame expected |
| Communities or stakeholders affected | How is the project progressing; How will the project impact them |
| Conservation community | Did the project achieve objectives and conservation results; What worked, what did not, and why |
| Academics and students | What is working, what is not, and why |
| Auditors, certifying entities | Is the project complying with laws and regulations; Is it following best practices |

*Source:* Conservation Measures Partnership, *Open Standards for the Practice of Conservation*. Version 3.0. 2013.

think about the range of information that they may be interested in. Some may be focused on the ecological outcomes, others may be most interested in human well-being aspects, and still others may want to know about the fund-raising and management of the project itself.

Different audiences also want information in different levels of detail. A project scientist, for example, may want to know the detailed demographic trends of a key target species, while the project director may only want to know some basic trends in the population. Not only do different audiences want different segments of monitoring information, but they also likely want it in different forms. FIGURE 12.6, for example, presents a visualization of impact measures for the Micronesia Challenge across five impact factors: ecological, management, finance, policy, and people (human well-being). The audience for this particular set of measures is primarily the board of directors and senior management team of a major conservation organization and secondarily the team managing the project.

FIGURE 12.7 contains two different indicators of progress on improving the Chesapeake Bay ecosystem in the eastern United States. These

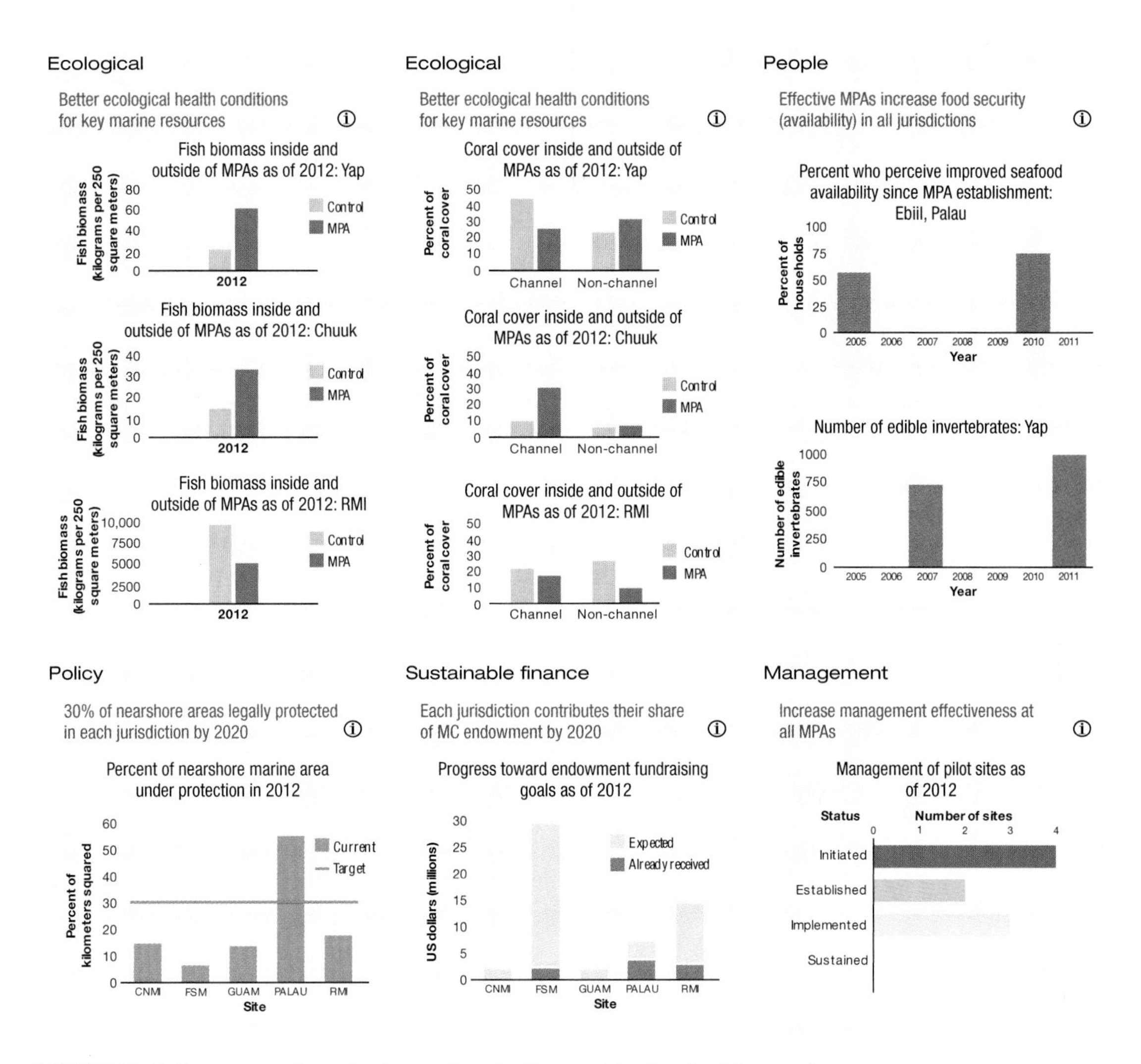


**FIGURE 12.6** Five types of monitoring and evaluation metrics for the Micronesia Challenge to assess progress in a priority project of a major conservation organization. These visualizations were designed with a specific audience in mind—the board of directors and a senior management team. (Data from The Nature Conservancy, 2013.)

graphs are intended to reach a broad public audience that is interested in the ecological recovery of Chesapeake Bay. For both Figure 12.6 and 12.7, considerably more information is available on the topic than is displayed in the graphs. Deciding what information to display, how much information, and in what form can be critical to conveying information to target audiences in a convincing manner. Doing so is not only an art but also a profession of graphic artists and other data visualization staff. The paper by David Speigelhalter and colleagues[20] on visualizing uncertainty is an excellent reference for designing visualizations that will work with different audiences. Stephen Few's *Show Me the Numbers: Designing Tables and Graphs to Enlighten* (2012) and Nathan

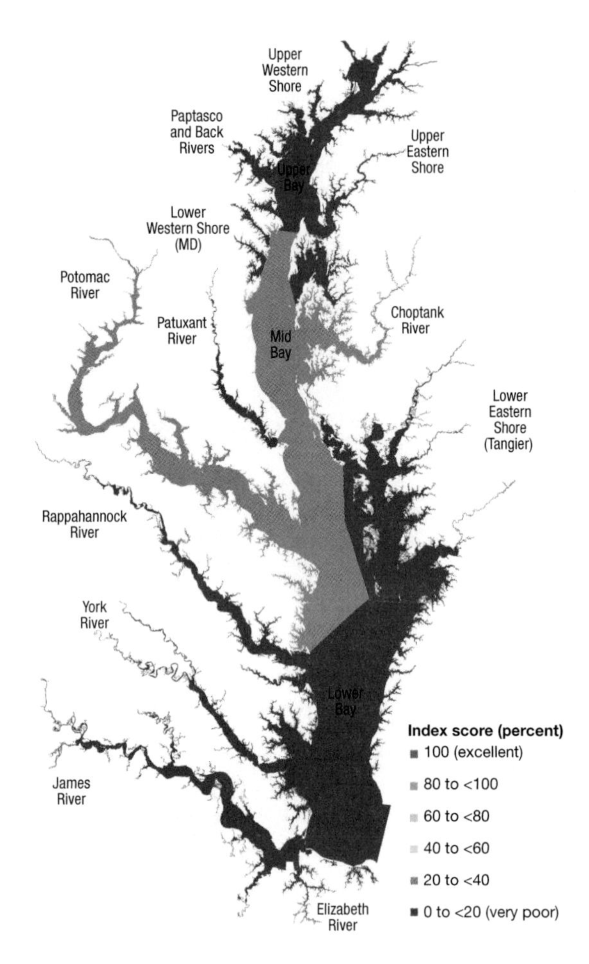


**FIGURE 12.7** Chesapeake Bay health–water quality thresholds. The different colors represent the percentage of time that water quality was above an acceptable threshold level for the period March–November 2013. (Green colors represent acceptable scores; red-orange colors, unacceptable.) This type of visualization was designed to reach a broad public audience and quickly demonstrates that water clarity remains poor throughout the ecosystem. (Data from *Chesapeake Bay Report Card.* 2014, Solomons, MD: University of Maryland Center for Environmental Science, Chesapeake Bay Laboratory.)

Yau's *Data Points: Visualization That Means Something* (2013) are standard texts on this topic that practitioners will also find useful.[21]

## B. Making Investment Choices

Thousands of conservation and natural resource management projects are being conducted around the globe every day, and managers rarely have enough resources to appropriately monitor each and every one. Managers may not invest in monitoring unless they have some guidance on how to invest, and know what sort of return they can expect from that investment. From a practical standpoint, not all of these projects will or should have M&E programs with rigorous statistical designs that infer the causality of management actions with some high degree of confidence. In some cases, for example, there may already be substantial evidence that an action will be successful if appropriately implemented and a low-rigor, inexpensive monitoring method may suffice. In other situations, the risks or consequences of an intervention failing may be so low that funding for monitoring could be better spent on other tasks.

In an insightful article on collaborative adaptive management, former undersecretary of the U.S. Interior Department Lynn Scarlett[22] makes the point that monitoring may not always be worth the investment in situations where it won't be accomplished in a time frame relevant to a management decision or the information gained, it isn't statistically valid, or it won't otherwise affect a management decision. Decision scientist Eve McDonald-Madden and others[23] made the same point in an article entitled "Monitoring Does Not Always Count"—and provided a decision tree for helping to make these strategic choices about monitoring (**FIGURE 12.8**).

Our colleague Jensen Montambault at The Nature Conservancy has developed and extensively tested a qualitative framework for

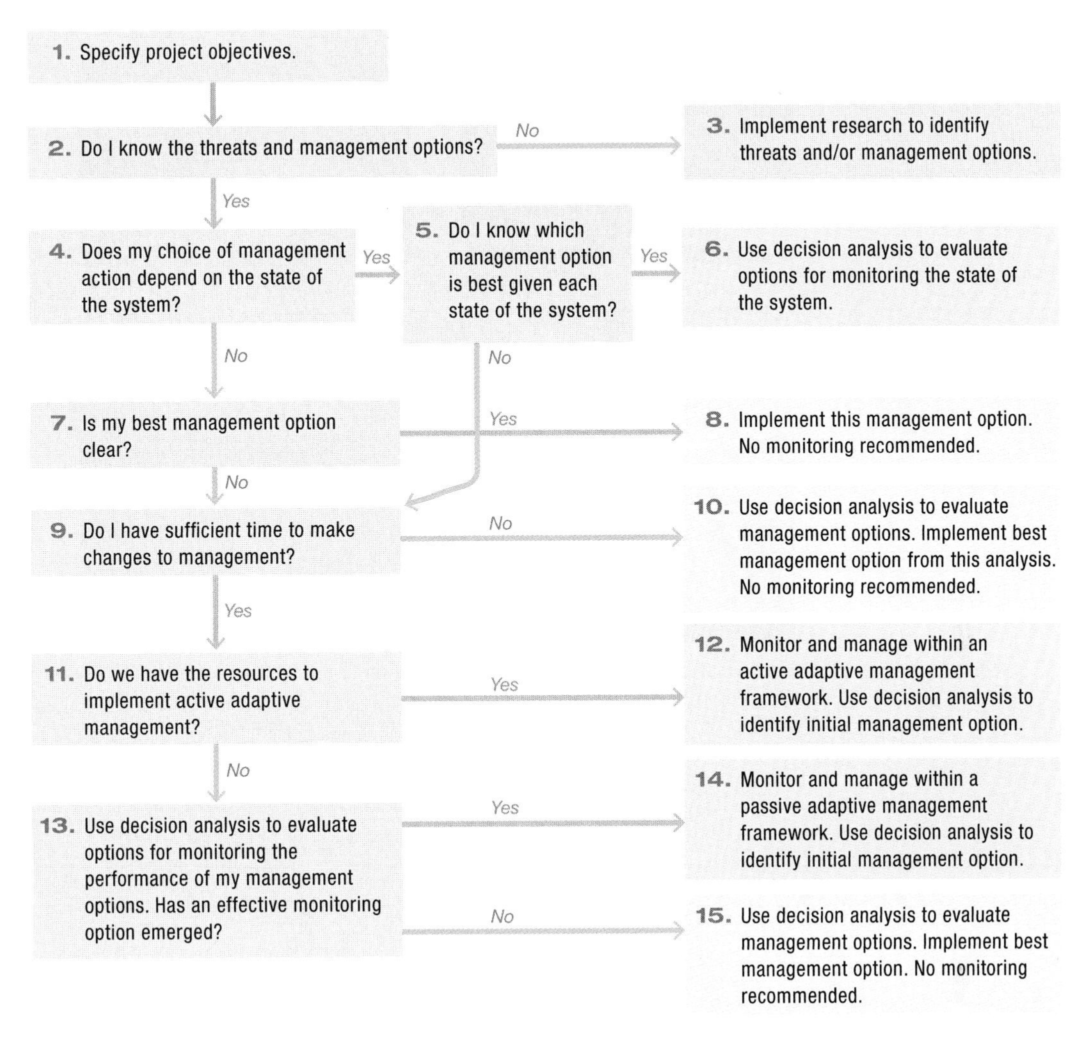


**FIGURE 12.8** Decision tree for helping to decide whether and in what form to invest in monitoring of conservation projects. (Data from McDonald-Madden et al., 2010.)

making decisions on M&E investments that applies three factors: risk, leverage, and conservation evidence.[24] Explicitly using these factors in guiding M&E investments is not typical of most adaptive management frameworks. Risk is the possibility of a negative effect on the entity being managed (often a target species or ecosystem, but also possibly a measure of human well-being) or the managing institution. Leverage is the replication or scaling up of the conservation impact of an intervention beyond any individual project or program (see Chapter 11 for more discussion on this topic). A demonstration or proof-of-concept project is one example of such leverage.[25] **Conservation evidence** is the existing body of scientific information that informs us as to the efficacy of different conservation interventions.[26] This framework demonstrates how risk, leverage, and existing evidence

FIGURE 12.9 Monitoring and evaluation investments in relation to risk and leverage. Risk and the potential for leverage are relative to the ecological, socioeconomic, political, institutional, and financial context in which the conservation action is implemented. Risk and leverage interact to determine relative monitoring investment, while scale and location influence actual monitoring cost. *Top left:* Using ecological considerations to modify the flows of water through hydropower facilities is a high-leverage conservation approach as demonstrated by this learning exchange between an engineer in the Zambezi River water authority and the director of a TNC freshwater program. *Top right:* The first third-part verified REDD project required exhaustive socioeconomic and ecological monitoring and evaluation by Fundación Amigos de la Naturaleza. *Bottom left:* U.S. Department of Defense adds simple postburn monitoring at little extra expense, allowing routine learning about what works and why in Fort Hood, Texas. *Bottom right:* Higher risk associated with The Nature Conservancy's promotion of small-scale marine conservation areas required a biological assessment of species such as this grouper in the western Pacific.

influence the value of additional information gained from M&E programs and how that value translates into decisions about differential investments in M&E. FIGURE 12.9 illustrates the use of this framework for risk and leverage across a range of conservation projects. In addition to risk and leverage, site-specific costs (**Box 12.2** and **Box 12.3** provide examples of cost-effective monitoring), feasibility, and previously existing monitoring protocols are also factors that influence the overall investment in M&E.

## Box 12.2 A Cost-Effective Monitoring Program

A monitoring program was designed for Australia's Environmental Stewardship Program, an incentive-based program to improve the stewardship of two endangered ecosystems on private land (e.g., Box Gum Grassy Woodland). The program has been able to demonstrate differences in biodiversity between stewardship and control sites while using a cost-effective investment scheme.

David Lindenmayer and colleagues,[a] who designed the monitoring program for the Australian Environmental Stewardship Program, came away with several important lessons. Three of them, highlighted here, can help practitioners make more efficient and effective investments in monitoring:

- Monitoring programs should be designed to meet the specific objectives of particular conservation projects and not necessarily apply or adapt these protocols from other conservation projects that have different objectives and may operate at different scales ("fit-for-purpose" monitoring design).
- In most situations, monitoring needs to begin with the implementation of strategies and actions, not as an afterthought. Doing so can lead to problems such as misinterpretation of "treatment" effects and inefficient sampling designs.
- Effective conservation projects will involve links between science-literate managers and management-literate scientists. For anyone who has been involved in implementing a conservation project, this lesson needs no further explanation.

[a] Lindenmayer, D. B. et al., *A novel and cost-effective monitoring approach for outcomes in an Australian biodiversity conservation incentive program.* PLoS ONE, 2012. **7**(12): e50872.

## Box 12.3 Using Systematic Monitoring to Evaluate and Adapt Management of a Tiger Reserve in Northern Lao PDR

Two major strategies were used to reduce threats to tigers in Nam Et-Phou Louey National Protected Area: law enforcement and outreach.[a] Monitoring was implemented at four spatial scales to discern the effectiveness of these two strategies and to track the status of tigers and their prey. Camera traps and scat DNA analysis were used to estimate tiger numbers. Monitoring the spatial coverage of law enforcement staff using a time series approach demonstrated that expansion of the law enforcement effort (foot patrols) reduced the sign of hunting in the protected area and that routine patrol of roads and markets was not as effective in reducing hunting and trade of tigers as was responding to informants.

Monitoring was also used to evaluate the change in behavior and knowledge of villagers, hunters, and government officials. Monitoring of the social marketing strategy included surveys with quasi-experimental designs, control sites, and a priori hypotheses. Results showed that social marketing, especially with villagers, had the greatest impact on reducing tiger hunting. Types of monitoring results that were important for making changes to strategies at the project level were quite different from those needed for donors and senior management as well as the broader conservation community.

As with many projects in developing countries where human resources are limited, staff underestimated the time and resources needed for providing technical training. This project demonstrated that ongoing, systematic, cost-effective monitoring could be used to improve the effectiveness of two strategies and that tigers and their prey could be detected at different spatial scales within the overall conservation project area.

---

[a] Johnson, A. et al., *Using systematic monitoring to evaluate and adapt the management of a tiger reserve in northern Lao PDR,* in *Working Paper Series.* 2012, New York: Wildlife Conservation Society.

Teams of managers, scientists, and other decision makers should discuss how the information from a conservation intervention's assessment will likely be used and decide on the minimum acceptable level of inference. We synthesize these approaches in a decision tree (FIGURE 12.10) to illustrate how risk, leverage, and evidence can influence the appropriate level of inference.

While randomized, controlled experiments are the gold standard of attributing cause and effect to a management action, it is not possible or desirable to use this design in most cases. Cases that involve endangered species will often have legal consequences and may lend themselves to experimental or quasi-experimental monitoring designs. When a conservation plan is enacted for such a species, it will likely be necessary to monitor this species under various management situations. In other cases, we may wish to replicate an action that is taking place in a demonstration project. For example, The Nature Conservancy and U.S. Army Corps of Engineers established a set of demonstration projects to improve the flows below Corps-operated dams to benefit freshwater resources (see Figures 11.6 and 12.9).[27] As part of this process, the feasibility of collecting baseline monitoring data was explored, as was setting up quasi-experimental and time series monitoring.

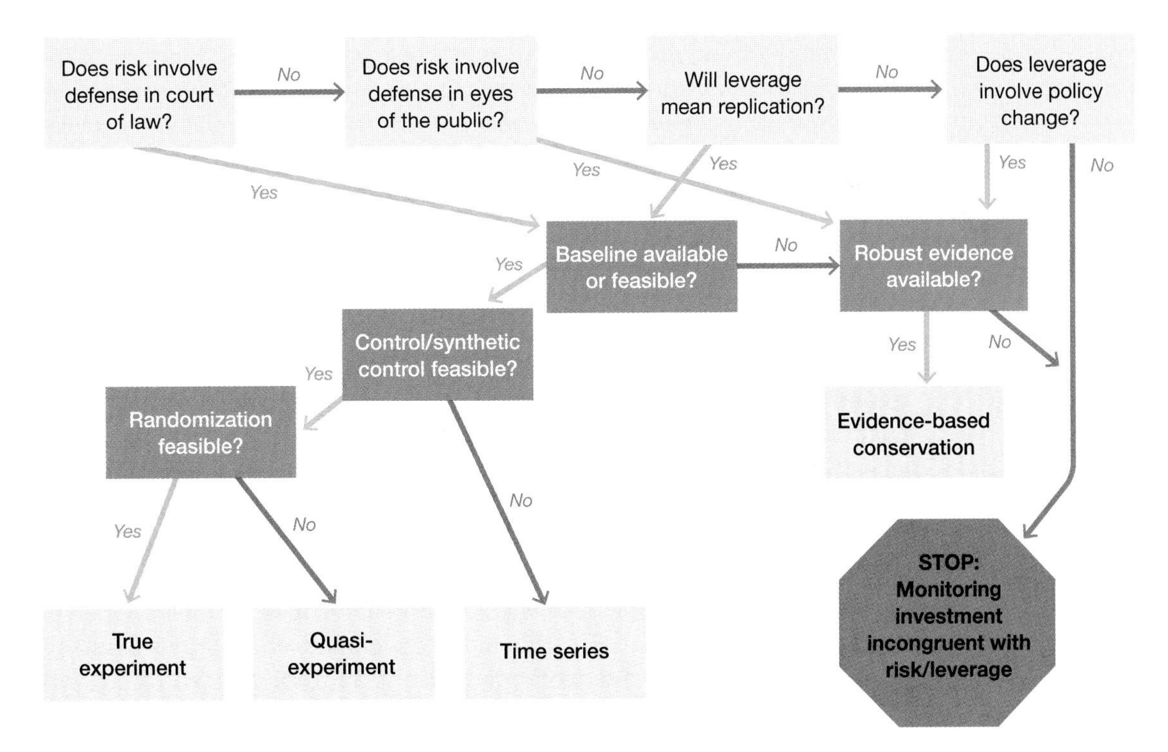


FIGURE 12.10 A decision tree for making investments in monitoring and evaluation based on potential risks, leverage, and available conservation evidence for strategies and actions being implemented in a conservation project. (See text for examples.)

Many actions in the day-to-day work of conservation organizations and natural resource agencies have relatively certain outcomes and limited risk and leverage. For example, this category includes many (but not all) land acquisitions that are made by **land trusts** and natural resource agencies to prevent natural areas from being developed. It is unlikely to be necessary or cost-effective to monitor all of these acquisitions for the effectiveness of the strategy because the cost of monitoring may exceed the value added by this information.[28] In these cases, managers should examine the evidence that the outcome is certain enough that future managers can make well-informed management decisions without monitoring information. In other cases, some interventions are risky and have wide application, but the weight of assembled conservation evidence is sufficiently convincing that the intervention's impact does not need further evaluation. For example, the biological effects of marine protected areas (MPAs) have been well established through formal meta-analyses,[29] and the only monitoring that may be necessary is that of management assessment to ensure that standard methods and practices for implementing MPAs have been followed and are resulting in the anticipated outcomes.

### C. Selecting Indicators

Once a decision has been made to invest in an M&E program and the appropriate level of investment is determined, the next step is to decide what to monitor. Making that decision requires selecting indicators that allow us to measure progress toward objectives. We first introduced the concept of indicators in Chapter 4 and noted the critical role that they play in allowing us to assess progress. We explained that a good indicator should be both *representative* and *measurable.* The Conservation Measures Partnership's *Open Standards* provide additional advice on what constitutes good indicators.[30]

One tendency we have observed in many projects is to measure too many indicators, many of which may tell us little about the effectiveness of our strategies and actions. To avoid this pitfall, project teams should give considerable thought to what indicators will be most relevant and useful to the objectives at hand. In most cases, the selection of indicators will be influenced by the cost-effectiveness of employing a particular indicator, the availability of existing data, its sensitivity to the action being taken, and how consistently it is defined and used. The availability of existing data is often a crucial factor because the need to gather new data can sometimes be too expensive. Many countries, especially the more developed ones, have existing human demographic and economic data at appropriate scales as well as ecological data that could prove useful as indicators.

Many practitioners find it useful to refer to their conceptual models and theories of change to help identify indicators (Chapter 6). FIGURE 12.11 and FIGURE 12.12 are examples of results chains with indi-

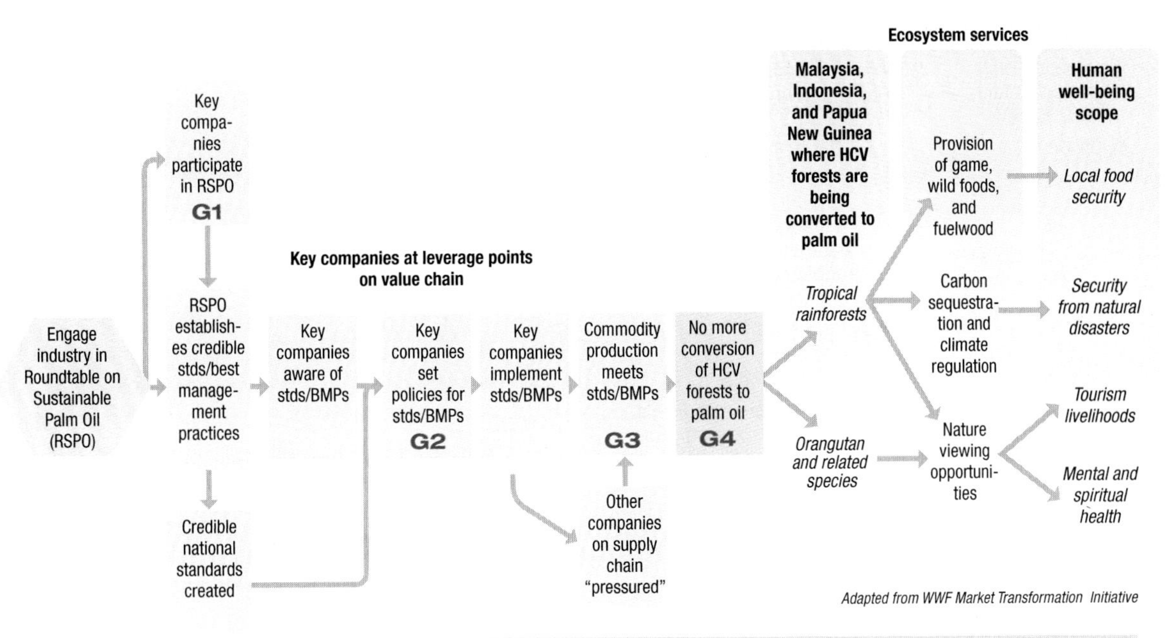


| Code | Objectives | Indicators |
|---|---|---|
| G1 | By 2013, at least 75% of global production of palm oil is represented in Roundtable of Sustainable Palm Oil (RSPO) membership. | Percentage of global production represented in RSPO membership |
| G2 | By 2014, 50% of targeted palm oil manufacturers and retailers have committed to procure 100% sustainable palm oil. | Percentage of targeted palm oil manufacturers and retailers committed to procure 100% sustainable palm oil |
| G3 | By 2015, all top 20 global palm oil companies procure 100% sustainable palm oil. | Number of global palm oil companies that procure 100% sustainable palm oil |
| G4 | All new plantings of palm oil since November 2005 have not replaced primary forest or any area containing more than one high conservation value (HCV) in World Wildlife Fund priority places (Heart of Borneo, etc.) | Number of hectares of HCV habitat lost to palm oil production each year in X, Y, Z priority places |

*Note: This results chain does not include a goal because, at this global scale, the ultimate measure of success of this strategy is the elimination or reduction of the threat of conversion of tropical forests to palm oil plantations.*

FIGURE 12.11 Results chain with objectives and indicators for engaging industry in sustainable palm oil production. Figure originally adapted for WWF Market Transformation Initiative. (Adapted from Conservation Measures Partnership, 2013.)

cators for two different types of projects: a policy strategy (a market transformation project in WWF) and a generic reintroduction project by a wildlife agency (freshwater mussels in this case). Both figures also contain sample objectives and indicators.

A report entitled *Measuring the Effectiveness of State Wildlife Grants* has many examples of different types of results chains, along with an appendix of objectives and indicators and how to develop them.[31] Indicators can be selected that will help evaluate progress toward an

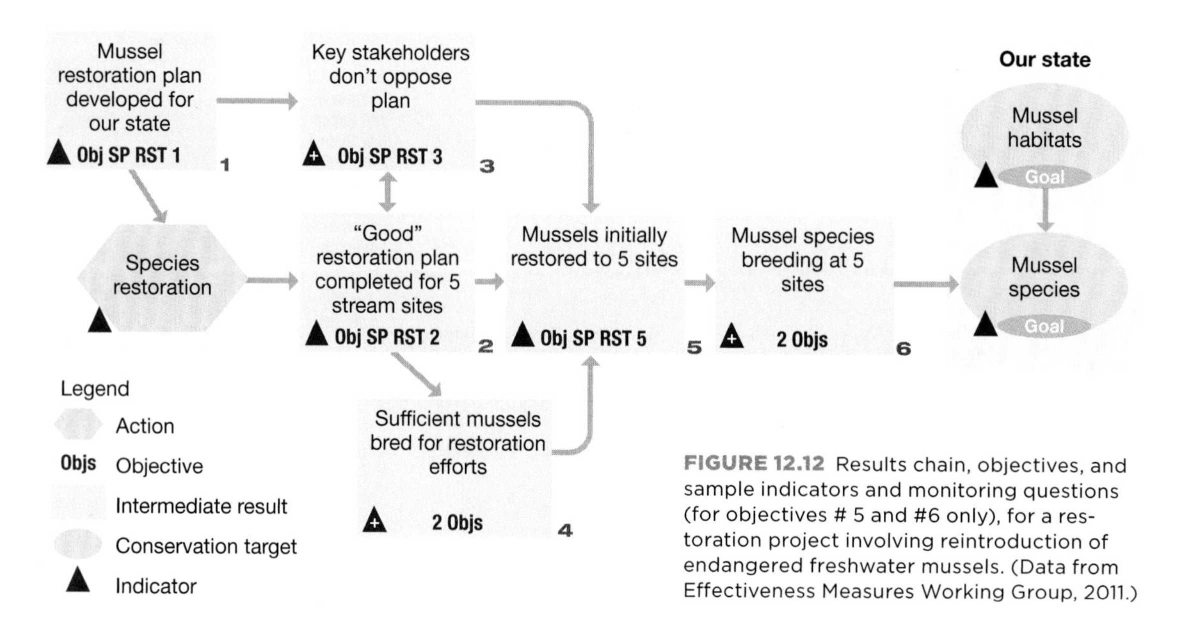


**FIGURE 12.12** Results chain, objectives, and sample indicators and monitoring questions (for objectives # 5 and #6 only), for a restoration project involving reintroduction of endangered freshwater mussels. (Data from Effectiveness Measures Working Group, 2011.)

intermediate objective (e.g., establishing a new protected area, developing a management plan for a protected area, or implementing a management action) as well as for measuring the effectiveness of strategies and actions for particular biological features (see Box 12.3 on tiger monitoring). Planners should pay particular attention to identifying indicators that may result in a go/no-go decision or serve as decision-making triggers for later actions in a conservation project.

Martin Nie and Courtney Schulz discuss such decision-making triggers in an adaptive management context where these types of indicators can be pre-identified to result in specific actions when the indicator reaches a particular threshold level.[32] For example, a trigger in a conservation project aimed at improving fish passage might be a certain number of anadromous fish that must pass certain points in their upstream movements through fish ladders or locks. After reviewing a range of cases in which species were monitored as they went extinct (cleverly referred to as "counting the books while the library burns"), David Lindenmayer and colleagues recommended including trigger points in monitoring plans and developing specific plans of action of what would be done when the trigger points were reached.[33]

As we indicated in Chapter 4, many conservation practitioners are likely to be somewhat familiar with ecological indicators, but likely less so for social and economic ones (see Box 4.1 for advice on social indicators). **TABLE 12.3** provides an example of social objectives for a conservation project in Gansu Province, China, aimed at reducing fuelwood consumption to conserve habitat for a rare primate species. **FIGURE 12.13** summarizes some of the indicators (metrics) used to evaluate progress toward these objectives.

| Label | Result | Objective | Measure (indicator) | Rolled-up measure | Monitoring questions |
|---|---|---|---|---|---|
| SP RST 05 | **Generic:** Species initially restored to sites (short-term) | By specified target date, the target number of units* have been introduced to Area(s) YYYY.<br>** Units could be individuals, breeding pairs, communities, pounds of fish fry, or other measures as appropriate* | Percent of target number of units that are released | Percent of projects that are able to release sufficient units, by taxa and by region | **9.** Has the project begun releasing species to restoration site(s)?<br>**10.** What percent of initial release work across all restoration sites has been completed?<br>**11.** What is the "unit" for measuring quantities of species released within restoration site(s)?<br>**12.** How many units of the species have been reintroduced? |
| | ***Mussel example:*** *Mussels initially restored to 5 sites* | *Within 2 years, more than 10,000 individuals of each species have been restored to each site.* | *Percent of 10,000 individuals that are released* | | |
| SP RST 06 | **Generic:** Species breeding at restoration sites | Within X years of introduction, the restored population is successfully breeding within the restoration site(s). | Percent of sites with restored population successfully breeding | Percent of all projects with restored species successfully breeding, by taxa and by region | **13.** Are the introduced populations breeding within the recovery site(s)?<br>**14.** What is the "unit" for measuring successful reintroduction of the species within restoration site(s)?<br>**15.** How many units of the species are present in the recovery site(s)? |
| | ***Mussel example:*** *Mussels breeding in 5 sites* | *Within 4 years, the mussel species are breeding at each of the 5 sites.* | *Percent of 5 sites with evidence of breeding* | | |
| N/A Conservation targets | **Generic:** Viability of species with greatest conservation need (SGCN) improved | **Goal:** Within X years of the start of the action, the species of interest have improved viability. | Species measure (e.g., population size, reproduction success) | Status measure—will not be rolled up | **16.** Are the introduced populations viable within the recovery site(s)?<br>**17.** Has the population goal for the target species within the restoration site(s) been achieved?<br>**18.** Has this project contributed to any changes regarding the conservation priority status (SGCN priority, threatened/endangered, etc.) of the target species in your state? |
| | ***Mussel example:*** *Viability of mussel population* | **Goal:** *Within 5 years, viable populations of mussels have doubled from 5 to 10 sites.* | *Number of viable populations* | | |

**FIGURE 12.12** *Continued*

As you may already be concluding, how to best incorporate social and economic objectives into conservation plans, establish indicators for these, and evaluate progress toward objectives is a topic of considerable current interest in conservation. There is no single best or right way to go about this work. The choice will likely depend on the local context—existing social assessments, existing data or resources

**TABLE 12.3 Socially oriented conservation objectives from a Rare Pride Campaign project in Gansu Province, China, aimed at reducing fuelwood use to conserve habitat for an endangered primate, the snub-nosed monkey**

| Type of objective | Objective |
|---|---|
| Knowledge | Pride Campaign increases community residents' awareness of the environment issue of fuelwood felling, the health issue of using traditional stoves, and the benefits of fuel-efficient stoves. |
| Attitudes | Pride Campaign increases communities' willingness in using fuel-efficient stoves.<br>Campaign improves communities' identification in adopting measures to reduce fuelwood felling. |
| Interpersonal communications | Pride Campaign stimulates discussions among target audiences about fuel-efficient stoves and environment protection. |
| Barrier removal | Cooperative partners technically support community in building fuel-efficient stoves.<br>Pride Campaign provides subsidies for fuel-efficient stoves. |
| Behavior change | Community residents utilize fuel-efficient stoves in their daily lives. |
| Threat reduction | The community fuelwood consumption starts to decline. |
| Conservation results | By October 2015, the biodiversity and forest quality of Yuhe Reserve will improve significantly (compared to pre-project, the biodiversity index will rise and the forest growing stock is increasing). |

*Source:* Data from DeWan, A. et al., *Using social marketing tools to increase fuel-efficient stove adoption for conservation of the golden snub-nosed monkey, Gansu Province, China.* Conservation Evidence, 2013. **10**: 32–36.

to obtain additional social and economic data—and on the preferences and concerns of stakeholders involved in the conservation projects. The most important point to keep in mind is that what we choose to measure needs to be related to the overall objectives of the project and within the sphere of what the project can influence.

### D. Designing a Monitoring Program and Analyzing Monitoring Data

Although it is essential to invest in designing an adequate monitoring program that will provide answers and evidence concerning science and management questions that conservation projects face, it is well beyond the scope of this book to provide advice on how to do that. Indeed, entire books have been written on this topic. If your project team lacks the scientific or statistical expertise to design monitoring programs, we highly recommend seeking that expertise through a local university or partner organization. A number of good books and publications on the topic provide recommendations as well more general advice about monitoring programs. **Box 12.4** highlights what we consider to be some of the best references.

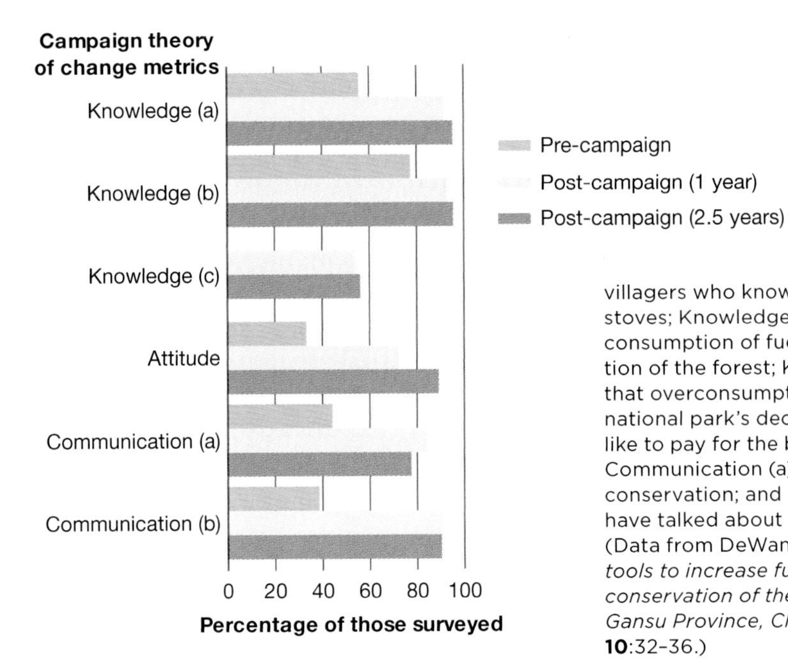


**FIGURE 12.13** Indicators used to evaluate progress on objectives in a Rare Pride (education and marketing) Campaign. This campaign was designed to reduce the use of fuelwood in Chinese villages in the province of Gansu to help prevent forest clearing and degradation. Indicators include Knowledge (a): villagers who know the advantages of fuel-efficient stoves; Knowledge (b): villagers who know that overconsumption of fuelwood is leading to the destruction of the forest; Knowledge (c): villagers who know that overconsumption of fuelwood is leading to the national park's decline; Attitude: villagers who would like to pay for the building of fuel-efficient stoves; Communication (a): villagers who talked about forest conservation; and Communication (b): villagers who have talked about the benefits of efficient stoves. (Data from DeWan, A. et al., *Using social marketing tools to increase fuel-efficient stove adoption for conservation of the golden snub-nosed monkey, Gansu Province, China.* Conservation Evidence, 2013. **10**:32–36.)

## Box 12.4 Recommended Reference Books and Publications on Designing Monitoring Programs

- Elzinga, C. et al., *Monitoring Plant and Animal Populations.* 2001, Malden, MA: Blackwell Press. 372.
- Gardner, T., *Monitoring Forest Biodiversity: Improving Conservation through Ecologically Responsible Management.* Earthscan Forest Library. 2010, London: Earthscan. 360.
- Gertler, P. J. et al., *Impact Evaluation in Practice.* 2011, Washington, DC: World Bank. 235.
- Lindenmayer, D. B. et al., *Improving biodiversity monitoring.* Austral Ecology, 2012. **37**(3): 285–294.
- Lindenmayer, D. B., and P. Gibbons, editors, *Biodiversity Monitoring in Australia.* 2012, Collingwood, Victoria, Australia: CSIRO Publishing. 224.
- Manly, B., *Statistics for Environmental Science and Management.* 2nd ed. 2008, Boca Raton, FL: CRC Press, Chapman and Hall. 295.
- Nichols, J. D., and B. K. Williams, *Monitoring for conservation.* Trends in Ecology & Evolution (personal edition), 2006. **21**(12): 668–673.
- Rosenbaum, P., *Design of Observational Studies.* Statistics Series 2010, New York: Springer. 384.

### E. Adjusting and Improving

Although this step seems simple and straightforward—taking what we have learned from a monitoring program and making adjustments to project strategies and actions—it doesn't get enough attention. Perhaps it is due to the difficulties that sometimes exist in getting scientists and program managers to work together, or maybe it is just a shortcoming in how a conservation project is managed, but too many conservation projects simply don't set aside the time to periodically evaluate feedback from monitoring reports or project evaluations. It is absolutely a critical step to improving project and program performance (a point discussed in more detail in Chapter 11). So even if an M&E program highlights some sobering results about the efficacy of a particular management action, our advice is not to procrastinate in getting this news to the project team. Decision-making trigger points that we discussed earlier are one mechanism for helping ensure that adjustments are made in conservation projects as a result of feedback from monitoring data. In our experience, too many projects keep pouring funding into strategies and actions that either just don't work or don't work well enough to merit continued investment.

Making adjustments to project strategies as a result of monitoring, whether done in the context of an adaptive management framework or not, is an example of how important learning is to improving project performance and impact. In this next section, we expand on the importance of learning for improving the implementation of conservation plans and the overall success of a conservation project.

## Learning

In this brief section we review the types of learning that are most relevant to advancing conservation projects and discuss a few conservation projects in which learning has been evaluated. We then offer a few key recommendations on how agencies and organizations can incorporate learning into the day-to-day management of conservation projects, what the likely benefits of doing so may be, and the risks of doing nothing.

There are many definitions of learning, but for our purposes a simple one may suffice: a change in an understanding of, or relationship, to the world.[34] From an environmental problem perspective, three types of learning have been defined: social, conceptual, and technical.[35] **Social learning** refers to learning that happens as the result of a dialogue among stakeholders or through engaging new collaborators in different ways. **Conceptual learning** involves redefining a conservation problem or developing new objectives in a conservation project. Finally, **technical learning** in conservation refers to undertaking new or revised conservation actions that will help in achieving a project's objectives. Learning experts tell us that social learning is the foundation for conceptual and technical learning. That is to say, if we don't have effective dialogue among stakeholders or strong collaborations, we aren't as likely to realize conceptual or technical learning.

A good example of social learning occurs in the U.S . Fire Learning Network (FLN), a cooperative program of the U.S. Forest Service, several Department of Interior agencies, and The Nature Conservancy.[36] The FLN uses a collaborative planning process based on the *Open Standards* to bring agencies, organizations, tribes, and private citizens together to "build trust and understanding, work together, and improve their community's capacity to safely co-exist with (wild) fire." The social learning occurs in the early stages of a set of workshops, when participants work together to identify issues and develop a mutually agreed upon vision and goals. These early stages of social learning then position the participants to develop a situation analysis (Chapter 5) or conceptual model of their conservation initiative—which, as you might guess, positions them to take on conceptual learning.

Another important aspect of learning is what is referred to as "single-loop" or "double-loop" learning (FIGURE 12.14).[37] Single-loop learning is easy for us to recognize—it's the familiar "plan-do-check-act" cycle that we have seen in the project cycle of the *Open Standards,* Conservation Action Planning, and other project management cycles that you may be familiar with (Chapter 1). We take a specific conservation action, learn from the outcomes of that action, and make adjustments as necessary. For example, we may use prescribed fire to help reduce invasive species but find that the particular intensity or timing wasn't sufficient and adjustments were needed in the burning protocol.

Double-loop learning, on the other hand, is not so much focused on learning from a particular action as it is on learning from a whole host of actions that are taken in a project, which in turn should cause a project team to rethink underlying assumptions and project goals. For example, the success of a REDD (Reduced Emissions from Deforestation and Forest Degradation) project in reducing greenhouse gas emissions might depend on the price of carbon on the global market, local government that regulates an industry for which forests are being cleared (palm oil, for example), and national government to regulate logging companies to achieve its objectives. If the price of carbon is too low to generate significant interest and revenue, or a particular government agency is failing to regulate timber harvest or the palm oil industry, then it might be necessary for the project team to rethink its goals and underlying assumptions. Such a situation requires a project team to engage in double-loop learning.

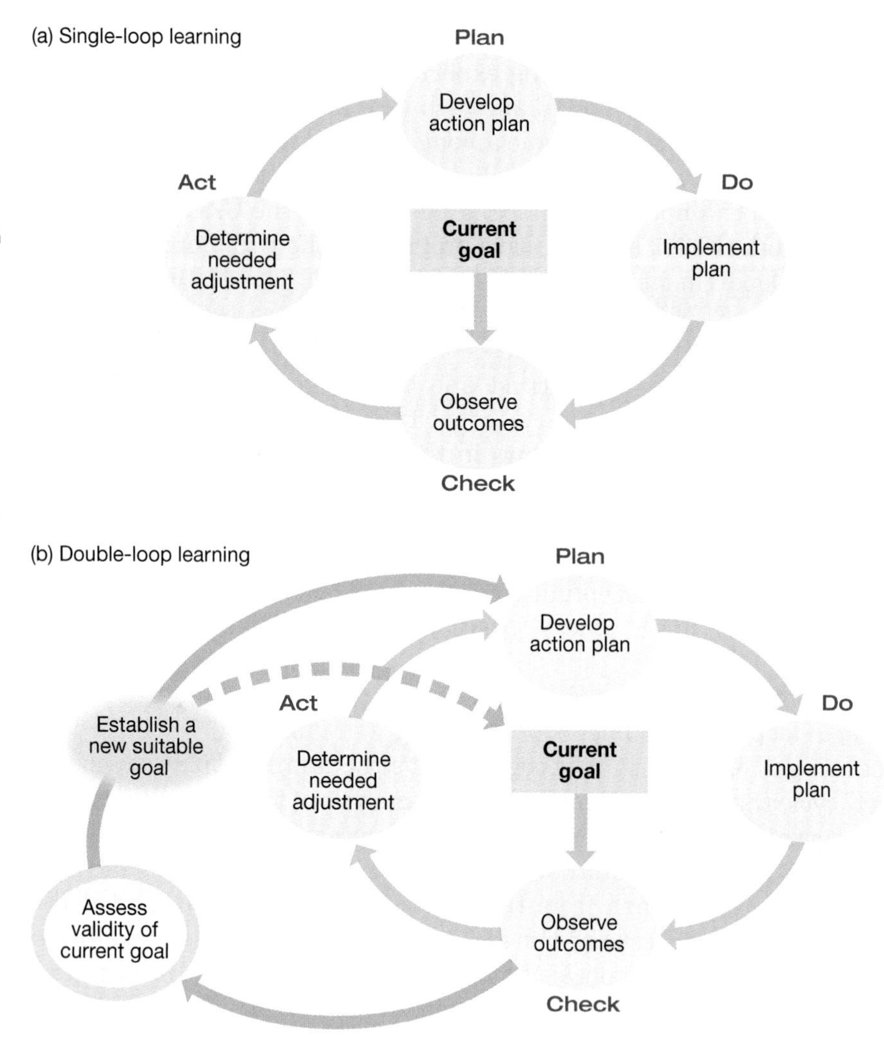


**FIGURE 12.14** Single-loop and double-loop learning in conservation projects. (a) Single-loop learning happens as the result of a traditional AM approach that evaluates a specific conservation action. (b) Double-loop learning occurs through the experience of all the different actions taken during a project. (Data from Dervitsiotis, 2004.)

One recent analysis offers insights into the importance of learning in conservation project performance. Anna Evely and colleagues in the United Kingdom analyzed the degree to which participation in a conservation project enhances learning[38] (i.e., social learning). They interviewed managers and participants in eight different conservation projects in the United Kingdom. Although they found that all participatory approaches to conservation projects fostered learning, learning outcomes were greatest when participants were involved in making project management decisions through open, participatory meetings of collaborators. The authors also suggested that higher levels of learning take place early in a project, which implies that projects need to invest in participatory processes at the outset.

Here are some additional recommendations we can make that will help foster learning both within and across projects and organizations:

- *Share information.* Both for the sake of institutional memory and for disseminating lessons learned to the broader conservation community, record information on your conservation project (e.g., where it is taking place; goals, strategies and actions; staff involved; data access) in a manner that will make it easy for others to find and learn from. In the USA, a conservation organization, Defenders of Wildlife, manages the Conservation Registry (http://www.conservationregistry.org)—an online database of over 100,000 conservation projects in all 50 states. The Conservation Measures Partnership is managing a similar database of projects that is international in scope.
- *Consider participating in a management review or "conservation audit" for your project.* A management review typically assesses a project against its stated goals and objectives, while a conservation audit evaluates a project against a standard set of procedures such as the *Open Standards.* For more information about audits, see the Conservation Measures Partnership website: http://www.conservationmeasures.org.
- *Consider joining a "community of practice"—groups of teams or individuals who work on similar conservation projects (e.g., projects designed to improve stream flows below operating dams[39]).* Some existing communities of practice have conservation planning as a centerpiece. One such example is the National Biodiversity Strategy and Action Plans (NBSAP) Forum, a community of practice organized around the development and improvements of NBSAPs (Chapter 1) that countries prepare under their obligations to the Convention on Biological Diversity.
- *Pay particular attention to the audience.* Different audiences learn in different fashions from many different communication media—publications, conferences, websites, blogs, podcasts, social networks, and TED talks, just to name a few.

## Getting Over the Hurdles

By now it's probably clear to most readers that monitoring—either for performance measurement, impact evaluation, or viewed through the lens of adaptive management—has been plagued by difficulties. In this section, we summarize a few key recommendations not made previously, based on years of experience in working with field projects, on how you can overcome some of the impediments to monitoring.

### Ask Management Relevant Questions

The best monitoring programs will be informed by a partnership of natural resource managers, project directors, planners, and scientists. Too often in conservation, scientists have gone about their job of designing monitoring programs that address questions of interest to scientists but not necessarily of value to managers.[40]

Conversely, managers and project directors need help from scientists and planners in framing questions in ways that make it possible to develop M&E programs designed to answer them. Carly Cook has examined this topic in some detail—a topic she and her colleagues refer to as science that bridges the knowledge-action boundary.[41] As they put it, "What is interesting (to scientists) is not always important (to managers), and what is important is not always interesting." In designing M&E programs, scientists may often strive for credibility in the form of objectivity, testing of hypotheses, and replication—in other words, rigor. But in effective conservation programs, this need for credibility and rigor must often be balanced with salience and legitimacy. Analyses that cost too much and take too long, for example, will not likely be legitimate for managers and decision makers.

According to Cook and colleagues, "boundary organizations" that understand how to best balance the needs of science and practicality are well suited to bridge the knowledge-action boundary, and in our case here, develop practical M&E programs that meet scientific and management needs. Examples of such boundary organizations include many major international conservation organizations, such as the Wildlife Conservation Society, World Wildlife Fund, and The Nature Conservancy, which collectively employ thousands of scientists and conservation practitioners in their field programs. In the United States, the Cooperative Fish and Wildlife Research units of the U.S. Geological Survey's Biological Resources Division were also established for this very purpose as a bridge among universities, federal natural resource agency research, and the management needs of state fish and wildlife agencies.

### Engage Leadership

Although we have all been in positions of having to report to funders about progress toward project outcomes, it has long been easy to report on means and not ends—in other words, to highlight the activities we have accomplished. Changing that mentality to one of a results-oriented culture is no easy task, and experience in several leading nonprofits suggests that leadership from the top (CEO, executive team, board) as well as from the leader of a measures or performance team really matters. Having senior leaders demonstrate the importance of being a results-focused organization seems to help organizations head faster down the path to taking performance measurement seriously.[42]

Several years ago, we brought over 20 project teams together in The Nature Conservancy from around the globe for peer reviews of the teams' performance measurement work. Having our then new CEO at the meeting to make a kickoff presentation and participate in a few sessions, followed by articles on this "measures summit" in key internal publications, has made some difference in building support for performance measures across The Nature Conservancy.[43]

### Peer Review

Peer review has a long history in many conservation organizations as a mechanism for improving performance. The Conservation Coaches Network—founded by Greening Australia, TNC, WWF, and Foundations of Success—is at its core a network of coaches who facilitate strategic planning and performance measures work in conservation organizations and with natural resource agency partners. It does this through peer-review workshops that follow the principles of CMP's *Open Standards.* Providing a "safe environment" for peer review and improvement of M&E programs is an incentive for many project teams to undertake such work, and participating in these workshops helps build broader support for M&E.[44] Sometimes that peer review will highlight the failings of a particular conservation effort. That's okay, but leaders of conservation organizations and agencies as well as funding institutions need to create an environment where it is safe to fail,[45] as long as we are learning and improving from those experiences.

### Don't Reinvent the Wheel

Learn and use evidence from other projects, sites, and analyses. Many of the same conservation interventions—establishing marine protected areas, controlling invasive species, or reintroducing species—are used every day, all over the world, by many different organizations. Increasingly, meta-analyses of some of these most important interventions are available in the published literature and through organizations like the Center for Evidence-Based Conservation (http://www.cebc.bangor.ac.uk) and online resource centers like Conservation Evidence.

Despite the emergency room atmosphere of conservation projects, doing a little more homework ahead of time on planned interventions and looking at their success in implementation by others may well lead to performance improvement. Recent research by some Cambridge University scientists suggests that the lack of access to scientific information and evidence may be a major impediment to implementing strategies and actions based on accumulated evidence.[46] When presented with synopses of techniques for reducing bird predations, conservation managers from Australia, New Zealand, and the United Kingdom were more likely to change the intervention they were currently using. They were also more likely to implement more effective actions and avoid ineffective ones. Similarly, doing cross-site comparisons of the same intervention through benchmarking,[47] a process that originated in the medical, business, and agricultural fields, may also lead to performance improvements in the implementation of conservation strategies and management actions.

### Better Communicate the Benefits

It is easier, and part of the ever-critical culture of science, to write about the limitations of monitoring or adaptive management. There is no shortage of publications that identify problems. On the other side of the coin, we need to publicize success stories about where M&E has gone well,[48] where it has improved the practice of conservation, and how it has been accomplished in a cost-effective manner. Many organizations have these important stories to tell, and more and more websites, blogs, and conservation newsletters are becoming available where they can be publicized.

### Make Smart Investment Choices

Perhaps the best advice we can give to practitioners is to think long and hard about whether to invest in M&E, why, and to what degree. There are simply not enough staff and financial resources to invest in monitoring results for every conservation project, and as we have outlined in this chapter, it makes little sense to do so. Being more strategic about M&E investments is likely to provide a better return on that investment than we have traditionally seen in conservation.

## KEY MESSAGES

- Distinguishing between what type of monitoring question a particular project needs to answer and between different types of M&E programs can help practitioners determine appropriate methods, match resources to those methods, and assess whether an adaptive management (AM) framework is needed.
- Defining who the key audiences are for monitoring and performance-related information is critical. Many of these audiences will want different levels of information and will want this information in different formats.
- Not all conservation projects merit the same staff or financial investment in monitoring the effectiveness of their interventions and actions. Examining a project through the lens of risk, leverage (the ability to scale up the results of a project and apply it

elsewhere), level of inference (if any) needed to demonstrate cause and effect, and existing conservation evidence for the interventions can help in determining how much to invest in M&E programs on a project-specific basis.

- M&E programs are only as successful as the project learning that accompanies them. Participating in communities of practice, engaging in management reviews and conservation audits, and sharing information across project boundaries are all useful practices that will encourage learning.
- The list of reasons for the limited success of AM and monitoring programs in conservation is long. It is important to have some understanding of these shortcomings to avoid perpetuating them. Considerable progress has been made in overcoming these limitations, and a number of good ideas can be implemented that will increase an organization or project's chances of successfully monitoring its results for improved performance.

## References

1. Richter, B. D. et al., *A collaborative and adaptive process for developing environmental flow recommendations.* River Research and Applications, 2006. **22**(3): 297–318.
2. Warner, A. T., L. B. Bach, and J. T. Hickey, *Restoring environmental flows through adaptive reservoir management: Planning, science, and implementation through the Sustainable Rivers Project.* Hydrological Sciences Journal, 2014. **59**(3–4): 770–785.
3. Muir, M. J., *Are we measuring conservation effectiveness? A survey of current results-based management practices in the conservation community.* 2010, Washington, DC: Conservation Measures Partnership.
4. Cook, C. N., M. Hockings, and R. W. Carter, *Conservation in the dark? The information used to support management decisions.* Frontiers in Ecology and the Environment, 2009. **8**(4): 181–186.
5. Hutto, R. L., and R. T. Belote, *Distinguishing four types of monitoring based on the questions they address.* Forest Ecology and Management, 2013. **289**: 183–189.
6. Mascia, M. B. et al., *Commonalities and complementarities among approaches to conservation monitoring and evaluation.* Biological Conservation, 2014. **169**: 258–267.
7. Carranza, T. et al., *Mismatches between conservation outcomes and management evaluation in protected areas: A case study in the Brazilian Cerrado.* Biological Conservation, 2014. **173**: 10–16.
8. Miteva, D. A., S. K. Pattanayak, and P. J. Ferraro, *Evaluation of biodiversity policy instruments: What works and what doesn't?* Oxford Review of Economic Policy, 2012. **28**(1): 69–92.

9. Cook, C. N., H. P. Possingham, and R. A. Fuller, *Contribution of systematic reviews to management decisions.* Conservation Biology, 2013. **27**(5): 902–915.
10. Pullin, A. S., and T. M. Knight, *Effectiveness in Conservation Practice: Pointers from Medicine and Public Health.* Conservation Biology, 2001. **15**(1): 50–54.
11. National Research Council, *Adaptive Management for Water Resources Project Planning.* 2004, Washington, DC: National Academy of Sciences. 138.
12. Williams, B. K., and E. D. Brown, *Adaptive Management: The U.S. Department of the Interior Applications Guide.* 2012, Washington, DC: U.S. Department of the Interior.
13. Hardesty, J. et al., *Simulating management with models.* Conservation Biology in Practice, 2000. **1**: 26–31.
14. Lemke, A. M. et al., *Evaluating agricultural best management practices in tile-drained subwatersheds of the Mackinaw River, Illinois.* Journal of Environmental Quality, 2011. **40**(4): 1215–1228.
15. Westgate, M. J., G. E. Likens, and D. B. Lindenmayer, *Adaptive management of biological systems: A review.* Biological Conservation, 2013. **158**: 128–139.
16. Keith, D. A. et al., *Uncertainty and adaptive management for biodiversity conservation.* Biological Conservation, 2011. **144**(4): 1175–1178.
17. Allen, C. R., and L. H. Gunderson, *Pathology and failure in the design and implementation of adaptive management.* Journal of Environmental Management, 2011. **92**(5): 1379–1384.
18. Westgate et al., *Adaptive management of biological systems: A review.* (See reference 15.)
19. Game, E. T. et al., *Conservation in a wicked complex world; challenges and solutions.* Conservation Letters, 2014. **7**(3): 271–277.
20. Spiegelhalter, D., M. Pearson, and I. Short, *Visualizing uncertainty about the future.* Science, 2011. **333**(6048): 1393–1400.
21. Few, S., *Show Me the Numbers: Designing Tables and Graphs to Enlighten.* 2012, Burlingame, CA: Analytics Press. 351; Yau, N., *Data Points: Visualization That Means Something.* 2013, Hoboken, NJ: John Wiley & Sons. 320.
22. Scarlett, L., *Collaborative adaptive management: Challenges and opportunities.* Ecology and Society, 2013. **18**(3).
23. McDonald-Madden, E. et al., *Monitoring does not always count.* Trends in Ecology & Evolution, 2010. **25**(10): 547–550.
24. Montambault, J., and C. R. Groves, *Investing in efficient, yet sufficient conservation monitoring.* (Manuscript in review, 2015).
25. Hall, J. A., and E. Fleishman, *Demonstration as a means to translate conservation science into practice.* Conservation Biology, 2010. **24**(1): 120–127.
26. Walsh, J. C., L. V. Dicks, and W. J. Sutherland, *The effect of scientific evidence on conservation practitioners' management decisions.* Conservation Biology, 2015: **29**(1): 88–98.

27. Konrad, C. P., A. Warner, and J. V. Higgins, *Evaluating dam re-operation for freshwater conservation in the Sustainable Rivers Project.* River Research and Applications, 2012. **28**(6): 777–792.

28. McDonald-Madden et al., *Monitoring does not always count.* (See reference 23.)

29. Selig, E. R., and J. F. Bruno, *A global analysis of the effectiveness of marine protected areas in preventing coral loss.* PLoS ONE, 2010. **5**(2): e9278.

30. Conservation Measures Partnership, *Open Standards for the Practice of Conservation 3.0.* 2013, Washington, DC: Conservation Measures Partnership.

31. Effectiveness Measures Working Group, *Measuring the Effectiveness of State Wildlife Grants.* 2011, Washington, DC: Association of Fish and Wildlife Agencies [accessed May 24, 2015]. http://www.fishwildlife.org/files/Effectiveness-Measures-Report_2011.pdf.

32. Nie, M. A., and C. A. Schultz, *Decision-making triggers in adaptive management.* Conservation Biology, 2012. **26**(6): 1137–1144.

33. Lindenmayer, D. B., M. P. Piggott, and B. A. Wintle, *Counting the books while the library burns: Why conservation monitoring programs need a plan for action.* Frontiers in Ecology and the Environment, 2013. **11**(10): 549–555.

34. Evely, A. C. et al., *High levels of participation in conservation projects enhance learning.* Conservation Letters, 2011. **4**(2): 116–126.

35. Lauber, T. B. et al., *Linking knowledge to action in collaborative conservation.* Conservation Biology, 2011. **25**(6): 1186–1194.

36. The Nature Conservancy, *Fire Learning Network: Working together to build resiliency in ecosystems and communities* [accessed January 22, 2015; 2-page fact sheet]. https://www.conservationgateway.org/ConservationPractices/FireLandscapes/FireLearningNetwork/Pages/fln-fact-sheet-feb-2013.aspx.

37. Dervitsiotis, K. N., *The design of performance measurement systems for management learning.* Total Quality Management & Business Excellence, 2004. **15**(4): 457–473; Petersen, B., J. Montambault, and M. Koopman, *The potential for double-loop learning to enable landscape conservation efforts.* Environmental Management, 2014. **54**(4): 782–794.

38. Evely et al., *High levels of participation in conservation projects enhance learning.* (See reference 34.)

39. Cundill, G. et al., *Soft systems thinking and social learning for adaptive management.* Conservation Biology, 2012. **26**(1): 13–20.

40. Lindenmayer, D. B., and G. E. Likens, *Adaptive monitoring: A new paradigm for long-term research and monitoring.* Trends in Ecology and Evolution, 2009. **24**(9): 482–486.

41. Cook, C. N. et al., *Achieving conservation science that bridges the knowledge–action boundary.* Conservation Biology, 2013. **27**(4): 669–678.

42. Nichols, J. D., and B. K. Williams, *Monitoring for conservation.* Trends in Ecology & Evolution (personal edition), 2006. **21**(12): 668–673.

43. Montambault, J., and C. R. Groves, *Making monitoring work for conservation: Lessons from The Nature Conservancy,* in *Biodiversity Monitoring in Australia,* D. B. Lindenmayer and D. W. Gibbons, editors. 2012, Collingwood, Victoria, Australia: CSIRO Publishing. 210.

44. Nichols and Williams, *Monitoring for conservation.* (See reference 42.)

45. Redford, K. H., and A. B. Taber, *Writing the wrongs: Developing a safe-fail culture in conservation.* Conservation Biology, 2000. **14**(6): 1567–1568.

46. Walsh et al., *The effect of scientific evidence on conservation practitioners' management decisions.* (See reference 26.)

47. Sutherland, W. J., and M. J. S. Peel, *Benchmarking as a means to improve conservation practice.* Oryx, 2011. **45**(1): 56–59.

48. Westgate et al., *Adaptive management of biological systems: A review.* (See reference 15.)

# 13

# Epilogue: Weaving Together the Futures of Conservation Planning and Nature Conservation

## Overview

The overarching focus of conservation planning in the early part of the twenty-first century will arguably be nature and people. Inherent to such a framing are many challenges, including our limited understanding of how nature conservation contributes to human well-being. Although many different ideologies underpin nature conservation, we have attempted to advance an approach to conservation planning in this book that is robust to these different ideologies. Future challenges remain for effective planning and implementation—and we have identified several of the most important ones here. At the same time, we have articulated a set of methods, tools, and examples in this book for the purpose of informing important conservation decisions, which we have asserted is the raison d'être for conservation planning.

## Topics

- *Phases of nature conservation*
- *Ecological restoration role*
- *Land use change*
- *Ecological thresholds*
- *Trans-boundary efforts*
- *Evidence-based planning*
- *Planning as decision making*

In a 2014 essay in *Science,* conservation biologist Georgina Mace traced the focus of nature conservation efforts through four phases.[1] These phases include conserving nature for its own values in the 1960s and '70s, nature despite people in the 1980s and '90s with an emphasis on the major threats to biodiversity, nature for people in the early 2000s with an increasing focus on ecosystem services, and today's conservation framing of "nature and people" with its underpinning of multidisciplinary science and necessary consideration of environmental change and adaptation. Not surprisingly, the history we provided of conservation planning in Chapter 1 closely parallels these phases of nature conservation. So as we conclude this book and reflect on the field of conservation planning, it is apropos to suggest that the future of conservation planning will almost certainly be closely tied to how the "nature and people" framing of twenty-first-century conservation plays itself out.

Along with Mace, we recognize that this framing of conservation is not without its tensions. There remain an array of ideologies and approaches to nature conservation that are being advocated by a wide range of actors. In this book, we have attempted to outline an approach to conservation planning that is robust to these different ideologies but at the same time recognizes the challenges of nature conservation in the twenty-first century. Paramount to that recognition is an understanding of the increasingly important role that global environmental change and adapting to that change will play in conservation. Equally important is appreciating that most conservation problems will be addressed only by understanding and working within the social-ecological systems in which conservation efforts take place, and by taking a multidisciplinary and multi-objective approach to conservation planning such as we have advocated throughout this book.

The approach to planning that we have outlined is based in part on our many years of experience in working with both nongovernmental conservation organizations and natural resource agencies in conducting conservation planning exercises. It is also based on our best thinking about the most appropriate planning methods and tools and—we cannot emphasize this enough—the thinking of a great many other conservation planners, scientists, and practitioners from a range of disciplines and institutions who have influenced us.

Like others, we remain concerned about the gap between knowing and doing[2] in conservation more broadly and conservation planning specifically. To close this gap, one of the principal means is to integrate planning methods that focus on where conservation should take place (spatial planning) with methods more focused on developing the strategies and actions necessary to get conservation done on the ground or through the policy arena. As a result, we have made the integration of spatial and strategic planning a major theme of this book. We firmly believe that undertaking planning in this manner will better inform conservation decisions that need to be made as a result of conservation planning.

Any approach to nature conservation—and in turn, conservation planning—faces many challenges; but a particularly difficult one may be the framing of conservation as integrating the needs of *nature and people*. For one, we have only a limited understanding of the ways that nature contributes to people and human well-being in particular. Moreover, adding people to the conservation planning equation[3] in any significant manner certainly increases the complexity of the planning process, and most conservation organizations and natural resource agencies are still stronger on the ecological aspects of conservation than on the social and economic. In addition, measuring outcomes in terms of monitoring and evaluation for people and nature can get difficult in a hurry. And as we have indicated in previous chapters, there are almost inevitably trade-offs in conservation efforts for people and nature. Indeed, one of the most significant challenges in conservation planning moving forward will be better understanding and estimating these trade-offs, and then making strategic decisions based on that understanding.

In the decade ahead, conservation planners will need to navigate a range of emerging challenges. Here are some of the most significant:

- Wrestling with the role that ecological restoration will need to play and the trade-offs with approaches focused on conserving more intact habitats.[4] Embedded within this challenge is a larger ongoing debate and discussion on the role of historical, hybrid, and novel ecosystems and the degree to which conservation plans will inevitably need to face up to these debates as objectives are established and targets are set for a range of conservation features.[5]
- Confronting the fact that rapid changes are occurring in many landscapes, seascapes, and watersheds as well as in patterns of land use and water use. It is also crucial to recognize that most conservation planning efforts will need to address the variety of such uses across any given planning context or situation and not necessarily just those areas that retain biodiversity or other conservation values.[6] How these landscapes and seascapes are planned for and managed in the future will have consequences not only for more traditional nature conservation values but almost certainly for issues like human health, including the spread of infectious disease.[7]

- Estimating ecological thresholds of human activities, beyond which the species, processes, and ecosystems that we care about are likely to be impaired. While significant efforts are being made to understand these thresholds, there are significant challenges to doing so: it is difficult to generalize from one species or ecosystem to another about these thresholds, and there are few examples of actually implementing thresholds derived from conservation or land use plans.[8] Nevertheless, pressures to do so will be part of the experience for many conservation planning efforts moving forward.
- Recognizing that most conservation efforts will need to take place across landscapes and seascapes that encompass a variety of geopolitical boundaries as well as institutional players and actors.[9] Recognition is of course only the first step; effective conservation planning efforts must incorporate approaches that counteract the natural human tendency to be parochial and focus inward on agency or other political boundaries. Working across or beyond boundaries (transboundary), whatever form they may take, is never easy; but it will be essential to twenty-first-century conservation efforts, especially in the context of climate change and necessary adaptations to it.
- Accumulating evidence for the need to take particular conservation interventions is likely to become an increasingly important factor that influences the practice of conservation planning.[10] Such transparency and accountability of the evidence has been critical in fields such as public health and medicine. Likewise, there are some signs that the resources for conservation in the future will become increasingly dependent on demonstrating the evidence that various strategies and actions are likely to succeed or that filling a critical gap in such evidence is needed to move forward. In turn, this need for a more evidence-based approach to conservation will place even greater importance on monitoring and evaluation practices that are needed to provide such evidence.

Fortunately, few conservation planning teams will face all of these challenges. But many will face at least one or more of them, and for that reason we have endeavored in this book to highlight the methods and tools that will allow planners and practitioners to successfully confront many of these challenges. For example, we have highlighted methods for better bringing social and economic data into the planning process and the types of social and economic data that are needed and generally available for conservation planning. We have emphasized participatory approaches that are essential to getting all the right stakeholders to the planning table. This helps ensure that from the outset, objectives are established for conservation projects that have buy-in from those who have a stake in the outcome. We have focused on the importance of evidence throughout the planning process—from setting conservation targets to evaluating a range of alternative interventions to feedback from monitoring and evaluation on which actions are most effective. Through numerous case studies and examples, we have demonstrated the importance of tessellating the entire landscape and seascape for conservation plans so that all land and sea uses are being taken into consideration in the planning process regardless of the specific objectives.

Nevertheless, we will certainly stop far short of suggesting that we have all the answers for the challenges we just articulated or that those are the only challenges conservation planners will face. Still, we hope that we have provided planners, scientists, and conservation practitioners with the tools they need to make informed decisions. Our view has been that conservation planning is fundamentally about making important decisions: what do we want to conserve, where should we do it, how should we do it, what resources will it take, and how will we know we have done it? Our hope and desire for this book is that the approach to conservation planning we have taken will inform these critical decisions and will indeed lead to conservation efforts that result in a healthier planet.

## References


1. Mace, G. M., *Whose conservation?* Science, 2014. **345**(6204): 1558–1560.
2. Hulme, P. E., Editorial: *Bridging the knowing–doing gap: Know-who, know-what, know-why, know-how and know-when.* Journal of Applied Ecology, 2014. **51**(5): 1131–1136.
3. Stephanson, S., and M. B. Mascia, *Putting people on the map through an approach that integrates social data in conservation planning.* Conservation Biology, 2014. **28**(5): 1236–1248.
4. Possingham, H. P., M. Bode, and C. J. Klein, *Optimal conservation outcomes require both restoration and protection.* PLoS Biol, 2015. **13**(1): e1002052.
5. Hobbs, R. J. et al., *Managing the whole landscape: Historical, hybrid, and novel ecosystems.* Frontiers in Ecology and the Environment, 2014. **12**(10): 557–564.
6. Morrison, S. A., *A framework for conservation in a human-dominated world.* Conservation Biology, 2015. **29**(3): 960–964.
7. Redford, K. H. et al., *Human health as a judicious conservation opportunity.* Conservation Biology, 2014. **28**(3): 627–629.
8. Johnson, C. J., *Identifying ecological thresholds for regulating human activity: Effective conservation or wishful thinking?* Biological Conservation, 2013. **168**: 57–65.

9. Dallimer, M., and N. Strange, *Why socio-political borders and boundaries matter in conservation.* Trends in Ecology & Evolution. **30**(3): 132–139.
10. Pullin, A. S., and T. M. Knight, *Doing more good than harm—Building an evidence-base for conservation and environmental management.* Biological Conservation, 2009. **142**(5): 931–934.

# Glossary

**abiotic.** The nonliving properties of the landscape.

**active adaptive management.** Management that involves deliberately implementing and testing a range of management options or actions in order to learn which is most effective.

**adaptive capacity.** The ability of a conservation feature or human community to adjust to the impacts of climate change.

**adaptive management.** The process of "learning by doing" by taking either a singular approach or intervention or a range of conservation interventions, monitoring the effectiveness of these approach(es), and using that feedback to make improvements in the management intervention(s).

**Aichi Biodiversity Targets.** In the 2011–2020 strategic plan of the Convention on Biological Diversity, five goals and 20 associated targets (the Aichi targets) related to addressing biodiversity loss for member nations to the convention.

**Analytic Hierarchy Process (AHP).** A frequently used and sophisticated way to establish criteria weights in MCDA, involving pairwise comparisons between criteria.

**apex consumer.** Refers to predators at the tops of food chains.

**attribute.** An alternative term for *indicator*.

**backcasting.** A professional planning method that starts with a plausible future condition or goal and then works backward to identify strategies, policies, or programs that could help achieve that future condition or goal.

**benefit transfer method.** A method used to estimate the monetary value of ecosystem services by using information already available in one place or situation and applying it to other places or situations.

**biodiversity.** The variety of life at all levels of genes, populations, species, and ecological communities.

**biodiversity offset.** A process for compensating for the loss of biodiversity in one place (usually because of development activities) by protecting another place.

**bundle.** Multiple, co-occurring ecosystem services within a landscape, seascape, or watershed that are grouped to help analyze and evaluate their value.

**business as usual (BAU).** The expected situation and outcomes if no additional, deliberate interventions are taken.

**coarse filter.** Conservation features intended to compensate for our incomplete knowledge of all biodiversity by reflecting diversity at a higher level of ecological organization that we can more readily observe, such as ecosystem types.

**cognitive bias.** Errors in judgment or reasoning that often occur from predictable cognitive processes. Also called *psychological biases*.

**community (ecological).** Relatively distinct assemblages of species that co-occur in space.

**community pool resource.** A principle according to which the natural assets of a community are held indivisibly rather than in the names of the individual members.

**conceptual learning.** Learning that involves placing ideas, events, and objects into mental categories to compare and contrast common features now and into the future. In conservation, this process allows participants to evaluate and revise project objectives.

**conceptual model.** A descriptive model of a system based on qualitative assumptions about its elements, their interrelationships, and system boundaries.

**conservation evidence.** The body of scientific information and data that can be brought to bear on a conservation problem or project and help managers make decisions on various courses of action.

**conservation feature.** A representation of biodiversity in a conservation plan.

**conservation planning problems.** Question(s) about the best way to achieve conservation objectives.

**constraint.** A condition that restricts potential options in a conservation plan (e.g., total budget or actions that have already been committed to).

**constructed scale.** An arbitrary scale used to reflect the possible range of an indicator.

**contingent valuation methods.** Survey-based economic methods for estimating the value of nonmarket goods. Also called stated preferences methods.

**cost.** The money, staff time, capital equipment, and other resources necessary to plan for and more broadly implement a strategy, action, or conservation plan.

**cost surface.** Generally, a geographic information systems (GIS) layer in conservation planning that accounts for costs in conservation interventions. Cost surface can vary spatially across a planning area.

**counterfactual.** A measure of the expected outcome in the absence of an intervention; used to evaluate the impact of an intervention.

**cultural ES (ecosystem service).** One of four main types of ecosystem services (ES) by which people receive intrinsic benefits, such as through recreation or cultural values.

**cultural keystone species.** Species that represent important components of a region's cultural heritage.

**decision science.** The application of analytical methods to help make better decisions. Also called operations research.

**decision theory.** A research field that includes economics, psychology, philosophy, mathematics, and computer science; concerned with values and uncertainties in making a decision, the rationale involved in that decision, and what factors may lead to the "best" or optimal decision.

**Delphi method.** A structured and systematic process for judgment and forecasting that relies on a panel of experts answering questions and revising judgments over two or more rounds.

**designed landscape.** Landscapes, seascapes, and watersheds whose configuration and function have been heavily influenced by people, generally for their benefit.

**diminishing returns.** A general rule whereby adding additional units of some factor (e.g., hectares of protection) yields lower marginal (incremental) per-unit returns (e.g., number of new species protected per hectare).

**downscale (climate modeling).** Statistical techniques used to produce climate models that have finer resolution than General Circulation Models. See also *General Circulation Model (GCM)*.

**ecological integrity.** The condition or health of an ecosystem or habitat; in particular, how intact it is.

**ecological network.** A network of conservation or protected areas that are ecologically connected, often through biological corridors that connect core conservation areas.

**ecological production function.** A mathematical function used to estimate the effect of changes in structure, function, and dynamics of ecosystems on ecosystem service outputs.

**ecological threshold.** The point at which a nonlinear or substantive change in the dynamics of an ecosystem is observed relative to some disturbance.

**ecosystem.** An ecological community and the physical environment it interacts with.

**ecosystem-based adaptation (EBA).** The conservation, sustainable management, and restoration of natural ecosystems to help people and nature adapt to climate change.

**ecosystem service.** The goods and services that natural ecosystems deliver to people.

**efficiency frontier.** All options or solutions that are Pareto optimal are considered to be on the efficiency frontier. Solutions not on this frontier are inefficient because they involve unnecessarily sacrificing achievement of one or more objectives; options along the efficiency frontier represent trade-offs that cannot be avoided.

**elicitation.** The process of gathering expert judgment.

**endemic.** A species whose distribution is limited to only one defined area.

**epistemic uncertainty.** The uncertainty associated with having imperfect knowledge of the state of a system and resulting from such things as insufficient data, poor measurement ability, and variability over time or space.

**expert judgment.** The opinion of an expert about a matter of fact (e.g., what is the population of tigers?) in contrast to a matter of value (e.g., how many tigers should there be?)

**exposure.** A measure of the degree of change likely to be experienced by a conservation feature or human community in the face of climate change.

**extinction debt.** The difference between the large number of species doomed to extinction and the smaller number of extinctions that have occurred to date.

**fine filter.** Important aspects of biodiversity unlikely to be well represented with coarse filters, such as individual species.

**flagship species.** Organisms of special charismatic or cultural value to the public.

**fundamental objectives.** Those goals that we ultimately want our actions to achieve; they are statements about the things we value.

**gap analysis.** The process whereby conservation features are mapped and overlaid with the distribution of conservation areas to determine which features are not adequately represented in the conservation network; those features that are not adequately represented are called *gaps*.

**General Circulation Model (GCM).** Global climate models that integrate thermodynamic terms, focus on Earth's atmosphere, and are used to predict changes in climate and weather forecasting.

**Global Environment Facility (GEF).** An independent organization that provides funding to biodiversity conservation, climate change, international water issues, land degradation, and ozone and other pollution problems.

**green economy.** An economy that aims for sustainable development without degrading the environment.

**habitat.** An ecosystem, often linked to particular species.

**heuristic algorithm.** An approach to mathematical optimization that uses rules to avoid the need to compare all possible solutions and still find one that is close to optimal.

**hotspot.** Locations that harbor unusually high concentrations of species.

**human well-being (HWB).** There are many different definitions, but HWB generally involves three components: people's needs are being met, they can act to pursue goals, and they have obtained a satisfactory quality of life.

**indicator.** In conservation, something that is reported on as evidence of how well the plan objectives are being achieved.

**Integrated Valuation of Ecosystem Services and Tradeoffs (InVEST).** A portfolio of spatially explicit software models developed by the Natural Capital Project to assist in the mapping and valuing of ecosystem services.

**intentionality.** The process of being deliberate and transparent in planning and implementing strategies and actions to abate and mitigate the impacts of climate change.

**Intergovernmental Science–Policy Platform on Biodiversity and Ecosystem Services (IPBES).** An intergovernmental body that operates under the auspices of the United Nations and its member countries to strengthen the interface between science and policy for biodiversity and ecosystem services.

**intermediate objective.** The desired outcomes that are part of the means of achieving fundamental objectives. Also called *means objectives* or *project objectives.*

**key biodiversity area (KBA).** A conservation site considered to contribute significantly on a global scale to the persistence of biodiversity.

**knowing-doing gap.** The gap existing between the number of conservation plans that are developed and those that are actually implemented.

**land facet.** Recurring units of land with relatively uniform characteristics of climate, topography, geology, and vegetation.

**land trust.** Usually nonprofit organizations whose missions are to conserve land through acquisition, conservation easement legislation, and/or ecological management activities.

**likelihood of success.** The estimate that a project will deliver the desired outcomes. This analysis includes assessing the likelihood that the project will be implemented effectively, and, if implemented, will be successful. The estimate might be on a linguistic scale or as a probability.

**linguistic uncertainty.** The uncertainty arising because the language or terminology used is underspecific, ambiguous, vague, context dependent, or exhibits theoretical indeterminacies.

**logic model.** The potential pathway of change that traces a logical sequence of steps from action to outcome.

**marginal and total value.** In economic terms, marginal value of a good or service is the value that an individual would attach to a specific increase or decrease in that service. Total value is the full worth of the amount of good or service being provided.

**marine spatial planning (MSP).** The process of analyzing and distributing the spatial arrangement of different types of human activities and uses of the marine environment.

**Marxan.** Software designed to support spatial prioritization by identifying the set of locations that most efficiently meet targets for a range of objectives.

**mathematical function.** A mathematical formula describing the relationship between two or more variables, typically the relationship between an input and an output.

**Micronesia Challenge.** A commitment by the Federated States of Micronesia, the Republic of the Marshall Islands, the Republic of Palau, Guam, and the Commonwealth of the Northern Marianas Islands to preserve the natural resources of Micronesia "that are crucial to the survival of Pacific traditions, cultures and livelihoods."

**Minimax.** A rule for determining the most robust solution by identifying the solution that minimizes the maximum potential loss in the worst-case scenario.

**minimum viable population (MVP).** The size at which a population's survival could be considered reasonably certain.

**motivational bias.** A situation where the ability to provide accurate expert judgment is impaired because the expert is likely to benefit from a particular decision related to that judgment.

**multi-criteria decision analysis (MCDA).** A process for incorporating multiple objectives into the evaluation of alternative options by identifying weights for different criteria/objectives. Also called *multi-criteria decision making* (*MCDM*) and *multi-criteria analysis* (*MCA*).

**natural scale.** An obvious and preexisting way of measuring an indicator.

**objective hierarchy.** A classification that groups similar objectives together to minimize the number of objectives at the top of the hierarchy.

**optimization (mathematical).** The process of identifying the option that delivers the best possible outcome for an objective. This is done by maximizing or minimizing a mathematical function that describes a performance measure for the objective.

**option.** An alternative strategy or action under consideration in a plan.

**Pareto optimal.** An option is Pareto optimal when no improvement in one objective can be made without simultaneously diminishing the achievement of at least one other objective.

**passive adaptive management.** Management in which only one management option or action is selected based on existing information, but results of that choice are monitored, and future management decisions are made based on this additional knowledge.

**Payment for Ecosystem Services (PES).** A type of conservation intervention in which incentives, usually in some financial form, are offered to one or more parties in exchange for actions that are beneficial to the conservation of one or more ecosystem services.

**planning context.** The circumstances in which a conservation plan is developed. This usually involves considering the purpose of a plan, decisions to be made, decision makers, constraints on a planning process, level of investment in the plan, and audience for the plan.

**population viability analysis (PVA).** A quantitative assessment of the probability that a population will become extinct during some specified time frame.

**precautionary principle.** The idea that lack of full scientific certainty shall not be used as a reason for postponing action.

**probability.** A measure of the likelihood that an event will occur, quantified as a number between 0 and 1 (where 0 indicates impossibility and 1 indicates certainty).

**process objective.** An objective related to how planning and decision making should take place.

**provisioning ES (ecosystem service).** One of four main types of ecosystem services (ES) in which ecosystems directly provide products such as food and water to human communities.

**proxy scale.** An approach to measuring an indicator by measuring an alternate indicator that is more easily measured on a natural scale and is likely to change in a similar fashion to the indicator of interest.

**quantitative.** An amount that can be measured and expressed numerically.

**refugia.** Areas with a relatively lower likelihood of being affected by climate change, often due to reasons related to topography or geology.

**regulating ES (ecosystem service).** One of four main types of ecosystem services (ES) in which ecological processes such as pollination or water filtration directly benefit humans.

**resilience.** The capacity of a system to resist or recover quickly from a perturbation.

**restricted-range endemic.** An endemic species with a highly restricted distribution.

**results chain.** A type of logic model that links an action to an outcome through a series of intermediate outcomes. Results chains are usually represented as polygons linked with lines.

**return on investment (ROI).** In conservation, a general term for prioritization approaches that explicitly consider the cost of the options being considered.

**revealed expenditures.** An approach to measuring the benefits of environmental quality by examining the value of environmental goods from other related market transactions.

**risk.** An uncertainty that might negatively affect the ability to achieve a project's objectives.

**robust solution.** A solution that would perform well despite potential changes due to uncertainty.

**rules of thumb.** The simple rules (or heuristics) that do a reasonable job of approximating the solutions found through formal optimization.

**scanning.** The process of systematically looking into the future, often with experts, to better understand what issues or factors may affect the development and implementation of a conservation plan or project later. Also referred to as horizon scanning.

**scenario analysis (or scenario).** The process of developing a set of possible futures (or scenarios) that describe how some of the main uncertainties—demographic trends, policies, markets, budgets, degree of climate change, stakeholder support—might behave, and then exploring how potential options would be expected to perform under these scenarios.

**scope.** The geographic area under consideration in a planning process and/or the breadth of strategies being considered and the time frame over which the plan is meant to be in use.

**sensitivity.** The degree to which a conservation feature or human community may be influenced by changes in climate.

**sensitivity analysis.** A mathematical or statistical technique for determining the degree to which any single variable, factor, or sets of factors influence the overall output of a model.

**site conservation planning.** Generally refers to conservation planning that takes place for a specific site on the ground or in the water to develop strategies and actions for conservation at that place.

**situation analysis.** The process of identifying and articulating how socioeconomic, political, institutional, and ecological factors drive change in the system of interest.

**sliding baseline.** The temptation to believe that memories of the near past reflect what natural abundance was, such that the perception of what is natural is progressively diminished.

**social adaptation to climate change.** A process or action taken by a group of people, often at the community level, to better prepare them for the risks and stresses associated with climate change.

**social-ecological system (SES).** An ecosystem and the set of social and institutional actors that interact regularly with it.

**social learning.** The learning that takes places through engagement and collaboration among stakeholders.

**spatial prioritization.** The process of solving conservation planning problems that involves identifying and selecting the best location to take conservation actions.

**species-area relationship (SAR).** A positive, saturating relationship between the number of species and the physical size of an island, habitat patch, or sampling space.

**species distribution model (SDM).** A model that combines the known locations of species with knowledge of the environmental conditions at those places to extrapolate where else across a landscape a species is likely to be found.

**stakeholder.** An individual, group, or organization that is interested in some aspect of a conservation plan or project and may be affected by, or will potentially affect, project activities.

**stakeholder analysis.** The process of identifying the individuals or groups that are likely to affect or be affected by a proposed strategy or action, and classifying them in some manner regarding their impact on the project or the project's impact on them.

**strategic foresight.** A systematic process for examining possible future conditions or states as part of a long-term planning process.

**strategic objective.** An objective related to how a plan and resulting decisions should align with an organization's mission and values.

**structured decision making.** An organized approach to developing and evaluating creative alternatives and making defensible choices.

**supporting ES (ecosystem service).** One of four main types of ecosystem services (ES) by which ecological processes such as nutrient cycling provide indirect benefits to people.

**summary indicator.** A single indicator made up of a small number of other indicators.

**surrogate.** A conservation feature used to represent another feature(s) and generally easier to observe, map, or measure than those features it is representing.

**SWOT analysis.** A strategic planning process in which the strengths, weaknesses, opportunities, and threats of a project or course of action are analyzed.

**systematic conservation planning (SCP).** An established conservation planning process, typically used to identify new conservation areas in order to meet targets for the representative protection of biodiversity.

**target.** Quantitative statements of the outcomes planners want to achieve for each objective.

**technical learning.** The learning that involves engaging in new tasks and objectives that require learning new competencies.

**theory of change (TOC).** An articulation of how an action is anticipated to achieve an objective, including a set of causal linkages and the assumptions underpinning them.

**trade-off.** A situation where achievement of one objective comes at the expense of achieving another objective.

**traditional ecological knowledge (TEK).** A special subset of expert knowledge that reflects the environmental knowledge held by members of indigenous or traditional communities.

**uncertainty.** A situation characterized by imperfect and/or unknown information.

**United Nations Framework Convention on Climate Change (UNFCCC).** An international treaty first adopted in 1992 and aimed at reducing greenhouse gas emissions to limit the impacts of global warming.

**utility function.** A mathematical function that describes an option's performance (or utility) against an objective.

**value of information.** The difference between the outcome of the decision you would make with the additional information and the outcome of the decision you would make without that information.

**vision statement.** A brief and inspirational statement about what the future of a conservation project, initiative, program, or area might look like.

**vulnerability.** The degree to which a system is susceptible to, or unable to cope with, adverse effects of change.

**vulnerability assessment.** A process undertaken to determine the relative susceptibility of ecological or human communities to adverse effects from climate change. Generally includes three components: sensitivity, exposure, and adaptive capacity.

**wave attenuation.** The process by which waves arriving on the coast are dissipated by either natural ecosystems or built infrastructure.

**willingness to pay.** The amount an individual is willing to pay to obtain a good or service, or to sacrifice to avoid something undesirable.

**Zonation.** Software designed to support spatial prioritization by indicating the relative conservation importance of every location in an entire landscape for an efficient solution.

# Credits

All illustrations credited to Emiko Paul, ECHO Medical Media, except where noted otherwise.

**CHAPTER OPENER BACKGROUND IMAGE**

© iStock.com/Qweek

**CHAPTER 1**

Figure 1.1 © 2009 Bridget Besaw /© The Nature Conservatory; Figure 1.2 Used with permission of John Wiley and Sons Limited, Adapted from Scott, J. M. et al., *Gap analysis: A geographic approach to protection of biological diversity.* Wildlife Monographs, 1993. 1993(123): 3–41; Figure 1.3 © 2008 Society for Conservation Biology. Used with permission of John Wiley and Sons Limited, Adapted from Pressey, R. L., and M. C. Bottrill, *Opportunism, threats, and the evolution of systematic conservation planning.* Conservation Biology 22(5):1340–1345; Figure 1.5 © 2002, with permission from Elsevier, Adapted from Sanderson, E. W. et al., *A conceptual model for conservation planning based on landscape species requirements.* Landscape and Urban Planning, 2002. 58(1): 41–56; Figure 1.6 Used with permission of Conservation Measures Partnership; Figure 1.7 © The Nature Conservancy; Figure 1.8 Used with permission of the Ecological Continuum Initiative of the Alps; Figure 1.9 Courtesy of USDA Forest Service; Figure 1.10 Used with permission of the Florida Natural Areas Inventory; Figure 1.12 Jessica Musengezi/© The Nature Conservancy

**CHAPTER 2**

Figure 2.1 Ron Gaetz/©The Nature Conservancy; Figure 2.2 © The Nature Conservancy; Figure 2.3 Pankaj Chandan/©WWF India; Figure 2.4 ©The Nature Conservancy, Data from The Nature Conservancy, *Conservation Business Planning Guidance*, Version 1.3. July 10, 2013, Arlington, VA [accessed March 4, 2015]. http:// www.conservationgateway.org /ConservationPlanning/BusinessPlanning /Documents/CBP_Guidance.pdf; Figure 2.5 ©Mark Anderson; Figure 2.6a Erika Nortemann/© The Nature Conservancy; Figure 2.6b Poster courtesy of California Bay Delta Authority; Figure 2.6c Photo courtesy of U.S. Forest Service; Figure 2.7 Used with permission of Dr. Charles Ehler/UNESCO Intergovernmental Oceanographic Commission; Figure 2.8 ©The Nature Conservancy; Table 2.1 Reprinted with permission of Elsevier; Table 2.2 Used with permission of John Wiley and Sons Limited; Table 2.3 Used with permission of Dr. Claudia Binder; Box 2.1 Miradi logo, used with permission of the Conservation Measures Partnership and the Sitka Technology Group

**CHAPTER 3**

Figure 3.2 (map) ©WWF Germany used under Creative Commons Attribution License; Figure 3.2 (photo) ©Erik Meijaard; Figure 3.3 ©Allen Alison; Figure 3.4 (photo) ©Timothy Boucher; Figure 3.4 (map) ©Government of Manitoba used under Creative Commons Attribution License; Figure 3.5 ©United States Geological Survey used under Creative Commons Attribution License; Figure 3.6 ©Commonwealth of Australia (Geoscience Australia) used under Creative Commons Attribution License; Figure 3.7a ©Alex Game; Figure 3.7b ©James Fitzsimons; Figure 3.8 ©The Nature Conservancy; Figure 3.9 ©Serge Andrefouet; Figure 3.10 Reprinted with permission of Wiley and Sons Limited, Adapted from Game, E.T., et al., *Incorporating climate change adaptation into national conservation assessments.* Global Change Biology, 2011. 17:3150-3160; Figure 3.11 ©Kent Redford; Figure 3.12 ©Ami Vitale/ The Nature Conservancy; Figure 3.13 ©Craig Groves; Figure 3.14 ©Edward Game

**CHAPTER 4**

Figure 4.1 ©Yongcheng Long/The Nature Conservancy; Figure 4.2a John Randall/©The Nature Conservancy; Figure 4.2b ©Richard Hamilton Smith/The Nature Conservancy; Figure 4.4 ©The Nature Conservancy; Figure 4.5a&b Reprinted with permission from Elsevier; Figure 4.5c John Dwyer; Figure 4.6 ©Alan Carmichael; Figure 4.7 Reprinted with permission from Elsevier; Figure 4.8a ©Kydd

Pollock/The Nature Conservancy; Figure 4.8b Adapted with permission from Kiki Dethmers; Figure 4.9a ©Jonathan Rossouw; Figure 4.9b ©Edward Game; Figure 4.10 ©Federal Government of Germany (Federal Maritime and Hydrographic Agency) used under Creative Commons Attribution License; Figure 4.11 ©Federal Government of Belgian (Science Policy Office) used under Creative Commons Attribution License; Figure 4.2.1 Reprinted with permission from Elsevier; Figure 4.2.2 Reprinted with permission from the Ecological Society of America

**CHAPTER 5**

Figure 5.1 ©The Nature Conservancy; Figure 5.2a ©Paula Deegan; Figure 5.2b ©Pierre Ibisch; Figure 5.3 Data from Selkoe, K.A., et al., *Evaluating anthropogenic threats to the Northwestern Hawaiian Islands.* Aquatic Conservation: Marine and Freshwater Ecosystems, 2008. 18: 1149–1165; Figure 5.4 and Figure 5.5 Data from Ban, N. C., et al., *Cumulative impact mapping: Advances, relevance and limitations to marine management and conservation, using Canada's Pacific waters as a case study.* Marine Policy, 2010. 34:876–886; Figure 5.6 Used with permission of University of Wisconsin Press, Adapted from Esselman, P.C., et al., *An index of cumulative disturbance to river fish habitats of the conterminous United States from landscape anthropogenic activities.* Ecological Restoration, 2011. 29: 133–151; Figure 5.7 © Wiley Periodical, Inc. Used under Creative Commons Attribution License; Figure 5.8 ©Cartoon Stock; Figure 5.9 ©Ami Vitale/The Nature Conservancy; Figure 5.10 ©United States Geological Survey used under Creative Commons Attribution License; Figure 5.11 ©The Nature Conservancy; Figure 5.12 ©The Nature Conservancy; Figure 5.13 Data from Theobald, D. M., *A general model to quantify ecological integrity for landscape assessments and US application.* Landscape Ecology, 2013. 28:1859–1874 used with permission from David Theobald; Figure 5.16 ©Public Library of Science used under Creative Commons Attribution License; Figure 5.17 ©Public Library of Science used under Creative Commons Attribution License; Figure 5.18 ©Public Library of Science used under Creative Commons Attribution License; Figure 5.21 ©Nate Peterson; Figure 5.22 ©Nicolas Houde; Figure 5.1.1 ©Michelle Lee; Figure 5.3.1 ©The Nature Conservancy; Figure 5.5.1 Used with permission of Dave Theobald, Adapted from Theobald, D. M., et al., *Ecological support for rural land-use planning.* Ecological Applications, 2005. 15:1906–1914; Figure 5.8.1 ©Nate Peterson; Figure 5.8.2 ©Nate Peterson

**CHAPTER 6**

Figure 6.2 ©The Nature Conservancy; Figure 6.3 ©The Nature Conservancy; Figure 6.4.1 © Elsevier used under Creative Commons Attribution License; Figure 6.5.1 ©Ian Shive/The Nature Conservancy

**CHAPTER 7**

Figure 7.1 Adapted with permission from Stephen Redpath; Figure 7.3 Wiley Periodical, Inc. used under Creative Commons Attribution License, Adapted from Goldstein, J. H., et al., *Using return-on-investment to guide restoration: A case study from Hawaii.* Conservation Letters, 2008 1:236–243 ©; Figure 7.4 ©Public Library of Science used under Creative Commons Attribution License, Data from Klein, C. J. et al., *Prioritizing Land and Sea Conservation Investments to Protect Coral Reefs.* PLoS One, 2010. 5:e12431; Figure 7.5 ©The Nature Conservancy; Figure 7.6 Reprinted with permission from Elsevier, Data from Smith, R. J. et al., *Designing a transfrontier conservation landscape for the Maputaland centre of endemism using biodiversity, economic and threat data.* Biological Conservation, 2008. 141:2127–2138; Figure 7.7 Adapted from Hermoso, V. et al., *Integrating multidirectional connectivity requirements in systematic conservation planning for freshwater systems.* Diversity and Distributions, 2012. 18:448–458 with permission from Virgilio Hermoso; Figure 7.8a ©Rod Salm; Figure 7.8b ©Robyn James/The Nature Conservancy; Figure 7.11 ©Heini Kujala; Figure 7.12 ©The Nature Conservancy; Figure 7.13 Reproduced from Moilanen, A. et al., *Spatial prioritization of conservation management.* Conservation Letters, 2011. 4:383–393 © Wiley Periodical, Inc. used under Creative Commons Attribution License; Figure 7.19 ©The Nature Conservancy; Figure 7.3.1 ©The Nature Conservancy; Figure 7.4.1, Figure 7.4.2 and Figure 7.4.3 ©Heini Kujala

## CHAPTER 8

Figure 8.1 ©Kenneth K. Coe/The Nature Conservancy; Figure 8.3 (map) ©United States Geological Survey used under Creative Commons Attribution License; Figure 8.3 (photo) ©Dave Lauridsen/The Nature Conservancy; Figure 8.5 ©United States Fish and Wildlife Service used under Creative Commons Attribution License; Figure 8.6 ©Intergovernmental Panel on Climate Change used under Creative Commons Attribution License; Figure 8.8 Data from Hill, M. J., and R. Olson, *Possible future trade-offs between agriculture, energy production, and biodiversity conservation in North Dakota.* Regional Environmental Change, 2013. 13:311–328; Figure 8.10 Adapted from Game, E. T. et al., *Subjective risk assessment for planning conservation projects.* Environmental Research Letters, 2013 8:045027 ©IOPscience used under Creative Commons Attribution License; Figure 8.11 ©IOP science used under Creative Commons Attribution License, Adapted from Game, E. T. et al., *Subjective risk assessment for planning conservation projects.* Environmental Research Letters, 2013. 8:045027

## CHAPTER 9

Figure 9.1 © Jack Stein Grove; Figure 9.3 Used with permission of the National Wildlife Federation, Data from Glick, P., B. A. Stein, and N. A. Edelson, *Scanning the Conservation Horizon: A Guide to Climate Change Vulnerability Assessment.* 2011, Washington, DC: National Wildlife Federation. 176; Figure 9.4 National Wildlife Federation; Figure 9.5 Used with permission of Nadine Marshall/CSIRO Land and Water, Data from Marshall, N. A. et al., *A Framework for Social Adaptation to Climate Change; Sustaining Tropical Coastal Communities and Industries.* 2009, Gland, Switzerland: IUCN; Figure 9.6 Reprinted with permission from Macmillan Publishers Ltd: *Nature* © 2013; Figure 9.7 © Taco Anema/IUCN, Adapted from Colls, A., N. Ash, and N. Ikkala, *Ecosystem-Based Adaptation: A Natural Response to Climate Change.* 2009, Gland, Switzerland: IUCN. 16; Figure 9.8 Used with permission of National Wildlife Federation, Adapted from Stein, B. A. et al., *Climate Smart Conservation: Putting Adaptation Principles into Practice.* 2014, Washington, DC: National Wildlife Federation. 262; Figure 9.9 Used with permission of John Wiley and Sons Limited; Figure 9.10 Adapted from Cross, M. S. et al., *The Adaptation for Conservation Targets (ACT) framework: A tool for incorporating climate change into natural resource management.* Environmental Management, 2012, 50(3): 341–351; Figure 9.11 Used with permission of GIZ Climate Protection Program; Figure 9.12 Courtesy of USDA Forest Service; Figure 9.13 Used with permission of the Ecological Society of America; Box 9.1 © Ben Hall with permission of Royal Society for Protection of Birds; Box 9.2 Adapted from Anthony, K. R. N. et al., *Operationalizing resilience for adaptive coral reef management under global environmental change.* Glob Change Biology, 2015. 21: 48–61. doi:10.1111/gcb.12700; Box 9.3 ©Steffen Reichle; Box 9.4 (all figures) © The Nature Conservancy; Box 9.5 Photo and permission by David Golden; Box 9.6 Used with permission of Oxford University Press, Data from Schwartz, M. W. et al., *Managed relocation: Integrating the scientific, regulatory,and ethical challenges.* BioScience, 2012. 62(8): 732–743; Box 9.7 © Mary O-Brien/Grand Canyon Trust; Table 9.1 Used with permission of Springer under Creative Commons Attribution License, Adapted from Ellison, J., *Vulnerability assessment of mangroves to climate change and sealevel rise impacts.* Wetlands Ecology and Management, 2015, 23: 115–137; Table 9.2 Used with permission of John Wiley and Sons Limited, Data from Cross, M. S. et al., *Accelerating adaptation of natural resource management to address climate change.* Conservation Biology, 2013. 27(1): 4–13; Table 9.3 Used with permission of UK Climate Impacts Programme; Table 9.4 Courtesy of U.S. Fish and Wildlife Service

## CHAPTER 10

Figure 10.1 Used with permission of Department of Agriculture & Food, Australia; Figure 10.2 Used with permission of PBL Netherlands Environmental Assessment Agency under Creative Commons License; Figure 10.3 Courtesy of Intergovernmental Platform on Biodiversity and Ecosystem Services; Figure 10.4a Chris Helzer/©The Nature Conservancy 2012; Figure 10.4b Chris Helzer/©The Nature Conservancy 2008; Figure 10.4c Mark Godfrey/© The Nature Conservancy 2013; Figure 10.5 Used with permission of Island Press, Data from Daily, G., ed., *Nature's Services: Societal Dependence on Natural Ecosystems.* 1997, Washington, DC: Island Press. 391; Figure 10.6 Used with permission of John Wiley and Sons Limited; Figure 10.8 Used with permission of Oxford University Press, Data from Egoh, B. N. et al., *Safeguarding biodiversity and ecosystem services in the Little Karoo, South Africa.* Conservation Biology, 2010. 24(4): 1021–1030; Figure 10.9 Courtesy of Heather Tallis/The Nature Conservancy; Figure 10.10 ©Wiley Periodical, Inc. used under Creative Commons Attribution License; Figure 10.11 © Michael Goulding; Box 10.1 Courtesy of USDA Forest Service; Box 10.2 Figures courtesy of the Natural Capital Project, Data from Guerry et al., *Modeling benefits from nature: using ecosystem services to inform coastal and marine spatial planning.* International Journal of Biodiversity Science, Ecosystem Services & Management, 2012. DOI:10.1080/21513732.2011.647835 (10.2.1), Clarke, C. et al., Belize Integrated Coastal Zone Management Plan. 2013, Belize City: Coastal Zone Management Authority and Institute. 423 (10.2.2); Box 10.3.1 Used with permission of K. Bagstad, U.S. Geological Survey; Table 10.2 Used with permission of Oxford University Press; Table 10.3 Used with permission of Ken Bagstad/U.S. Geological Survey; Table 10.4 ©Wiley Periodical, Inc. used under Creative Commons Attribution License; Data from Schröter, M. et al., *Ecosystem services as a contested concept: A synthesis of critique and counter-arguments.* Conservation Letters, 2014: 7(6):514–523

**CHAPTER 11**

Figure 11.1 © Mike Flanagan; Figure 11.2 ©Randy Glasbergen; Figure 11.3 © The Nature Conservancy; Figure 11.4 © Andrew Kornylak/ The Nature Conservancy; Figure 11.5 Used with permission of Elsevier, Data from Knight, A. T. et al., *Walking in STEP: Lessons for linking spatial prioritizations to implementation strategies.* Biological Conservation, 2011. 144(1): 202–211; Figure 11.6 Photo courtesy of U.S. Army Corps of Engineers; Figure 11.7 Courtesy of European Commission Environment [accessed November 11, 2015], http://ec.europa.eu/environment/nature/natura2000/barometer/index_en.htm; Figure 11.8 Used with permission of Frank W. Davis, University of California, Santa Barbara; Figure 11.9 © Michael Nichols, National Geographic; Box 11.3 Reprinted with permission from Elsevier; Figure 11.4.1 Courtesy of illustrator Anders Busse Nielsen, and the Government of Denmark, Ministry of Environment and Food; Table 11.1 Courtesy of Conservation Measures Partnership; Table 11.2 © WWF International, Data from Mikov, M. et al., *Step 3.1 Work plans and budgets,* in *Resources for Implementing the WWF Project and Programme Standards.* 2007, Gland, Switzerland:World Wildlife Fund; Table 11.3 ©WWF International; Table 11.4 © WWF International, Data from Mikov, M. et al., *Step 3.1 Work plans and budgets,* in *Resources for Implementing the WWF Project and Programme Standards.* 2007, Gland, Switzerland: World Wildlife Fund

**CHAPTER 12**

Figure 12.1 Erika Nortemann/©The Nature Conservancy; Figure 12.2 Used with permission of L. Greig; Figure 12.3 Photo courtesy of Patrick Kolar, U.S. Geological Survey; Figure 12.4 © Society American Foresters; Figure 12.5 Andreas Mujaldi/ ©The Nature Conservancy; Figure 12.6 Michael Nakomoto/© The Nature Conservancy; Figure 12.7 © University of Maryland Center for Environmental Science; Figure 12.8 Reprinted with permission of Elsevier, Data from McDonald-Madden, E. et al., *Monitoring does not always count.* Trends in Ecology & Evolution, 2010. 25(10): 547–550; Figure 12.9 (bottom left) Rich Kostecke/© The Nature Conservancy, (bottom right) Daniel and Robbie Wisdom/© The Nature Conservancy, (top right) Andy Drumm/©The Nature Conservancy, (top left) Brian Richter/© The Nature Conservancy; Figure 12.10 © The Nature Conservancy; Figure 12.11 Courtesy of Conservation Measures Partnership; Figure 12.12 Courtesy of Conservation Measures Partnership; Box 12.2 © Matt Looby; Box 12.3 © WCS-Laos PDR / NEPL NPA; Table 12.1 Reprinted with permission from Elsevier, Data from Mascia, M. B. et al., *Commonalities and complementarities among approaches to conservation monitoring and evaluation.* Biological Conservation, 2014. 169: 258–267; Table 12.2 Courtesy of Conservation Measures Partnership

# Index

**Page numbers in bold** indicate figures.